T0320381

Biomass Energy for Sustainable Development

The potential future fluctuations in energy security and potential climate change impacts require an emphasis on clean and renewable energies to safeguard the environment as well as economic livelihoods. The current recalcitrant nature of biomass processing has led researchers to find the most suitable technique for its depolymerization, as well as various strategies to pretreat the biomass which include physical, thermochemical, and biochemical methods and a combination of these. *Biomass Energy for Sustainable Development* examines how optimal biomass utilization can reduce forest management costs, help mitigate climate change, reduce risks to life and property, and help provide a secure, competitive energy source into the future.

Features:

- Provides a comprehensive review of biomass energy and focuses on in-depth understanding of various strategies to pretreat biomass including physical, chemical, and biological.
- Explores multidisciplinary, novel approaches including AI for furthering the understanding and generation of models, theories, and processes in the field of bioenergy.
- Covers the sustainable development goals for bioenergy, including the related concepts of bioeconomy and the potential environmental impact from reliance on bioenergy.

Maulin P. Shah is an active researcher and scientific writer in his field for over 20 years. He received a BSc degree (1999) in Microbiology from Gujarat University, Godhra (Gujarat), India. He also earned his PhD degree (2005) in Environmental Microbiology from Sardar Patel University, Vallabh Vidyanagar (Gujarat), India. His research interests include biological wastewater treatment, environmental microbiology, biodegradation, bioremediation, and phytoremediation of environmental pollutants from industrial wastewater. He has published more than 250 research papers in national and international journals of repute on various aspects of microbial biodegradation and bioremediation of environmental pollutants. He is the editor of 110+ books of international repute.

Pardeep Kaur received her BSc degree (2005) in Biological Sciences from Punjabi University, Patiala, Punjab, India. She received her MSc degree (2007) in

Microbiology and PhD degree (2017) in Microbiology from Punjab Agricultural University, Ludhiana. Currently, she is an Assistant Professor at the University Institute of Biotechnology, Chandigarh University, Punjab, India. She has more than 12 publications in reputed peer-reviewed journals. Her research interests include bioprocess development and optimization, biofuel production, and fermentation technology.

Biomass Energy for Sustainable Development

Edited by
Maulin P. Shah and Pardeep Kaur

CRC Press
Taylor & Francis Group
Boca Raton London New York

CRC Press is an imprint of the
Taylor & Francis Group, an **informa** business

Cover image: © Shutterstock

First edition published 2024
by CRC Press
6000 Broken Sound Parkway NW, Suite 300, Boca Raton, FL 33487-2742

and by CRC Press
4 Park Square, Milton Park, Abingdon, Oxon, OX14 4RN

CRC Press is an imprint of Taylor & Francis Group, LLC

© 2024 Taylor & Francis Group, LLC

ISBN: 978-1-032-52400-9 (hbk)
ISBN: 978-1-032-52403-0 (pbk)
ISBN: 978-1-003-40650-1 (ebk)

DOI: 10.1201/9781003406501

Typeset in Times
by codeMantra

Contents

Chapter 3   Thermochemical Conversion: An Approach for the Production
of Energy Materials from Biomass 35

*Chitra Devi Venkatachalam, Mothil Sengottian,
Sathish Raam Ravichandran, Sarath Sekar,
and Premkumar Bhuvaneshwaran*

Chapter 4   Biochemical Approach to Biomass Conversion: Biofuel
Production 62

Nishu Sharma, Lovepreet Kaur, and Shiwani Guleria Sharma

Chapter 5 Microbial Approach for Biofuel Production by Biomass
Conversion

Khalida Bloch and Sougata Ghosh

Chapter 6 Anaerobic Digestion: A Sustainable Biochemical Approach
to Convert Biomass to Bioenergy

Gaganpreet Kaur, Nisha Yadav, and Sachin Kumar

Chapter 13   Process Design of Various Biomass Gasification Processes Using
Aspen Plus and Its Effects on Syngas and Hydrogen Production........... 260

Deepanshu Awasthi, Bhautik Gajera, Rakesh Godara,
Arghya Datta, Nikhil Gakkhar, and Tapas Kumar Patra

Chapter 14   Implementing Targeted Total Soluble Product Recovery during
Food Waste Biomethanation for Enhanced Recovery of Energy
and Value-Added Products...279

S. Hemapriya and P. Sankar Ganesh

Chapter 15 Torrefaction of Agriculture Residues and Municipal Solid
Waste for Char Production 299

Sugali Chandra Sekhar, Bukke Vani, and Sridhar Sundergopal

Chapter 16 The Circular Bioeconomy Concept........................ 313

Larissa Echeverria, Carina Contini Triques,
Keiti Lopes Maestre, Jacqueline Ferandin Honório,
Veronice Slusarski-Santana, Gabriela Eduarda Zeni,
Fabio Augusto Gubiani, Leila Denise Fiorentin-Ferrari,
and Mônica Lady Fiorese

Chapter 17 A Circular Economy Approach to Valorisation of
Lignocellulosic Biomass-Biochar and Bioethanol Production......... 335

Sai Shankar Sahu and Subodh Kumar Maiti

Chapter 18 Global Research Trends in Biomass as Renewable Energy 355

N. Premalatha and S. R. Saranya

*Kamla Malik, Monika Kayasth, Sujeeta Yadav, Nisha Arya,
Shweta Malik, Rekha Chahar, Kashish Sharma,
Dandu Harikarthik, Shikha Mehta, and Meena Sindhu*

About the Editors

Maulin P. Shah is an active researcher and scientific writer in his field for over 20 years. He received a BSc degree (1999) in Microbiology from Gujarat University, Godhra (Gujarat), India. He also earned his PhD degree (2005) in Environmental Microbiology from Sardar Patel University, Vallabh Vidyanagar (Gujarat), India. His research interests include biological wastewater treatment, environmental microbiology, biodegradation, bioremediation, and phytoremediation of environmental pollutants from industrial wastewater. He has published more than 250 research papers in national and international journals of repute on various aspects of microbial biodegradation and bioremediation of environmental pollutants. He is the editor of 110+ books of international repute (published by RSC, Wiley, Nova Sciences, De Gruyter, Elsevier, Springer, and CRC Press).

Pardeep Kaur is an Assistant Professor in the University Institute of Biotechnology (UIBT), Chandigarh University, India. Her PhD and master's degrees are from Punjab Agricultural University, Ludhiana. She holds a teaching experience of more than 6 years. She has published research and review articles in many journals of repute, and has delivered oral and poster presentations in numerous platforms. Her areas of interest are bioprocessing, waste valorisation, microbial enzymes, and fermentation technology.

Preface

The book entitled *Biomass Energy for Sustainable Development – Opportunities and Challenge* explores the feasibility of biomass as a potential substrate for bioenergy generation. Lignocellulosic biomass holds immense potential as a green, sustainable, and renewable feedstock for next-generation advanced bioenergy production. The widespread use of biomass will help in the replacement of conventional petroleum-derived energy sources that form the basis of a contemporary bioeconomy. The positive attributes of biomass for energy generation, especially biofuels, can go a long way in contributing to a healthy environment and economy. Biomass utilization can help reduce forest management costs, mitigate climate change, reduce risks to life and property, and provide a secure, competitive energy source. Shifting to a homegrown, renewable energy economy provides opportunities for growth and expansion, especially for rural communities as these renewable feedstocks are directly connected to the land, primarily agricultural and forestry lands. Currently, biomass-related technologies are being used for the designing of feedstock, biorefining, and waste valorization, besides the production of green and sustainable products, e.g., biofuels, bioenergy, and biomaterials. This book covers the broader aspect of biomass-generated energy and provides an in-depth insight into production technology, advancements, economy, impact, and current status.

Maulin P. Shah

Pardeep Kaur

Contributors

Priya Agarwal
School of Biology
Indian Institute of Science Education
  and Research Thiruvananthapuram
  (IISER TVM)
Thiruvananthapuram, India

Shiksha Arora
Indian Institute of Maize Research
ICAR, Ludhiana
Ludhiana, India

Nisha Arya
Department of Textile and Apparel
  Designing
CCS Haryana Agricultural University
Hisar, India

Deepanshu Awasthi
Sardar Swaran Singh National Institute
  of Bio-Energy
Kapurthala, India

Premkumar Bhuvaneshwaran
Department of Food Technology
Kongu Engineering College
Perundurai, India

Manisha Bisht
Department of Chemistry
L.S.M Govt. P.G-College
Pithoragarh, India

Sourav Debsarma Biswas
Environmental Biotechnology
  Laboratory, Department of
  Biological Sciences
Birla Institute of Technology and
  Science, Pilani, Hyderabad Campus
Hyderabad, India

Khalida Bloch
Department of Microbiology, School of
  Science
R.K. University
Rajkot, India

Rekha Chahar
Department of Microbiology
CCS Haryana Agricultural University
Hisar, India

Sugali Chandra Sekhar
Membrane Separations Laboratory
Chemical Engineering and Process
  Technology Department
CSIR-Indian Institute of Chemical
  Technology, Hyderabad
Telangana, India
and
Academy of Scientific and Innovative
  Research (AcSIR)
Ghaziabad, India

Arghya Datta
Sardar Swaran Singh National Institute
  of Bio-Energy
Kapurthala, India

Priyanka Devi
Lovely Professional University
Phagwara, India

Palvi Dogra
Department of Agronomy, School of
  Agriculture
Lovely Professional University
Phagwara, India

Larissa Echeverria
Postgraduate Program in Chemical
    Engineering
State University of West Parana
Toledo, Brazil

Leila Denise Fiorentin-Ferrari
Postgraduate Program in Chemical
    Engineering
State University of West Parana
Toledo, Brazil

Mônica Lady Fiorese
Postgraduate Program in Chemical
    Engineering
State University of West Parana
Toledo, Brazil

Bhautik Gajera
Sardar Swaran Singh National Institute
    of Bio-Energy
Kapurthala, India

Nikhil Gakkhar
Ministry of New & Renewable Energy
Atal Akshay Urja Bhavan
New Delhi, India

Sougata Ghosh
Department of Microbiology, School of
    Science
R.K. University
Rajkot, India
Department of Physics, Faculty of
    Science
Kasetsart University
Bangkok, Thailand

Rakesh Godara
Sardar Swaran Singh National Institute
    of Bio-Energy
Kapurthala, India

Fabio Augusto Gubiani
Postgraduate Program in Chemistry
Federal University of Santa Catarina
Florianópolis, Brazil

Dandu Harikarthik
Department of Microbiology
CCS Haryana Agricultural University
Hisar, India

S. Hemapriya
Environmental Biotechnology
    Laboratory, Department of
    Biological Sciences
Birla Institute of Technology and
    Science, Pilani, Hyderabad Campus
Hyderabad, India

Jacqueline Ferandin Honório
Postgraduate Program in Chemical
    Engineering
State University of West Parana
Toledo, Brazil

K. Jagadeesh Chandra Bose
Chandigarh University
Mohali, India

Gaganpreet Kaur
Biochemical Conversion Division
Sardar Swaran Singh National Institute
    of Bioenergy
Kapurthala, India

Gurkanwal Kaur
Department of Biochemistry
Punjab Agricultural University
Ludhiana, India

Jaspreet Kaur
School of Agricultural Biotechnology
Punjab Agricultural University
Ludhiana, India

Lovepreet Kaur
School of Bioengineering and
    Biosciences
Lovely Professional University
Phagwara, Punjab

Pardeep Kaur
University Institute of Biotechnology
Chandigarh University
Mohali, India

Simran Jot Kaur
Department of Biotechnology
Sri Guru Granth Sahib World University
Fatehgarh Sahib, India

Monika Kayasth
Department of Microbiology
CCS Haryana Agricultural University
Hisar, India

Prasann Kumar
Department of Agronomy, School of
    Agriculture
Lovely Professional University
Phagwara, India

Sachin Kumar
Biochemical Conversion Division
Sardar Swaran Singh National Institute
    of Bioenergy
Kapurthala, India

Sunaina Kumari
Department of Microbiology, College of
    Basic Sciences and Humanities
CCS Haryana Agricultural University
Hisar, India

Keiti Lopes Maestre
Postgraduate Program in Chemical
    Engineering
State University of West Parana
Toledo, Brazil

Subodh Kumar Maiti
Department of Environmental Science
    and Engineering
Indian Institute of Technology (Indian
    School of Mines)
Dhanbad, India

Kamla Malik
Department of Microbiology
CCS Haryana Agricultural University
Hisar, India

Shweta Malik
Department of Agronomy
CCS Haryana Agricultural University
Hisar, India

Shikha Mehta
Department of Microbiology
CCS Haryana Agricultural University
Hisar, India

Sunaina Nag
Environmental Biotechnology
    Laboratory, Department of
    Biological Sciences
Birla Institute of Technology and
    Science, Pilani, Hyderabad Campus
Hyderabad, India

Mithila V. Nair
Chandigarh University
Mohali, India

Manoj K. Pal
Department of Microbiology
Graphic Era (Deemed to be University)
Dehradun, India

Tapas Kumar Patra
Sardar Swaran Singh National Institute
    of Bio-Energy
Kapurthala, India

N. Premalatha
Department of Biotechnology
Rajalakshmi Engineering College
Chennai, India

Roshni Raj
Environmental Biotechnology
 Laboratory, Department of
 Biological Sciences
Birla Institute of Technology and
 Science, Pilani, Hyderabad Campus
Hyderabad, India

Sathish Raam Ravichandran
Department of Chemical Engineering
Kongu Engineering College
Perundurai, India

Sai Shankar Sahu
Department of Environmental Science
 and Engineering
Indian Institute of Technology (Indian
 School of Mines)
Dhanbad, India

Anita Saini
Department of Microbiology, School of
 Basic and Applied Sciences
Maharaja Agrasen University
Solan, India

P. Sankar Ganesh
Environmental Biotechnology
 Laboratory, Department of
 Biological Sciences
Birla Institute of Technology and
 Science, Pilani, Hyderabad Campus
Hyderabad, India

S. R. Saranya
Department of Chemical Engineering,
 Alagappa College of Technology
Anna University
Chennai, India

Loveleen Kaur Sarao
Department of Plant Breeding and
 Genetics
Punjab Agricultural University
Ludhiana, India

Jyoti Sarwan
Chandigarh University
Mohali, India

Sarath Sekar
Department of Food Technology
Kongu Engineering College
Perundurai, India

Mothil Sengottian
Department of Chemical Engineering
Kongu Engineering College
Perundurai, India

Ankush Sharma
Indian Institute of Maize Research
ICAR, Ludhiana
Ludhiana, India

Kashish Sharma
Department of Microbiology
CCS Haryana Agricultural University
Hisar, India

Nishu Sharma
Department of Microbiology, COBS&H
Punjab Agricultural University
Ludhiana, Punjab

Shiwani Guleria Sharma
Department of Microbiology, COBS&H
Punjab Agricultural University
Ludhiana, Punjab

Smile Sharma
Chandigarh University
Mohali, India

Meena Sindhu
Department of Microbiology
CCS Haryana Agricultural University
Hisar, India

Joginder Singh
Department of Microbiology, School of
    Bioengineering and Biosciences
Lovely Professional University
Phagwara, India

Veronice Slusarski-Santana
Postgraduate Program in Chemical
    Engineering
State University of West Parana
Toledo, Brazil

Sridhar Sundergopal
Membrane Separations Laboratory
Chemical Engineering and Process
    Technology Department
CSIR-Indian Institute of Chemical
    Technology, Hyderabad
Telangana, India
and
Academy of Scientific and Innovative
Research (AcSIR)
Ghaziabad, India

Monica Sachdeva Taggar
Department of Renewable Energy
    Engineering
Punjab Agricultural University
Ludhiana, India

Yamini Tripathi
Helmholtz Centre for Infection
    Research
Braunschweig, Germany

Carina Contini Triques
Postgraduate Program in Chemical
    Engineering
State University of West Parana
Toledo, Brazil

Nazim Uddin
Chandigarh University
Mohali, India

Bukke Vani
Membrane Separations Laboratory
Chemical Engineering and Process
    Technology Department
CSIR-Indian Institute of Chemical
    Technology, Hyderabad
Telangana, India
and
Academy of Scientific and Innovative
    Research (AcSIR)
Ghaziabad, India

Chitra Devi Venkatachalam
Department of Food Technology
Kongu Engineering College
Perundurai, India

Nisha Yadav
Biochemical Conversion Division
Sardar Swaran Singh National Institute
    of Bioenergy
Kapurthala, India
Centre for Energy and Environment
Dr B R Ambedkar National Institute of
    Technology
Jalandhar, India

Sujeeta Yadav
Department of Microbiology, College of
    Basic Sciences & Humanities
Chaudhary Charan Singh Haryana
    Agricultural University
Hisar, India

Gabriela Eduarda Zeni
Postgraduate Program in Chemical
    Engineering
State University of West Parana
Toledo, Brazil

1 Biomass Energy
An Introduction

Jaspreet Kaur
Punjab Agricultural University

Pardeep Kaur
Chandigarh University

Monica Sachdeva Taggar
Punjab Agricultural University

1.1 INTRODUCTION

The expansion of the world's population and industrialization has accelerated global energy consumption over the last few decades (Arumugam et al., 2022). The worldwide global energy consumption was reported to be increased by 2.3% in 2018 as compared to that reported in 2010 (Zhao et al., 2022). Currently, crude oils are the main source of energy that contributes approximately 83% of the global fuel energy utilization during the year 2020 (Holechek et al., 2022). However, these crude oil reserves are unrenewable and definable, and exhausting annually at a rate of 4 billion tonnes. It was estimated that these oil reserves will be exhausted by 2060 if their depletion continues at the present rate (Saleem, 2022). The growing consumption of crude oils at the global level has also increased the emissions of CO_2 up to 33.1 Gt, which causes consequential environmental issues including changes in climate and global warming (BP, 2019). The finite accessibility of crude oils and environmental issues are effective indicators to produce alternative sources of renewable energy. Lignocellulosic feedstock represents a promising fuel alternative to overcome energy disasters at the world level in a viable and environmentally safe manner (Fatma et al., 2018). Currently, 5×10^{19} kJ of energy was produced by biomass, which represents 10% of the global energy depletion per annum. This value could reach 150×10^{19} kJ by the year 2050 depending on the type of feedstock (Alper et al., 2020).

Lignocellulosic biomass is an abundant, sustainable, and renewable reserve on the Earth. In contrast to fossil fuels, it is a carbon-neutral biomass for bioenergy production with vast resources (Zhao et al., 2022). It is constantly formed via the photosynthesis process using atmospheric CO_2, soil H_2O, and energy from the sun.

Approximately 200 billion metric tonnes of lignocellulosic biomass is generated every year (Diaz & Blandino, 2022). Various types of waste including agricultural wastes, industrial wastes (wood, paper, pulp), energy crops, forest residues, and municipal solid waste are the main source of lignocellulosic feedstocks (Ashokkumar et al., 2022) (Figure 1.1). The utilization of lignocellulosic feedstock evades the direct competition between fuel and food, and therefore, presents reliable feedstock for energy production (Lu et al., 2022). Traditionally, lignocellulosic biomass is dumped in landfills. But the major disadvantages associated with its dumping are the high price of transport and ignition problems owing to its high water content and low calorific value. The use of these forest and agricultural feedstock as fodder has been

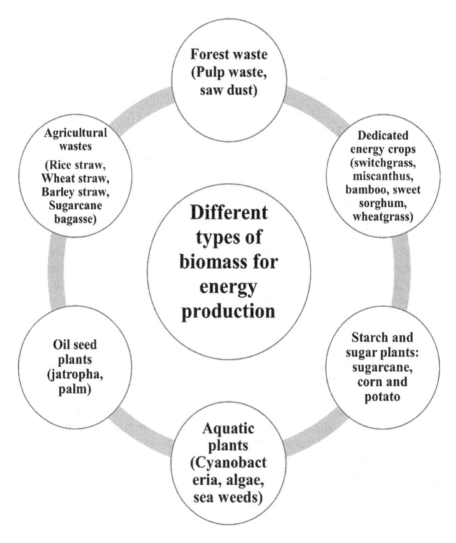

FIGURE 1.1 Types of biomasses used for energy production.

tried, but its consumption by animals on a large scale is limited (Liguori et al., 2013). Hence, its divergence toward innovative green processes instead of setting it ablaze will add value to the waste and help in conserving the natural resources of the ecosystem (Roy et al., 2021). The energy produced from lignocellulosic biomass has a number of advantages including zero-carbon emission, CO_2-neutral nature, waste reduction, agricultural diversification, and reduction in greenhouse gases emissions and global warming, and it also does not influence the food chain (Fatma et al., 2018; Prajapati et al., 2021).

This chapter provides a detailed insight into the potential of various biomasses used for energy production. This chapter involves a detailed discussion of lignocellulosic biomass structure, pretreatment techniques, and conversion technologies that can be barely utilized for energy production from biomass.

1.2 STRUCTURE OF LIGNOCELLULOSIC BIOMASS

Lignocellulosic biomass is chiefly formed of plant cell walls, i.e., primary and secondary. Cellulose and pectin are the two major constituents of the primary cell wall, whereas the secondary cell wall is mainly formed of a complex polymeric matrix, named lignocellulose (McCann & Carpita, 2015). Lignocellulose, as the name indicates, is formed by two main groups of biopolymers, i.e., holocellulose (cellulose plus hemicellulose) and lignin. These compounds play important roles requisite for plant viability survival such as structural integrity, water resistance, and resistance to microbial attack (Cosgrove & Jarvis, 2012).

Cellulose is an ample component (~40% by mass) of the lignocellulose biomass and acts as a backbone of the secondary plant cell wall. Hemicellulose is the third major constituent of biomass and forms 20%–30% of the total biomass (Sannigrah et al., 2010; Wegner & Jones 2009). Lignin is the second utmost polymer of lignocellulosic feedstock and provides shape and rigidity to the plant cell wall (Poovaiah et al., 2014). The structure and composition of the agricultural biomass highly depend on the type of source, i.e., hardwood, softwood, or grasses (Anwar et al., 2014). Lignocellulosic biomass contains cellulose (35%–40%), hemicellulose (20%–35%), and lignin (10%–25%) as their major elements, and the remaining part is made up of proteins, oils, and ash (Wei et al., 2017). Table 1.1 summarizes the chemical composition (cellulose, hemicellulose, and lignin content) of distinctive agricultural biomass.

Besides three main constituents, lignocellulosic biomass also contains inorganic compounds and organic extractives. These components are mainly responsible for flavor, color, smell, and natural resistance to rotting of some species and are known as nonstructural constituents of biomass. Organic extractives include various components, viz. fats, waxes, proteins, terpenes, simple sugars, resins, alkaloids, gums, phenolics, pectins, saponins, and essential oils. The extraction of these nonstructural components could be done by using polar (water and alcohol) and nonpolar solvents (toluene and hexane). The inorganic components form biomass weight by less than 10% which converts into ash during pyrolysis process (Daful & Chandraratne, 2020; Wang et al., 2017).

TABLE 1.1

Composition of Diverse Agricultural Biomass

Agricultural Biomass	Cellulose (%)	Hemicellulose (%)	Lignin (%)	References
Sugarcane bagasse	39.55 ± 0.25	30.20 ± 0.12	20.05 ± 0.11	Kaur et al. (2023)
Coconut shell	25.2 ± 1.0	27.7 ± 0.3	46.0 ± 0.8	Anuchi et al. (2022)
Coconut husk	37.6 ± 1.1	15.2 ± 0.9	41.3 ± 1.6	Anuchi et al. (2022)
Corn stover	34.50	21.95	25.50	García-Negrón and Toht (2022)
Corn cobs	34.7 ± 0.48	33.8 ± 0.64	18.4 ± 0.32	Sunkar and Bhukya (2022)
Eucalyptus sawdust	42.67 ± 0.80	16.05 ± 0.24	23.73 ± 0.1	Tavares et al. (2022)
Rice straw	$35.8 \pm 2.1\%$	17.5 ± 1.4	14.4 ± 0.4	Valles et al. (2021)
Rice husk	31.13	17.71	28.25	Bazargan et al. (2020)
Oil palm fiber	38.67	30.22	23.76	Tareen et al. (2020)
Switchgrass	39.42 ± 0.51	20.25 ± 0.07	21.22 ± 0.63	Wang et al. (2020)
Wheat straw	33.7 ± 1.62	19.1 ± 1.25	19.8 ± 1.50	Zheng et al. (2018)
Sorghum straw	37.74 ± 1.61	28.07 ± 1.12	21.48 ± 0.63	Dong et al. (2019)
Pine sawdust	39.85	20.51	31.41	Kruyeniski et al. (2019)
Soybean straw	39.8 ± 0.6	22.6 ± 1.0	10.5 ± 0.7	Martelli-Tosi et al. (2017)

1.3 LIGNOCELLULOSIC BIOMASS COMPONENTS

1.3.1 CELLULOSE

Cellulose was invented by Anselme Payen in 1838. It is abundantly present on the Earth and forms 40%–60% weight of the biomass (Sharma et al., 2019). Cellulose is composed of β-D-glucopyranose units linked by β-(1,4) glycosidic bonds, with their fundamental repeating disaccharide unit "cellobiose" (Figure 1.2) (Kaur et al., 2020). The chains of cellulose are formed of 500–1,400 units of D-glucose, which are packed cooperatively to form the microfibrils. These microfibrils are arranged simultaneously to construct the cellulose fibrils (Robak & Balcerek, 2018). The large ordered crystalline region and the small disordered amorphous region form these cellulose fibrils (Seddiqi et al., 2021). The degradation of the amorphous region of cellulose by chemicals and enzymes is easier due to the poor organization of hydrogen bonds as compared to a crystalline structure. The degree of polymerization (DP) of cellulose determines the recalcitrance nature of lignocellulosic biomass and varies greatly depending upon the biomass source (Zoghlami & Paës, 2019). About half of the organic carbon exists in cellulose form in the biosphere, and therefore, its conversion into liquid fuels and valuable chemicals has immense importance (Isikgor & Becer, 2015).

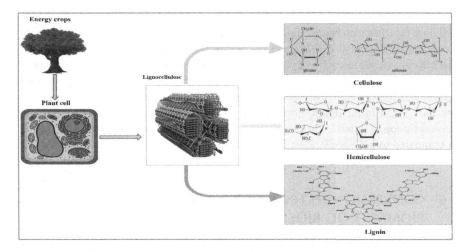

FIGURE 1.2 Organization of lignocellulosic biomass components.

1.4 HEMICELLULOSE

Hemicellulose is the least (by mass) abundant component of lignocellulosic biomass corresponding to 20%–35% of the total weight (Chandel et al., 2018). It is a heteropolysaccharide and chiefly formed of hexose sugars (mannose, glucose, and galactose), pentose sugars (xylose and arabinose), and uronic acids (D-glucuronic acid, 4-O-methyl-D-glucuronic acid, and D-galacturonic acid) (Figure 1.2) (Ahmad & Zakaria, 2019). The DP of hemicelluloses is found to be considerably lower, i.e., 100–200 units as compared to cellulose (Mota et al., 2018). Hemicelluloses are easily degraded by hemicellulases as well as by dilute acids or bases (Isikgor & Becer, 2015). These act as physical obstacles that limit the availability of enzymes. The elimination of hemicelluloses by steam explosion or dilute acid pretreatment could enhance the availability of cellulose to the cellulase enzymes (Herbaut et al., 2018; Santos et al., 2018). Hemicelluloses form a composite network of bonds in plant cell walls by cross-linking with cellulose fibers and lignin and give structural strength to the plant (Isikgora & Becer, 2015). The hemicelluloses also play a role in ameliorating the mechanical properties of plant cell walls (rigidity and flexibility) by their relation with cellulose and lignin.

1.5 LIGNIN

Lignin is the second most abundantly present aromatic polymer following cellulose and forms 15%–40% weight of the lignocellulosic feedstock (Ragauskas et al., 2014). The term "lignin" was coined by Candolle in the year 1819 and is obtained from the Latin word "Lignum" (wood). It is a complex amorphous hetero-polymer consisting of three phenylpropane elements guaiacyl-(G), syringyl-(S), and p-hydroxyphenyl (H) derived from three precursor aromatic alcohols, viz. coniferyl, sinapyl, and p-coumaryl alcohols, respectively (Figure 1.2) (Haq et al., 2021). The composition and organization of each unit in lignin depend upon the type of source. Lignin acts like a glue that holds cellulose together with hemicellulose and makes the matrix structure more recalcitrant to enzymatic degradation (Thapa et al., 2020). It provides structural strength to the plant fibers to resist insects' and pathogens' attack (Frei, 2013). Hardwood lignin is predominantly comprised of guaiacyl-(G) and syringyl-(S) elements, whereas lignin from softwood primarily contains guaiacyl-(G) units (~90%). However, all three precursor units are found in the lignin of nonwoody biomass (Den et al., 2018). The microbial decomposition of lignin is not easy owing to its composite structure and multiple bonds as compared to polysaccharides. However, beneficial chemicals, fuels, and fuel additives can also be generated by efficient depolymerization of lignin (Alper et al., 2020).

1.6 CONVERSION TECHNOLOGIES FOR BIOMASS INTO BIOENERGY

The conversion of biomass into energy depends upon the processing of residue by various pathways. Various parameters including biomass quality and quantity, availability, the overall operational cost, choice of end-products, and environmental problems affect the selection of a conversion technique to be employed to the biomass. The conversion technologies are mainly of three types, i.e., thermochemical conversions, biochemical conversions, and chemical conversions (Figure 1.3; Haq et al., 2021).

1.7 THERMOCHEMICAL CONVERSION

The thermochemical transformation process utilizes heat to generate energy as heat. It's a chemical reformation process performed at an elevated temperature that involves the fragmentation of bonds and organic matter into biochar (solid), bio-oil

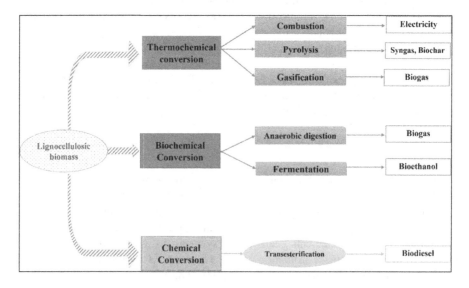

FIGURE 1.3   Types of processes used for conversion of biomass into bioenergy.

(liquid), and syn gas (Lee et al., 2019). Thermochemical conversion processes are mainly of three types, i.e., pyrolysis, gasification, and liquefaction (Jha et al., 2022).

1.8 COMBUSTION

Combustion is a high-temperature exothermic technique that involves the burning of biomass process in the presence of air (oxygen). The oxidation of hydrogen and carbon present in biomass into H_2O and CO_2 produces heat through the combustion method, which can be utilized as electricity (Kaushika et al., 2016). This approach is mostly utilized for the transformation of lignin-rich residues. The combustion process is mainly of two types: (i) direct combustion, which involves the complete conversion of the residue, and (ii) biochemical combustion, in which a certain part of the biomass is not undergoing the conversion process (Garba, 2020). The energy content of biomass and reaction conversion efficiency are the two main parameters that determine the production of heat in the combustion process (Vassilev et al., 2013). Biomass residue constitutes a high amount of volatile matter that significantly affects the thermal degradation of residue along with the combustion performance of solid fuels. Several factors, viz. biomass properties, particle size, combustion temperature, and atmosphere, affect the combustion of feedstock. The temperature for the combustion process is mainly used in the range of 800°C–1,000°C (Tursi, 2019). The byproducts of the combustion process involve char (organic carbon) and ash (inorganic oxides and carbonates).

1.9 PYROLYSIS

Pyrolysis is the most favorable process to transform biomass into solid (biochar) and liquid products (bio-oil). It is a thermochemical conversion method that involves the irreversible degeneration of the organic material of biomass to produce biofuels. Pyrolysis conversion processes are categorized into three types, i.e., slow pyrolysis (300°C–500°C temperature, <50°C/min heating time, >30 min residence time), intermediate pyrolysis (400°C–600°C temperature, 200–300°C/min heating time, 10 min residence time), and fast pyrolysis (400°C–900°C temperature, 10–1,000°C/min heating time, 10 min residence time) (Jha et al., 2022). In the fast pyrolysis process, fast liquefaction of the volatile hydrocarbon vapors into bio-oil occurs due to the rapid rate of heating along with the short vapor residence time. However, the slow rate of heating and longer time of vapor residence result in the production of biochar due to the slower carbonization of biomass in a slow pyrolysis process (Zaimes et al., 2015). The temperature of the pyrolysis process generally varies between 300°C and 700°C (Chandraratne & Daful, 2022). The characteristics of the final pyrolysis product depend on the various parameters including the temperature of the pyrolysis process, rate of heating, biomass composition, and residence time (Jha et al., 2022).

1.10 GASIFICATION

Gasification is a thermochemical technique that transforms carbonaceous material into gaseous products (syngas or producer gas) (Makwana et al., 2019). It is an exothermic reaction that involves partial biomass oxidation for high yields of gaseous products rich in CO, H_2, CH_4, and CO_2 (Tanger et al., 2013). Syn gas can be used to produce different types of energy such as heat, power, biofuel, biomethane, chemicals, and hydrogen (Centi, 2020). The gasification process also results in the formation of bio-chemicals and condensable liquids rich in water. The major objection in the gasification process is the condensation of higher molecular weight volatiles into tars. These tars' polycyclic aromatic hydrocarbons are a major threat to the environment (Milne et al., 1998). Thermal gasification of biomass occurs at high temperatures, i.e., 800–1,200°C. However, relatively average temperatures (400°C–700°C) can be employed for hydro-thermal gasification. Clean fuels and value-added products can also be produced from syngas by the Fischer–Tropsch catalysis (dos Santos et al., 2020).

1.11 BIOCHEMICAL CONVERSIONS

Biochemical conversion processes involve the use of yeast and/or bacteria to convert the waste into useful energy (Lee et al., 2019). Different types of fuels (biogas, hydro-gen, ethanol), chemicals (butanol, acetone), and organic acids can be produced by using biochemical technologies (Garba, 2020). These techniques are more clean and efficient in contrast to other transformation processes (Chen et al., 2016). Biochemical conversion processes can be categorized into two types: (i) anaerobic digestion (AD) and (ii) microbial fermentation (Brethauer & Studer, 2015). The major advantages of biochemical processes are that it mainly occurs at lower temperatures and lower reaction rates as compared to thermochemical conversion.

1.12 ANAEROBIC DIGESTION

AD is a multistep biological process that involves the transformation of lignocellulosic biomass into methane (Wang et al., 2021). The microbial processes involve four steps, viz. hydrolysis, acidogenesis, acetogenesis, and methanogenesis, that act sequentially to break down the organic polymers into simple polymers to produce the biogas (mix-ture of carbon dioxide and) and digestate (Venkiteshwaran et al., 2015). This process involves a sequence of chemical reactions accomplished by microorganisms in an anaerobic environment by using natural metabolic pathways (Anukam et al., 2019). Several types of biomasses, viz. sewage sludge, Municipal solid waste (MSW), ani-mal manure, and agricultural residues can be utilized in this process (Kasinath et al., 2021). The digestates generated throughout this process are nutrient-rich and can be utilized as fertilizer for various agricultural goals (Czekała et al., 2022).

1.13 FERMENTATION

Fermentation is a biochemical approach that is used to break down simple sugars (hexose and pentoses) into ethanol and CO_2 by using microorganisms (yeast) under anaerobic conditions (Sharma et al., 2020). The fermentation of hexose and pentose sugars is carried out as per the following reactions below:

$$\text{Glucose: } C_6H_{12}O_6 \rightarrow 2C_2H_6O + 2CO_2$$

$$\text{Xylose: } 3C_5H_{10}O_5 \rightarrow 5C_2H_6O + 5CO$$

The most common yeast *Saccharomyces cerevisiae* is used for the fermentation process. The feedstock used for the fermentation process is divided into three main groups, i.e., sugar crops, starchy crops, and lignocellulosic wastes (Zabed et al., 2016). From 100 g of hexoses and pentoses, it is calculated that 51.14 g of ethanol and 48.86 g of CO_2 can be produced theoretically. Several co-products, viz. CO_2, glycerol, and carboxylic acids, are also produced along with ethanol during the fermentation process. Several parameters, viz. type of biomass, temperature, inoculum, pH, and fermentation time, affect the quality along with the yield of the formation process (Strezov et al., 2014). Depending upon the starting material, different metabolic pathways could be utilized for the transformation of sugars into ethanol. For example, hexose and pentose sugars are converted into ethanol through glycolysis or the Embden–Meyerhof pathway (EMP) and the pentose phosphate pathway (PPP), respectively (Tursi, 2019). The crude alcohol produced at the end of the process must be processed through distillation and dehydration steps to obtain the concentrated alcohol (Bibi et al., 2017). The solid part obtained after fermentation could be utilized as livestock feed or as fuel for the generation of gas in the boilers (Mariod, 2016).

1.14 PHYSICOCHEMICAL CONVERSION PROCESS

Transesterification/esterification is a chemical process that converts biomass into high-density biodiesel. During the transesterification process, lipids are transformed into fatty acid alkyl esters (FAAE) by reacting with alcohol (Mandari & Devarai, 2022). Different types of vegetable oils (rapeseed oil and sunflower oil) and animal fats could be utilized for biodiesel generation. Biodiesel can also be yielded from waste cooking oil and microbial oil such as algal oil and categorized as second- and third-generation biodiesel, respectively (Tursi, 2019). During the transesterification process, triglycerides present in oils are broken down into monomeric units, i.e., fatty acids and glycerol. The lipid on reaction with an excess amount of methyl or ethyl alcohol in the presence of a catalyst (i.e., NaOH or KOH) yields methyl or ethyl esters (biodiesel) (Kumar et al., 2019). The temperature needed for the transesterification of triglycerides is generally low, i.e., 50°C–70°C.

$$\text{Oil} + \text{Short chain alcohol} \rightarrow \text{Biodiesel} + \text{glycerol}$$

This process involves an interchange of alcohol with another alcohol and is also known as alcoholysis. The reaction time, pressure, catalyst type, concentration, alcohol:oil ratio, and type of feedstock oil are the main factors that influence the biodiesel yield during the transesterification process (Mandari & Devarai, 2022).

1.15 ENERGY FROM BIOMASS

Biomass can be transformed into energy by utilizing different pretreatment methods and conversion techniques. The physiochemical properties, viz. moisture content, fixed carbon proportions and volatile substances, ash content, caloric value, cellulose to lignin ratio, and alkali metal content, act as important parameters for the choice of the biomass for generation of energy (Tursi, 2019). The intrinsic nature of lignocellulosic biomass makes it indestructible without any pretreatment to break down the lignin layer of biomass (Den et al., 2018). Pretreatment is, therefore, a precondition and an essential step for the conversion of lignocellulose into bioenergy and value-added products (Deng et al., 2023). The primary objective of the pretreatment is to disrupt the polymeric lignin structure and reduction in cellulose crystallinity. The pretreatment process increases the accessibility of cellulose to saccharification enzymes to aid in the breakdown of complex sugars into simple fermentable sugars (Verma et al., 2011). Several types of pretreatments have been utilized for the transformation of lignocellulosic waste into liquid fuels. Pretreatment methods are categorized into four groups depending upon their mode of action: (i) physical methods (e.g., milling, grinding, and irradiation); (ii) chemical methods (e.g., alkali, acid, oxidizing agents, organic solvents, and deep-eutectic solvents); (iii) physicochemical methods (e.g., steam pretreatment, wet-oxidation, and hydro thermolysis); and (iv) biological methods (Galbe & Wallberg, 2019) (Table 1.2). The significance of all the pretreatment methods is to reduce the size of biomass and to open its physical structure.

Different types of energy like biodiesel, bioethanol, biobutanol, biogas, bioelectricity, etc. could be generated from lignocellulosic biomass. Table 1.3 highlights the different forms of energy produced from different types of biomasses (Table 1.3). Bioenergy obtained from the fermentation of feedstock sugar such as corn grains (starch) and sugarcane (sucrose) is known as first-generation biofuel, while bioenergy derived from biomass like lignocellulosic biomass, vegetable oil, cooking oil, animal fats, wood, and pulp comes under the category of second-generation biofuels (Zabed et al., 2017). Second-generation bioenergy provides several advantages as compared to first-generation energy as it overcomes issues of food versus fuel competition. Energy derived from algae is termed as third-generation biofuel (Ganguly et al., 2021; Zabed et al., 2017). The main advantage of using algae is that it can grow on waste water at a rapid rate as compared to other feedstocks. Bioelectricity is known as the advanced form of energy produced by microbial fuel cells. It directly transforms the chemical energy into electricity by using the catalyst, i.e., biofilm (Singh et al., 2022).

1.16   CONCLUSION

The consumption of fossil energy has increased due to unprecedented population growth and industrialization during the past few decades. The substitution of fossil fuels with biomass energy has enormous positive benefits from viewpoints of health, economics, and environment (climate change, global warming). The abundantly available and unutilized biomass has become an important chemical raw material that could be transformed into various forms of energy. Different types of wastes such as agriculture residues, woody materials, grasses, and other plant materials could also be managed by their conversion into energy. The plant-derived renewable lignocellulosic wastes have immense potential for the generation of 2-G energy. It can be used as an energy source in several forms, i.e., liquid fuels, gaseous fuels, combustible fuel, and fuel pellets. Technological innovations are the main problem to increase productivity and reduce the costs of such biomass-derived renewable energy. Efforts should be made for the advancement of highly user-friendly and cost-efficient technologies for the production of biomass energy.

TABLE 1.2

Different Pretreatment Processes Available for Conversion of Lignocellulosic Wastes into Bioenergy

Pretreatment	Type	Definition	Advantages	Disadvantages	Reference
Physical	Milling	Mechanical pulverization of biomass to decrease the particle size	• Simple and easy to operate • Pre-treatment of immense amounts of biomass.	• High consumption of power	Yu et al. (2019)
	Irradiation	Involves the use of irradiations of high energy, viz. X-rays and gamma rays disrupt the structure of lignocellulosic biomass	• Reduction in crystallinity of cellulose and increase in the pore and surface area	• Additional steps of pretreatment are required	Kassim et al. (2016)
	Ultrasound	Use of ultrasound energy to generate the highly reactive radicals for lignin degradation and cellulose conversion into sugars	• No utilization of chemicals		Flores et al. (2021)
Chemical	Alkali	Solubilization of lignin along with some part of hemicellulose by using different alkaline agents, viz., NaOH, Ca(OH)$_2$, KOH, and NH$_4$OH to ameliorate the cellulose availability to hydrolytic enzymes	• The availability of cellulose increased • Degradation of the structure of lignin	• High cost of chemicals • Equipments are related to corrosion issues	Kim et al. (2016)
	Acid	Destruction of the covalent bonds, van der Waals' forces, and hydrogen bonds among biomass constituents results in the hydrolysis of hemicellulose component, especially xylan, by using different types of acids (H$_2$SO$_4$, H$_3$PO$_4$, HNO$_3$, HCl)	• Hydrolyses hemicellulose into various sugar fractions	• Formation of liable inhibitory byproducts • Requires long residence time (time-consuming)	Amin et al. (2017)
	Oxidizing agents	Use of oxidation agents such as organic peroxide (H$_2$O$_2$, C$_2$H$_4$O$_3$) to degrade the ring structure of lignin			Lee et al. (2014)
	Organic solvents	Extraction of lignin as well as recovery of high-purity cellulose from lignocellulosic wastes by using organic solvents or their aqueous solutions			Borand and Karaosmanoğlu (2018)

(Continued)

TABLE 1.2 (Continued)
Different Pretreatment Processes Available for Conversion of Lignocellulosic Wastes into Bioenergy

Pretreatment	Type	Definition	Advantages	Disadvantages	Reference
	Deep-eutectic solvents	A new class of ionic fluids that breaks supramolecular hydrogen bonds between hemicellulose and lignin matrix in lignocellulosic biomass, thus rendering the covalent bonds to degrade more easily			Scelsi et al. (2021)
Physicochemical	Steam pretreatment	The lignocellulosic biomass along with saturated steam is introduced into a batch or continuous reactor at high pressure. A sudden drop in pressure leads to vapor expansion inside fibers, resulting in the degradation of the fibrous structure	• Causes degradation of hemicellulose and lignin • Economically viable	• In complete degradation of the lignin and carbohydrate complex	Ziegler-Devin et al. (2021)
	Wet-oxidation	Involves pretreatment of biomass in water and oxygen-rich environment at high temperatures (120°C–238°C) and pressure (45–480 psi)		• Production of inhibitory byproducts	Zhou et al. (2023)
	Hydro thermolysis	Involves the degradation of biomass by auto-ionization of water under pressure at elevated temperatures and forms the acetic acid from acetyl groups present in the hemicellulose part of the biomass. Also termed as pressure-cooking in water, hydrothermal treatment, and autohydrolysis			Galbe and Wallberg (2019)
Biological		Degradation of lignin and hemicellulose components of biomass by using microorganisms (mainly fungi white-rot, brown-rot, and soft-rot fungi)	• Effective disruption of lignocellulosic components, i.e., cellulose and hemicellulose • Appropriate for wastes with high and low moisture contents • Less energy requirement	• Disruption of lignin is insignificant • Slow hydrolysis rate • Requires long residence times and vast space	Weng et al. (2021)

TABLE 1.3

Various Forms of Energy Produced from Biomass Wastes

Biomass	Conditions	Conversion Process	Type of Energy	References
Paper waste and cow dung (1:1)	Retention time—45 days and mesophilic temperature range of 26°C–43°C	Biological (anaerobic digestion)	Biogas $(9.34 \pm 0.11 \text{ dm}^3/$ kg slurry)	Ofoefule et al. (2010)
Switchgrass	Alkali pretreatment (1.0% NaOH) for 12 h at 50°C	Biological (fermentation)	Bioethanol	Xu et al. (2010)
Sugarcane	Alkali pretreatment [2.75% (w/v) NaOH, 10% (w/v)] solid loading microwave pretreatment for 20 min at 500 W power	Biological (fermentation)	Bioethanol	Singh et al. (2013)
Food wastes	Digestion period—70 days, mesophilic temperature—30°C–37°C, 8% total solid concentration	Biological (anaerobic digestion)	Biogas (1,090–8,016.67 mL/day)	Ojikutu and Osokoya (2014)
Jatropha oil	Catalyst ratio, i.e., 0.02:1 (w/w) calcium oxide, reaction time—133.1 min, and 5.15:1 mol/mol of methanol to the pretreated oil	Chemical (transesterification)	Biodiesel (95.8% yield)	Reddy et al. (2016)
Food waste	Mesophilic temperature (37°C $\pm$ 0.1°C; MDi), feeding time—6 days, hydraulic retention time (HRT)—20 days, and organic loading rate (OLR)—3 g VS/L/d for all the digesters	Biological (anaerobic digestion)	Biogas (480 mL/g VS)	Zamanzadeh et al. (2017)
Rice straw	Alkali pretreatment (1 M NaOH 8% solid loading ratio, autoclave at 121°C for 1 h)	Biological (fermentation)	Bioethanol (30.5 g/L)	Takano and Hoshino (2018)
Waste cooking oil	50°C temperature, methanol to oil ratio—8:1 WCO, 1% calcium oxide nanocatalyst, and 90 min reaction time	Chemical (transesterification)	Biodiesel (96% yield)	Degfie et al. (2019)
Wheat straw	Alkali pretreatment (1% w/v potassium hydroxide followed by 24 h room temperature incubation, autoclaved at 121°C and 15 psi)	Biological (fermentation)	Bioethanol (14.70 g/L)	Sharma et al. (2021)

REFERENCES

Alper, K., Tekin, K., Karagöz, S., & Ragauskas, A. J. (2020). Sustainable energy and fuels from biomass: A review focusing on hydrothermal biomass processing. *Sustainable Energy and Fuels, 4,* 4390–4414. doi:10.1039/D0SE00784F

Amin, F. R., Khalid, H., Zhang, H., Rahman, S. U., Zhang, R., Liu, G., & Chen, C. (2017). Pretreatment methods of lignocellulosic biomass for anaerobic digestion. *AMB Express, 7,* 72. doi:10.1186/s13568-017-0375-4

Anuchi, S. O., Campbell, K. L. S., & Hallet, J. P. (2022). Effective pretreatment of lignin-rich coconut wastes using a low-cost ionic liquid. *Scientific Reports, 12,* 6108. doi:10.1038/s41598-022-09629-4

Anukam, A., Mohammadi, A., Naqvi, M., & Granström, K. (2019). A review of the chemistry of anaerobic digestion: Methods of accelerating and optimizing process efficiency. *Processes, 7,* 504. doi:10.3390/pr7080504

Anwar, Z., Gulfraz, M., & Irshad, M. (2014). Agro-industrial lignocellulosic biomass a key to unlock the future bio-energy: A brief review. *Journal of Radiation Research and Applied Sciences, 7,* 163–173. doi:10.1016/j.jrras.2014.02.003

Ashokkumar, V., Venkatkarthick, R., Jayashree, S., Chuetor, S., Dharmaraj, S., Kumar, G., Chen, W. H., & Ngamcharussrivichai, C. (2022). Recent advances in lignocellulosic biomass for biofuels and value-added bioproducts: A critical review. *Bioresource Technology, 344,* 126195. doi:10.1016/j.biortech.2021.126195

Bazargan, A., Wang, Z., Barford, J. P., Saleem, J., & McKay, G. (2020). Optimization of the removal of lignin and silica from rice husks with alkaline peroxide. *Journal of Cleaner Production, 260,* 120848. doi:10.1016/j.jclepro.2020.120848

Bibi, R., Ahmad, Z., Imran, M., Hussain, S., Ditta, A., Mahmood, S., & Khalid, A, (2017). Algal bioethanol production technology: A trend towards sustainable development. *Renewable and Sustainable Energy Reviews, 71,* 976–985. doi:10.1016/j.rser.2016.12.126

Borand, M. N., & Karaosmanoğlu, F. (2018). Effects of organosolv pretreatment conditions for lignocellulosic biomass in biorefinery applications: A review. *Journal of Renewable and Sustainable Energy, 10,* 033104. doi:10.1063/1.5025876

BP. (2019). *Statistical Report: Statistical Review of World Energy,* 68th edn. Bioenergy Europe

Brethauer, S., & Studer, M. H. (2015). Biochemical conversion processes of lignocellulosic biomass to fuels and chemicals: A review. *CHIMIA International Journal for Chemistry, 69,* 572–581. doi:10.2533/chimia.2015.572

Centi, G., & Perathoner, S. (2020). Chemistry and energy beyond fossil fuels: A perspective view on the role of syngas from waste sources. *Catalysis Today, 342,* 4–12. doi:10.1016/j.cattod.2019.04.003

Chandel, A. K., Garlapati, V. K., Singh, A. K., Antunes, F. A. F., & da Silva, S. S. (2018). The path forward for lignocellulose biorefineries: Bottlenecks, solutions, and perspective on commercialization. *Bioresource Technology, 264,* 370–381. doi:10.1016/j.biortech.2018.06.004

Chandraratne, M. R., & Daful, A. G. (2022). Recent advances in thermochemical conversion of biomass. In M. Bartoli & M. Giorcelli (Eds.), *Recent Perspectives in Pyrolysis Research.* IntechOpen. doi:10.5772/intechopen.100060

Chen, H., & Wang, L. (2016). *Technologies for Biochemical Conversion of Biomass.* Academic Press

Cosgrove, D., & Jarvis, M. (2012). Comparative structure and biomechanics of plant primary and secondary cell walls. *Frontiers in Plant Science, 3,* 204. doi:10.3389/fpls.2012.00204

Czekała, W. (2022). Digestate as a source of nutrients: Nitrogen and its fractions. *Water, 14,* 4067. doi:10.3390/w14244067

Daful, A. G., & Chandraratne, M. R. (2020). Biochar production from biomass waste-derived material. In S. Hashmi & I. A. Choudhury (Eds.), *Encyclopedia of Renewable and Sustainable Materials* (pp. 370–378). Elsevier

Degfie, T. A., Mamo, T. T., & Mekonnen, Y. S. (2019) Optimized biodiesel production from waste cooking oil (WCO) using calcium oxide (CaO) nano-catalyst. *Scientific Reports*, 9, 18982. doi:10.1038/s41598-019-55403-4

Den, W., Sharma, V. K., Lee, M., Nadadur, G., & Verma, R. S. (2018). Lignocellulosic biomass transformations via greener oxidative pretreatment processes: Access to energy and value-added chemicals. *Frontiers in Chemistry*, 6, 141. doi:10.3389/fchem.2018.00141

Deng, W., Feng, Y., Fu, J., Guo, H., Guo, Y., Han, B., Jiang, Z., Kong, L., Li, C., Liu, H., Nguyen, P. T. T., Ren, P., Wang, F., Wang, S., Wang, Y., Wang, Y., Wong, S. S., Yan, K., Yan, N., Yang, X., Zhang, Y., Zhang, Z., Zeng, X., & Zhou, H. (2023). Catalytic conversion of lignocellulosic biomass into chemicals and fuels. *Green Energy & Environment*, 8, 10–114. doi:10.1016/j.gee.2022.07.003

Diaz, A. B., & Blandino, A. (2022). Value-added products from agro-food residues. *Foods*, 11, 766. doi:10.3390/foods11050766-

Dong, M., Wang, S., Xu, F., Wang, J., Yang, N., Li, Q., Chen, J., & Li, W. (2019). Pretreatment of sweet sorghum straw and its enzymatic digestion: Insight into the structural changes and visualization of hydrolysis process. *Biotechnology for Biofuels and Bioproducts*, 12, 276. doi:10.1186/s13068-019-1613-6

dos Santos, R. G., & Alencar, A. C. (2020). Biomass-derived syngas production via gasification process and its catalytic conversion into fuels by Fischer Tropsch synthesis: A review. *International Journal of Hydrogen Energy*, 45, 18114–18132. doi:10.1016/j.ijhydene.2019.07.133

Fatma, S., Hameed, A., Noman, M., Ahmed, T., Shahid, M., Tariq, M., Sohail, I., & Tabassum, R. (2018). Lignocellulosic biomass: A sustainable bioenergy source for the future. *Protein & Peptide Letters*, 25, 48–163. doi:10.2174/0929866525666180122144504

Flores, E. M. M., Cravotto, G., Bizzi, C. A., Santos, D., & Iop, G. D. (2021). Ultrasound-assisted biomass valorization to industrial interesting products: State-of-the-art, perspectives and challenges, *Ultrasonics Sonochemistry*, 72, 105455. doi:10.1016/j.ultsonch.2020.105455

Frei, M. (2013). Lignin: Characterization of a multifaceted crop component. *The Scientific World Journal*, 2013, 1–25. doi:10.1155/2013/436517

Galbe, M., & Wallberg, O. (2019). Pre-treatment for biorefineries: A review of common methods for efficient utilization of lignocellulosic materials. *Biotechnology for Biofuels and Bioproducts*, 12, 1–26. doi:10.1186/s13068-019-1634-1

Ganguly, P., Sarkhel, R., & Das, P. (2021). The second-and third-generation biofuel technologies: Comparative perspectives. In S. Dutta & C. M. Hussain (Eds.), *Sustainable fuel Technologies Handbook* (pp. 29–50). Elsevier. doi:10.1016/B978-0-12-822989-7.00002-0

Garba, A. (2021). Biomass conversion technologies for bioenergy generation: An introduction. In T. P. Basso, T. O. Basso, & L. C. Basso (Eds.), *Biotechnological Applications of Biomass*. IntechOpen. doi:10.5772/intechopen.93669

García-Negrón, V., & Toht, M. J. (2022) Corn stover pretreatment with Na2CO3 solution from absorption of recovered CO2. *Fermentation*, 8, 600. doi:10.3390/fermentation8110600

Haq, I. U., Qaisar, K., Nawaz, A., Akram, F., Mukhtar, H., Zohu, X., Xu, Y., Mumtaz, M. W., Rashid, U., Ghani, W. A. K., & Choong, T. S. Y. (2021). Advances in valorization of lignocellulosic biomass towards energy generation. *Catalysts*, 11, 309. doi:10.3390/catal11030309

Herbaut, M., Zoghlami, A., Habrant, A., Falourd, X., Foucat, L., Chabbert, B., & Paes, G. (2018). Multimodal analysis of pretreated biomass species highlights generic markers of lignocellulose recalcitrance. *Biotechnology for Biofuels*, 11, 52. doi:10.1186/s13068-018-1053-8

Holechek, J. L., Geli, H. M. E., Sawalhah, M. N., & Valdez, A. R. (2022). Global assessment: Can renewable energy replace fossil fuels by 2050? *Sustainability*, 14, 4792. doi:10.3390/su14084792

Isikgora, F. H., & Becer, C. R. (2015). Lignocellulosic biomass: A sustainable platform for the production of bio-based chemicals and polymers. *Polymer Chemistry*, 6, 4497. doi:10.1039/C5PY00263J

Jha, S., Nanda, S., Acharya, B., & Dalai, A. K. (2022). A review of thermochemical conversion of waste biomass to biofuels. *Energies*, *15*, 6352. doi:10.3390/en15176352

Kasinath, A., Fudala-Ksiazek, S., Szopinska, M., Bylinski, H., Artichowicz, W., Remiszewska-Skwarek, A., & Luczkiewicz, A. (2021). Biomass in biogas production: Pretreatment and codigestion. *Renewable and Sustainable Energy Reviews*, *150*, 111509. doi:10.1016/j.rser.2021.111509

Kassim, M. A., Khalil, H. P. S. A., Serri, N. A., Kassim, M. H. M., Syakir, M. I., Aprila, N. A. S., & Dungani, R. (2016). Irradiation pretreatment of tropical biomass and biofiber for biofuel production. In W. A. Monteiro (Eds.), *Radiation Effects in Materials*. InTech. doi:10.5772/62728

Kaur, J., Taggar, M. S., Kalia, A., Sanghera, G. S., Kocher, G. S., & Javed, M. (2023). Valorization of sugarcane bagasse into fermentable sugars by efficient fungal cellulolytic enzyme complex. *Waste and Biomass Valorization*, *14*, 963–975. doi:10.1007/s12649-022-01918-3

Kaur, P., Taggar, M. S., & Kaur, J. (2020). Cellulolytic microorganisms: Diversity and role in conversion of rice straw to bioethanol. *Cellulose Chemistry and Technology*, *54*, 613–634. doi:10.35812/CelluloseChemTechnol.2020.54.61

Kaushika, N. D., Reddy, K. S., & Kaushik, K. (2016). Biomass energy and power systems. In N. D. Kaushika, K. S. Reddy, & K. Kaushik (Eds.), *Sustainable Energy and the Environment: A Clean Technology Approach* (pp. 123–137). Springer. doi:10.1007/978-3-319-29446-9_9

Kim, J. S., Lee, Y. Y., & Kim, T. H. (2016). A review on alkaline pretreatment technology for bioconversion of lignocellulosic biomass. *Bioresource Technology*, *199*, 42–48. doi:10.1016/j.biortech.2015.08.085

Kruyeniski, J., Ferreira, P. J. T., da Graça, M. V. S. C., Vallejos, M. E., Felissia, F. E., & Area, M. C. (2019). Physical and chemical characteristics of pretreated slash pine sawdust influence its enzymatic hydrolysis. *Industrial Crops and Products*, *130*, 528–536. doi:10.1016/j.indcrop.2018.12.075

Kumar, N. S., Geetha, R. S., Vasu, A. T., & Ramu, D. V. (2019). Performance assessment of biodiesel transesterified from *Labeo rohita* visceral oil using IC engine. *Current Trends in Biotechnology and Pharmacy*, *13*, 348–356

Lee, H. V., Hamid, S. B. A., & Zai, S. K. (2014). Conversion of lignocellulosic biomass to nanocellulose: Structure and chemical process. *Scientific World Journal*, 2014, 1–20. doi:10.1155/2014/631013

Lee, S. Y., Sankaran, R., Chew, K. W., Tan, C. H., Krishnamoorthy, R., Chu, D. T., & Show, P. L. (2019). Waste to bioenergy: A review on the recent conversion technologies. *BMC Energy*, *1*, 4. doi:10.1186/s42500-019-0004-7

Liguori, R., Amore, A., & Faraco, V. (2013). Waste valorization by biotechnological conversion into added value products. *Applied Microbiology and Biotechnology*, *97*, 6129–6147. doi:10.1007/s00253-013-5014-7

Lu, H., Yadav, V., Bilal, M., & Iqbal, H. M. N. (2022). Bioprospecting microbial hosts to valorize lignocellulose biomass- Environmental perspectives and value-added bioproducts. *Chemosphere*, *288*, 132574. doi:10.1016/j.chemosphere.2021.132574

Makwana, J. P., Pandey, J., & Mishra, G. (2019). Improving the properties of producer gas using high temperature gasification of rice husk in a pilot scale fluidized bed gasifier (FBG). *Renewable Energy*, *130*, 943–951. doi:10.1016/j.renene.2018.07.011

Mandari, V., & Devarai, S. K. (2022). Biodiesel production using homogeneous, heterogeneous, and enzyme catalysts via transesterification and esterification reactions: A critical review. *Bioenergy Research*, *15*, 935–961. doi:10.1007/s12155-021-10333-w

Mariod, A. A. (2016). Extraction, purification, and modification of natural polymers. In O. Olatunji (Ed.), *Natural Polymers* (pp. 63–91). Springer. doi:10.1007/978-3-319-26414-1_3

Martelli-Tosi, M., Assis, O. B. G., Silvaa, N. C., Espostoa, B. S., Martins, M. A., & Tapia-Blácidoa, D. R. (2017). Chemical treatment and characterization of soybean straw and soybean protein isolate/straw composite films. *Carbohydrate Polymers, 157*, 512–520. doi:10.1016/j.carbpol.2016.10.013

McCann, M. C., & Carpita, N. C. (2015). Biomass recalcitrance: A multi-scale, multi-factor, and conversion-specific property. *Journal of Experimental Botany, 66*, 4109–4118. doi:10.1093/jxb/erv267

Milne, T. A., Abatzoglou, N., & Evans, R. J. (1998). Biomass gasifier "tars": Their nature, formation, and conversion. *National Renewable Energy Laboratory*, NREL/TP-570-25357. doi:10.2172/3726

Mota, T. R., Oliveira, D. M., Rogério Marchiosi, O., & Ferrarese-Filho Santos, W. D. (2018). Plant cell wall composition and enzymatic deconstruction. *Bioengineering, 5*, 63–77. doi:10.3934/bioeng.2018.1.63

Ofoefule, A. U., Nwankwo, J. I., & Cynthia, N. (2010). Biogas production from paper waste and its blend with cow dung. *Advances in Applied Science Research, 1*, 1–8

Ojikutu, A. O., & Osokoya, O. O. (2014). Evaluation of biogas production from food waste. *The International Journal of Engineering and Science, 3*, 1–7

Poovaiah, C. R., Nageswara-Rao, M., Soneji, J. R., Baxter, H. L., & Stewart, Jr., C. N. (2014). Altered lignin biosynthesis using biotechnology to improve lignocellulosic biofuel feedstocks. *Plant Biotechnology Journal, 12*, 1163–1173. doi:10.1111/pbi.12225

Prajapat, P., Varjani, S., Singhania, R. R., Patel, A. K., Awasthi, M. K., Sindhu, R., Zhang, Z., Binod, P., Awasthi, S. K., & Chaturvedi, P. (2021). Critical review on technological advancements for effective waste management of municipal solid waste: Updates and way forward. *Environmental Technology & Innovation, 23*, 101749. doi:10.1016/j.eti.2021.101749

Ragauskas, A. J., Beckham, G. T. Biddy, M. J., Chandra, R., Chen, F., Davis, M. F., Davison, B. H., Dixon, R. A., Gilna, P., Keller, M., Langan, P., Naskar, A. K., Saddler, J. N., Tschaplinski, T. J., Tuskan, G. A., & Wyman, C. E. (2014). Lignin valorization: Improving lignin processing in the biorefinery. *Science, 344*, 1246843. doi:10.1126/science.1246843

Reddy, A. N. R., Saleh, A. A., Islam, M. S., Hamdan, S., & Maleque, M. A. (2016). Biodiesel production from crude jatropha oil using a highly active heterogeneous nanocatalyst by optimizing transesterification reaction parameters. *Energy & Fuels, 30*, 334–343. doi:10.1021/acs.energyfuels.5b01899

Robak, K., & Balcerek, M. (2018). Review of second-generation bioethanol production from residual biomass. *Food Technology and Biotechnology, 56*, 174–187. doi:10.17113/ftb.56.02.18.5428

Roy, S., Dikshit, P. K., Sherpa, K. C., Singh, A., Jacob, S., & Rajak, R. C. (2021). Recent nanobiotechnological advancements in lignocellulosic biomass valorization: A review. *Journal of Environmental Management, 297*, 113422. doi:10.1016/j.jenvman.2021.113422

Saleem, M. (2022). Possibility of utilizing agriculture biomass as a renewable and sustainable future energy source. *Heliyon, 8*, e08905. doi:10.1016/j.heliyon.2022.e08905

Sannigrahi, P., Ragauskas, A. J., & Tuskan, G. A. (2010). Poplar as a feedstock for biofuels: A review of compositional characteristics. *Biofuels, Bioproducts and Biorefining, 4*, 209–226. doi:10.1002/bbb

Santos, V. T. O., Siqueira, G., Milagres, A. M. F., & Ferraz, A. (2018). Role of hemicellulose removal during dilute acid pretreatment on the cellulose accessibility and enzymatic hydrolysis of compositionally diverse sugarcane hybrids. *Industrial Crops and Products, 111*, 722–730. doi:10.1016/j.indcrop.2017.11.053

Scelsi, E., Angelini, A., & Pastore, C. (2021). Deep eutectic solvents for the valorisation of lignocellulosic biomasses towards fine chemicals. *Biomass, 1*, 29–59. doi:10.3390/biomass1010003

Seddiqi, H., Oliaei, E., Honarkar, H., Jin, J., Geonzon, L. C., Bacabac, R. G., & Klein-Nulend, J. (2021). Cellulose and its derivatives: Towards biomedical applications. *Cellulose*, *28*, 1893–1931. doi:10.1007/s10570-020-03674-w

Sharma, H. K., Xu, C., & Qin, W. (2019). Biological pretreatment of lignocellulosic biomass for biofuels and bioproducts: An overview. *Waste and Biomass Valorization*, *10*, 235–251. doi:10.1007/s12649-017-0059-y

Sharma, R., Garg, P., Kumar, P., Bhatia, S. K., & Kulshrestha, S. (2020). Microbial fermentation and its role in quality improvement of fermented foods. *Fermentation*, *6*, 106. doi:10.3390/fermentation6040106

Sharma, S., Jha, P. K., & Panwar, A. (2021). Production of bioethanol from wheat straw via optimization of co-culture conditions of *Bacillus licheniformis* and *Saccharomyces cerevisiae*. *Discover Energy*, *1*, 5. doi:10.1007/s43937-021-00004-4

Singh, S., Sarkar, P., & Dutta, K. (2022). Bioenergy: An overview of bioenergy as a sustainable and renewable source of energy. In P. Verma & M. P. Shah (Eds.), *Bioprospecting of Microbial Diversity* (pp. 483–502). Elsevier. doi:10.1016/B978-0-323-90958-7.00006-6

Strezov, V. (2014). Properties of biomass fuels. In V. Strezov & T. J. Evans (Eds.), *Biomass Processing Technologies* (pp. 1–32). CRC Press

Sunkar, B., & Bhukya, B. (2022). An approach to correlate chemical pretreatment to digestibility through biomass characterization by SEM, FTIR and XRD. *Frontiers in Energy Research*, *10*, 802522. doi:10.3389/fenrg.2022.802522

Takano, M., & Hoshino, K. (2018). Bioethanol production from rice straw by simultaneous saccharification and fermentation with statistical optimized cellulase cocktail and fermenting fungus. *Bioresources and Bioprocessing*, *5*, 16. doi:10.1186/s40643-018-0203-y

Tanger, P., Field, J. L., Jahn, C. E., DeFoort, M. W., & Leach, J. E., (2013). Biomass for thermochemical conversion: Targets and challenges. *Frontiers in Plant Science*, *4*, 218. doi:10.3389/fpls.2013.00218

Tareen, A. K., Punsuvon, V., & Parakulsuksatid, P. (2020). Investigation of alkaline hydrogen peroxide pretreatment to enhance enzymatic hydrolysis and phenolic compounds of oil palm trunk. *Biotech*, *10*, 179. doi:10.1007/s13205-020-02169-6

Tavares, D., Cavali, M., de Oliveira, V. A. T., Torres, L. A. Z., Rozendo, A. S., Filho, A. Z., Soccol, C. R., & Woiciechowski, A. L. (2022). Lignin from residual sawdust of eucalyptus spp.- isolation, characterization, and evaluation of the antioxidant properties. *Biomass*, *2*, 195–208. doi:10.3390/biomass2030013

Thapa, S., Mishra, J., Arora, N., Mishra, P., Li, H., O'Hair, J., Bhatti, S., & Zhou, S. (2020). Microbial cellulolytic enzymes: Diversity and biotechnology with reference to lignocellulosic biomass degradation. *Reviews in Environmental Science and Biotechnology*, *19*, 621–648. doi:10.1007/s11157-020-09536-y

Tursi, A. (2019). A review on biomass: Importance, chemistry, classification, and conversion. *Biofuel Research Journal*, *22*, 962–979. doi:10.18331/BRJ2019.6.2.3

Valles, A., Capilla, M., Álvarez-Hornos, F. J., García-Puchol, M., San-Valero, P., & Gabaldón, C. (2021). Optimization of alkali pretreatment to enhance rice straw conversion to butanol. *Biomass and Bioenergy*, *150*, 106131. doi:10.1016/j.biombioe.2021.106131

Vassilev, S. V., Baxter, D., & Vassileva, C. G. (2013). An overview of the behavior of biomass during combustion: Part I. Phase-mineral transformations of organic and inorganic matter. *Fuel*, *112*, 391–449. doi:10.1016/j.fuel.2013.05.043

Venkiteshwaran, K., Bocher, B., Maki, J., & Zitomer, D. (2015). Relating anaerobic digestion microbial community and process function. *Microbiology Insights*, *8*, 37–44. doi:10.4137/MBI.S33593

Verma, A., Kumar, S., & Jain, P. K. (2011). Key pre-treatment technologies on cellulosic ethanol production. *Journal of Scientific Research*, *55*, 57–63

Wang, F., Shi, D., Han, J., Zhang, G., Jiang, X., Yang, M., Wu, Z., Fu, C., Li, Z., Xian, M., & Zhang, H. (2020). Comparative study on pretreatment processes for different utilization purposes of switchgrass. *ACS Omega*, *5*, 21999–22007. doi:10.1021/acsomega.0c01047

Wang, S., Dai, G., Yang, H., & Luo, Z. (2017). Lignocellulosic biomass pyrolysis mechanism: A state-of-the-art review. *Progress in Energy and Combustion Science*, *62*, 33–86. doi:10.1016/j.pecs.2017.05.004

Wegner, T. H., & Jones, E. P. (2009). A fundamental review of the relationships between nanotechnology and lignocellulosic biomass. In L. A. Lucia & O. J. Rojas (Eds.), *The Nanoscience and Technology of Renewable Biomaterials* (pp. 1–41). Blackwell. doi:10.1002/9781444307474.ch1

Wei, H., Yingting, Y., Jingjing, G., Wenshi, Y., & Junhong, T. (2017). Lignocellulosic biomass valorization: Production of ethanol. *Encyclopedia of Sustainable Technologies*, *3*, 601–604. doi:10.1016/B978-0-12-409548-9.10239-8

Weng, C., Peng, X., & Han, Y. (2021). Depolymerization and conversion of lignin to value-added bioproducts by microbial and enzymatic catalysis. *Biotechnology for Biofuels*, *14*, 84. doi:10.1186/s13068-021-01934-w

Xu, J., Cheng, J. J., Sharma-Shivappa, R. R., & Burns, J. C. (2010). Sodium hydroxide pretreatment of switchgrass for ethanol production. *Energy & Fuels*, *24*, 2113–2119. doi:10.1021/ef9014718

Yu, H., Xiao, W., Han, L., & Huang, G. (2019). Characterization of mechanical pulverization/phosphoric acid pretreatment of corn stover for enzymatic hydrolysis. *Bioresource Technology*, *282*, 69–74. doi:10.1016/j.biortech.2019.02.104

Zabed, H., Sahu, J. N., Boyce, A. N., & Faruq, G. (2016). Fuel ethanol production from lignocellulosic biomass: An overview on feedstocks and technological approaches. *Renewable and Sustainable Energy Reviews*, *66*, 751–774. doi:10.1016/j.rser.2016.08.038

Zabed, H., Sahu, J. N., Suely, A., Boyce, A. N., & Faruq, G. (2017). Bioethanol production from renewable sources: Current perspectives and technological progress. *Renewable and Sustainable Energy Reviews*, *71*, 475–501. doi:10.1016/j.rser.2016.12.076

Zaimes, G. G., Soratana, K., Harden, C. L., Landis, A. E., & Khanna, V. (2015). Biofuels via fast pyrolysis of perennial grasses: A life cycle evaluation of energy consumption and greenhouse gas emissions. *Environmental Science & Technology*, *49*, 10007–10018. doi:10.1021/acs.est.5b00129

Zamanzadeh, M., Hagen, L. H., Svensson, K., Linjordet, R., & Horn, S. J. (2017). Biogas production from food waste via co-digestion and digestion- effects on performance and microbial ecology. *Scientific Reports*, *7*, 17664. doi:10.1038/s41598-017-15784-w

Zhao, L., Sun, Z. F., Zhang, C. C., Nan, J., Ren, N. Q., Lee, D. J., & Chen, C. (2022). Advances in pretreatment of lignocellulosic biomass for bioenergy production: Challenges and perspectives. *Bioresource Technology*, *343*, 126123. doi:10.1016/j.biortech.2021.126123

Zheng, Q., Zhou, T., Wang, Y., Cao, X., Wu, S., Zhao, M., Wang, H., Xu, M., Zheng, B., Zheng, J., & Guan, X. (2018). Pretreatment of wheat straw leads to structural changes and improved enzymatic hydrolysis. *Scientific Reports*, *8*, 1321. doi:10.1038/s41598-018-19517-5

Zhou, Z., Ouyang, D., Liu, D., & Zhao, X. (2023). Oxidative pretreatment of lignocellulosic biomass for enzymatic hydrolysis: Progress and challenges. *Bioresource Technology*, *367*, 128208. doi:10.1016/j.biortech.2022.128208

Ziegler-Devin, I., Chrusciel, L., & Brosse, N. (2021). Steam explosion pretreatment of lignocellulosic biomass: A mini-review of theoretical and experimental approaches. *Frontiers in Chemistry*, *9*, 705358. doi:10.3389/fchem.2021.705358

Zoghlami, A., & Paës, G. (2019). Lignocellulosic biomass: Understanding recalcitrance and predicting hydrolysis. *Frontiers in Chemistry*, *7*, 874. doi:10.3389/fchem.2019.00874

2 Physical Approach to Biomass Conversion

Monika Kayasth, Kamla Malik, Sunaina Kumari,
Meena Sindhu, and Shikha Mehta
CCS Haryana Agricultural University

2.1 INTRODUCTION

The need of energy is enhanced by developments and growing advancements in the world. The fossil fuels and conventional or traditional energy resources are the major sources of power that are the main causes for the environmental changes and global warming. The fossil fuels like coal, petroleum, etc. are the fundamental energy sources producing near about 80% of the total world's energy needs (Tumuluru et al., 2019). These non-renewable energy resources are having the probability of getting exhausted in the near future. In contrast to this, renewable sources of energy such as biomass, wind, solar, tidal, and geothermal are plentiful in the nature and can be replenished by themselves. Among these renewable energy resources, biomass contributes as the third largest source of energy in the entire world (Vaish et al., 2022). It is also the most prevalent energy source for heating and cooking for three-fourths of the developing countries' population and accounts for about 14% usage of the total world-wide energy (Baqir et al., 2018). Biomass is the organic matter that includes agricultural waste, crops waste, agricultural crop residues, wood wastes, animal wastes, aquatic plants, municipal wastes, etc., which contribute to the major production source of energy (Sivakumar & Mohan, 2010). In addition to this, biomass can be available on a recurring basis.

The non-renewable resources such as oil, coal, natural gas, and fossil fuels have been diminished by continuous consumption, which will lead to fuel scarcity (Sirajudin et al., 2013). Large quantities of organic waste such as fruit wastes and vegetable wastes left unutilised or allowed to decompose in an improper way are the major causes of environmental pollution. Various studies have been reported for finding the solutions to manage the organic matter or agricultural wastes by converting them into something valuable for people around the world. The utilisation and exploitation of biomass resources in producing bio-briquettes that are economic, renewable, and eco-friendly can solve the problems of disposal caused by a huge quantity of agricultural waste production annually and can reduce the potential environmental pollution problem (Tan et al., 2017).

Large quantity of biomass waste can be transformed to briquettes through the process of densification. Compaction or densification of biomass, also known as briquetting of agro wastes, has been practiced in several countries for many years. Some recent studies have reported the rise in combustion properties by 20% after the biomass compaction into solid densified briquettes and the emission of greenhouse gases (GHGs) such as SO_2 and NOx was also less, i.e., one-tenth to that of coal (Chen et al., 2015).

The utilisation and production of biomass briquettes are considered eco-friendly and environmentally safe. Several studies have reported that biomass is abundant in nature and its energy has the significant benefits of being almost carbon neutral in comparison to other renewable energy resources. However, the biomass resource's carbon neutrality depends upon the emission of toxic and harmful GHGs across the entire life cycle processes. The CO_2 emitted through the utilisation of biomass, burning, and exploitation methods does not create a boom in atmospheric CO_2 but alternatively leads to a faster transfer of CO_2 into the atmosphere. This is recycled through plants to supply biomass again (Tursi, 2019). This eco-friendly aspect of biomass makes it an excellent renewable and sustainable source for production of briquette. Recent research has explored numerous types of biomass substances in combination with some non-biomass substances utilised in briquetting power (Danjuma et al., 2013).

Plant origin biomass is lignocellulosic in nature as it is made up of cellulose, hemicellulose, and lignin which include some organic components like lipids and extractives (Ramamoorthy et al., 2020). The lignocellulosic biomass nature makes it rich in energy content. In developing nations, huge quantities of biomass residues are produced annually as by-products of the commercial agricultural, forestry, and business sectors (Njenga et al., 2008).

Biomass residues, which are non-woody, have a major drawback of having low bulk densities for their effective utilisation that results in requirements of vast storage space, trouble in handling, and greater transportation charges, which make them inefficient and expensive as a marketable property. Also, low bulk densities and the free nature of available biomass are related to quicker combustion of fuels resulting in major losses of fuels (lower running thermal efficiencies) and emissions in fly ash form or particulates within the atmosphere. These types of emissions make them poor-quality biomass fuels.

Densification of biomass may be executed by using the pelletising technology or briquetting that compresses unfastened biomass into more densified or compacted forms. This reduces the storage and transportation costs and improves the efficacy of biomass for utilisation as a fuel that is combustible. Recent studies revealed that the combustion properties had been expanded by way of 20% after the biomass became moulded into strong briquettes and the emission of GHGs such as SO_2 and NOx have been one-ninth, one-fifth, and one-tenth to that of coal. Briquettes are used industrially and domestically for heat and power generation. The utilisation of renewable energy from biomass is one of the few confirmed, economical, available, and eco-friendly technologies that can lower CO_2 emissions.

2.2 RAW MATERIALS FOR BRIQUETTING

Agro residues which include rice husk, saw dust, groundnut shell, stalks of pigeon pea, cotton stalks, soybean stalks, sugar cane bagasse, mustard stalks, wooden chips, pods of tamarind, castor husk, dried tapioca stick, coconut shell powder, and espresso husk are the raw materials that are commonly used for densification or briquetting in India. All these residues may be briquetted individually and in combination with or without the use of binders.

Raw materials selection depends upon the factors that are particle size, moisture content, ash content, flow characteristics, and their availability in the local market. About 10%–15% moisture content is generally preferred because moisture content higher than this range will create issues in grinding and greater energy is needed for drying. The biomass ash content affects its slagging behaviour collectively with the ash mineral composition and operating conditions. However, feedstock biomass having as much as 4% of ash content material is utilised for briquetting. The granular homogeneous substances that can easily flow in conveyors, garage silos, and bunkers are appropriate for utilisation in briquetting.

2.3   FACTORS REQUIRED FOR BRIQUETTING

There are various factors required for utilisation of biomass residue for briquetting. Some factors are discussed in the following.

2.3.1   MOISTURE CONTENT

High moisture content affects grinding and makes drying operations more energy-intensive. As a result, biomass with a lower moisture content, i.e., below 10%–15%, is preferred.

2.3.2   ASH CONTENT

The majority of biomass wastes, with the exception of rice husk, which contains 20% ash, have low ash contents but greater alkaline mineral contents (particularly potash), which lower the sintering temperature and cause ash deposition. The likelihood of slagging increases with rising ash concentration and gets more severe when biomass contains more than 4% ash.

2.3.3   FLOW CHARACTERISTICS

The fine material of similar size can flow easily inside the garage bins/silos and fuel hoppers. Therefore, sawdust, coffee husk, mustard stalk (pulverised), cotton sticks, and groundnut shell are appropriate biomass substances for briquetting, which can be ground to make fine material of uniform size.

2.4   PRE-TREATMENT/PRE-PROCESSING

Pre-treatment methods including drying, cleaning, size reduction, and addition of binders are being utilised to make the biomass conversion process easy with the manufacturing of a high-quality product.

2.4.1 CLEANING

The first steps in the densification of biomass resources into briquettes, which are required to improve the fabric's performance for burning, are cleaning and sorting of the feedstock. This procedure, also known as sieving, is used to remove any unwanted items while ensuring that all of the feedstock is of appropriate size. To get the cleanest feedstocks possible, screening equipment including sieves and magnetic conveyors are used to remove contaminants like soil, dust, metals, plastic threads, etc. These unwelcome substances are produced during the collection and storage of leftovers. Washing these impure materials with solvents is another method of eliminating impurities produced by the use of chemicals, alkali oxide, and fertiliser in agricultural operations.

2.4.2 DRYING

If the Feedstock is moist, its drying is essential to enhance its efficiency. However, adding a small amount of moisture helps in aggregation of the biomass particles. Forced drying methods and herbal evaporation methods are being utilised in various industries. There are two types of forced drying methods such as direct drying (e.g., in hot air oven) and indirect drying (heat passed via some metallic surface). Drying method is the most energy-consuming and out of total energy utilised, 70% is consumed within the process of densification of biomass residues.

2.4.3 SIZE REDUCTION

Reduction of size of biomass residues is a very essential process prior to biomass briquetting. Recent studies revealed that size reduction partially hydrolyses the lignin content material of biomass and enhances the overall surface area, which result in greater inter-particle bonding. In biomass, size reduction also enhances the bulk density, which improves the flow of biomass during the process of densification. There are several techniques for reducing the particle size, including chopping, chipping, hammer milling, crushing, shredding, and grinding. Biomass that has been reduced in size was classified as chopped (50–250 mm), chipped (8–50 mm), or ground (8 mm). The use of a sieve, either with a vibratory screen or an oscillating display screen, is another method of lowering the size of the biomass before densification. Tumuluru and Heikkila (2019) described a two-stage process for grinding biomass feedstock made of both wood and herbaceous material. While the second stage comprises grinding to a smaller size to make the biomass suitable for thermochemical and biochemical conversion techniques, the first stage involves the grinder breaking the biomass bundles into a large size material to improve its transportation within the conveyors.

2.4.4 BINDER ADDITION

Another key factor is binder addition. Addition of binder with biomass materials has increased the inter-particle bonding strongly forming a bridge. Addition of binders can be done before densification and during the preparation of mixture of the

BIOMASS FEEDSTOCK PRE-PROCESSING

Step 1 CLEANING	Step 2 DRYING	Step 3 SIZE REDUCTION	Step 4 BINDER ADDITION
The first step of briquetting is sorting and cleaning of feedstock. This procedure is also called sieving, which is done to remove all unwanted materials ensuring that all the feedstock is of the required size.	Drying of feedstock increases its efficiency but should not be excessively dried. Allowing a small amount of moisture helps in binding the biomass particles.	Size reduction in biomass increases bulk density, which improves the flow of biomass during densification. It partially breaks down the lignin content of biomass and increases the total surface area leading to greater inter-particle bonding	it partially breaks down the lignin content of biomass and increases the total surface area leading to greater inter-particle bonding of the product. It forms a bridge to enhance strong inter-particle bonding with biomass components.

FIGURE 2.1   Steps showing biomass feedstock pre-processing.

different feedstock/substrate or after the carbonisation of feedstock. Compaction technique such as low pressure is generally used for the agglomeration of some biomass materials with the addition of binder. Addition of binder to the biomass as a feedstock is a common preparation, which generally helps in the densification of biomass or enhances thermal and mechanical properties of the desired product.

Briquette production utilises three types of binders such as inorganic and organic and compound mixing in biomass. General examples of different categories of inorganic binders are lime, clay, plaster, cement, and Na_2SiO_3. On the other hand, these binders are further sub-classified into biomass used as binder, viz. cow dung, molasses, paste of cassava, wood pulp and paper waste, starch, etc., and other binders are pitch, petroleum bitumen, polymers, tar, lignosulphonate, etc. However, two or more organic and inorganic binders can be combined for compound binder formation.

However, briquettes made up of inorganic binders have generally more compaction ratio, high compressive strength, and hydrophobically in nature as compared to the briquettes prepared from organic binder. For efficient production of fuel, the binder which is utilised must be plastic and elastic due to the addition of biomass-based binder, which generally improved the durability, density, and resistance to shearing (Figure 2.1).

2.5 BINDING MECHANISMS

The briquette-making technologies are divided on the basis of compaction as follows:

- Low-pressure compaction
- Medium-pressure compaction
- High-pressure compaction

Generally, medium- as well as high-pressure compaction did not require binders but sometimes biomass pre-heating is utilised to increase the process of compaction. All the compaction processes include pressing of the individual particles mixing in a limited volume. In the low-pressure compaction method, the highly viscous bonding media used as binders, e.g., cow dung and other molecular organic liquid, tar, etc., are used to more strengthen the adhesion between biomass particles by creation of solid–liquid bridges. Biomass also contained lignin, which is also used as a binder particle facilitating bonding due to its adsorption on solid particle layers and softening at higher temperatures. These types of interaction and material properties or characteristics utilised in the agglomeration are responsible for its strength.

2.6   COMPACTION OF BIOMASS RESIDUES

Densification or compaction is the general process, where the raw biomass residues or feedstocks are transformed into a dense form known as briquettes and pellets. Generally, biomass used for pellets/briquetting making machine are essential for compaction process. This compaction process relies on the type of densification technology adopted.

For this, three types of different compacting pressures are measured for the process for compaction of briquettes and maximum compaction pressure is achieved at 150 kPa, which is good for improved quality of briquettes when cassava has been used as a binder.

The compaction quality and densification of biomass depends on the durability and strength of the particle bonding, and also influences various process parameters such as die temperature, die diameter, pressure, binder's types, and pre-heating of the biomass/feedstock components mixture. The particle bonding processes are divided into five sub-categories: (i) attraction of forces between solid particles, (ii) interfacial forces and capillary pressure occurring in liquid movable surfaces, (iii) adhesion and cohesion forces not freely moving between binder bridges, (iv) solid bridges, and (v) mechanical interlocking.

Kaliyan and Morey (2010) investigated the densification process of switchgrass and corn stover using scanning electron microscopes (SEMs) for understanding the arrangement of solid-type bridges. A SEM image revealed that the bonding between the biomass particles was created generally through solid-type bridges. These solid-type bridges were formed between the biomass particles due to the natural binders used in the process of densification. According to UV auto-fluorescence analysis, the briquettes and pellets were further examined and observed that the creation of solid-type bridges was made due to the addition of lignin and protein as a natural binder. However, that makes strong inter-particle bonding between them. It was also observed that the activation of natural binders occurs when temperature and moisture are essential in the range of the glass transition phase. Tumuluru et al. (2011) reported that more studies on using different techniques like SEM and TEM at a micro level have been useful for knowing the material properties, intra-particles cavities, and variable interactions processes on the quality traits of densified feedstock/biomass.

2.7 TECHNOLOGIES OF DENSIFICATION/BRIQUETTING

The biomass/crops residues compaction is referred to as densification or briquetting process/technology. It is also known for technology for biomass particles/residues utilisation under the process of compaction with different levels of pressure for making a solid biomass fuel. Extruder, screw press machine, hydraulic piston press, piston-type machine, roller press machine, and pellet press (ring & flat die) machines by applying an appropriate pressure are used to making a briquettes/pellet. For solid biomass fuels, the quality of briquettes/pellets depends upon the mechanical durability, strength, density, moisture content, calorific values, etc. that play an important role during storage, handling, and transportation. During the densification process, binders are the materials that give the strength and solidification of the ground biomass residues. Major binders which are generally used for densification include starch, molasses, cassava, corn, gelatin, bio-solids, macro- and microalgae, etc. Densification process of biomass residues depends on various factors such as the moisture content of the raw residue material and steam under high pressure, which easily hydrolyses lignin and hemicellulose and converts them into simple carbohydrates, i.e., sugar polymers, lignin by-products, and other derivatives. These types of products are prepared under heat and pressure in the die of machines, and also used as glues/gums which act as a strong binder of the particles mixed together (Chen et al., 2015).

The main types of densification process/techniques that are usually used for manufacturing the briquettes or pellets (solid biomass fuels) are described in the following.

2.7.1 Screw Compaction/Extruder Machine

Screw extruder compaction or extruder machine has a screw which rotates to exert a force on crop residues or biomass for compression which are used as a feedstock. Operating a screw inside the compaction machine requires mechanical drive. To minimise the particle friction, a significant quantity of energy is required as the biomass residues are temporarily compressed. At initial temperature and high temperature (2,000°C–2,500°C), biomass particles dissipate their elasticity characteristics and they become softer (Chen et al., 2015). This leads to the formation of local bonds and interlinking of particles. A particle ideal for extrusion period must have a size range between 2 and 4 mm, as they are responsible for optimum binding of biomass residue particles. When biomass enters in the tapered die section, moisture which is remaining gets evaporated again because of high temperature of 2,800°C. This process aids to strengthen the compression of biomass residue materials. The heavy compression and high temperature make the biomass material more suitable for co-firing and burning process. A well-defined diagram of biomass screw-type extruder used for briquettes or pellets is mentioned in Figure 2.2.

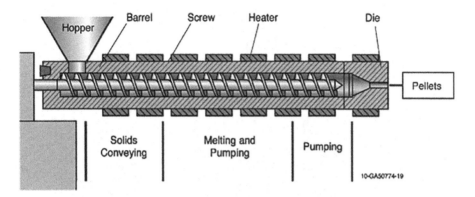

FIGURE 2.2 Biomass screw-type extruder. (Adapted from Tumuluru et al., 2011.)

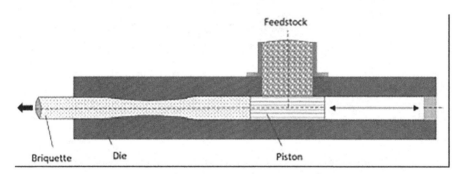

FIGURE 2.3 Piston press machine. (Adapted from Tumuluru et al., 2011.)

2.7.2 PISTON PRESS OR PUMP-TYPE MACHINE

Piston press or pump machine is used with some hydraulic mechanism or flywheel for the densification of biomass into briquettes/pellets for fuel production. The biomass residue materials generally come from crop residue as a feedstock converted into the cylindrical-shaped piston drive that could be mechanically/hydraulically done. During compression process, the temperature ranges between 15,000°C and 30,000°C for heating of biomass residue materials. The compression of biomass residue material coming from the conical-shaped die through the piston press and the briquettes become extruded from the opening face or front of the die (Tumuluru et al., 2011). Biomass-based piston press/pump-type machine is depicted in Figure 2.3.

2.7.3 ROLLER PRESS/MILL MACHINE

A press/mill biomass roller type machine is illustrated in Figure 2.4. The roller mill main unit consists of two rollers rotating in the opposite direction of each other. In this

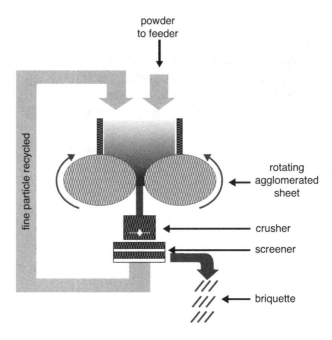

FIGURE 2.4   Roller press machine. (Adapted from Tumuluru et al., 2011.)

machine, the biomass residue material is fed through the hopper. The biomass residue material flow is optimised by the control mechanism. The two rollers exert forces on pre-compressed feedstock material by continuously rotating in the opposite direction (Danjuma et al., 2013). The biomass residues or feedstock material can be withdrawn from one side, and from the other side, the densified product is extruded. The gap between the two rollers depends on biomass residue material type which is utilised, ratio of binders, size of particle, content of moisture, and several other elements.

2.7.4 Pellet Mill Machine

A press-type biomass pellet machine is illustrated in Figure 2.5. A pellet mill machine is generally used for producing pellets from biomass residue material having a length of 13–19 mm and whose diameter is 6.3–6.4 also having 1,125–1,190 kg/m^3 density range and cylindrical shape. This press-type pellet mill machine has a circular matrix and roller. The residue material of the biomass, which is to be pelletised, is compacted between the roller mechanism and circular matrix. The circular matrix consists of a huge number of orifices whose sizes are pre-determined. The biomass residue materials which get densified or pelletised are extruded from these circular matrix orifices automatically. The pellets are cut in the desired length by the knife or cutter that is installed at the machine, and the final pelletised product is obtained (Danjuma et al., 2013) in the form of pellets.

Process of briquette formation is depicted in Figure 2.6. Densified briquettes/pellets manufactured from biomass are quite good alternatives/substitutes for lignite, coal, and firewood as they have various advantages.

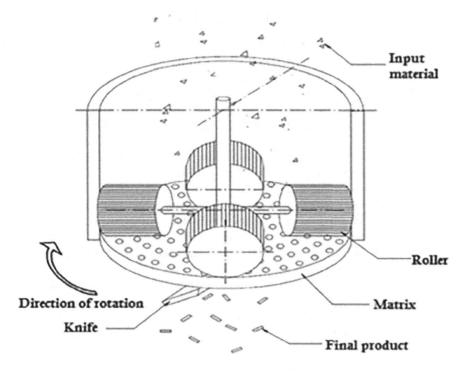

FIGURE 2.5   Biomass pellet press-type machine. (Adapted from Holubcik et al., 2012.)

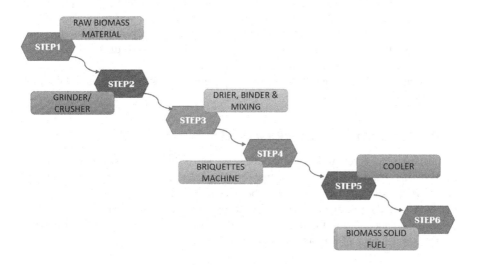

Biomass solid fuels (Briquettes/pellets) manufacturing process

FIGURE 2.6   Process of briquette formation.

2.8   ADVANTAGES IN BRIQUETTING OF BIOMASS

1. The briquetting or the densification process helps to resolve the issue concerned with generation of agricultural and agro-industrial waste, which in turn reduces the environmental pollution.
2. The process enhances the net heating value per unit volume.
3. The fuel which is produced from densification is uniform in size and quality.
4. Sulphur emission is negligible during burning of densified product. Moreover, no toxic gases are released and even no odour is generated as the briquettes are more volatile as compared to conventional fuels.
5. Briquette formation reduces the biomass volume and enhances compaction that makes the transportation and storage of densified product very easy.
6. Minimum risk of fire in loose storage of biomass.
7. The densification process generates fuel having high thermal value which can be attributed to their low moisture content and their ability to produce more heat than other fuels with lower ash content (2%–10%) in comparison to coal (20%–40%), so it is also known as white coal.
8. The briquettes produced from densification have lower ignition temperature compared to coal and are easy to burn.
9. During burning of the briquettes, gas is produced which fastens the burning efficiencies and inhales CO_2, and oxygen is released into the atmosphere.

2.9   DISADVANTAGES ASSOCIATED WITH BRIQUETTING

1. Manual feeding of briquettes to the furnace is a labour-intensive process that leads to high energy consumption input and high investment cost.
2. Fire doors through which feeding is done sometimes allows heat to escape into the atmosphere that generates black smoke which in turn reduces the efficiency of manual briquette fired boilers.
3. Briquettes tend to get loosen when exposed to water or in the humid environmental conditions.

2.10   APPLICATIONS OF BRIQUETTES

Briquettes being a renewable source of energy have several industrial and domestic applications. They can act as a substitute for charcoal, firewood, or other solid fuels and can serve as development intervention because hike in fuel prices and current scarcities have forced consumers to search for reasonable, energy-efficient, and economical resources (Oladeji, 2015). Figure 2.7 depicts the various forms of pellets and briquettes.

The several uses of briquettes are depicted in Table 2.1

FIGURE 2.7 (a) Pellets, (b) singular pellets, and (c and d) biomass briquettes.

TABLE 2.1
Potential Applications of Briquettes

S. No.	Industry/Sector	Possible Application
1.	Domestic use	For heating and cooking purpose
2.	Commercial and institutional catering	Water heating, cooking and for grilling purpose
3.	Industrial boilers	Generation of steam and heat
4.	Hospitality	Heating (outdoor dining areas) and cooking purpose
5.	Crop processing	For tea drying, tobacco curing, and oil milling purpose
6.	Textiles	For bleaching and dyeing purpose
7.	Food processing	For bakeries, distilleries, restaurants, canteens, and for drying purpose
8.	Ceramic production	For making tiles, brick, pots, etc.
9.	Gasification	For utilisation as a fuel to generate electricity
10.	Charcoal production	For initiating pyrolysis necessary to make efficient charcoal
11.	Poultry	For incubation of chicks and for heating purpose

2.11 CHALLENGES AND PROSPECTS

As reported earlier, in African nations, the briquetting is innovative, but this technology is in advanced stage in America, Europe, and Asia. Successes have been observed in the utilisation and production of briquettes in these advanced nations, but this advancement cannot be possible in Africa or other developing countries.

Basically, the expansion of densification of biomass depends on three factors, such as adequate technologies, residue availability, and production market for briquettes.

The biomass residue availability does not present a problem for countries in the developing world; however, for most of the numerous feedstocks, there is requirement of optimisation of the mechanical and chemical treatments which remains a challenge. According to the communities that are rural where power is very less, an accurate means of pre-processing is a must and will need less input energy. As reviewed, most of the technologies generating high-quality briquettes are most expensive and also require high input energy. However, large-scale briquettes production will require major investment. This causes difficulty in the expansion of densification of biomass. Therefore, efforts should be made towards the development of energy-effective and more cost-effective technologies that are user-friendly at various scales to attract a great investment in areas lacking high energy input and adequate financial capacity. Finally, the market for efficient utilisation of briquette as an alternate to conventional biomass (fuelwood) and fossil fuel exists. However, some challenges should be addressed to carry the full potential of this briquetting technology.

2.12 SUMMARY AND CONCLUSION

Tremendous increase in population has led to increase in standard of living that eventually increases the demand of energy. There are several reports indicating the pressure on conventional fuels that are formed from non-renewable sources and are decreasing at an alarming rate and their vast utilisation by the rising population has detrimental effects on global climate. To anticipate future crisis of energy and to mitigate change in climate, it has become vital to focus on renewable and sustainable energy sources. Biomass residue seems to be a good alternative to reduce our reliance on fossil fuels.

Storage, handling, and transportation of biomass are typically complicated by its large bulk and poor energy density. This is a significant obstacle. However, densification, or briquetting, solves this problem and enhances biomass density, burn time, and calorific value. The technical and financial implications of biomass densification, or briquetting, were covered in this chapter. This chapter demonstrated how the kind of biomass material (feedstock), pre-processing, briquetting technique, and technology all affect the quality of briquettes. Briquetting or densification can be carried out using either a high- or low-pressure method. But building more robust and high-energy-density briquettes requires technology that makes use of high temperatures and high compaction pressures. These days, briquetting requires the use of roller presses, screw press extruders, and piston presses (either mechanical or hydraulic). Furthermore, as the densified products are meant to replace current fuels, successful briquetting requires financial and commercial sustainability. Analysing the numerous expenses, including economic metrics like Payback period (PBP), Net present value (NPV), Internal rate of return (IRR), and Benefit cost ratio (BCR), is another way to evaluate it. Ultimately, it is in the best interests of producers and customers alike for briquette costs to be more reasonable and efficient than those of the fuels they are anticipated to replace. For residential heating, biomass briquettes may be utilised in both rural and urban settings. They can also be used in industrial settings for energy-generating processes like gasification and heating. High-pressure technologies and conducive environments are not widely accessible in many rural communities in poor nations. To increase the quality of fuel briquettes produced at lower temperatures and pressures, more research must be done. But a suitable briquetting equipment that can produce these high-quality briquettes at affordable prices that are suited for local populations has to be created.

REFERENCES

Bajwa, D.S., Peterson, T., Sharma, N., Shojaeiarani, J., Bajwa, S.G. (2018). A review of densified solid biomass for energy production. *Renew. Sustain. Energy Rev*, 96, 296–305.

Baqir, M., Kothari, R., Singh, R.P. (2018). Fuel wood consumption, and its influence on forest biomass carbon stock and emission of carbon dioxide: A case study of Kahinaur, district Mau, Uttar Pradesh, India. *Biofuels*, 1–10.

Chen, H. (2015). Lignocellulose biorefinery product engineering. In: *Lignocellulose Biorefinery Engineering*, 1st edn. Woodhead Publishing Limited, Cambridge, pp. 125–165.

Danjuma, M.N., Maiwada, B., Tukur, R. (2013). Disseminating biomass briquetting technology in Nigeria: A case for briquettes production initiatives in Katsina State. *Int. J. Emerg. Technol. Adv. Eng.*, 3, 2–20.

Holubcik, M., Nosek, R., Jandacka, J. (2012). Optimization of the production process of wood pellets by adding additives. *Int. J. Energy Optimiz. Eng.* 1(2), 20–40.

Kaliyan, N., Morey, R.V. (2010). Natural binders and solid bridge type binding mechanisms in briquettes and pellets made from corn stover and switchgrass. *Bioresour. Technol.*, 101, 1082–1090.

Njenga, M., Karanja, N., Prain, G., Malii, J., Munyao, P., Gathuru, K. (2009). *Community-Based Energy Briquette Production from Urban: Organic Waste at Kahawa Soweto Informal Settlement*, Nairobi, Urban Harvest Working Paper Series No. 5. International Potatoe Center, Lima, Peru

Oladeji, J. (2015). Theoretical aspects of biomass briquetting: A review study. *J. Energy Technol. Policy*, 5, 72–82.

Ramamoorthy, N.K., Sahadevan, R. (2020). Production of bioethanol by an innovative biological pre-treatment of a novel mixture of surgical waste cotton and waste card board. *Energy Sour. A Recover Util. Env. E* 42, 942–953.

Sirajudin, N., Jusoff, K., Yani, S., Ifa, L., Roesyadi, A. (2013). Biofuel production from catalytic cracking of palm oil. *World Appl. Sci. J.*, 26, 67–71. doi:10.5829/idosi.wasj.2013.26. nrrdsi.26012

Sivakumar, K., Mohan, N.K. (2010). Performance analysis of downdraft gasifier for agriwaste biomass materials. *Indian J. Sci. Technol.* 3, 58–60.

Tan, X., Liu, S., Liu, Y., Gu, Y., Zeng, G., Hu, X., Wang, X., Liu, S., Jiang, L.H. (2017). Biochar a potential sustainable precursor for activated carbon production: Multiple applications in environmental protection and energy storage. *Bioresour. Technol.* 227, 359–372.

Tumuluru, J.S., Heikkila, D.J. (2019). Biomass grinding process optimization using response surface methodology and a hybrid genetic algorithm. *Bioengineering*, 6(1), 12.

Tumuluru, S.J., Wright, C.T., Hess, J.R., Kenney, K.L. (2011). A review of biomass densifi cation systems to develop uniform feedstock commodities for bioenergy application. *Biofuels Bioprod. Bioref.*, 5, 683–707.

Tursi, A. (2019). A review on biomass: Importance, chemistry, classification, and conversion. *Biofuel Res. J.*, 22, 962–979.

Vaish, S., Sharma, N.K., Kaur, G. (2022). A review on various types of densification/briquet-ting technologies of biomass residues. *IOP Conf. Ser. Mater. Sci. Eng.* 1228, 012019.

3 Thermochemical Conversion

An Approach for the Production of Energy Materials from Biomass

Chitra Devi Venkatachalam, Mothil Sengottian,
Sathish Raam Ravichandran, Sarath Sekar,
and Premkumar Bhuvaneshwaran
Kongu Engineering College

3.1 INTRODUCTION

Biomass energy is derived from organic matter, such as plants, crops, trees, agricultural and forestry wastes, and organic trash. Biomass is considered renewable because the plants and trees that are used to create biomass can be replanted and regrown. The process of converting biomass into energy is known as bioenergy [1,2].

Bioenergy can be used in a variety of forms, including solid biomass, liquid biofuels, and biogas [3]. Solid biomass includes wood pellets, wood chips, and other forms of biomass that can be burned to produce heat or electricity [4,5]. For example, many homes and businesses use wood pellets or chips as a fuel source for heating. Liquid biofuels: Liquid biofuels are made from biomass, such as corn or sugarcane, and can be used in vehicles or other engines in place of fossil fuels [6,7]. Ethanol and biodiesel are two popular liquid biofuels. Biogas is created when organic waste in landfills, agricultural waste, or livestock manure decomposes. Biogas can be utilized to power heating systems, generate electricity, and transport vehicles [8].

Biomass has several advantages as one of the renewable sources of energy. It is abundant, widely available, and can be produced locally, reducing dependence on foreign energy sources. In addition, using biomass for energy can reduce greenhouse gas emissions and provide a valuable use for organic waste materials [9]. However, the use of biomass can have environmental consequences, such as deforestation or soil deterioration, and should be managed in a sustainable manner to confirm its sustainability as a renewable energy source [10].

Biorefining involves the conversion of biomass into multiple useful products, such as biofuels, chemicals, catalysts, and energy storage materials [11,12]. Biomass can be directly used as fuel but to enhance the productivity and quality of the fuel source

DOI: 10.1201/9781003406501-3

and also to diversify the products produced, the biomass is subjected to conversion techniques that enable the production of multiple kinds of bioproducts. The conversion methods include physical, chemical, biological, and thermochemical methods.

Thermochemical conversion methods are a collection of different types which are classified into dry and wet conversion processes based on the feedstock condition and preparation. Dry thermochemical conversion generally requires the least amount of moisture in the feedstock to be able to get the highest quality and yield. But the products obtained through this have less oxygenated compounds and high hydrogenated compounds. The wet thermochemical conversion process can handle feedstocks with higher moisture levels than the dry process and thus the products obtained through it have higher oxygenated compounds and are generally lower in calorific value [13]. This chapter looks into the thermochemical conversion of biomass into important value-added products such as biochar, biocrude, and biogas, as well as their applications in fuels, energy storage, catalysts, carbon nanocomposites, and so on.

3.2 THERMOCHEMICAL CONVERSION

Thermochemical conversion is one of the energy production processes which converts a large amount of biomass into valuable energy products that involve biochar, biocrude, and biogas. It requires a wide range of heating and reduction or oxidation processes for the degradation of biomass structure. The superheated water is used to form biofuels from the organic matter present in the feedstock during the conversion process [14,15]. Chemical energy accumulated in the biomass can be released during the breakdown of bonds joined between the oxygen, carbon, and hydrogen molecules. The heat energy can be absorbed and released within the biomass during the conversion. Endothermic (absorbed energy) and exothermic (released energy) are the major thermochemical conversions that are classified based on their reaction environment, temperature gradient, absorption and rejection of heat, heating rate, particle size, and presence and absence of oxygen [16,17].

In addition, the conversion process, such as pyrolysis, torrefaction, carbonization, liquefaction, and gasification, which is held in the aqueous and non-aqueous environment are considered as wet and dry thermochemical conversions, respectively. The aforementioned process can be regarded as endothermic and exothermic reactions depending on their operating temperature [18–21]. A detailed discussion is given about various types of the thermochemical conversion process in the following section.

3.2.1 DRY THERMOCHEMICAL CONVERSION

A dry thermochemical conversion process is performed for various biomass such as wood, algae, plastic, municipal waste, and agricultural and forest residue in non-aqueous operating conditions [18,22–24]. When increasing the temperature of these conversion processes, the chemical structure present in the biomass waste could be degraded to a condensable (liquid state) and non-condensable (gaseous molecules) vapor state. The volatile matter is released from deep inside the biomass at an operating temperature of around 300°C, which is considered a torrefaction process, and the reaction happens near the surface of the biomass when the temperature is between 300°C and 900°C, which is considered for pyrolysis and carbonization processes. The aforementioned process produces the flame over the surface of the biomass that leads to formation of the pyrolyzed vapor and biochar. Likewise, a dry liquefaction process is carried out at 150°C–420°C to form an organic liquid product, which also requires higher pressure (1–24 MPa) than pyrolysis and carbonization processes [17,25,26]. The mentioned process is discussed in detail in the following section.

3.2.1.1 Pyrolysis

Direct pyrolysis is a reduction-type biomass degrading process that uses heat as an energy source. During pyrolysis, the biomass is heated to temperatures ranging from 250°C to 700°C in an inert environment at varying speeds. According to research, nitrogen flow is mostly employed to create an inert atmosphere [17,27]. Pyrolysis is normally classified into slow, fast, and flash pyrolysis based on their operating parameters including heating rate, temperature, residence time, and feed rate. In situ or slow pyrolysis takes higher residence time (in hours) for heating the biomass to the preferred temperature. As a result, it provides a large amount of solid remnant and high-quality liquid products than the hydrothermal liquefaction (HTL) process [28–30]. Fast pyrolysis often requires a higher heating rate (15°C/s–220°C/s) and produces better bio-liquids than the traditional process. Thus, the heating rate ranges between 1,200°C/s and 9,800±200°C/s are needed for flash pyrolysis to obtain a very high liquid yield of 83±5 wt.%. As the heating rate increases, bio-oil yield increases and biochar yield decreases due to the cracking of char under secondary reaction [28,31,32]. The percentage of bio-product yield is varied for different pyrolysis processes which are briefly discussed in the following sections.

3.2.1.1.1 Slow Pyrolysis

Slow pyrolysis is used to produce biofuels from a variety of feedstock, including agricultural and vegetable residue, municipal waste, wood, and micro and macro algae. To obtain maximal and enhanced biochar, optimal process parameters such as temperature, pressure, residence time, and operating environment are necessary [18,33,34]. Moreover, physio-chemical properties like moisture content, fixed carbon, ash, and other volatile substances are also important for better conversion. Slow pyrolysis is performed on various kinds of agricultural residues that include corn, barley, wheat straw, coco peat, vine shoots, sugar beet, and paddy straw [35–39]. These residues have a chemical composition consisting of cellulose, hemicellulose, and lignin which has a wide range of fixed carbon (from 10 to 25 wt.%), moisture content (from 7 to 12 wt.%), volatile substance (from 74 to 87 wt.%), and ash content (from 1 to 3 wt.%) in dried conditions [40]. Whence, the high amount of fixed carbon and less ash content are predominant conditions to attain the maximum yield of biochar. Especially, lignin is an important factor in producing biochar because of its highly branched biopolymer. This process is conducted with an operating temperature between 350°C and 780°C, which increases the pressure from 0.1 to 1 MPa and residence time from hours to days. During slow pyrolysis, the biochar yield shows indirect proportionality with the operating pressure and temperature and it shows linear proportionality with the particle size. Regardless of the type of agricultural residue, the carbon stability in biochar is increased when the operating temperature is maintained from 400°C to 590°C. Thus, the process provides a maximum of 30–50 wt.% of biochar yield from the agricultural residues [33,41–43]. Therefore, the simple reaction mechanisms followed for slow pyrolysis are given by:

1. Biomass → Unreacted remnant + H_2O
2. Unreacted remnant → (Volatile matter)$_1$ + (Char)$_1$ + (Incondensable liquid)$_1$
3. Char → (Volatile matter)$_2$ + (Char)$_2$ + tar

Slow pyrolysis is also implemented for various types of wood particles like sawdust, birch and rubber wood, pine and bamboo barks, beech and poplar wood, etc. It is performed at a heating rate of 5–30°C/min [43]. The chemical compound present in the wood bark is similar to the agricultural wastes. Likewise, lignin plays a vital role to produce biochar. Moreover, particle size and peak temperature are important parameters to obtain higher biochar yield. Generally, slow pyrolysis is operated within the temperature ranges from 300°C to 800°C with a residence time of 2–24 hours for wood biomass [26,44]. The biochar yield is increased with the increasing heating rate for particle size more than 245 μm and there is no significant change of biochar yield for a tiny particle concerning heating rate. The large-size particle remains intact for slow and medium heating rates. Approximately, 24–28 wt.% of biochar is produced from the wood biomass of tiny particles and the biochar yield ranges from 16 to 20 wt.% for particle sizes more than 235 μm at a heating rate of 5–30°C/min. A maximum biochar yield from 35 to 55 wt.% is obtained at a temperature between 300°C and 345°C due to easy degradation of cellulose and hemicellulose in wood barks, and it starts to decrease when temperature increases beyond 350°C in slow pyrolysis [36,42,45–47].

Micro and macro (up to 80 meters long) algae are used for biofuel preparation under slow pyrolysis. Algae-based biofuel reduces global warming and also is environment friendly. It is a group of eukaryotic organisms that has better carbon sources than other biomass. Mostly, microalgae like *Spirulina maxima*, *Chlorella vulgaris*, *Chlorella pyrenoidosa*, *Nannochloropsis gaditana*, *Dunaltella tertiolecta*, *Cladophora glomerata*, *Botryococcus braunit*, and some macroalgae like *Enteromorpha clathrata*, *Fucus serratus*, *Laminaria digitata*, and *Macrocystis pyrefera* which have the fixed carbon from 10 to 17 wt.% are used to perform slow pyrolysis [38,44,48]. Whence, *Dunaltella tertiolecta* has a higher amount of fixed carbon (27 wt.%) than other micro- and macro-organisms. The chemical composition of algae consists of proteins, lipids, and carbohydrates. Microalgae have comparatively higher lipid content than macroalgae, which have carbohydrates as its major components. In general, lipids and carbohydrates are important components for the production of biofuels. Process parameters including algae type and size, moisture content, chemical composition, pressure, temperature, residence time, and heating rate are the optimizing parameters to get better bio-product yield in slow pyrolysis. It is observed that the yield of biochar is gradually increasing for temperatures up to 350°C in algal biomass. The maximum amount of bio-oil yield is obtained at a temperature between 380°C and 550°C.

Interestingly, a temperature of more than 550°C leads to secondary cracking (breaking down of larger molecules into smaller ones) of algae, which reduces the bio-oil quantity. Further increasing the temperature promotes the carbon content in the final products. In the case of particle size, a larger one increases the temperature gradient, which promotes the biochar yield and reduces the bio-oil yield. Likewise, higher residence time encourages polymerization and increases the biochar yield. Hence, a lower heating rate (0.1–1 K/s), higher temperature (550°C–950°C), and residence time are suitable for getting better biochar yield from slow pyrolysis [35,36,38,49].

3.2.1.1.2 Fast Pyrolysis

Fast pyrolysis is the process of heating biomass at a higher temperature and at a faster rate in the absence of oxygen. It also decreases the possibility of secondary reactions occurring due to the very short residence period and rapid heating rate. As previously stated, the extent of chemical compound degradation is determined by characteristics such as temperature, pressure, heating rate, and residence time. A mixture of organic compounds present in the biomass has broken out at different temperatures and mechanisms. Whence, lignin has a broader range of decomposing temperatures and better thermal stability when compared to other chemical compounds which degrade at a limited temperature range. Therefore, fast pyrolysis provides a higher bio-oil yield than biochar due to the lack of secondary reactions of lignin. In general, the maximum amount of bio-oil is acquired at a temperature near 500°C with a residence time of fewer than 2.5 seconds followed by rapid condensation of vapor during fast pyrolysis [28,50,51].

Temperatures between 495°C and 550°C are chosen for wood biomass to get a bio-oil output of 78 wt.% (in dry form) with a residence period of less than 1 second. According to research, a very short residence time (250 ms) leads to inappropriate lignin degradation and the formation of heterogeneous bio-oil, which is utilized as a food and chemical additive; in addition, bio-oil derived with a residence period of up to 2 seconds is used as a fuel. An extended residence period (greater than 5 seconds) stimulates secondary chemical reaction and lowers bio-oil output and quality [27,48]. The hydrothermal pretreatment of the biomass before fast pyrolysis also enhanced the thermal stability and yield of the bio-oil due to improved amount of levoglucosan and reduction of hemicellulose (from 21.5 to 3.5 wt.%), ketones, and acids [50,51]. Agricultural residue has more hemicellulose (34–42 wt.%) and minimum lignin content (15–28 wt.%) than softwood biomass. It also provides a lesser bio-oil yield (44–53 wt.%) with higher aqueous solution and gaseous product containing high alkali metals when compared to wood biomass [50].

Microalgae may be cultivated to produce 68 weight percent of lipids, which can then be transformed into a greater amount of bio-product. It has been discovered that quick pyrolysis of *Chlorella vulgaris* at 800°C yields the best value bio-oil (61 wt.%) [30,52,53]. Similarly, when compared to other algae species at the same temperature, the fast pyrolysis of *Botryococcus braunii* yields the highest bio-oil production of 69 wt.% at 500°C. The bio-oil production of macroalgae ranges between 36 wt.% and 46 wt.% at operating temperatures ranging from 400°C to 550°C, which is smaller than the output of microalgae. Thus, for temperatures above 550°C, the bio-oil output of macroalgae gradually decreases. As a result, temperatures ranging from 380°C to 800°C are acceptable for increasing bio-oil yield from microalgae, while temperatures ranging from 400°C to 550°C are favored for macroalgae. Interestingly, the yield of bio-oil gradually increased for a feeding rate up to 0.6 kg/hour, and it gradually decreased for further heating during fast pyrolysis. Irrespective of the biomass, the yield of both biochar and bio-yield is reduced owing to insufficient secondary reactions when increasing the flow rate of carrier gas and lowering the residence time [48,52,53].

$$\text{Biochar yield}, \eta = \frac{\text{mass of biochar-mass of ash in biochar}}{\text{mass of dry feedstock-mass of ash in dry feedstock}} \times 100$$

3.2.1.2 Torrefaction

Torrefaction is the process of converting different types of biomass into high-quality solid fuels by performing suitable heat treatment under inert conditions. It can be used for diverse feedstock such as wood, agro-residue, algal biomass, and sewage sludge [54–56]. Pre-heating and drying, intermedial heating, post-drying, torrefaction, and annealing are the important stages of the torrefaction process. Pre-heating is carried out at 100°C to remove the moisture content in the biomass. Then, the biomass is heated to around 200°C to vaporize the light fractions in the second stage of the aforementioned process. Torrefaction starts at a temperature of 200°C and is heated up to 300°C where a large amount of mass is lost (complete decomposition of hemicellulose and partial decomposition of lignin and cellulose), and it cools down to the initial temperature. At last, an annealing process is carried out to cool down the solid fuel to ambient temperature [57,58].

In the case of the torrefaction mechanism, the decomposition of biomass into a carboxyl group happens with the emission of CO_2 and is followed by an aromatization reaction. Whence, the nature of biomass can change from hydrophilic to hydrophobic with dark brown solid fuel during torrefaction with inert conditions. The mass of feedstock can be deduced to around 29 wt.% based on operating parameters and composition, later the solid fuels can be developed with 89.5% of energy content from the initial biomass. So, there are a few advantages of the biofuels obtained from the torrefaction process including more homogeneity, fine particle, calorific value, better bulk density, and high durability. It is observed that the solid products obtained from dry torrefaction have higher alkali content and slightly less calorific value than another type of torrefied product. Therefore, dry torrefaction is well-suited for less moisture biomass because most of the decarboxyl reaction starts after the removal of moisture content [55,56,59,60].

Carbonization, oxidation, and degasification are the important processes followed in oxidative torrefaction under liquid or gaseous environments. Whence, torrefaction and oxidation are simultaneously performed processes without involving each other. Oxidative torrefaction uses oxygen as a carrier gas for developing a durable torrefied product. It also improves the particle density and provides the solid product with lighter hydrocarbon. It also reduces the hydrophobicity because of the higher surface area and also improves the hydroxyl group during the process. There is no separate process for removing the nitrogen from the torrefied biomass which also reduces the overall cost of the process. But, the yield of torrefied solid obtained from oxidation conditions is comparatively less than the yield of non-oxidative torrefaction [59,60].

3.2.1.2.1 Enhancement of Biomass Properties under Torrefaction

The solid product obtained from oxidative torrefaction has higher elemental content, enhanced hydrophobicity, and comparatively less hardness than conventional torrefaction due to the simultaneous process of both oxidation and torrefaction. This combined process leads to the disintegration of O–H bonds and makes the biomass resist moisture. The hydrophobic nature of the biomass can be improved by increasing the operating time and temperature of the torrefaction. The increasing residence time of the biomass also improves the amount of fixed carbon and calorific value within the biomass because of releasing a higher amount of hydrogen and oxygen molecules than the carbon molecules [56,60,61]. In addition, fine particle size and their surface area distribution are the enhanced properties for the torrefied biomass than the raw material.

During torrefaction, the particle becomes more brittle and smaller with uniform structure due to its improved C/O and C/H ratios. Moreover, the size of the torrefied biomass is important for the upgrading the process because higher energy is required for upgrading the large-size particle. Instead of using raw biomass, torrefied biomass is used for the upgrading process due to lesser energy consumption. More amount of mass can be released during torrefaction which gives a more porous solid product. As a result, the biomass's bulk density gradually drops, resulting in more than 60% rise in energy density. Torrefied biomass also reduces the emission of CO and CO_2 to 1/3rd of that emitted from regular biomass [56,59,60].

3.2.1.3 Carbonization

Carbonization is an earlier kind of slow pyrolysis that can provide charcoal (a higher amount of carbon than char) from the feedstock. It is performed at the temperature ranges between 320°C and 900°C under inert conditions for a residence time of more than 1 hour in the retort or kiln furnace. It takes comparatively more residence time than slow pyrolysis [19,62]. It is the simultaneous process of hydrogen removal, condensation, and isomerization with a faster reaction rate. To begin with, heat is applied to add hydrogen to the naphthenic ring and stabilize an unstable hydrocarbon [62].

In addition, the weaker C-C bond present in the biomass has been broken to form fine lighter hydrocarbon vapor. Thereby, further heating of stable vapor tends to crack reactive molecules to form a high polymeric closed carbon chain (coke), condensate liquid, and volatile gases. Carbonization also provides a mostly equal amount of bioproducts (biochar, bio-liquid, biogas) from 30% to 35% as compared to another thermal conversion process. Normally, higher pressure and a slower heating rate are preferred for getting a larger yield of cracked biochar during carbonization. Charcoal resulting from carbonization under CO_2 conditions has a more porous and high surface area than compared to nitrogen atmosphere [63,64].

3.2.1.3.1 Carbonization Chemistry for Different Biomass

Carbonization is performed for different biomass such as wood, agricultural animal residue, algae, sewage sludge, food, fruit waste, etc. [23]. As aforementioned in pyrolysis, the yield of bioproducts is based on the thermal conversion of chemical components present in the biomass. In the case of both wood and vegetable biomass, dehydration is the primary reaction with a slight extension of decarboxylation. As a result, the decomposition of ligin, hemicellulose, and cellulose occurs at 240°C–350°C, 200°C–325°C, and 285°C–525°C, respectively. Consequently, dry carbonization reduces the ratios of H/C and O/C for biochar to 0.65 and 0.13, respectively, which is better than hydrothermal carbonization.

Here, the dehydration process consists of the separation of glyceride from the lipids and the removal of both amino and carboxyl groups from the protein which reduces the H/C and O/C ratios. Thus, the strong volatile content is released from the protein as the carbonization temperature attains 320°C. Increasing the temperature above 355°C tends to reduce the yield of charcoal to 60% and also decrease the amount of hydrocarbon than oxygen [19,65]. In the case of algal biomass, carbonization is carried out for four different stages between 250°C and 700°C. Firstly, the dried algae are heated to 250°C for removing the moisture. In addition, the volatilization of CO_2 from protein, carbohydrates, and lipids is occurring for further heating of algae between 260°C and 480°C. Therefore, a low temperature is required for the decomposition of algae when compared to lignocellulosic biomass. Finally, the degradation of the C-C bond and loss of mineral residue is observed when the temperature is increased to 720°C. Hence, a large amount of mesoporous carbon is derived from the carbonization of algae at 900°C due to the evaporation of impure atoms [64].

3.2.1.4 Liquefaction

Dry liquefaction is the high-pressure (up to 18 MPa) and low-temperature (250°C–500°C) thermal conversion process with dried biomass which provides a large amount of biocrude with a fraction of biochar and biogas. The water becomes a supercritical state during liquefaction and acts as a solvent which has higher solubility of organic compounds present in the biomass [66]. As aforementioned in carbonization, the pathway for decomposition of biomass follows the degradation of polymer into monomer, removal of moisture, decarboxylation, and deamination process [32,66]. Moreover, a tiny fragment is converted into unstable molecules, thereby forming a final product by altering those unstable molecules through condensation and polymerization. The O/H ratio of the biocrude can be deduced by removing both the carboxyl group and CO_2. A biocrude produced from the dry liquefaction has a high calorific value with more amount of aldehyde and phenol. In addition, the carbon present in the biocrude can be improved by increasing the liquefaction temperature [17,49]. The heavy oil produced from the liquefaction process has 18% more oxygen with a higher heating value [29].

The yield of both heavy soluble oils increased for temperature ranges from 285°C to 345°C. It also decreases when the temperature rises from 300°C to 380°C due to the breaking of liquid into the gaseous product. The sub-critical temperature is preferred for dry liquefaction rather than the supercritical temperature. But, the yield of soluble oil decreased for residence time after 30 min irrespective of the temperature, with similar decrease in heavy oil with increase in temperature. On the other side, the yield of the solid product decreases up to 30 min and it is gradually increased to 30% at a residence time of 60 min especially for 380°C because of the condensation of the gaseous product. The bio-oil yield shows direct proportionality with the operating pressure as reported in Le Chatelier's law. The solid product gradually decreases for the higher solvent-to-biomass ratio. On the other side, the bio-liquid provides maximum yield, especially for the ratio between 3 and 5 [67].

3.2.1.5 Gasification

Gasification is the combined process of slow pyrolysis, combustion, and reduction of biomass into valuable syngas products consisting of steam, hydrogen, hydrocarbons, methane, carbon monoxide, and carbon dioxide. It is similar to the combustion process but differs in the energy present between the chemical bonds in the final product. The predominant component of the biochar obtained from the pyrolysis process decides the reduction rate and operating time of the gasification process [68,69]. Most of the biomass consists of 68%–85% of volatile matter, which can be easily converted into a gaseous product during gasification. Initially, the biomass is dried and heated at a temperature ranging from 200°C to 720°C followed by the combustion started at a temperature of about 700°C without oxygen.

Then, the gasification is performed at a temperature between 790°C and 980°C for obtaining a higher gaseous product. It also provides the product gas with a predominant amount of H/C ratio by cracking of hydrocarbons present in the biomass. In addition, gasification also uses air, CO_2, and steam as the gasifying agent for improving the calorific value, H/C ratio as well as energy efficiency in the final product [20,70]. Likewise, the presence of hydrogen and carbon monoxide in the syngas can be improved by air gasification. The injection of oxygen as a gasification agent increases the amount of O_2 in CO_2 and reduces the O_2 in carbon monoxide [70,71].

3.2.1.5.1 Influence of Process Conditions on Biogas during Gasification

Gasification is mostly influenced by process parameters like reaction temperature, pressure, and flow rate of both biomass and gasification agent. The use of a combined mixture of biomass provides a wide range of behavior and heterogeneity in their syngas properties. The composition of an ideal syngas can be obtained when the system becomes thermodynamically equilibrium above 880°C [72,73]. The syngas with higher calorific value and hydrogen is attained when gasification is carried out under steam condition.

In addition, ash, moisture, heat conductivity, calorific value, organic and inorganic matter of the biomass, and the size of biochar obtained from the pyrolysis are the important properties to enhance both the quality and ranges of the biogas. Especially, the small-sized biochar produced at the pyrolysis process increases the amount of hydrogen and carbon monoxide, and promotes the homogeneous distribution of the gas flow. The ratio of fuel to gasification agent improves the effectiveness of the gasifier as well as the yield of syngas. In the case of the gasifier type, the efficient bio-product yield (tar, hydrogen, CO_2, methane, and nitrogen) is obtained in the circulating fluidized bed reactor than in other reactors like updraft, downdraft, and bubbling fluidized bed reactor [20,74–76].

3.2.2 Wet Thermochemical Conversion

The biomass with more than 31% moisture is considered wet biomass, which can be used for the wet thermal conversion process. Moreover, there is no requirement for energy for drying the biomass that can also provide more amount of biocrude at high pressure and nominal temperature than the dry conversion process [77,78]. In hydrothermal conversion, the reaction starts with the hydrolysis (homogeneous reaction) of biomass into glucose and phenol products, which are different from dry conversion. It also follows the anhydrous distillation process to make high-quality heating fuel as well as advanced bio-oil. It can also be classified into hydrothermal carbonization, liquefaction, and gasification [79–81]. A moderate temperature is followed for hydrothermal carbonization to produce more biochar under sub-critical conditions [78]. In the case of gasification, it is operated at high temperatures to develop hydrogen, carbon dioxide, and gaseous hydrocarbons under supercritical conditions [82].

3.2.2.1 Wet Pyrolysis

Wet pyrolysis is the thermal degradation of organic matter in the presence of water or another aqueous medium. In this process, the organic matter like microorganisms present in the biomass is heated with water, which leads to the higher production of biochar with a fraction of bio-oil and biogas. The water content in the biomass prevents the formation of a harmful product during pyrolysis, but it also reduces the overall energy efficiency of the process. The presence of water in the process medium can influence the reaction kinetics, control operating temperature, and also lead to a steam reforming process [83,84]. Wet pyrolysis is operated at temperature ranges from 350°C to 600°C and pressure between 0.5 and 2 MPa that prevents the release of volatile compounds present in the biomass.

The reaction is considered to be endothermic and forms a mixture of carbon dioxide and steam when the biomass is heated up to 315°C [85]. Consequently, the biomass is pyrolyzed and a large amount of flammable matter is released for temperature between 315°C and 500°C. In addition, the optimum temperature depends on various factors such as the type of feedstock and desired bio-product. [85,86].

3.2.2.2 Hydrothermal Carbonization

Hydrothermal carbonization involves the treatment of organic matter with pressurized hot water in anaerobic conditions. It is operated at temperature ranges from 160°C to 240°C with saturated steam pressure up to 15 MPa. It requires only mild temperature (less than 250°C) and minimum residence that can provide higher conversion efficiency than other biomass conversion methods. The temperature and pressure can also be increased to 375°C and 22 MPa (supercritical conditions) to get the maximum amount of bio-liquid and biogas yield [78,79,87]. The energy required for heating the reaction medium in hydrothermal carbonization (HTC) is comparatively lower than in the dry conversion process. In the case of bio-product, high-pressure carbonization can result in carbon-rich crystalline biochar with fine particle size, and lower-pressure carbonization provides biochar with more amorphous as well as hydrophilic in nature. In addition, the produced biogas like carbon dioxide, oxides of nitrogen, and sulfur are converted into corresponding salts and acids when they are dissolved in water. It is also a promising method to manage municipal water waste as it reduces the weight of the waste and makes it easier to transport [88–90].

3.2.2.2.1 Reaction Mechanism during Hydrothermal Carbonization

Dehydration plays a more vital role in HTC than other degradation process like carboxylation, liquefaction, polymerization, and aromatization during sub-critical conditions. In HTC, the properties of the water are changed under the sub-critical conditions that occur between 175°C and 240°C at 5 MPa. Consequently, the water loses its dielectric constant and converts the hydrogen molecules into hydronium (H$^+$) and hydroxide (OH$^-$) ions during the alienation of water. These excessive H$^+$ ions provide a suitable medium for reacting with the organic matter to form an improved carbon structure with acid-catalyzed without adding further catalyst [91–94].

3.2.2.3 Hydrothermal Liquefaction

HTL is a thermochemical conversion process that involves heating wet biomass, such as agricultural residue, municipal waste, algae, and lignocellulosic biomass in the occurrence of a solvent (water) under high pressure and temperature. It breaks down the complex organic molecules (higher molecular weight) in the biomass into simpler compounds with a lower molecular weight that includes a large amount of biocrude with a fraction of bio-solids and biogas (CO, H$_2$, and CH$_4$). It typically operates at temperature ranges from 250°C to 325°C with pressure increases from 10 to 25 MPa [80,95,96].

During the HTL process, the acid formation may corrode the reaction chamber which can be reduced by adding a weak base into the reaction. In the case of bio-product, the yield of bio-oil decreases gradually at temperatures more than 315°C due to the re-polymerization of biocrude into bio-solid. In addition, alkali catalysts like sodium carbonate, potassium carbonate, sodium hydroxide, potassium hydroxide, and calcium hydroxide can be used for improving the yield of biocrude without compromising its chemical composition. The major advantage of HTL is to produce high-quality biofuels that are compatible with existing petroleum-based infrastructure including engines, pipelines, and refineries [21,96–98].

3.2.2.3.1 Reaction Mechanism during Hydrothermal Liquefaction

The reaction mechanism of HTL consists of depolymerization, degradation, and re-polymerization of biomass. In depolymerization, there is a sequence of dissolving carried out for biomass to convert large-chain biopolymer to small-chain hydrocarbons. As aforementioned in the previous conversion process, the decomposition consists of dehydration, decarboxylation, and deamination. The removal of oxygen from the biomass is observed through water and CO$_2$ during dehydration and decarboxylation. The hydrogen present in the water and biomass breaks down into polar monomers. In addition, macromolecules present in the biomass are converted into oligomers and monomers through hydrolysis. After the complete reaction of the hydrogen ions, the hydrogen present in the organic matter is re-polymerized to bio-char with higher molecular weight [96,97,99].

3.2.2.4   Hydrothermal Gasification

Wet organic materials in the presence of water at supercritical conditions turn into gaseous organic compounds comprising methane, CO_2, and hydrogen. Wet organic material is fed into a high-pressure reactor vessel and heated to temperatures ranging from 320°C to 600°C under a pressure of 15–24 MPa [82,100]. This process requires a significant amount of energy input to maintain the high pressure and temperature conditions, which can make it less efficient than other energy conversion processes.

Gasification may be divided into three categories. Aqueous Phase Reforming (APR), catalyzed near critical gasification (NCG), and supercritical water gasification (SWG) are examples of these processes. In APR, hydrogen and carbon dioxide are produced by gasifying biomass compounds at temperatures ranging from 220°C to 270°C in the presence of a heterogeneous catalyst. In the case of SWG, combustible hydrogen and CO_2 are primarily produced in the presence of a carbon catalyst at supercritical temperatures. Furthermore, the gas generated by this process may be utilized to create heat and refined to a higher-quality biogas and bio-liquid [81,101].

3.2.2.4.1   Reaction Mechanism during Hydrothermal Gasification

The reaction mechanism is classified into hydrolysis, acidogenesis, acetogenesis, and methanogenesis. The organic matter is hydrolyzed into smaller molecules like glucose, amino acids, and fatty acids in the presence of water and a catalyst. These smaller molecules are broken down into volatile fatty acids and alcohols by an acidogenic process. Then, the volatile matter is converted into acidic acids and hydrogen by the acetogenic process. Finally, these products are reacted with the water to form volatile hydrocarbons like methane and carbon dioxide. The water can dissolve and convert organic matter, which can reduce solid waste and prevent the formation of greenhouse gases [81,102,103].

Figure 3.1 shows the types of thermochemical conversion and its process conditions.

3.3   BIOMASS-BASED CARBON IN ENERGY APPLICATIONS

The growth and demand for renewable energy materials with high efficiency, ease of manufacturing, and cost-efficiency have turned the attention toward biomass, the largest resource of carbon that is environmentally safe, cost-efficient, and tailored. These biomasses can be converted into highly efficient fuels or highly porous carbon for energy applications. There are challenges to be overcome in converting these biomasses into sustainable and eco-friendly substitutes for fossil fuels and energy storage applications. This section provides insight into the applications of carbon generated through the conversion process described in the previous section and also the current and future scope for these carbon materials.

Wood and other residues were collected and burned directly to produce heat, which is eventually used to cook food, stay warm, and even deter predators during the early civilizations. In the event of modern civilization, people discovered fossil fuels which reduced the dependence on biomass for fuel applications. But the discovered fossil fuel was finite and it could become zero soon. Also, the methods by which the fossil fuel was extracted and the procedures that converted it into fuels

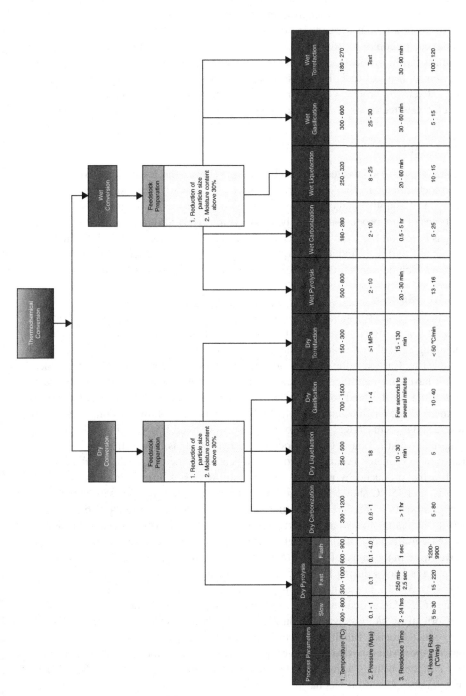

FIGURE 3.1 Thermochemical conversion types and process conditions.

and subsequent burning of that fuel increased the emissions of pollutants into the atmosphere, which led to various ill effects on the environment.

So, to reduce the dependence on fossil-based fuels like gasoline, diesel, CNG, LPG, etc., people are finding alternates such as solar, wind, geothermal, and biomass-based fuel sources. Out of these renewables, biomass is much closer in comparison with fossil fuels in terms of energy, and chemical and physical characteristics. Even the products can be tailor-made for application as there is no need to develop new technologies to extract energy out of fuels produced through biomass.

3.3.1 Fuel

Biomass can be converted into fuels through conversion techniques like thermal, thermochemical, and biological processes. Out of these, the thermochemical process is the most adaptive and closely related to natural conversion techniques to produce all types of fuel products. Typically, products are referred to as biochar, bio-oil, and biogas based on the yield and process. The produced fuel has similar characteristics to that of fossil fuels and it may require further processing if needed. Applications of biofuels include transportation, steam generation, city heating, etc.

3.3.1.1 Biochar

Biochar is a solid substance produced during the thermochemical conversion of biomass at low to high temperatures. Biochar can be made from a variety of feedstocks, including agricultural residue, lignocellulosic biomass, algal biomass, and municipal solid waste. Carbon is the primary constituent, although depending on the feedstock, it may have varying mineral or chemical compositions. The biochar's properties vary depending on the feedstock utilized and the process parameters under which it is created. However, biochar refers to the solid products obtained through pyrolysis, whereas hydrochar refers to the solid products obtained by hydrothermal synthesis. The major distinction between biochar and hydrochar is that hydrochar has higher ratios of oxygenated functional groups and is acidic, whereas biochar is alkaline.

Biochar produced by pyrolysis of agricultural wastes at temperatures of 250°C, 350°C, and 450°C showed increased porosity and thermal stability with varied mineral composition. The greater heating values of the biochar were 24,23.6 and 23.08 MJ/kg, indicating that when the temperature is raised, the heating values of the biochar decrease marginally [104]. The higher heating values for coconut fiber and eucalyptus leaves were in the range of 24.7–30.6 MJ/kg for coconut fiber and 25.3–29.4 MJ/kg for eucalyptus leaves during hydrothermal carbonization at temperatures ranging from 200°C to 375°C. These increased heating values are extremely similar to the lignite value of 25 MJ/kg, demonstrating that these solid products can be used as fuel [105].

Corn straw biomass was carbonized in an inert atmosphere at 350°C and 700°C to produce biochar which was further formed into briquettes using NovoGro as the binder; the results showed that the briquettes based on the biochar produced at 350°C has a higher heating value between 18.45 and 21.46 MJ/kg compared 6.96 MJ/kg of briquettes based on biochar produced at 700°C. It was noted that when the carbonization temperature was increased from 350°C to 700°C, the percentage of ash increased exponentially while the percentage of volatiles and fixed carbon reduced significantly showing that the volatiles could have vaporized, thus reducing the calorific value of the biochar [106].

Slow pyrolysis of walnut shells at 375°C produced a biochar with a higher heating value of 18.4 MJ/kg confirming that the char is hydrophobic, aromatized, and carbonized and could be used in the thermal conversion process in place of coal [107]. Torrefaction of cassava rhizome using nitrogen and carbon dioxide as a sweeping gas with various flow rates and composition produced a biochar with a maximum higher heating value of 24.54 MJ/kg at 300°C with a nitrogen flow rate of 50 mL/min [108].

3.3.1.2   Bio-oil/Biocrude

Based on the process of conversion, the liquid product obtained through thermochemical conversion is referred to as biocrude or bio-oil. Bio-oil is a liquid by-product of pyrolysis that has similar properties to biocrude, which is created through another thermochemical conversion, specifically hydrothermal liquefaction. Bio-oil is primarily composed of water in the 30%–40% range, depending on the moisture content of the feedstock used in the conversion process. The composition of bio-oil is highly dependent on process parameters and biomass composition, and it commonly consists of phenols, hydrocarbons, and ester combinations.

Pyrolysis of lignocellulosic biomass produces certain bio-oils that have furans, aldehydes, acids, and oxygenated compounds. On the other hand, bio-crudes generally have a higher degree of oxygenated and sulfur-based compounds and are generally corrosive, toxic, and have higher moisture content than bio-oil. Both bio-oils and biocrude certainly need upgradation to be used in refineries for fuel purposes. These remedies include the usage of catalysts, post-treatment to remove toxic substances, and reduction of moisture.

Fast pyrolysis of sawdust in a circulating fluidized bed produced bio-oil with a maximum yield of 60% at 500°C and a higher heating value of 24.1 MJ/kg; to use this bio-oil as an energy source, additional research on unit operations to transform it into a fuel source is required [109]. Co-pyrolysis of sawdust and waste polystyrene foam (WPSF) at 500°C with two different combinations yielded a biocrude with 48.33% (only sawdust) and 63.31% (25% sawdust + 75% of WPSF) with 17.81 MJ/kg (only sawdust) and 39.65 MJ/kg (25% sawdust + 75% of WPSF). This showed that the quality and energy capacity of the biocrude can be increased when there is an addition of polymer compounds which also reduces waste plastics from accumulating [110].

During the Maillard reaction between protein and carbohydrate during the HTL of microalgae, a biocrude with a mass yield of 47.6% was formed, with a higher heating value of 38.07 MJ/kg at 320°C. The findings revealed that the Maillard reaction plays an important role in the conversion of proteins and carbohydrates during the HTL of microalgae [111]. A multi-batch process was optimized to improve the yield and calorific value of the biocrude oil, and at temperatures ranging from 280°C to 310°C and residence times ranging from 15 to 45 minutes, the tobacco waste feedstock generated a maximum yield of 52% with a calorific value of 31.9 MJ/kg [112].

3.3.1.3   Syngas

One of the most important intermediates in the fuel processing area, syngas comprises hydrogen, carbon-di-oxide, carbon monoxide, and methane, and a few trace hydrocarbons. Though there is some generation of syngas during pyrolysis and few other thermochemical conversions, the highest yields of syngas are produced by gasification and hydrothermal gasification processes. Syngas is one of the types of products that can be directly used to fuel gas turbines or other combustion power plants. Production of syngas has its limitations owing to the higher temperature and pressure of the process and large-scale storage and distribution of the gas similar to natural gases.

SWG, also known as hydrothermal gasification, is the process of converting biomass into hydrogen and methane in the presence of a substantial amount of water at temperatures exceeding 370°C. The process parameters for hydrothermal gasification are critical for increasing hydrogen and methane output. Various catalysts are added to the mixture to increase the production of the syngas. The most common ingredient in syngas, hydrogen, has numerous potential applications like fuel for transportation.

Microwave-assisted gasification is a modified pyrolysis technique where the feedstock is uniformly heated, able to handle large particle sizes of biomass, and cost-effective and enables cleaner syngas production. The potential of this technique is yet to be tapped as it's still under the development stage and further research is needed to understand the effect of microwaves in different conditions.

3.3.2 ELECTRODE MATERIAL

Biomass-based carbon electrodes are promising alternatives due to the increase in, surface area, excellent electrical conductivity, and surface porosity due to the thermochemical conversion process making them a superior alternative and a strong candidate for energy storage applications. The electrode material from biomass can be used for supercapacitors (SCs), ion batteries, and fuel cells.

3.3.2.1 Ion Batteries

Carbon electrodes synthesized through the carbonization of hemp obtained a capacity of 463 mAh/g at 0.04 A/g of current density and a capacity retention of 97% after 6 months of usage. The carbon was porous, added with phosphorous nanoparticles by gas deposition, and was assembled into a potassium-ion battery that obtained around 294.2 Wh/kg of energy density [113]. Porous carbon nanospheres were generated through high-temperature carbonization of corn straw, and the generated carbon was activated using calcium chloride, which preserved the spherical shape in addition to increasing the porosity. The electrochemical performance of the carbon sphere was 546 mAh/g at the end of 100 cycles [114].

In the test, a porous interlocked nanostructure with a nitrogen content of 19% and a specific capacity of 523.6 mAh/g was produced by H_2SO_4-assisted hydrothermal treatment and subsequent pyrolysis using maize silk, demonstrating that a superior carbon material from biomass can be derived through thermochemical conversion and used for advanced energy applications [115].

Magnolia grandiflora Lima leaf was pyrolyzed to obtain a hard carbon compound, which when used as anode material exhibited a capacity of 315 mAh/g in sodium-ion battery and 263.5 mAh/g in potassium-ion battery [116].

3.3.2.2 Direct Carbon Fuel Cell

Fuel cells operate at comparatively high efficiency and can help if renewable fuels are used, which in turn will reduce greenhouse gas emissions. In the recent past, a direct carbon fuel cell (DCFC) have been used for the conversion of a molten carbonaceous solid fuel into electricity coal and biochar. Direct carbon solid oxide fuel cells derived from the biochars of wheat straw, corncob, and bagasse showed improved performances which closely resemble the feedstock characteristics.

The wheat straw and corn cob fuel cells peaked at 187 and 204 mW/cm^2, respectively, whereas the biochar fuel cell through bagasse produced high output of 260 mW/cm^2 [117]. The output performance of a DCFC using biochar derived from wheat straw was 197 mW/cm^2 and increased to 258 mW/cm^2 when 5% calcium was added as a catalyst; this showed that carbon derived from biochar can be used in preparing the solid-state fuel cell [118]. When biochar material was utilized to build a DCFC in the presence of a eutectic electrolyte, a maximum power density of 32.8 mW/cm^2 was obtained [119]. Forestry waste and sawmill residue were used to prepare microbial fuel cell electrodes with outputs of 532 and 457 mW/m^2, respectively. These electrodes were 90% less expensive than commercial electrodes made from activated carbon, demonstrating that biochar can be a cost-effective alternative for electrode manufacturing [62].

Despite its enormous potential and substantial research, the commercialization of biochar as an energy storage material is hampered by its lower performance when compared to standard energy storage materials. More research will be needed to understand the relationships between the properties of biochar and energy storage, as well as to enhance overall efficiency to increase energy storage efficiency.

3.3.2.3 Supercapacitors

SCs have identical anode and cathode electrodes that are both coupled to electrolytes (bilayers). The effect of charging in the double layer controls the charging process of SCs. SCs use three types of electrode materials: carbon-based materials, transition metal oxide, and conducting polymers. Because of their availability, ease of processing, non-toxicity, high chemical stability, and broad temperature range of use, carbon nanotubes, carbon aerogels, graphite, activated carbon, carbon nanofibers, and nano-sized carbon have been a long time SC material. Recently, novel carbon electrodes based on biomass-based materials and wastes have been created as a potential energy and environmental solution. A range of biomasses, including wood, plant tissues, industrial biomass wastes, and agricultural waste can be used to manufacture electrodes due to their low cost.

Waste aluminum soda cans were converted into alumina nanoparticles through calcination at higher temperatures. The produced alumina when tested using sodium sulfate electrolyte solution produced exceptional specific capacitance of around 1,297 F/g at 0.7 A/g while retaining 92% of the capacity for around 5,000 cycles [120]. The biochar material produced had an exclusive porous microrod- like structure and was nitrogen-doped. It was mixed in a ratio of 1:3 with KOH solution and activated at 900°C. The biochar has a high specific surface area of 2757.63 m^2/g with a total of 1.47 cm^3/g pore volume. Because of these specifics, it exhibited higher stability, while retaining the capacity of about 97% at around 5,000 cycles, and 390 F/g of specific capacitance at current density of 1 A/g [121].

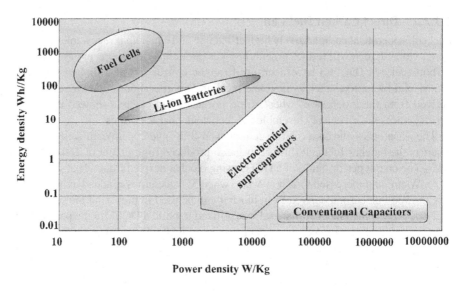

FIGURE 3.2   Ragone plot for carbon-based energy storage devices.

Figure 3.2 shows the Ragone plot for energy storage materials based on energy vs power density.

3.3.3 GRAPHENE-BASED COMPOUNDS

The single-layered graphitic sheet made of sp^2 hybridized carbon atoms is known as graphene. It is suitable for SC application owing to its huge surface area, excellent conductivity, and thermal steadiness. High diffusion of electrolyte ions at an ideal spacing between graphene nanosheets produces a high specific capacitance. The graphene compound was synthesized from biowaste through high-temperature conversion at 900°C using KOH as pore former and as a graphitization agent, the graphene sheets with a d-spacing of 0.345 nm were formed with 2,308 m²/g of the surface area [65]. Graphene was synthesized from carbonization and subsequent graphitization of glucose at 400°C, through both chemical and physical activation using NH_4Cl and CO_2, the graphene formed has a surface area of 3,657 m²/g [122].

Multilayered graphene was synthesized from corn husk through pyrolysis and activation. The graphene compound was turned into a zinc composite using zinc oxide and was used as a photocatalyst for dye degradation [123]. Porous graphite carbon sheets were synthesized from tissue paper through a facile one-step activation process, the produced graphite sheets have unique properties with interlayers to enhance ionic transfer suitable to be used in SCs [124].

3.3.4 CARBON NANOFIBERS AND NANOTUBES

Carbon nanofibers and nanotubes are nanostructures of carbon mainly graphene in different shapes. If they form a perfect cylinder, they are called carbon nanotubes. These advanced graphene structures have a higher surface area and porosity.

The carbon nanostructures obtained from biomass have multiple advantages and are very much feasible in energy storage devices [125]. Recent advances in microwave pyrolysis to convert biomass into carbon nanostructures by controlling the reaction conditions showed that microwave irradiation helps in improving the pore structure of the compound and helps in the formation of the carbon nanostructures [126].

Lignin is an attractive compound in fabricating a functional carbon composite; poplar wood biomass was used to prepare lignin-based carbon nanocomposite fiber, which was used in the SC application and showed a greater surface area of 1062.5 m^2/g producing a specific capacitance of 350 F/g [127]. Carbon nanofibers were synthesized from waste walnut shells through direct liquefaction and subsequent electrospinning followed by carbonization at elevated temperatures. These nanofibers were electro-spun using polyvinyl alcohol before carbonization. The carbon nanofiber composites were converted into the electrode to be tested in a Li-ion battery. The results showed that the highest specific capacity of 380 mAh/g was obtained with 55% efficiency [128].

3.4 CONCLUSION

The biomasses from multiple sources can be converted into useful materials through various conversion processes, yet the thermochemical process, when compared with other methods, produces quality products that can be tailored-based on the application and the process parameters can be altered to suit the product need. Conversion process like pyrolysis produces good quality solid products with a better pore distribution and requires limited activation for further processing, and the bio-oil generated has fewer oxygenated compounds than the hydrothermal process due to low moisture feeds. This can be a drawback for some feedstocks where removal of moisture is difficult; instead, hydrothermal process can be used in this case to produce products in all physical forms. Further, the products can be modified based on the applications through activation, extraction, etc. The applications of these products are typically carbon-based and are fuel, activated carbon for energy applications, fertilizers, and catalysts. This shows that renewable energy materials like biomass can be easily converted into useful products through a thermochemical conversion process, which is cost-efficient and easy to control, and produces higher-quality products.

REFERENCES

1. Chitra Devi V, Mothil S, Sathish Raam R, Senthilkumar K. Thermochemical conversion and valorization of woody lignocellulosic biomass in hydrothermal media. In: R Praveen Kumar, B Bharathiraja, R Kataki and VS Moholkar, editors. *Biomass Valorization to Bioenergy*. Singapore: Springer Singapore; 2020. p. 45–63. doi:10.1007/978-981-15-0410-5_4
2. Yusuf AA, Inambao FL. Characterization of Ugandan biomass wastes as the potential candidates towards bioenergy production. *Renew. Sust. Energy Rev.* 2020;117:109477. doi:10.1016/j.rser.2019.109477
3. Lee SY, Sankaran R, Chew KW, Tan CH, Krishnamoorthy R, Chu D-T, et al. Waste to bioenergy: A review on the recent conversion technologies. BMC Energy 2019;1:4. doi:10.1186/s42500-019-0004-7

4. Hamzah N, Tokimatsu K, Yoshikawa K. Solid fuel from oil palm biomass residues and municipal solid waste by hydrothermal treatment for electrical power generation in malaysia: A review. *Sustainability* 2019;11:doi:10.3390/su11041060

5. Singh S, Chakraborty JP, Mondal MK. Torrefaction of woody biomass (Acacia nilotica): Investigation of fuel and flow properties to study its suitability as a good quality solid fuel. *Renew. Energy* 2020;153:711–724. doi:10.1016/j.renene.2020.02.037

6. Gin AW, Hassan H, Ahmad MA, Hameed BH, Mohd Din AT. Recent progress on catalytic co-pyrolysis of plastic waste and lignocellulosic biomass to liquid fuel: The influence of technical and reaction kinetic parameters. *Arab. J. Chem.* 2021;14:103035. doi:10.1016/j.arabjc.2021.103035

7. Wang C, Zhang X, Liu Q, Zhang Q, Chen L, Ma L. A review of conversion of lignocellulose biomass to liquid transport fuels by integrated refining strategies. Fuel Process. Technol. 2020;208:106485. doi:10.1016/j.fuproc.2020.106485

8. Mlonka-Mędrala A, Evangelopoulos P, Sieradzka M, Zajemska M, Magdziarz A. Pyrolysis of agricultural waste biomass towards production of gas fuel and high-quality char: Experimental and numerical investigations. *Fuel* 2021;296:120611. doi:10.1016/j.fuel.2021.120611

9. Yue D, You F, Snyder SW. Biomass-to-bioenergy and biofuel supply chain optimization: Overview, key issues and challenges. *Comput. Chem. Eng.* 2014;66:36–56. doi:10.1016/j.compchemeng.2013.11.016

10. Ochieng R, Gebremedhin A, Sarker S. Integration of waste to bioenergy conversion systems: A critical review. *Energies* 2022;15:doi:10.3390/en15072697

11. Sengottian M, Venkatachalam CD, Ravichandran SR, Sekar S, Thirumoorthi A, Selvakumar KA, et al. Bisphenol A adsorption using hydrothermal carbonization derived biochar resulting from Casuarina equisetifolia L. and Wrightia tinctoria. *AIP Conf. Proc.* 2020;2240:130003. doi:10.1063/5.0010998

12. Sengottian M, Venkatachalam CD, Ravichandran SR. Optimization of alkali catalyzed hydrothermal carbonization of Prosopis juliflora woody biomass to biochar for copper and zinc adsorption and its application in supercapacitor. *Int. J. Electrochem. Sci.* 2022;17:2. doi:10.20964/2022.09.22

13. Venkatachalam CD, Sengottian M, Ravichandran SR. Hydrothermal conversion of biomass into fuel and fine chemicals. In: M Jerold, S Arockiasamy and V Sivasubramanian, editors. *Bioprocess Engineering for Bioremediation: Valorization and Management Techniques.* Cham: Springer; 2020. p. 201–224. doi:10.1007/698_2020_583

14. Tanger P, Field JL, Jahn CE, DeFoort MW, Leach JE. Biomass for thermochemical conversion: Targets and challenges. *Front. Plant Sci.* 2013;4:doi:10.3389/fpls.2013.00218

15. Zhang J, Zhang X. The thermochemical conversion of biomass into biofuels. *Thermochem. Convers. Biomass Biofuel.* 2019;327–368. doi:10.1016/B978-0-08-102426-3.00015-1

16. Awasthi MK, Sarsaiya S, Chen H, Wang Q, Wang M, Awasthi SK, et al. Global status of waste-to-energy technology. *Neur. Comput. Appl.* 2019;31–52. doi:10.1016/B978-0-444-64083-3.00003-8

17. Peters B, Bruch C. Drying and pyrolysis of wood particles: Experiments and simulation. *J. Anal. Appl. Pyrolysis.* 2003;70:233–250. doi:10.1016/S0165-2370(02)00134-1

18. Vilas-Boas ACM, Tarelho LAC, Kamali M, Hauschild T, Pio DT, Jahanianfard D, et al. Biochar from slow pyrolysis of biological sludge from wastewater treatment: Characteristics and effect as soil amendment. *Biofuel Bioprod. Biorefin.* 2021;15:1054–1072. doi:10.1002/bbb.2220

19. Ronsse F, Nachenius RW, Prins W. *Carbonization of Biomass.* In: *Recent Advances in Thermo-Chemical Conversion of Biomass.* Kent: Universiteit Gent; 2015. p. 293–324. doi:10.1016/B978-0-444-63289-0.00011-9.

20. Sheth PN, Babu BV. Production of hydrogen energy through biomass (waste wood) gasification. *Int. J. Hydrog. Energy* 2010;35:10803–10810. doi:10.1016/j.energy.2018.09.119

21. Alper K, Tekin K, Karagöz S. Hydrothermal liquefaction of lignocellulosic biomass using potassium fluoride-doped alumina. *Energy Fuels* 2019;33:3248–3256. doi:10.1021/acs.energyfuels.8b04381

22. Zhang W, Lin N, Liu D, Xu J, Sha J, Yin J, et al. Direct carbonization of rice husk to prepare porous carbon for supercapacitor applications. *Energy* 2017;128:618–625. doi:10.1016/j.energy.2017.04.065

23. Yoshida T, Antal MJ. Sewage sludge carbonization for terra preta applications. *Energy Fuels* 2009;23:5454–5459. doi:10.1021/ef900610k

24. Demirba Ş A, Balat M, Bozba Ş K. Direct and catalytic liquefaction of wood species in aqueous solution. *Energy Sour.* 2005;27:271–277. doi:10.1080/00908310490441971

25. Ratte J, Marias F, Vaxelaire J, Bernada P. Mathematical modelling of slow pyrolysis of a particle of treated wood waste. *J. Hazard. Mater.* 2009;170:1023–1040. doi:10.1016/j.jhazmat.2009.05.077

26. Shaaban A, Se S-M, Mitan NMM, Dimin MF. Characterization of biochar derived from rubber wood sawdust through slow pyrolysis on surface porosities and functional groups. *Proc. Eng.* 2013;68:365–371. doi:10.1016/j.proeng.2013.12.193

27. Wang G, Dai Y, Yang H, Xiong Q, Wang K, Zhou J, et al. A review of recent advances in biomass pyrolysis. *Energy Fuels* 2020;34:15557–15578. doi:10.1021/acs.energyfuels.0c03107

28. Quispe I, Navia R, Kahhat R. Energy potential from rice husk through direct combustion and fast pyrolysis: A review. *Waste Manag.* 2017;59:200–210. doi:10.1016/j.wasman.2016.10.001

29. Sand U, Sandberg J, Larfeldt J, Bel Fdhila R. Numerical prediction of the transport and pyrolysis in the interior and surrounding of dry and wet wood log. *Appl. Energy* 2008;85:1208–1224. doi:10.1016/j.apenergy.2008.03.001

30. Piloni RV, Daga IC, Urcelay C, Moyano EL. Experimental investigation on fast pyrolysis of freshwater algae. Prospects for alternative bio-fuel production. *Algal. Res.* 2021;54:102206. doi:10.1016/j.algal.2021.102206

31. Vîˆjeu R, Gerun L, Tazerout M, Castelain C, Bellettre J. Dimensional modelling of wood pyrolysis using a nodal approach. *Fuel* 2008;87:3292–3303. doi:10.1016/j.fuel.2008.06.004

32. Porteiro J, Míguez JL, Granada E, Moran JC. Mathematical modelling of the combustion of a single wood particle. *Fuel Process. Technol.* 2006;87:169–175. doi:10.1016/j.fuproc.2005.08.012

33. Arnold S, Rodriguez-Uribe A, Misra M, Mohanty AK. Slow pyrolysis of bio-oil and studies on chemical and physical properties of the resulting new bio-carbon. *J. Clean. Prod.* 2018;172:2748–2758. doi:10.1016/j.jclepro.2017.11.137

34. Lu J-S, Chang Y, Poon C-S, Lee D-J. Slow pyrolysis of municipal solid waste (MSW): A review. *Bioresour. Technol.* 2020;312:123615. doi:10.1016/j.biortech.2020.123615

35. Chaiwong K, Kiatsiriroat T, Vorayos N, Thararax C. Study of bio-oil and bio-char production from algae by slow pyrolysis. *Biomass Bioenergy* 2013;56:600–606. doi:10.1016/j.biombioe.2013.05.035

36. Manyà JJ, Azuara M, Manso JA. Biochar production through slow pyrolysis of different biomass materials: Seeking the best operating conditions. *Biomass Bioenergy* 2018;117:115–123. doi:10.1016/j.biombioe.2018.07.019

37. Şensöz S. Slow pyrolysis of wood barks from Pinus brutia Ten. and product compositions. *Bioresour. Technol.* 2003;89:307–311. doi:10.1016/S0960-8524(03)00059-2

38. Pourkarimi S, Hallajisani A, Alizadehdakhel A, Nouralishahi A. Biofuel production through micro- and macroalgae pyrolysis: A review of pyrolysis methods and process parameters. *J. Anal. Appl. Pyrolysis* 2019;142:104599. doi:10.1016/j.jaap.2019.04.015

39. Lee Y, Park J, Ryu C, Gang KS, Yang W, Park Y-K, et al. Comparison of biochar properties from biomass residues produced by slow pyrolysis at 500°C. *Bioresour. Technol.* 2013;148:196–201. doi:10.1016/j.biortech.2013.08.135

40. Fagernäs L, Kuoppala E, Tiilikkala K, Oasmaa A. Chemical composition of birch wood slow pyrolysis products. *Energy Fuels* 2012;26:1275–1283. doi:10.1021/ef2018836

41. Brown TR, Wright MM, Brown RC. Estimating profitability of two biochar production scenarios: Slow pyrolysis vs fast pyrolysis. *Biofuel Bioprod. Biorefin.* 2011;5:54–68. doi:10.1002/bbb.254

42. Guerrero MRB, Salinas Gutiérrez JM, Meléndez Zaragoza MJ, López Ortiz A, Collins-Martínez V. Optimal slow pyrolysis of apple pomace reaction conditions for the generation of a feedstock gas for hydrogen production. *Int. J. Hydrog. Energy* 2016;41:23232–23237. doi:10.1016/j.ijhydene.2016.10.066

43. Xiong Z, Wang Y, Syed-Hassan SSA, Hu X, Han H, Su S, et al. Effects of heating rate on the evolution of bio-oil during its pyrolysis. *Energy Convers. Manag.* 2018;163:420–427. doi:10.1016/j.enconman.2018.02.078

44. Ronsse F, van Hecke S, Dickinson D, Prins W. Production and characterization of slow pyrolysis biochar: Influence of feedstock type and pyrolysis conditions. *GCB Bioenergy* 2013;5:104–115. doi:10.1111/gcbb.12018

45. Sahoo SS, Vijay VK, Chandra R, Kumar H. Production and characterization of biochar produced from slow pyrolysis of pigeon pea stalk and bamboo. *J. Clean. Prod.* 2021;3:100101. doi:10.1016/j.clet.2021.100101

46. Toloue Farrokh N, Suopajärvi H, Mattila O, Umeki K, Phounglamcheik A, Romar H, et al. Slow pyrolysis of by-product lignin from wood-based ethanol production - A detailed analysis of the produced chars. *Energy* 2018;164:112–123. doi:10.1016/j.energy.2018.08.161

47. Yorgun S, Yıldız D. Slow pyrolysis of paulownia wood: Effects of pyrolysis parameters on product yields and bio-oil characterization. *J. Anal. Appl. Pyrolysis* 2015;114:68–78. doi:10.1016/j.jaap.2015.05.003

48. Yang C, Li R, Zhang B, Qiu Q, Wang B, Yang H, et al. Pyrolysis of microalgae: A critical review. *Fuel Process. Technol.* 2019;186:53–72. doi:10.1016/j.fuproc.2018.12.012

49. Chintala V. Production, upgradation and utilization of solar assisted pyrolysis fuels from biomass: A technical review. *Renew. Sustain. Energy Rev.* 2018;90:120–130. doi:10.1016/j.rser.2018.03.066

50. Khosravanipour Mostafazadeh A, Solomatnikova O, Drogui P, Tyagi RD. A review of recent research and developments in fast pyrolysis and bio-oil upgrading. *Biomass Convers. Biorefin.* 2018;8:739–773. doi:10.1007/s13399-018-0320-z

51. Xu F, Ming X, Jia R, Zhao M, Wang B, Qiao Y, et al. Effects of operating parameters on products yield and volatiles composition during fast pyrolysis of food waste in the presence of hydrogen. *Fuel Process. Technol.* 2020;210:106558. doi:10.1016/j.fuproc.2020.106558

52. Neto CJD, Letti LAJ, Karp SG, Vítola FMD, Soccol CR. Production of biofuels from algae biomass by fast pyrolysis. *Biofuel Bioprod. Biorefin.* 2019;461–473. doi:10.1016/B978-0-444-64192-2.00018-4

53. Ly HV, Kim S-S, Choi JH, Woo HC, Kim J. Fast pyrolysis of Saccharina japonica alga in a fixed-bed reactor for bio-oil production. *Energy Convers. Manag.* 2016;122:526–534. doi:10.1016/j.enconman.2016.06.019

54. Tsalidis GA, Tsekos C, Anastasakis K, de Jong W. The impact of dry torrefaction on the fast pyrolysis behavior of ash wood and commercial Dutch mixed wood in a pyroprobe. *Fuel Process. Technol.* 2018;177:255–265. doi:10.1016/j.fuproc.2018.04.026

55. Sarker TR, Nanda S, Dalai AK, Meda V. A review of torrefaction technology for upgrading lignocellulosic biomass to solid biofuels. *Bioenergy Res.* 2021;14:645–669. doi:10.1007/s12155-020-10236-2

56. Phusunti N, Phetwarotai W, Tekasakul S. Effects of torrefaction on physical properties, chemical composition and reactivity of microalgae. *Korean J. Chem. Eng.* 2017;35:503–510. doi:10.1007/s11814-017-0297-5

57. Acharya B, Dutta A, Minaret J. Review on comparative study of dry and wet torrefaction. *Sustain. Energy Technol. Assess.* 2015;12:26–37. doi:10.1016/j.seta.2015.08.003

58. Acharya B, Sule I, Dutta A. A review on advances of torrefaction technologies for biomass processing. *Biomass Convers. Biorefin.* 2012;2:349–369. doi:10.1016/j.seta.2015.08.003

59. Das P, V.P C, Mathimani T, Pugazhendhi A. Recent advances in thermochemical methods for the conversion of algal biomass to energy. *Sci. Total Environ.* 2021;766:144608. doi:10.1016/j.scitotenv.2020.144608

60. Barskov S, Zappi M, Buchireddy P, Dufreche S, Guillory J, Gang D, et al. Torrefaction of biomass: A review of production methods for biocoal from cultured and waste lignocellulosic feedstocks. *Renew. Energy* 2019;142:624–642. doi:10.1016/j.renene.2019.04.068

61. Olugbade TO, Ojo OT. Biomass torrefaction for the production of high-grade solid biofuels: A review. *Bioenergy Res.* 2020;13:999–1015. doi:10.1007/s12155-020-10138-3

62. Huggins T, Wang H, Kearns J, Jenkins P, Ren ZJ. Biochar as a sustainable electrode material for electricity production in microbial fuel cells. *Bioresour. Technol.* 2014;157:114–119. doi:10.1016/j.biortech.2014.01.058

63. Tian Z, Xiang M, Zhou J, Hu L, Cai J. Nitrogen and oxygen-doped hierarchical porous carbons from algae biomass: Direct carbonization and excellent electrochemical properties. *Electrochim. Acta* 2016;211:225–233. doi:10.1016/j.electacta.2016.06.053

64. Tian Z, Qiu Y, Zhou J, Zhao X, Cai J. The direct carbonization of algae biomass to hierarchical porous carbons and $CO2$ adsorption properties. *Mater. Lett.* 2016;180:162–165. doi:10.1016/j.matlet.2016.05.169

65. Nanaji K, Sarada BV, Varadaraju UV, N Rao T, Anandan S. A novel approach to synthesize porous graphene sheets by exploring KOH as pore inducing agent as well as a catalyst for supercapacitors with ultra-fast rate capability. *Renew. Energy* 2021;172:502–513. doi:10.1016/j.renene.2021.03.039

66. Behrendt F, Neubauer Y, Oevermann M, Wilmes B, Zobel N. Direct liquefaction of biomass. *Chem. Eng. Technol.* 2008;31:667–677. doi:10.1002/ceat.200800077

67. Peterson AA, Vogel F, Lachance RP, Fröling M, Antal JMJ, Tester JW. Thermochemical biofuel production in hydrothermal media: A review of sub- and supercritical water technologies. *Energy Environ. Sci.* 2008;1:32. doi:10.1039/B810100K

68. Hu Y, Cheng Q, Wang Y, Guo P, Wang Z, Liu H, et al. Investigation of biomass gasification potential in syngas production: Characteristics of dried biomass gasification using steam as the gasification agent. *Energy Fuels* 2019;34:1033–1040. doi:10.1021/acs.energyfuels.9b02701

69. Bellouard Q, Abanades S, Rodat S, Dupassieux N. Solar thermochemical gasification of wood biomass for syngas production in a high-temperature continuously-fed tubular reactor. *Int. J. Hydrog. Energy* 2017;42:13486–13497. doi:10.1016/j.ijhydene.2016.08.196

70. Kaewluan S, Pipatmanomai S. Potential of synthesis gas production from rubber wood chip gasification in a bubbling fluidised bed gasifier. *Energy Convers. Manag.* 2011;52:75–84. doi:10.1016/j.enconman.2010.06.044

71. Mazaheri N, Akbarzadeh AH, Madadian E, Lefsrud M. Systematic review of research guidelines for numerical simulation of biomass gasification for bioenergy production. *Energy Convers. Manag.* 2019;183:671–688. doi:10.1016/j.enconman.2018.12.097

72. Weerachanchai P, Horio M, Tangsathitkulchai C. Effects of gasifying conditions and bed materials on fluidized bed steam gasification of wood biomass. *Bioresour. Technol.* 2009;100:1419–1427. doi:10.1016/j.biortech.2008.08.002

73. Hannula I, Kurkela E. A parametric modelling study for pressurised steam/O2-blown fluidised-bed gasification of wood with catalytic reforming. *Biomass Bioenergy* 2012;38:58–67. doi:10.1016/j.biombioe.2011.02.045
74. Couto N, Rouboa A, Silva V, Monteiro E, Bouziane K. Influence of the biomass gasification processes on the final composition of syngas. *Energy Proc.* 2013;36:596–606. doi:10.1016/j.egypro.2013.07.068
75. Pio DT, Tarelho LAC, Pinto RG, Matos MAA, Frade JR, Yaremchenko A, et al. Low-cost catalysts for in-situ improvement of producer gas quality during direct gasification of biomass. *Energy* 2018;165:442–454. doi:10.1016/j.energy.2018.09.119
76. Van der Meijden CM, Veringa HJ, Rabou LPLM. The production of synthetic natural gas (SNG): A comparison of three wood gasification systems for energy balance and overall efficiency. *Biomass Bioenergy* 2010;34:302–311. doi:10.1016/j.biombioe.2009.11.001
77. Inoue S. Hydrothermal carbonization of empty fruit bunches. *J. Chem. Eng. Jpn.* 2010;43:972–976. doi:10.1252/jcej.10we027
78. Simsir H, Eltugral N, Karagoz S. Hydrothermal carbonization for the preparation of hydrochars from glucose, cellulose, chitin, chitosan and wood chips via low-temperature and their characterization. *Bioresour. Technol.* 2017;246:82–87. doi:10.1016/j.biortech.2017.07.018
79. Hoekman SK, Broch A, Felix L, Farthing W. Hydrothermal carbonization (HTC) of loblolly pine using a continuous, reactive twin-screw extruder. *Energy Convers. Manag.* 2017;134:247–259. doi:10.1016/j.enconman.2016.12.035
80. de Caprariis B, De Filippis P, Petrullo A, Scarsella M. Hydrothermal liquefaction of biomass: Influence of temperature and biomass composition on the bio-oil production. *Fuel* 2017;208:618–625. doi:10.1016/j.fuel.2017.07.054
81. Kruse A. Hydrothermal biomass gasification. *J. Supercrit. Fluids* 2009;47:391–399. doi:10.1016/j.supflu.2008.10.009
82. Schmieder H, Abeln J, Boukis N, Dinjus E, Kruse A, Kluth M, et al. Hydrothermal gasification of biomass and organic wastes. *J. Supercrit. Fluids* 2000;17:145–153. doi:10.1016/S0896-8446(99)00051-0
83. Alves SS, Figueiredo JL. A model for pyrolysis of wet wood. *Chem. Eng. Sci.* 1989;44:2861-2869,. doi:10.1016/0009-2509(89)85096-1
84. Libra JA, Ro KS, Kammann C, Funke A, Berge ND, Neubauer Y, et al. Hydrothermal carbonization of biomass residuals: A comparative review of the chemistry, processes and applications of wet and dry pyrolysis. *Biofuels* 2011;2:71–106. doi:10.4155/bfs.10.81
85. He Q, Raheem A, Ding L, Xu J, Cheng C, Yu G. Combining wet torrefaction and pyrolysis for woody biochar upgradation and structural modification. *Energy Convers. Manag.* 2021;243:114383. doi:10.1016/j.enconman.2021.114383
86. Bilbao R, Mastral JF, Ceamanos J, Aldea ME. Modelling of the pyrolysis of wet wood. *J. Anal. Appl. Pyrolysis* 1996;36:81–97. doi:10.1016/0165-2370(95)00918-3
87. Başakçılardan Kabakcı S, Baran SS. Hydrothermal carbonization of various lignocellulosics: Fuel characteristics of hydrochars and surface characteristics of activated hydrochars. *Waste Manag.* 2019;100:259–268. doi:10.1016/j.wasman.2019.09.021
88. Hoekman SK, Broch A, Robbins C. Hydrothermal carbonization (HTC) of lignocellulosic biomass. *Energy Fuels* 2011;25:1802–1810. doi:10.1021/ef101745n
89. Khan TA, Saud AS, Jamari SS, Rahim MHA, Park J-W, Kim H-J. Hydrothermal carbonization of lignocellulosic biomass for carbon rich material preparation: A review. *Biomass Bioenergy* 2019;130:105384. doi:10.1016/j.biombioe.2019.105384
90. Sermyagina E, Saari J, Kaikko J, Vakkilainen E. Hydrothermal carbonization of coniferous biomass: Effect of process parameters on mass and energy yields. *J. Anal. Appl. Pyrolysis* 2015;113:551–556. doi:10.1016/j.jaap.2015.03.012
91. Tremel A, Stemann J, Herrmann M, Erlach B, Spliethoff H. Entrained flow gasification of biocoal from hydrothermal carbonization. *Fuel* 2012;102:396–403. doi:10.1016/j.fuel.2012.05.024

92. Zheng Q, Morimoto M, Takanohashi T. Production of carbonaceous microspheres from wood sawdust by a novel hydrothermal carbonization and extraction method. *RSC Adv.* 2017;7:42123–42128. doi:10.1039/C7RA07847A

93. Ahmed Khan T, Kim HJ, Gupta A, Jamari SS, Jose R. Synthesis and characterization of carbon microspheres from rubber wood by hydrothermal carbonization. *J. Chem. Technol. Biotechnol.* 2018;94:1374–1383. doi:10.1002/jctb.5867

94. Funke A, Ziegler F. Heat of reaction measurements for hydrothermal carbonization of biomass. *Bioresour. Technol.* 2011;102:7595–7598. doi:10.1016/j.biortech.2011.05.016

95. Yua J, Biller P, Mamahkel A, Klemmer M, Becker J, et al. Catalytic hydrotreatment of bio-crude produced from hydrothermal liquefaction of aspen wood: A catalyst screening and parameter optimization study. *Sustain. Energy Fuels* 2017;1:832–841. doi:10.1039/C7SE00090A

96. Akhtar J, Amin NAS. A review on process conditions for optimum bio-oil yield in hydrothermal liquefaction of biomass. *Renew. Sustain. Energy Rev.* 2011;15:1615–1624. doi:10.1016/j.rser.2010.11.054

97. Gollakota ARK, Kishore N, Gu S. A review on hydrothermal liquefaction of biomass. *Renew. Sustain. Energy Rev.* 2018;81:1378–1392. doi:10.1016/j.rser.2017.05.178

98. Sintamarean IM, Grigoras IF, Jensen CU, Toor SS, Pedersen TH, Rosendahl LA. Two-stage alkaline hydrothermal liquefaction of wood to biocrude in a continuous bench-scale system. *Biomass Convers. Biorefin.* 2017;7:425–435. doi:10.1007/s13399-017-0247-9

99. Braz A, Mateus MM, Santos RGd, Machado R, Bordado JM, Correia MJN. Modelling of pine wood sawdust thermochemical liquefaction. *Biomass Bioenergy* 2019;120:200–210. doi:10.1016/j.biombioe.2018.11.001

100. Okolie JA, Nanda S, Dalai AK, Kozinski JA. Optimization studies for hydrothermal gasification of partially burnt wood from forest fires for hydrogen-rich syngas production using Taguchi experimental design. *Environ. Pollut.* 2021;283:117040. doi:10.1016/j.envpol.2021.117040

101. Elliott DC. Catalytic hydrothermal gasification of biomass. *Biofuel Bioprod. Biorefin.* 2008;2:254–265. doi:10.1002/bbb.74

102. Zhang Y, Cui Y, Chen P, Liu S, Zhou N, Ding K, et al. Gasification technologies and their energy potentials. In: MJ Taherzadeh, K Bolton, J Wong, and A Pandey, editors. *Gasification Technologies and their Energy Potentials.* Amsterdam: Elsevier; 2019. p. 193–206. doi:10.1016/B978-0-444-64200-4.00014-1

103. Luterbacher JS, Fröling M, Vogel F, Maréchal F, Tester JW. Hydrothermal gasification of waste biomass: Process design and life cycle asessment. *Environ. Sci. Technol.* 2009;43:1578–1583. doi:10.1021/es801532f

104. Waqas M, Aburiazaiza AS, Miandad R, Rehan M, Barakat MA, Nizami AS. Development of biochar as fuel and catalyst in energy recovery technologies. *J. Clean. Prod.* 2018;188:477–488. doi:10.1016/j.jclepro.2018.04.017

105. Liu Z, Quek A, Kent Hoekman S, Balasubramanian R. Production of solid biochar fuel from waste biomass by hydrothermal carbonization. *Fuel* 2013;103:943–949. doi:10.1016/j.fuel.2012.07.069

106. Wang T, Li Y, Zhi D, Lin Y, He K, Liu B, et al. Assessment of combustion and emission behavior of corn straw biochar briquette fuels under different temperatures. *J. Environ. Manag.* 2019;250:109399. doi:10.1016/j.jenvman.2019.109399

107. Alfattani R, Shah MA, Siddiqui MIH, Ali MA, Alnaser IA. Bio-char characterization produced from walnut shell biomass through slow pyrolysis: Sustainable for soil amendment and an alternate bio-fuel. *Energies* 2021;15:1. doi:10.3390/en15010001

108. Nakason K, Khemthong P, Kraithong W, Chukaew P, Panyapinyopol B, Kitkaew D, et al. Upgrading properties of biochar fuel derived from cassava rhizome via torrefaction: Effect of sweeping gas atmospheres and its economic feasibility. *Case Stud. Therm. Eng.* 2021;23:100823. doi:10.1016/j.csite.2020.100823

109. Park JY, Kim J-K, Oh C-H, Park J-W, Kwon EE. Production of bio-oil from fast pyrolysis of biomass using a pilot-scale circulating fluidized bed reactor and its characterization. *J. Environ. Manag.* 2019;234:138–144. doi:10.1016/j.jenvman.2018.12.104

110. Van Nguyen Q, Choi YS, Choi SK, Jeong YW, Kwon YS. Improvement of bio-crude oil properties via co-pyrolysis of pine sawdust and waste polystyrene foam. *J. Environ. Manag.* 2019;237:24–29. doi:10.1016/j.jenvman.2019.02.039

111. Qiu Y, Aierzhati A, Cheng J, Guo H, Yang W, Zhang Y. Biocrude oil production through the maillard reaction between leucine and glucose during hydrothermal liquefaction. *Energy Fuel* 2019;33:8758–8765. doi:10.1021/acs.energyfuels.9b01875

112. Saengsuriwong R, Onsree T, Phromphithak S, Tippayawong N. Conversion of tobacco processing waste to biocrude oil via hydrothermal liquefaction in a multiple batch reactor. *Clean Technol. Envir.* 2023;25:397–407. doi:10.1007/s10098-021-02132-w

113. Wang P, Gong Z, Zhu K, Ye K, Yan J, Yin J, et al. Nano-phosphorus supported on biomass carbon by gas deposition as negative electrode material for potassium ion batteries. *Electrochim. Acta* 2020;362:137153. doi:10.1016/j.electacta.2020.137153

114. Yu K, Wang J, Wang X, Liang J, Liang C. Sustainable application of biomass by-products: Corn straw-derived porous carbon nanospheres using as anode materials for lithium ion batteries. *Mater. Chem. Phys.* 2020;243:122644. doi:10.1016/j.matchemphys.2020.122644

115. Yang X, Zheng X, Yan Z, Huang Z, Yao Y, Li H, et al. Construction and preparation of nitrogen-doped porous carbon material based on waste biomass for lithium-ion batteries. *Int. J. Hydrog. Energy* 2021;46:17267–17281. doi:10.1016/j.ijhydene.2021.02.131

116. Zhu Z, Zhong W, Zhang Y, Dong P, Sun S, Zhang Y, et al. Elucidating electrochemical intercalation mechanisms of biomass-derived hard carbon in sodium-/potassium-ion batteries. *Carbon Energy* 2021;3:541–553. doi:10.1002/cey2.111

117. Qiu Q, Zhou M, Cai W, Zhou Q, Zhang Y, Wang W, et al. A comparative investigation on direct carbon solid oxide fuel cells operated with fuels of biochar derived from wheat straw, corncob, and bagasse. *Biomass Bioenergy* 2019;121:56–63. doi:10.1016/j.biombioe.2018.12.016

118. Cai W, Liu J, Liu P, Liu Z, Xu H, Chen B, et al. A direct carbon solid oxide fuel cell fueled with char from wheat straw. *Int. J. Energy Res.* 2019;43:2468–2477. doi:10.1002/er.3968

119. Kacprzak A, Kobyłecki R, Włodarczyk R, Bis Z. The effect of fuel type on the performance of a direct carbon fuel cell with molten alkaline electrolyte. *J. Power Sour.* 2014;255:179–186. doi:10.1016/j.jpowsour.2014.01.012

120. Mostafa MMM, Alshehri AA, Salama RS. High performance of supercapacitor based on alumina nanoparticles derived from Coca-Cola cans. *J. Energy Storage* 2023;64:107168. doi:10.1016/j.est.2023.107168

121. Wu F, Gao J, Zhai X, Xie M, Sun Y, Kang H, et al. Hierarchical porous carbon microrods derived from albizia flowers for high performance supercapacitors. *Carbon* 2019;147:242–251. doi:10.1016/j.carbon.2019.02.072

122. Jung S, Myung Y, Kim BN, Kim IG, You I-K, Kim T. Activated biomass-derived graphene-based carbons for supercapacitors with high energy and power density. *Sci. Rep. UK* 2018;8:1–8. doi:10.1038/s41598-018-20096-8

123. Sebuso DP, Kuvarega AT, Lefatshe K, King'ondu CK, Numan N, Maaza M, et al. Corn husk multilayered graphene/ZnO nanocomposite materials with enhanced photocatalytic activity for organic dyes and doxycycline degradation. *Mater. Res. Bull.* 2022;151:111800. doi:10.1016/j.materresbull.2022.111800

124. Harimohan E, Nanaji K, Rao BVA, Rao TN. A facile one-step synthesis of bio-inspired porous graphitic carbon sheets for improved lithium-sulfur battery performance. *Int. J. Energy Res.* 2022;46:4339–4351. doi:10.1002/er.7430

125. Azwar E, Wan Mahari WA, Chuah JH, Vo D-VN, Ma NL, Lam WH, et al. Transformation of biomass into carbon nanofiber for supercapacitor application - A review. *Int. J. Hydrog. Energy* 2018;43:20811–20821. doi:10.1016/j.ijhydene.2018.09.111

126. Omoriyekomwan JE, Tahmasebi A, Dou J, Wang R, Yu J. A review on the recent advances in the production of carbon nanotubes and carbon nanofibers via microwave-assisted pyrolysis of biomass. *Fuel Process. Technol.* 2021;214:106686. doi:10.1016/j.fuproc.2020.106686

127. Du B, Zhu H, Chai L, Cheng J, Wang X, Chen X, et al. Effect of lignin structure in different biomass resources on the performance of lignin-based carbon nanofibers as supercapacitor electrode. *Ind. Crop. Prod.* 2021;170:113745. doi:10.1016/j.indcrop.2021.113745

128. Tao L, Huang Y, Zheng Y, Yang X, Liu C, Di M, et al. Porous carbon nanofiber derived from a waste biomass as anode material in lithium-ion batteries. *J. Taiwan Inst. Chem. E* 2019;95:217–226. doi:10.1016/j.jtice.2018.07.005

4 Biochemical Approach to Biomass Conversion
Biofuel Production

Nishu Sharma
Punjab Agricultural University

Lovepreet Kaur
Lovely Professional University

Shiwani Guleria Sharma
Punjab Agricultural University

4.1 BACKGROUND

In earlier times, industrialization and urban society were directly or indirectly linked to energy production in a variety of sectors such as scientific and economic sectors. But as of now, the rise in the energy demand has led to the inevitable destruction of the surrounding nature, which ultimately led to global warming and various climatic changes. The main concern that is alarming now is the exhaustion of fossil fuel sources and the generation of a large volume of gases that are toxic in nature such as greenhouse gases (Prasad et al. 2007). Because of this, there is a need for alternative options to deal with the energy crisis globally. For this purpose, the need of the hour is to determine and study the environmental concerns and the strategies for their mitigation. Modernization has shifted the focus of the generation of fuel by various other sources (Boateng et al. 2006). The fuel can be generated with the help of various other sources which are sustainable in nature like wind, solar, geothermal, marine, and hydropower sources. But this is not suitable for the substitute fossil-based fuels. So for this purpose, biomass is required (Sun et al. 2002). In nature, biomass is categorized as non-lignocellulosic or lignocellulosic and is present in several forms either as a by-product of agricultural material or as aquatic leftovers and woody materials. The non-lignocellulosic biomass comprises some proteins, lipids, and a few forms of saccharides. Moreover, they have inorganic components, which may contain little fractions of lignin and cellulose (Jin et al. 2007), whereas lignocellulosic biomass contains cellulose more than 30% in each of its types. The leftover agricultural material such as corn stalks, sugarcane bagasse, sorghum straw, and grasses accounts for lignocellulosic biomass (Himmel et al. 2007).

DOI: 10.1201/9781003406501-4

All the products that constitute carbon in an organic form are known as form biomass. With the help of various processes, this form of carbon which is organic in nature can be transformed into the inorganic form. This inorganic form of carbon along with various other compounds such as hydrogen and oxygen can be utilized as solar energy (Chen et al. 2012). During the photosynthesis process, by a combination of the energy from the sun's rays and carbon dioxide, chemical energy can be generated in the form of carbohydrates from the biomass. This is due to the carbon dioxide that is stored during the process of photosynthesis, which is released when it burns; it has the ability to be an energy source (Taherzadeh et al. 2008). It is one of the most economic and readily available forms which can be used from the easily available residues of forests and agricultural waste. There are enormous methods for the conversion of biomaterial into essential products, contingent upon the type of biomass and the end product. Mostly, the technologies that can be adopted in biomass conversion are mainly categorized as thermochemical or biological methods (Prasad et al. 2007).

In the future, energy resources will be very critical to factor in the economic growth of nations. One of the main leading forms of energy is biomass with a significant worldwide load of energy of more than 10%. However, in remote and rural areas, biomass is responsible for approximately 90% of the whole energy required in various nations (Miao et al. 2011). Several types of biomass can be employed for biochemical conversion and biofuel production. Some of them are discussed as wood and agricultural waste and solid waste. Wooden residues and agricultural waste account of about more than 40% of biomass waste. They are abundantly present and used for household purposes. They comprise logs of wood, chips, bark, and sawdust. Such waste can be used for the production of electricity (Wyman et al. 2005). For past thousands of years, people have been using wood to heat and cook their homes and it was the prime energy source in the United States and the rest of the nations. In the United States, 80% of wood-based fuel is consumed by various businesses and electric producers. In 2021, 2.1% of the total energy consumption was from wood and wood residues in the United States. Solid waste forms energy by burning the trash into a usable form of energy (Serapiglia et al. 2013). It has been reported that one tonne (2,000 pounds) of garbage constitutes approximately similar to 500 pounds of coal which is heat energy. Garbage is composed of a number of components including biomass. Possibly, 50% of its energy proportion is derived from plastics which further produced from petroleum and natural gas. The solid waste is composed of the following organic matter which is about 57%, paper and other cardboard materials are about 13%, while the metals, glass, and wood are approximately 7%, 4%, and 2%, respectively, in the solid waste composition that is collected from the industries as well as residential areas (Davis et al. 2020).

Biomass is divided into distinct forms such as woody biomass, herbaceous biomass, and aquatic and animal waste. The conversion of biomass can be done by various approaches but for attaining the best results and proper fuel production initially the pretreatment of the biomass is carried out (Anukam et al. 2017). In the case of the pretreatment, it can be done either by physical, chemical, or biological means. It relies upon the type and the recalcitrant nature of the biomass that is used. Following pretreatment, different techniques are implemented to convert the biomass into biofuel that are either physical-chemical or thermochemical in nature (Liu et al. 2015). The thermochemical approaches are further classified as per the method adopted for the conversion process like pyrolysis, gasification, combustion, and liquefaction (Habert et al. 2010).

4.2 CLASSIFICATION OF BIOFUELS

Biofuels are primarily classified into two groups (Chakraborty et al. 2012).

4.2.1 PRIMARY BIOFUELS

These types of biofuels are used in a raw form. Their uses include heating, cooking, and electricity generation such as wood chips and pellets, fuelwood, etc. Release of toxic environmental pollutants and energy inefficiency are the two major limitations of these biofuels (Chakraborty et al. 2012).

4.2.2 SECONDARY BIOFUELS

The biofuels produced by processing biomass are secondary. For example, bioethanol and biodiesel are produced by processing of lignocellulosic biomass and transesterification of vegetable oils and fats, respectively. Their usage includes running vehicles and various kinds of industrial processes. They are further divided into three generations described below (Rodionova et al. 2016):

First-generation biofuels: These involve the use food or feed materials, sugar-containing materials, starch, oil crops, or animal fats as substrates. Technologies employed for their production include fermentation, esterification, and distillation. Vegetable oils (Roy 2017), biodiesel (Hoekman et al. 2012), biogas (Balat and Balat 2009), and solid biofuels (Dafnomilis et al. 2017) are some examples. Various sugar-containing substances are used as substrates for their production. For example, ethanol production utilize corn as feedstock in the United States, wheat and corn in Canada, and sugarcane in Brazil while potatoes, wheat, and sugar beets are used in Europe (Tse et al. 2021). Although these biofuels are economical, they constitute a menace to biodiversity conservation and demand agricultural lands, thus restricting their commerciality (Sorigue et al. 2016). Their production is limited by the following issues:

1. Their cultivation requires land and water, which are otherwise used for food production (Searchinger et al. 2008).
2. Production process is expensive and requires subsidies from the government to rival with petroleum-derived products (Doornbusch and Steenblik 2007).

Second-generation biofuels: As long as food crops are being employed as substrates for first-generation biofuels, the food vs fuel problem will scourge the biofuel industry. To meet this end, extensive research is being conducted on biofuel production from non-food biomass, and fuels produced from such sources are called second-generation biofuels (Chakraborty et al. 2012). The use of non-edible substrates for second-generation biofuel production has ended the food versus fuel debate. They are produced either by biological or thermochemical methods. Substrates used are mainly whole plant lignocellulosic biomass or non-edible residue of a food crop (Nigam and Singh 2011). These are more advanced industrial versions as compared to first-generation biofuels and feedstock, such as wood, agricultural waste, food waste, and some biomass crops are prime substrates for production (Berla et al. 2013). Lignocellulosic biomass constitutes primarily cellulose, hemicellulose, and lignin, which are very cumbersome to degrade. This recalcitrant nature is the prime cause of feedstock resistance to degradation treatments such as chemical and biological methods, thus necessitating the need for pretreatment methods (Baruah et al. 2018). Second-generation biofuels are less viable economically as they yield lower oil per acre of culturable land (Hossain 2019). Unlike first-generation biofuels, these do not necessitate the use of food crops but their production process requires sophisticated technologies for processing of samples (Juodeikiene et al. 2014). In comparison to first-generation biofuels, second-generation biofuels yield higher productivity per unit of land, thus enabling efficient use of land (Nigam and Singh 2011).

Third-generation biofuels: The prime source of production of third-generation biofuels is algae. Microalgae are photosynthetic micro-organisms, either autotrophic or heterotrophic, and can be found thriving in freshwater or marine water environments. Richmond (2004) reported that out of 50,000 prevailing species of microalgae, only 30,000 have been discerned. More than 50% of oil is present in algal biomass, which produces 10–100 times higher yields of biomass and fuels as compared to energy crops such as soybean, sunflower, and rapeseed. Cultivation of algae can be done under conditions that are incongruous with conventional crop cultivation (AL-Rajhia et al. 2012). A very prominent yield (90,000 L/ha) of biodiesel has been observed in algae as compared to other oilseeds such as soybean (450 L/ha), rapeseed (1,200 L/ha), and oil palm (6,000 L/ha) (Haag 2007). Algal biomass can be converted to biofuels through biochemical, chemical, or thermochemical conversion processes (Behera et al. 2015). Algae has many commendatory attributes necessary for biofuel conversion (Hu et al. 2008): (i) expeditious growth rate, resulting in high biomass productivity, (ii) great amount of lipid accumulation for biodiesel production (Scott et al. 2010), (iii) marginal land tolerance to avoid rivalry with agricultural lands (Day et al. 2011), (iv) greenhouse gases confiscation property (Packer 2009), and (v) production of valuable co-products of commercial importance (Raja et al. 2008). The process of consolidation of microalgae cultivation and biofuel generation is a distinctive symbiosis and it could be one of the alternatives for the prospective biorefinery sources (Ichsan et al. 2014). In the industrial and academic zone, lipid production and its use for mass production of biodiesel is the major center of attraction for research purposes (Kim et al. 2017).

Biodiesel production from microalgae is a multiphase process: selection of oil-producing microalgal strain, mass cultivation, harvesting of biomass, cell lysis, extraction of lipids, conversion of lipids to biodiesel, utilization of residual biomass for valuable bio-products (Seo et al. 2018). Mass production of oil-producing microalgae is the major hurdle for the commercial success of biodiesel as it costs approximately 40% of the total cost (Kim et al. 2016). Other obstacles such as low lipid production, and unstable microalgae culture outdoors in untreated water are also some of the limiting constraints (Cho et al. 2013). Microalgae can accumulate vast amount of lipids under nutrient-deficient conditions in laboratory conditions. Nitrogen is of utmost importance for the synthesis of proteins. An increased amount of nitrogen source in growth media results in enhanced growth while nitrogen deficiency creates conditions under which a substantial amount of photosynthetically fixed carbon is used for lipid synthesis. For example, *Neochlorisoleo abundans* produced two-fold and *Nannochloropsis* sp. F&M-M24 produced onefold, increased lipids under nitrogen-limiting environment (Rodolfi et al. 2009). An increase in the concentration of nitrogen causes increased microalgal growth. Another sought-after nutrient is phosphorus, which affects many intracellular processes. It was reported that under sufficient conditions 25%–28% of lipids were present in *Scenedesmus* sp. LX1 while under phosphorus-limiting conditions 53% lipids were produced (Xin et al. 2010). However, at the industrial level, the nutrient deprivation process for lipid generation is not feasible as it requires vast amounts to produce for commercialization purposes. Therefore, scientists have been prompted to think about an alternate strategy for mass cultivation and for the induction of a large number of lipids to accumulate.

Another issue to address is that residual biomass left after lipid extraction contains many other valuable compounds like carbohydrates, proteins, antioxidants, and phenolic compounds which can be purified and have many industrial applications (Chen et al. 2017). Despite these promising features of algal-derived biofuels, biomass conversion technology needs further refinement for the sustainable commercialization of these biofuels (Hossain 2019). Mass production of these biofuels for commercial production is paralyzed by the following constraints (Chakraborty et al. 2012):

1. Search for the most robust and efficient microalgal strain.
2. Optimization of cultivation conditions of selected strains for the highest growth rates.
3. Genetic modification of the metabolic pathways for lipid production optimization by algal cells at the desired level.
4. Designing an economical and optimized protocol for lipid extraction from microalgal cells.

4.3 CLASSIFICATION OF BIOMASS

Biomass is classified depending on the following criteria (Tursi 2019):

1. Based on biomass type prevailing in nature.
2. Based on application of biomass feedstock.

Therefore, the first category biomass is divided into the following five types.

4.3.1 WOOD AND WOODY BIOMASS

It is the prime renewable energy resource. Carbohydrates and lignin are the main ingredients of wood and woody biomass types. Examples include residues of trees and roots, leaves, and barks of woody shrubs (Vassilev et al. 2012). Usually, four types of biomass are used for energy generation: (i) production residues; (ii) non-merchant timber residues; (iii) wood wastes after combustion; and (iv) urban and agricultural wastes.

4.3.2 HERBACEOUS BIOMASS

According to European standard EN 14961-1, "Herbaceous biomass is from plants that have a non-woody stem and which die back at the end of the growing season. It includes grains or seeds crops from the food processing industry and their by-products such as cereal straw" (Chum et al. 2011). It belongs either (i) to agricultural residues which are by-products of food industries or fibers or (ii) to energy crops (Tursi 2019).

4.3.3 AQUATIC BIOMASS

Macro- or microalgae and emerging plants come under the aquatic biomass category (Di Benedetto 2011). Microalgae are microscopic, unicellular or multicellular, prokaryotic, or eukaryotic micro-organisms with astounding photosynthetic ability and are capable of surviving in a stressful environment (Greenwell et al. 2010). It can accumulate more than 50% oil in its biomass which has application in biodiesel production (Al-Rajhia et al. 2012). Macroalgae are mainly involved in food production and hydrocolloid extraction, and can reach up to a height of 60 m. Marshes and swamps are the main habitats for emerging plants. They metabolize sunlight, water, and CO_2, and algal biomass is produced (Vassilev and Vassileva 2016).

4.3.4 HUMAN AND ANIMAL WASTE BIOMASS

Bones, meat, animal manure, and human excreta are prominent examples of human and animal waste biomass (Vassilev et al. 2012). Earlier, these wastes were used as fertilizers but the due implementation of strict rules by regulatory agencies has stopped this practice. Biogas technology can be used to digest these wastes anaerobically to syngas, i.e., biogas having application in cooking, generating electricity, and in internal combustion engines (Horan 2018).

4.3.5 BIOMASS MIXTURES

When the substrates described above are in mixed form, they are called biomass mixtures (Tursi 2019).

4.4 PRETREATMENT OF BIOMASS

As the structure and composition of the biowaste are very complex and it is recalcitrant in nature, it requires a pretreatment process step (Figure 4.1). Pretreatment can be done either by physical method, chemical method, or biological means (Shen et al. 2010).

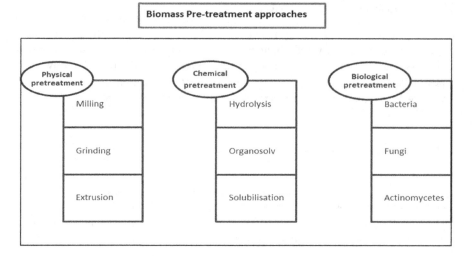

FIGURE 4.1   Biomass pretreatment approaches.

In the case of the physical approach, initially the milling, chipping, briquetting, pelleting, and torrefaction are done, whereas in the case of the biological means the bacteria, fungi, or actinomycetes are used (Alizadeh et al. 2005). In case of the chemical pretreatment, it can be attained by the process of hydrolysis with the help of some enzymes, solubilization, and by using organic solvents (Dayton et al. 1999). The most significant challenges that current pretreatment technologies face are huge cost and generating a pre-treated product with minimal destruction or elimination of the vital components.

4.4.1   PHYSICAL PRETREATMENT

This pretreatment process of the biomass focuses on improvising the surface area and carrying out the modifications in the size of the pore by mechanically reducing the particle size. Physical preparation for lignocellulosic biomass materials reduces the following characteristics, i.e., cellulose crystallinity and the degree of polymerization (Zhu et al. 2006). This is the main step before the metabolic and thermochemical conversion of biomass. As per reviews and researches it has been noted that there are very few details and knowledge regarding how physical pretreatment methods work, particularly in terms of how biomass's chemical constituents as well as structural forms are altered. The biomass used in the process relies upon the type of physical pretreatment process to be utilized (Liang et al. 2010). For elaboration, in the case of size reduction, milling will be necessary for the biochemical conversion of lignocellulose biomaterial to enhance the component's ability to be broken down by the various forms of enzymes. Pretreatment techniques like densification, pelletization, and torrefaction need the process of milling to prepare biomass for thermochemical conversion (Ananthi et al. 2019). In the case of the thermochemical and biochemical pretreatment conversion methods, the initial step of the reduction in the size is important to get the mass and transfer of the heat constraints; chipping is also a widely adopted physical pretreatment technique used in place of the biomass as a feedstock used for the thermochemical conversion processes (Kumar et al. 2009). This is due to most thermochemical conversion systems that require feedstock with a diameter between 50 and 50 mm due to their size-specific nature. The physical pretreatments are effectively used as an aid in enhancing the yields of enzymatic hydrolysis (Bak et al. 2009). Increased surface area by mechanical refining or decreased particle size by mechanical breakdown is the major principle. The only drawback of all the methods of the physical pretreatment is that they require an input of energy for their proper functioning (Himmel et al. 2007).

4.4.1.1   Some of the Physical Pretreatment Approaches Are as Follows

Milling: It is being widely used to make lignocelluloses more available to cellulases depending on the existing cellulose structure and their extent of crystallinity (Robert et al. 2018). Cellulases are enzymes that facilitate the disintegration of cellulose, but for maximum performance and catalysis, the availability of the substrate must be increased (Ananthi et al. 2019). The lignocellulosic material needs to be minimized in size before being subjected to enzymatic hydrolysis. Milling with ball or colloid, vibro-energy milling, hammer milling, and two-roll milling are some of the specific sorts of grinding processes (Kumar et al. 2009). Colloid crushers, dissolvers, and fibrillation work well with wet materials, while the mills such as hammers, cryogenic, roller, and extruders are widely used for dry ingredients (Liang et al. 2010).

Microwave: Irradiation by microwave is a major treatment process for crop residues. This pretreatment approach provides many benefits, including ease of pretreatment, maximum heating capacity, rapid working, little inhibitor production, and low use of energy (Himmel et al. 2007). A few academics from Kyoto University, Japan, in 1984 have published the study on microwave irradiation in an enclosed container. With the addition of the water, they microwave-treated rice hulls, rice straws, and sugarcane bagasse. Fifty milliliters of glass jars, 2,450 MHz energy, and more than 2.0 kW of microwave radiation are the requirements for microwave therapy (Demirbas et al. 2009).

Extrusion mechanically: In this case of physical pretreatment method, the materials slide through a die that is already having a predefined cross-section and come out with a fixed, definite profile. This extrusion method is well-known for recovering sugar from biomass (Liang et al. 2010). A few benefits of the mechanical extrusion pretreatment process are adaptability to alterations, the elimination of products that are damaged, environment management, and high output. It has two types: single screws and twin screws (Kumar et al. 2009).

4.4.2   Chemical Pretreatment

The formation of specific bonds of the primary organic components of biomass leads to physical instability of the biomass. Biomass is resistant to chemical breakdown, in particular when it comes to lignocellulosic biomass (LCB) (Yu et al. 2010). Complex anatomy, type of biomass, crystallinity of its cellulose components, the degree of lignification are some elements contributing to its recalcitrance. To break down the biomass components, various chemicals are used that can be alkali, acids, or wet oxidation (Schacht et al. 2008). Increased yields and less time consumption are

TABLE 4.1

Classification of Chemicals in Accordance with the Different Varieties of the Biomass

S. No.	Classification of Chemical	Chemical	Type of Biomass	References
1.	Green solvents	Ethanol, acetone, and propanol	Wheat straws, cotton, switchgrass, bagasse, corn stover, and hardwoods of different hardness	Yu et al. (2010)
2.	Acid	Dilute sulfuric acid, phosphoric acid, and nitric acid	Switchgrass, softwood, and cottonwood	Kumar et al. (2009)
3.	Wet oxidation	Sodium carbonate and ammonium persulfate	Fodder, corn stover, peanuts, rye, and baggase of sugarcane	Shogo et al. (2021)
4.	Alkali	Sodium hydroxide	Roughage of corn, bagasse, and fodders	Demirbas et al. (2009)

two main benefits of chemical pretreatment. Chemical precipitation, neutralization, adsorption, disinfection (chlorine, ozone, ultraviolet (UV) light), and ion exchange are the chemical treatment procedures that are used most frequently (Shogo et al. 2021). Some of the chemicals along with their classification are as follows (Table 4.1):

4.4.3 BIOLOGICAL PRETREATMENT

In this process of the pretreatment of the biomaterial initially, the biomass is incubated with specific microbes that create extracellular enzymes that ultimately alter the biomass, enhancing its suitability for use in either biological or thermochemical processing. Biological pretreatment requires little energy input (Shruti et al. 2021). These treatment methods amend the composition of the chemicals as well as the anatomy of lignocellulosic substrate to make the biomass less resistant to enzymatic assault. These biological pathogens involve white-rot fungi, brown-rot fungi, soft-rot fungi, and bacteria. The white-rot fungi specifically target lignin, while brown-rot and soft-rot fungi typically target cellulose and mildly alter lignin (Kim et al. 2015). All of the physicochemical parameters encouraging optimum growth and enzyme expression are therefore crucial for efficient pretreatment since biological pretreatment depends on the right micro-organisms and their enzymes (Robert et al. 2018).

4.5 BIOMASS TO BIOFUEL CONVERSION APPROACHES

4.5.1 PHYSICOCHEMICAL APPROACH

Diverse strategies are used in the physicochemical transformation of the waste to enhance the physical and chemical features of solid waste (Kobayashi et al. 2004). It is primarily used to degenerate this resistance, remove the cellulose from the matrix polymers, and increase the cellulose's accessibility for enzymatic hydrolysis. Steam is generated by conversion of combustible portion of waste high-energy fuel pellets (Fernandez et al. 2014). The pretreatment can increase sugar yields for biomass like wood, grass, and corn to more than 90% of the theoretical yield. Drying is done for reduction of the high amount of moisture in wastes. Before compaction and turning into fuel pellets, the waste is physically separated from sand, grit, and other incombustible material (Lizasoain et al. 2017). The fuel pellets have several distinct benefits over coal and wood as they are clean, rid of non-flammables, and have lower ash and content moisture. Furthermore, the fuel pellets are also consistent in size. They are cost-effective and environmentally safe.

4.5.2 THERMOCHEMICAL APPROACH

Several thermochemical techniques such as combustion, torrefaction, hydrothermal liquefaction, pyrolysis, and gasification have been studied and used to transform liquid biomass into fuels (Duff et al. 1996). Biofuel is generated by the decomposition of biomass under precise operating conditions, which results in solid, liquid, and gas that require an added catalytic enhancement method to biomass as a feedstock for biomaterials. Some of the other approaches that are used in thermochemical approaches generate liquid fuels (Qiao et al. 2013). Main advantage is its ability to use any form of method to convert biomass into biofuels, such as pyrolysis, gasification, liquefaction, and combustion (Sadhwani et al. 2018).

Pyrolysis: Thermal degradation of organic material biomass under oxygen-deficient conditions is known as pyrolysis. Generally, it is done at a very high temperature at or above 500°C as it provides enough heat to degrade the robust bio-polymers (Wang et al. 2007). Biochar, bio-oil, and gases like methane, hydrogen, carbon monoxide, and carbon dioxide are its by-products. It is classified into three different categories: conventional/slow, fast, and ultra-rapid/flash (Ishizawa et al. 2007). Every method that is adopted has variations in terms of temperature, residence time, heating rate, and end products. In this approach, a variety of biomass feedstock can be utilized (Sheth et al. 2004).

Gasification: It is a technological approach that transforms biomass or leftover agricultural waste into hydrogen and other products without burning using various methods (Cara et al. 2008). The process of gasification provides a more well-established process for producing energy and other useful products from feedstocks like coal, biomass, and some waste streams in comparison to other procedures (Viola et al. 2008). In the forthcoming years, gasification may play a bigger role in the global energy and industrial sectors as a result of its benefits in particular situations and applications, particularly in the clean generation of electricity from coal (Wyman et al. 2005). Coal will likely continue to be the primary feedstock choice for gasification systems in the coming decades due to its low price and plentiful supply everywhere in the world. Five stages of gasification of biomass are (Figure 4.2).

Any carbonaceous raw material, such as coal, can be gasified to yield fuel gas, usually alluded to as synthesis gas. Gasification occurs in a gasifier, which is typically a

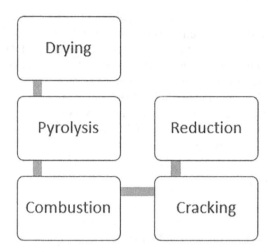

FIGURE 4.2   Flowchart representing stages of gasification.

high-temperature/pressure vessel where air or oxygen comes into direct contact with coal or some other feed material (Basu et al. 2010). The feed material is converted into syngas and ash/slag (mineral residues) by these chemical processes.

Liquefaction: It is a process used to transform biomass into charcoal, gases, and the target product bio-oil. The procedure is typically accomplished at temperature and pressure ranging from 200°C to 400°C and 5–25 MPa, respectively, in water or any other suitable solvent (Mofijur et al. 2013). Various catalysts are used in the process of the liquefaction of biomass. They depend upon the type of biomass that is utilized for the processing. Studies have already been done and it has been observed that both homogeneous and heterogeneous catalysts can be utilized in liquefaction process (Chen et al. 2014). Indirect and direct liquefactions are the two main techniques used in this process of biomass conversion. Apart from this, there are two primary forms of thermodynamic liquefaction of biomass, namely, pyrolysis liquefaction and hydrothermal liquefaction. These liquefaction protocols depend and are done on the operating circumstances which rely on the form of the biomass utilized (Narayanaswamy et al. 2013).

Combustion: The only carbon-based renewable fuel is biomass, so with the exhaustion of fossil fuel resources its use is becoming more and more crucial for preserving the environment (Alencar et al. 2017). Combustion is the only thermochemical conversion method that has been successfully used to generate both heat and electricity (the others being gasification and pyrolysis). The burning of organic material is termed as combustion of biomass. Various systems for burning biomass are present in a size that extends from a few kW to over 100 MW. Generally, it has been noted that around more than 10% of the primary energy used globally is generated by burning biomass (Tanjore et al. 2015).

There are various applications of the combustion of biomass. Some of them are as follows. Lignocellulosic fiber (wood, straws, stalks, nuts shells, etc.) is the major biomass used.

- The heat is used for daily living (stoves) (Singh et al. 2015).
- A community heating system is focused on a specific kind of heat network that is concerned with providing heat and hot water to a single basic apartment of a building that has multiple heat customers (Neves et al. 2016).
- Use at the industrial level for heat and electricity generation. Most widely utilized in the pulp and paper industry for providing combined heat and power (CHP) supply (Cormier et al. 2006).
- The combustion approach of biomass is widely implemented in the generation of forest products and sugarcane-processing industries (James et al. 2015).

4.5.3 Biochemical Approaches

Biochemical conversion for biofuel production is an efficient way to handle agricultural biomass. Biomass waste can be transformed to biofuel either by anaerobic digestion technology or by fermentation technology elaborated below.

4.5.3.1 Anaerobic Digestion Technology

It is the major type of dynamic biochemical biomass conversion technology suitable for high moisture harboring biomass. It is a spontaneous process that occurs in oxygen-deficient conditions producing a mixture of gases called biogas, which is composed of 40%–70% methane (CH_4), 20%–30% carbon dioxide (CO_2), 100–3,000 ppm hydrogen sulfide (H_2S) and water, trace gases and other impurities (Ramaraj et al. 2016) and can be used to produce heat, electricity, and biofuels for vehicles. Generally, brooks, sediments, damp soils, unprocessed materials wastewater from industries and municipalities, and wastes generated from food and agro-industrial activities are the main habitats for biogas production (Kythreotou et al. 2014, Ward et al. 2008). The operation of anaerobic digestion is divided into three types: (i) wet digestion: when the substrate undergoing digestion has dry matter less than 10% and is employed for the management of liquid waste with a low organic loading rate (OLR); (ii) dry digestion: when the substrates undergoing digestion have dry matter more than 20%; and (iii) dry matter digestion intermediate (semi-dry digestion) is mainly applied for treating organic dry wastes with very low water level (Gkamarazi 2015). The process of anaerobic digestion involves complex interactions of microbiological, biochemical, and physical-chemical processes. Biogas and biofertilizer (Digestate) are two major end products of anaerobic digestion. Biofertilizer is a wet, inert substance retaining many important nutrients and humus for plant growth (Adams et al. 2018).

4.5.3.1.1 Microbiology of Anaerobic Digestion Process

The process is divided into four consecutive phases catalyzed by four trophic groups of micro-organisms (Kwietniewska and Tys 2014) (Figure 4.3):

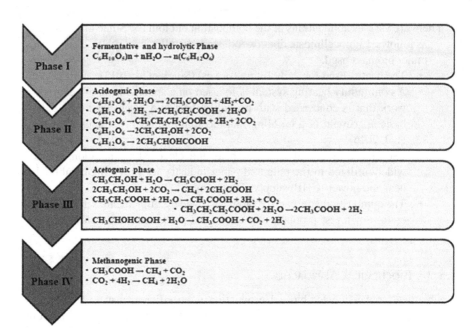

FIGURE 4.3 Different phases of anaerobic digestion technology. (*Source*: Deepanraj et al. 2014.)

Phase I or fermentative and hydrolytic phase: This is the foremost and long-winded phase of anaerobic digestion technology wherein complex polymeric organic matter is degraded to simpler products by the action of fermentative and hydrolytic micro-organisms. A different group of facultative or obligate fermentative bacteria is involved in this phase (Jingquan 2002). Fermentative bacteria of the genus *Micrococci, Bacteroides, Clostridia, Butyrivibrio, Fusobacterium, Selenomonas,* and *Streptococcus* release exoenzymes like cellulase, cellobiase, xylanase, amylase, protease, lipase which are used by some hydrolytic bacteria for conversion of complex substrates including proteins, amino acids, and carbohydrates into mono- and disaccharides and lipids into long-chain fatty acids and glycerin along with many unwanted side products (Christy et al. 2014). Many pretreatment methods in the form of biological, chemical, and mechanical processes, etc. are used to intensify the hydrolysis rate of substrates allowing the release of intracellular contents to microflora and thus lessening the digester retention time of the substrates (Ferrer et al. 2008). Sugars and proteins present in the feedstock are degraded by saccharolytic and proteolytic microbes, respectively (Adekunle and Okolie 2015).

Phase II or acidogenic phase: After the completion of the fermentative and hydrolytic phase, commences phase II or the acidogenic phase. Hydrolysis products of phase I transformed into volatile fatty acids (C1–C5), alcohols, carbon dioxide, hydrogen, ammonia, and H_2S aided by acidogenic fermentative micro-organisms. Acetic acid (CH_3COOH), propionic acid (CH_3CH_2COOH), butyric acid ($CH_3CH_2CH_2COOH$), and valeric acid ($CH_3CH_2CH_2CH_2COOH$) are prime acids produced in this stage (Shefali 2002). *Streptococcus, Lactobacillus, Bacillus, Escherichia coli,* and *Salmonella* are some prominent instances of acidogenic fermentative micro-bacteria (Christy et al. 2014). Acid production brings on a drop in the pH to 4.5–5.5 which is favorable for furnishing acidogenic and acetogenic micro-organisms. Acetic and butyric acids are the two principal products formed in this phase as these two are preferential progenitor for methane generation in the last step (Hwang et al. 2001). Acetic acid produced in this phase is directly metabolized in the fourth stage while other products formed are used by acetogenic bacteria in the third stage. The contents of the final products formed also depend on the amount of hydrogen produced in this stage. A lesser amount of reduced compounds will be formed if the partial pressure of hydrogen is too high (Gerardi 2003).

Phase III or acetogenic phase: This is called the acetogenesis phase of anaerobic digestion. Acetogenic bacteria metabolize the fatty acids (>C2) produced in the acidogenic phase and transform them into hydrogen, carbon dioxide, and acetate. These bacteria are very much sensitized to environmental repercussions (Christy et al. 2014). Depending upon the hydrogen present in the system, a symbiotic relationship exists between anaerobes performing oxidation and methanogens. Protons act as final electron acceptors under oxygen-free conditions producing hydrogen molecules. Oxidation can occur only under conditions of low pressure of hydrogen which is provided by methanogens by metabolizing hydrogen to produce methane (Adekunle and Okolie 2015)

Phase IV or methanogenic phase: This phase is also known as methanogenesis wherein methanogenic bacteria thriving in a slightly alkaline medium (Kothari et al. 2014) convert the components of phase III to methane, the final production of the anaerobic digestion process. Methanogenic archaea fall into two categories. First is the acetoclastic group, which metabolizes acetic acid or methanol to methane, and the second one is the hydrogenotrophic group, which combines hydrogen and carbon dioxide to produce methane molecule (Gkamarazi 2015). Acetic acid alone accounts for 66% of total methane production via acetate decarboxylation reaction, while carbon dioxide reduction reaction produces only 34% methane (Nayono 2009).

4.5.3.1.2 Factors Affecting the Anaerobic Digestion

Anaerobic digestion is a steady operation and the concerned microbial population takes about three weeks to acclimatize to new environmental conditions (Deublein and Steinhauser 2008). Favorable conditions are required to be maintained to balance the microbial population involved in the anaerobic digestion process. The most important variables affecting the system are temperature, moisture, total solids, H^+ or OH^- ions, retention time, OLR, C/N ratio, and mixing (Aslanzadeh 2014) discussed below:

Temperature: Temperature plays the most pivotal factor in the smooth functioning of the anaerobic digestion system. Based on their temperature optima, methanogens are partitioned into three classes: thermophilic at 45°C–60°C, mesophilic at 20°C–45°C, and psychrophilic below 20°C. However, only mesophilic and thermophilic temperature ranges are considered in anaerobic digestion because the system generally halts below 10°C (Nijaguna 2011). Thermophiles are very much sensitized to temperature and environmental changes as compared to mesophiles. Although they require higher temperatures, they have more efficacies in terms of retention time, biogas yield, and loading rate compared to mesophiles (Nijaguna 2011). Lei et al. (2020) observed that *Methanoculleus, Methanosarcina, Tepidiphilus, Sedimentibacter*, and *Gelria* are the genus predominantly present during the thermophilic anaerobic digestion of high cold tolerant energy crop *Arundo donax.*

Moisture: Water or moisture is of utmost importance for microbial degradation of organic matter. It performs various important functions: (i) acts as a universal solvent, (ii) diffusion and mass transfer of microbial cells, and (iii) permitting efficient reaction microbe and its substrate (Bollon et al. 2011; Vavilin et al. 2003). For the anaerobic digestion process to occur spontaneously, water activity >0.91 is essential (NáthiaNeves et al. 2018).

Total solids: Biogas yield is affected greatly by the total solid content of the digesting feedstock. It should neither be too thick nor too thin. Depending on the type of waste, the optimum total solid concentration should vary from 7% to 25% (Abbasi et al. 2012). Depending upon the solid content, anaerobic digestion is divided into three categories (Kothari et al. 2014): (i) low solids content (<15% TS), (ii) medium solids content (15%–20% TS), and (iii) high solids content (20%–40% TS)

H^+ or OH^- Ions: Each microorganism requires an optimum pH at which it shows the highest growth characteristics (Montañés et al. 2014). Based on the type of substrate and technique utilized, the optimum pH for the anaerobic digestion process ranges from 6.8 to 7.2 (Ward et al. 2008; Zhai et al. 2015). Methanogens, for example, show growth inhibition below pH 6.6 while around pH 8 leads cell lysis to occur and the overall process stops. As compared to other micro-organisms, acetogenins and methanogens do not easily acclimatize pH changes because former bacteria acidify the media by producing volatile acids that are converted into acetic acid, hydrogen, and carbon dioxide, which further cause negative effects on methanogens (Montañés et al. 2014). The optimum pH for hydrogen, and butyric acid production are 5, 4 respectively while for acetic and propionic acid production pH of 8 is optimum (Stavropoulos et al. 2016; Appels et al. 2008). Fatty acid produced in Phase II is much more prominent than biomethane production in Phase IV, which causes the pH to become variable during the entire digestion process further causing the pH to drop below the optimum range which inhibits the activity of methanogens as they are not capable to thrive in acidic conditions (Nijaguna 2011; Khanal 2008). This problem of pH drop can be alleviated by adding certain chemicals like sodium hydroxide, sodium carbonate, ammonia, sodium bicarbonate, ammonium hydroxide, lime, and potassium (Khanal 2008). The addition of ammonia increases basicity levels, which result in switching the gas content to methane, thereby producing purified biogas with more methane content (Hossain 2019).

Retention time: The total period spent by digesting organic substances inside the digester for the formation of final products, i.e., biogas is called retention time. It depends upon two factors: (i) type of feedstock and (ii) temperature (Rai 2010). The term SRT is used for the total time that bacteria (solids) reside inside the digester, while hydraulic retention time (HRT) is used for substrate retention time. HRT is the total time for which the feedstock stays inside the reactor. It is affected by the temperature and feedstock components. It ranges from 10 to 40 days and 14 days, for mesophiles and thermophiles, respectively (Kothari et al. 2014). They can be calculated by the following expression (Hossain 2019).

$$HRT = V / Q$$

$$SRT = \frac{1}{4} \frac{VX}{QXi}$$

where V is denoted as the volume of the tank, Q is the influent flow rate, X is the concentration of suspended solids, and Xi is the inlet concentration. For methanogens to be retained in digesters, values of both HRT and SRT must be high. Methane yield will be the highest at a low loading rate and high HRT while it will be lower at a high loading rate and low HRT. Shi et al. (2017) reported that methane content and stability were higher when the retention time of anaerobic digestion of wheat straw was increased from 20 to 60 days.

Organic loading rate: OLR denotes the biological conversion efficiency of anaerobic digestion. It can be expressed in terms of the quantity of feedstock fed to the digester per unit volume per day. It is a very important process parameter in a continuous culture system and an excess load of feedstock into the digester affects the biogas production process through acid accumulation (Deepranjan et al. 2014). It can be calculated by the expression given below:

$$OLR(kg\,VS\,(m^3 day^{(-1)})) = Q \times VS/V$$

where V (m^3) is the volume, Q (kg/day) is defined as the daily flow rate, and VS (kg VS (kg)$^{-1}$) is volatile solids (Kothari et al. 2014). Biogas production increases by increasing the feed rate up to a definite limit. Beyond this limit, excess volatile acids generation results in acidification of the environment which further halts methanogenic activity, thereby decreasing the biogas yield (Kwietniewska and Tys 2014; Mao et al. 2015).

C:N ratio: The amount of carbon and nitrogen present in the organic matter is defined as the C/N ratio. It depends upon the type and structural complexity of the substrate (Kothari et al. 2014). The carbohydrate content of the feedstock has a very important part in anaerobic digestion as it enhances the activity of the protease enzyme (Mao et al. 2015). Another important element of utmost importance is nitrogen, which is required for protein synthesis. Nitrogen-containing compounds are degraded to release ammonia which further maintains the pH of the system (Khalid et al. 2011). Therefore, any kind of imbalance in the carbon and nitrogen content of the digesting mixture leads to a disturbed C/N ratio and causes hindrance in the smooth functioning of the whole process. Carbohydrates (carbon sources) and proteins and ammonia nitrates (nitrogen) are metabolized by anaerobic bacteria to fulfill their growth requirements (Rai 2010). Bacteria utilize carbon for growth 20–30 times more rapidly than nitrogen so the optimum C: N for bacterial growth range from 20 to 30 (Bardiya and Gaur 1997). At a higher than optimal C:N, the rate of degeneration will be slower while at a lower than optimal ratio accumulation of ammonia in the digester causes inhibition of methanogens which further leads to a volatile acid build-up (Montingelli et al. 2015). This whole process causes a significant drop in the biogas yield. However, this situation can be effectively alleviated by co-digestion of a lignocellulosic substrate having a high C:N ratio with a substrate having a lower C:N (Ajeej et al. 2015; Ward et al. 2014). Table 4.2 depicts various studies showing the relationship between the C:N ratio of different substrates and their corresponding biogas yield.

Mixing: For maintaining process stability and homogeneity of the reaction mixture, mixing is necessary (Kaparaju et al. 2008). It can be done by mechanical stirrers or recirculating digesting slurry by centrifugal pumps (Nayono 2009). It not only facilitates the amalgamation of fresh feedstock with the microbial inoculum already present in the digester but also fends off the formation of thermal strata and scum (Karim et al. 2005).

TABLE 4.2
Studies Show the Relationship between the C:N Ratio and Biogas Yield of Different Substrates

Substrate	C:N Ratio	Biogas Yield	References
Rice husks	47:1	0.28 m³/kg	Dioha et al. (2013)
Waste activated sludge and perennial ryegrass	17:1	310 mL/g VS	Dai et al. (2016)
Carica solid waste	25:1	1.7825 mL/g TS	Jos et al. (2018)
Mixed waste	25.12:1	$2,371 \pm 20.10$ m³/day	Getahun et al. (2014)
Bagasse and water hyacinth	30:1	339 L/kg COD removed	Hadiyarto et al. (2019)

4.5.3.1.3 Types of Digesters Used in Anaerobic Digestion Technology

Continuous stirred tank reactors (CSTRs): A CSTR is a reaction tank whereby reactants and reagents glide inside the vessel with simultaneous removal of products formed. An agitator is installed for continuous or intermittent aeration, maximizing contact between microflora and effluvium, preventing solid sedimentation, and reducing mass transfer resistance. Due to the constant stirring of the reaction mixture, maintenance of a prominent amount of biomass is not possible which further leads to a reduction in biogas yield (Show et al. 2011). Energy needs for the constant stirring of the agitator also add to the overall cost of the reactor. Another limitation of constant stirring is microbial cell lysis and decreased oxidation of volatile fatty acids which leads to their amassing and significant pH drop leading to lesser biogas yield (Mao et al. 2015; Ward et al. 2008).

Fixed bed reactors: In this reactor, reactions occur on a stationary or fixed microbial support or bed made up of a solid catalyst. These solid catalysts or supports are made up of materials like activated carbon, ceramics, etc., and are immobile or biodegradable. Besides, cost and energy requirement, another major constraint of this type of reactor, is the low hydraulic turbulence property which causes mass transfer problems resulting in low substrate conversion efficiency by microbial population and hence lesser hydrogen yield. It also hinders maintaining uniform pH throughout the media and hence unequal distribution of microbial population. However, this problem can be alleviated by biomass recirculation which further reduces the resistance to mass transfer (NáthiaNeves et al. 2018)

Upflow anaerobic sludge blanket (UASB) reactor: The UASB reactor is simple, stable, cheap, and commonly used for wastewater remediation. It is an adaptable reactor used in biogas production wherein biomass to substrate ratio is kept constant. A thick bed made of sludge settled at the base of the reactor maintains the contact between biomass and wastewater. Longer start time, high wastewater microbial load, and longer HRT are its drawbacks (Arimi et al. 2015). The development of anammox UASB has shown the removal of nitrogen content in wastewater to a greater extent (Rikmann et al. 2014).

Fluidized bed reactor: Digesters of this type of configuration make certain the growth and upward flow of contents by involving the use of particles of small size like sand or alumina. It is mainly employed for soluble and biodegradable substrates. These not only help to keep the microbial cells in suspension but also allow high SRT, efficient mixing, and microbial mass transfer process and are expensive (Mao et al. 2015; Arimi et al. 2015).

4.5.3.2 Fermentation

Fermentation is the process whereby sugar units are broken down to ethanol and carbon dioxide by the action of fermenting micro-organisms, particularly yeast. In this conversion process, simple fermentable sugars are produced by the breakdown of complex polysaccharide compounds. These sugars are consumed by fermenting micro-organisms and ethanol is generated as the end product of their metabolism (Balat 2011). Then, ethanol can be recovered from the fermentation broth by distillation (FAO 2004). Bioethanol production is achieved through the fermentation of carbohydrates or sugary substances from plant sources or microalgae, e.g., sugarcane, wheat, lignocellulosic material, etc., through the action of microbes. It can be used in modified spark engines or a certain percentage of ethanol is mixed with petrol (Gasoline) and used as fuel. Countries like Australia, Sweden, and the United States employ high blends of ethanol (85% ethanol mixed with 15% gasoline called E85) to run FFVs abbreviated as 'flexible-fuel vehicles'. In Brazil, a few cars use 100% bioethanol as fuel, which is produced from sugarcane (Blottnitz and Curran 2007). The advantage of blending ethanol with petrol is that it improves the combustion properties of fuels as a result of which lesser amounts of Green House Gases (GHGs) are emitted. It contains a very minute amount of sulfur in its composition; hence, blending it with petrol reduces overall sulfur content and a lesser release of sulfur oxide occurs from such engines (Chakraborty et al. 2012). Sugarcane is regarded as a superior crop for ethanol production as it is sugar-rich and cheap. After sugarcane, maize is the major source of ethanol production in the United States, while (Hertel et al. 2010) wheat and sugarbeet are used in Europe.

4.5.3.2.1 Fermentation Process of Biomass to Biofuel Conversion

Generally, the bioethanol production process is divided into the following steps: (i) pretreatment, (ii) hydrolysis, (iii) fermentation, and (iv) biofuel recovery (Figure 4.4).

Pretreatment: Pretreatment of the feedstock for bioethanol production is the most high-cost step accounting for about $0.30/gallon of ethanol produced (Edeh 2020). It is necessary to break the recalcitrant structure of the biomass and make the fermentable sugars content available for the action of microbes to carry out the fermentation process. Pretreatment remarkably influences fermentation and significantly enhances ethanol yield as it releases fermentable sugars in excessive amounts (Srichuwong et al. 2009). Pretreatment methods include traditional ones including chemical, physical, physicochemical, and biological methods while advanced pretreatment is classified into acid-based fractionation and ionic liquid-based fractionation (Maurya et al. 2015). Physical methods of pretreatment involve comminution, extrusion, torrefaction, and irradiation (ultrasound, microwaves, gamma rays, and electron beam) whereby the size of substrate particles is reduced by milling or grinding (Soudham 2015). On the other hand, chemical methods include ozonolysis, acid hydrolysis, alkaline hydrolysis (Wang et al. 2009), and organosol-based treatment (Zhao et al. 2009). Physicochemical methods comprise ammonia fiber explosion (Prasad et al. 2007) and steam treatment (Kurian et al. 2013). The biological treatment uses certain microbes, particularly white, brown, and soft-rot fungi, which are capable of degrading lignocellulosic material (Narayanaswamy et al. 2013; Zhang et al. 2007). These microbes can only degrade hemicellulose and lignin, not

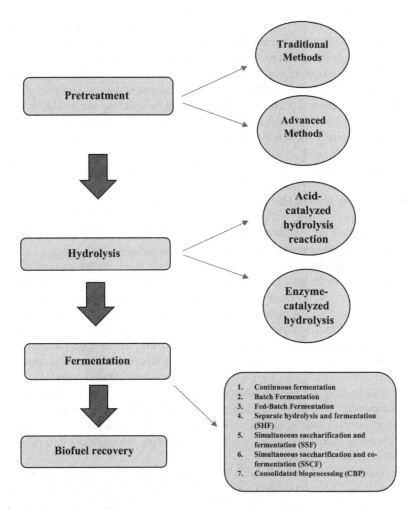

FIGURE 4.4   Flow chart representing biofuel production from fermentation. (*Source*: Edeh 2020; Azhar et al. 2017.)

cellulose that is intact, hence digestibility of substrate is enhanced (Sánchez 2009; Nanda et al. 2014). *Phanerochaete chrysosporium* when cultivated under carbon- or nitrogen-limited conditions releases lignin-degrading enzymes such as lignin peroxidases, and manganese-dependent peroxidases (Blanchette 1991). Cellulose is the substrate for biofuel production by a fermentation process, which is primarily made up of glucose units linked through β-1-4 glycosidic linkages. The polymer is embedded deep into a matrix made of lignin and hemicellulose. The major goal of each pretreatment method is to disrupt the original structure and release the cellulose polymers from the matrix so that cellulose molecules become easily accessible for the action of enzymes released by micro-organisms responsible for ethanol production. Pretreatment of feedstock before fermentation is a very advantageous process for enhancing ethanol yield by 90% or more while without pretreatment microbial enzymes instead of binding on the surface of cellulose, get adhered to lignin molecules, thereby hampering the hydrolysis reaction (Ceballos 2018). Parameters affecting the availability of cellulose in biomass include direct factors (approachable surface area) and indirect factors including anatomy-related features (size, volume of pore, particle size, and specific surface area), chemical compositions (lignin, hemicelluloses, etc.), and cellulose structure-related characteristics (crystallinity and polymerization) (Zhao et al. 2012).

Hydrolysis or Saccharification: The pretreatment step is succeeded by the most important step called hydrolysis whereby polymeric substances, i.e., celluloses or hemicelluloses are hydrolyzed to monomers. It can either be acid-mediated or enzyme-mediated.

Acid-catalyzed hydrolysis reaction: Acid hydrolysis is the conventional method of hydrolysis (Jeffries and Jin 2000). It can be performed in two ways: dilute acid at low concentration and high temperature or through concentrated acid at high concentration and low temperature. However, the former technique is used commonly but its major drawback is the formation of a vast amount of inhibitors. Lignocellulosic feed degradation by acid hydrolysis is a two-step process. Step 1 where diluted acid is used to hydrolyze hemicellulose and Step 2 where concentrated acid is used for cellulose hydrolysis. The use of concentrated acid hydrolysis involves lower temperature and high concentration, resulting in 90% sugar recovery in a short period (Joshi et al. 2011). However, this method is expensive because of trouble in acid recovery, its disposal, recycling, and formation of inhibitors (Azhar et al. 2017).

Enzyme-catalyzed hydrolysis: It uses enzymes for carrying out hydrolysis reactions. The reaction proceeds at a temperature of 45°C–50°C and pH of 4.8–5.0. Unlike acid-mediated hydrolysis, this method has more efficacy, a high sugar recovery rate is more efficient, does not form inhibitors, and does not cause corrosion of reactors. This reaction is sensitive to pH, time, temperature, enzyme loading, and substrate concentration (Canilha et al. 2012). Hydrolysis of celluloses can be increased by adding substances such as Polyethylene glycol (PEG) or Tween 20, which act as surfactants and reduce cellulose adsorption on lignin molecules (Joshi et al. 2011). Three types of enzymes are responsible for this process, namely, endo-1,4-β-glucanases, cellobiohydrolases, and β-glucosidases. Some bacterial and fungal spp. secreting cellulolytic enzymes are *Clostridium, cellulosomes, Erwinia, Thermonospora, Bacteriodes, Bacillus, Ruminococcus, Acetovibrio, Streptomyces, Trichoderma, Penicillium, Fusarium, Phanerochaete, Humicola*, and *Schizophillum*. But these enzymes have several problems like instability and inhibition by substrates or products. But with the advent of recombinant DNA technology that is robust and economic, these problems have been alleviated to a great extent (Edeh 2020).

Mechanism of hydrolysis: Cellulose hydrolysis to glucose units occurs in the following three steps (Edeh 2020):

1. Initially, the β-1,4 linkages of the cellulose are bonded with molecules of water. This step is catalyzed by the enzyme endoglucanase (1,4-β-Dglucanohydrolase) and cellodextrins are formed (Edeh 2020).
2. The cellodextrins are then degraded to cellobioses by exoglucanase (1,4-β-D-glucan cellobiohydrolase) enzymes (Edeh 2020).
3. Then, products of the second step are hydrolyzed to glucose units with the aid of enzyme β-glucosidases (Edeh 2020).

Hemicellulose hydrolysis involves the following steps:

Hemicellulose is mainly comprised of 10%–15% xylan in softwoods and 10%–35% xylan in hardwoods. The main chain of xylan is hydrolyzed by enzyme endo-β-1,4-xylanase (EC 3.2.1.8) and β-xylosidase (EC 3.2.1.37) to shorter xylan oligosaccharides, which are further degraded to xyropyranose by enzyme β-xylosidase while enzymes such as feruloyl esterase (EC 3.1.1.73), α-L-arabinofuranosidase (EC 3.2.1.55), α-glucuronidase (EC 3.2.1.139), and acetylxylan esterase (EC 3.1.1.72) are responsible for the hydrolysis of outer chains of xylan molecules (Edeh 2020).

Fermentation: Fermentation is an oxidation-reduction operation, in which an organic compound acts as an electron donor and electron acceptor yielding ethanol and carbon dioxide as end products along with releasing energy for ATP synthesis. Two micro-organisms capable of performing fermentation are yeast and bacteria *Zymomonas mobilis*. Organic substances are fermented according to the equation:

$$C_6H_{12}O_6 \rightarrow 2C_2H_5OH + 2CO_2 + ATP$$

(Glucose)   (Ethanol)   (Carbon dioxide)

Glucose is fermented to pyruvate by the enzyme zymase released by yeast in the fermentation reaction. Enzyme pyruvate decarboxylase then decarboxylates pyruvate acetaldehyde, which is then reduced to ethanol by the reductant $NADH_2$. A total of 2 mol ATP per mol glucose substrate is yielded in the fermentation reaction while 38 mol ATP per mol glucose is formed in the aerobic reaction. At the industrial scale, production of ethanol requires a robust strain of yeast that is capable of producing high ethanol yield because ethanol is inhibitory to yeast sp. at higher concentrations. In *Z. mobilis*, fermentation reaction occurs by Entner–Doudoroff (ED) pathway producing 2 mol pyruvate per mol glucose, which is decarboxylated to acetaldehyde via pyruvate decarboxylase and then reduced to ethanol (Müller 2008).

4.5.3.2.2 Types of Fermentation Processes for Biofuel Production

Different types of fermentation operations like continuous fermentation, batch fermentation, and fed-batch fermentation process exist but primarily three main fermentation processes used for bioethanol production are separate hydrolysis and fermentation (SHF), simultaneous saccharification and fermentation (SSF), and simultaneous saccharification and co-fermentation (SSCF) (Azhar et al. 2017). Many workers reported bioethanol production from lignocellulosic biomass (Table 4.3)

Continuous fermentation: In this type of fermentation process, substrates, inoculum, and nutrient medium are added to the fermentor continuously while maintaining the culture volume constant, and products of the fermentation are recovered continuously. Continuous fermentation is more economical, high yielding, requiring small-volume fermenters (Jain and Chaurasia 2014). However, yeast instability due to longer retention time and contamination of products formed are some of the limitations (Chandel et al. 2007). Ethanol productivity can be increased in continuous fermentation at higher dilution, but it decreases ethanol yield because at higher dilution substrate consumption by yeasts is incomplete (Sanchez and Cardona 2008). Therefore, this process is not suitable for the commercial production of ethanol.

Batch fermentation: In batch fermentation, substrates, inoculum, and nutrient medium are added to the fermentor at the initiation of the fermentation process for a predetermined period, and products are withdrawn only after the completion of the fermentation time. The culture environment of the batch culture continuously changes as micro-organisms continue to metabolize the nutrients which are added only once during the start of the process (Yang and Sha 2019). A standard growth curve comprising all four phases of microbial growth, namely, lag phase, log phase, stationary phase, and death phase, can be seen in this type of fermentation (Tse et al. 2021). It is an easy-to-control and adaptable process but it suffers from problems like low product yield, longer retention time, and laboriousness which render it unsuitable for large-scale bioethanol production (Liu et al. 2019). Due to periodic washing and sterilization of the fermentor in-between fermentation batches, there is also enhanced interlude. This type of fermentation is frequently employed for long-term or solid-state fermentations (Wang et al. 2013).

Fed-batch fermentation: It combines the attributes of both batch and continuous mode which is being used to handle the substrate inhibition issue encountered during the batch process. This process is the same as the batch process except that the nutrient medium is fed to the fermentor exponentially throughout the fermentation process (Yang and Sha 2019). Adding nutrients constantly leads to increased cell density in the log phase of the growth cycle, thereby increasing the product yield. Culture volume in fed-batch culture varies. At low substrate concentrations, sugar molecule conversion to ethanol is enhanced; hence, productivity increases (Jain and Chaurasia 2014). This has a shorter retention time, high yield, and lower toxin production (Chandel 2007). But feed rate and concentration of cell mass limit its use for ethanol production at an industrial scale (Liu et al. 2019).

TABLE 4.3

Studies Showing Bioethanol Production from Different Types of Lignocellulosic Waste

Substrate	Technology Used	Bioethanol Uield	References
Fruit waste	Saccharification with *Phoma sp.* (fungus) followed by fermentation with yeast *Saccharomyces cerevisiae* at 25°C for 48 hours	0.024% g	Senthilguru et al. (2011)
Sugarcane and maize waste (sugarcane bagasse, sugarcane bark, corncob, corn stalk, corn husk)	Pretreatment with acid with 5% H_2SO_4 at 121°C for 5 minutes, enzymatic saccharification (termamyl enzyme), hydrolysis at 45°C for 30 minutes followed by fermentation with yeast *Saccharomyces cerevisiae* at 30°C for 5 days	Sugarcane stipite, sugarcane bark, cornstalk, corncob, and cornhusk gave ethanol yields of 6.72%, 6.23%, 6.17%, 4.17%, and 3.45%, respectively	Braide et al. (2016)
Municipal solid waste (waste newspaper)	Pretreatment with 1.5% H_2SO_4 at 121°C for 45 minutes, saccharification with bacteria *Cytophaga huchnosonni* followed by fermentation with (i) yeast *Saccharomyces cerevisiae* and (ii) an isolated microorganism	6.849% v/v with *Cytophaga huchnosonni* and 6.031% (v/v) with isolated microorganism	Byadgi and Kalburgi (2016)
Energy cane	Pretreatment with 1.4 M, 0.8 M, and 0.16 M sulfuric acid followed by fermentation with *Klebsiella oxytoca* for fermentation of cellulose and *Pichia stipites* for fermentation of hemicellulose under simultaneous saccharification and fermentation (SSF) and separate hydrolysis and fermentation (SHF) conditions	Under SHF, Cellulosic ethanol produced was 6995 mg/L and hemicellulosic ethanol was 3624 mg/L	Shields and Boopathy (2011)
Paddy straw	Pretreatment with 2% sodium hydroxide for 1 hour at 15 psi, enzymatic hydrolysis (commercial cellulase) at 50°C for 2 hours followed by fermentation with yeast *Saccharomyces cerevisiae* at 30°C + 2°C for 48 hours	20–30 g/L	Wati et al. (2007)
Recycled paper sludge	Acid pretreatment with 1% H_2SO_4 at 120°C for 2 hours, followed by fermentation under separate hydrolysis and fermentation (SHF) using *Pichia stipites*	3.73 ± 0.16 g/L	Dubey et al. 2012

Separate hydrolysis and fermentation (SHF): In this process, hydrolysis reaction and fermentation reactions are performed independently in separate reactors. The supremacy of the process lies in the fact that it allows enzymes and micro-organisms to perform at their optimum temperature in separate reactors for better performance. Usually, enzymatic hydrolysis and saccharification occur at the optimum temperature of ~50°C and 30°C, respectively (Soudham 2015). But, the cost of operating two separate fermentors for each reaction and inhibition of the activity of cellulase enzymes by sugars released are major constraints of this process (Azhar et al. 2017).

Simultaneous saccharification and fermentation (SSF): Inhibition of cellulase enzyme by sugars is the limitation challenge in the SHF process. This problem is remediated by the SSF process in which hydrolysis and fermentation reactions are performed simultaneously in the same fermentor. In addition, fermenting microbes also transform the enzyme inhibitory compounds to less inhibitory compounds (Almeida et al. 2007). Thus, this SSF is considered a more effective substitute for SHF. As the same reactor is used for the two separate reactions, the high-temperature requirement of the cellulase enzyme might reduce the performance of fermenting organisms (Azhar et al. 2017). Thermotolerant and filamentous fungi such as *Trichoderma* and *Aspergillus* or bacteria like *Streptomyces* and yeast sp. like *Saccharomyces cerevisiae* are commonly used for this process as they can tolerate high temperature employed during the enzymatic hydrolysis step (Hsu et al. 2011; Prasad et al. 2007). It is more economical to substitute microbes for sugar hydrolysis instead of high-priced enzymes, but the major challenges are a longer incubation period and maintaining the optimum process conditions (Tse et al. 2021).

Simultaneous saccharification and co-fermentation (SSCF): Hydrolysis and saccharification reactions are carried out in the same fermentor while carrying out the fermentation of pentose sugars concomitantly. As yeast *S. cerevisiae* cannot metabolize pentoses, genetically engineered strains of *S. cerevisiae* are used (Edeh 2020).

Consolidated bioprocessing (CBP): Consolidated bioprocessing (CBP) involves the usage of biomass to produce value-added compounds in a single step in a single vessel. It uses a single reactor for the simultaneous production of enzymes, hydrolysis, and saccharification. The microorganism used can either be native, genetically engineered, or a consortium of microbes. This system is not only eco-friendly but also very cheap as it decreases the inputs for infrastructure and the number of chemicals used in the processing of samples. But its commercialization is hindered by inefficiency in the conversion process (Singhania et al. 2022). Many micro-organisms are used for this process such as cellulolytic bacteria *Ruminococcus* isolated from the stomach of ruminants and *Clostridium* isolated from insects. Bacteria *Clostridium thermocellum*, *C. phytofermentans*, and the yeast *S. cerevisiae* are the most widely studied and efficient CBP micro-organisms (Weimer et al. 2009). *C. thermocellum* is an obligate anaerobic bacteria that show the highest growth rate when crystalline cellulose is used as a substrate. Consortium containing fungi or yeast in synergy with bacteria employing a co-cultivation system could also be used (Minty et al. 2013; Zuroff et al. 2013)

Biofuel recovery: The last step of biofuel production is ethanol recovery from the medium. First anhydrous ethanol (99.5%) is produced by evaporating the water content. Ethanol can be recovered from the fermentation broth by different methods like adsorption distillation, azeotropic distillation, diffusion distillation, extractive distillation, vacuum distillation, membrane distillation, and chemical dehydration. Some of the newer methods of ethanol recovery are pervaporation and salt distillation (Nitsche and Gbadamosi 2017; Nagy et al. 2015).

4.6   CONCLUSION AND KEY CHALLENGES

Population boom, rapid industrialization, and development of sophisticated technologies demand greater amounts of energy which resulted in issues like energy crisis and environmental contamination. Rapid use of exhaustible fuels for fulfilling the energy demands of a growing population has caused their shortage. Moreover, billions of years are required for their production and they emit toxic substances in the environment. Use of abundantly available biomass as substrate for biofuel production is contemplated as solution for this major problem. Biomass can be metamorphosed to biofuels by three approaches, i.e., physicochemical, thermochemical conversion, and biochemical conversion. In biochemical approach, anaerobic digestion and fermentation technology are the main methods for biomass to biofuel conversion. However, due to complex and recalcitrant nature of biomass, pretreatment is required which is the most exorbitant step in biofuel production process. Another challenge is large-scale production of biofuel at the industrial level. Therefore, these challenges need to be handled carefully before commercialization.

REFERENCES

Abbasi T, Tauseef SM and Abbasi SA (2012) Anaerobic digestion for global warming control and energy generation: An overview. *Renew. Sust. Energy Rev.* 16:3228–42.

Adams P, Bridgwater T, Lea-Langton A, Ross A and Watson I (2018) Biomass conversion technologies. In: *Greenhouse Gases Balances of Bioenergy Systems* (pp. 107–39). doi:10.1016/b978-0-08-101036-5.00008-2.

Adekunle KF and Okolie JA (2015) A review of biochemical process of anaerobic digestion. *Adv. Biosci.* Biotechnol. 6:205–12.

Ajeej A, Thanikal JV, Narayanan CM and Kumar RS (2015) An overview of bioaugmentation of methane by anaerobic co-digestion of municipal sludge along with microalgae and waste paper. *Renew. Sust. Energy Rev.* 50:270–76. doi:10.1016/j.rser.2015.04.121

Alencar BRA, Rocha JMTS, Rocha GJM and Gouveia ER (2017) Effect of tween-80 addition in dilute acid pretreatment of waste office paper on enzymatic hydrolysis for bioethanol production by SHF and SSF processes. *Cellulose Chem. Technol.* 51:121–26. doi:10.17648/sinaferm-2015-33868.

Alizadeh H, Teymouri F, Gilbert TI and Dale BE (2005) Pretreatment of switchgrass by ammonia fiber explosion (AFEX). *Appl. Biochem. Biotechnol.* 124:1133–41. doi:10.1385/ ABAB:124:1-3:1133.

Almeida JRM, Modig T, Petersson A, Hähn-Hägerdal B, Liden´ G and Gorwa-Grauslund MF (2007) Increased tolerance and conversion of inhibitors in lignocellulosic hydrolysates by Saccharomyces cerevisiae. *J. Chem. Technol. Biotechnol.* 82:340–49.

Al-Rajhia S, Raut N, AL- Qasmi F, Qasmi M and Al Saadi A (2012) Treatments of industrial wastewater by using microalgae. *Int. Conf. Environ. Biomed. Biotechnol. IPCBEE.* 41:217–21.

Ananthi V, Prakash GS, Chang SW, Ravindran B, Nguyen DD, Vo D-N, Lag DD, Bachh Q-V, Wongi JWC, Gupta SK, Selvaraj A and Arun A (2019) Enhanced microbial biodiesel production from lignocellulosic hydrolysates using yeast isolates. *Fuel* 256:1–13. doi:10.1016/j.fuel.2019.115932.

Anukam A, Mamphweli S, Okoh O and Reddy P (2017) Influence of torrefaction on the conversion efficiency of the gasification process of sugarcane bagasse. *Bioengineering* 4:22–23

Appels L, Baeyens J, Degrève J and Dewil R (2008) Principles and potential of the anaerobic digestion of waste-activated sludge. *Prog. Energy Combust. Sci.* 34:755–81. doi:10.1016/j. pecs.2008.06.002.

Arimi MM, Knodel J, Kiprop A, Namango SS, Zhang Y and Geißen S-U (2015) Strategies for improvement of biohydrogen production from organic-rich wastewater: A review. *Biomass Bioenergy* 75:101–18. doi:10.1016/j.biombioe.2015.02.011.

Aslanzadeh S (2014) Pretreatment of cellulosic waste and high rate biogas production. Ph.D. thesis, University of Borås, Borås.

Azhar SHM, Abdulla R, Jambo SA, Marbawi H, Gansau JA, Faik AAM and Rodrigues KF (2017) Yeasts in sustainable bioethanol production: A review. *Biochem. Biophys. Rep.* 10:52–61

Bak JS, Ko JK, Han YH, Lee BC, Choi IG and Kim KH (2009) Improved enzymatic hydrolysis yield of rice straw using electron beam irradiation pretreatment. *Bioresour. Technol.* 100:1285–90.

Balat M (2011) Production of bioethanol from lignocellulosic materials via the biochemical pathway: A review. *Energy Convers. Manag.* 52:858–75.

Balat M and Balat H (2009) Biogas as renewable energy source: A review. *Energy Sour. A* 31:1280–93.

Bardiya N and Gaur A (1997) Effects of carbon and nitrogen ratio on rice straw biomethanation. *J Rural Energy* 4:1–16.

Baruah J, Nath BK, Sharma R, Kumar S, Deka RC, Baruah DC and Kalita E (2018) Recent trends in the pretreatment of lignocellulosic biomass for value-added products. *Front. Energy Res.* 6:1–19.

Basu P (2010) *Biomass Gasification and Pyrolysis: Practical Design and Theory.* Academic Press: Cambridge, MA.

Behera S, Singh R, Arora R, Sharma NK, Shukla M and Kumar S (2015) Scope of algae as third generation biofuels. *Front. Bioeng. Biotechnol.* 2:1–13. doi:10.3389/fbioe.2014.00090.

Berla BM, Saha R, Immethun CM, Maranas CD, Moon TS and Pakrasi HB (2013) Synthetic biology of cyanobacteria: Unique challenges and opportunities. *Front Microbiol.* 4:1–14. doi:10.3389/fmicb.2013.00246.

Blanchette RA (1991) Delignification by wood-decay fungi. *Annu. Rev. Phytopathol.* 29:381–98.

Blottnitz HV and Curran MA (2007) A review of assessments conducted on bio-ethanol as a transportation fuel from a net energy, greenhouse gas and environmental life cycle perspective. *J. Clean. Prod.* 15:607–19.

Boateng AA, Hicks KB and Vogel KP (2006) Pyrolysis of switchgrass (*Panicum virgatum*) harvested at several stages of maturity. *J. Anal. Appl. Pyrolysis* 75:55–64. doi:10.1016/j. jaap.2005.03.005

Bollon J, Le-hyaric R, Benbelkacem H and Bufere P (2011) Development of a kinetic model for anaerobic dry digestion processes: Focus on acetate degradation and moisture content. *Biochem. Eng. J.* 56:212–18. doi:10.1016/j.bej.2011.06.011

Braide W, Kanu IA, Oranusi US and Adeleye SA (2016) Production of bioethanol from agricultural waste. *J. Fundam. Appl. Sci.* 8:372–86.

Byadgi SA and Kalburgi PB (2016) Production of bioethanol from waste newspaper. *Proc. Environ. Sci.* 35:555–62.

Canilha L, Chandel AK, dos Santos Milessi TS, Antunes FAF, da Costa Freitas WL, das Gracas Almeida Felipe M and da Silva SS (2012) Bioconversion of sugarcane biomass into ethanol: An overview about composition, pretreatment methods, detoxification of hydrolysates, enzymatic saccharification and ethanol fermentation. *J. Biomed.* Biotechnol. 1–15. doi:10.1155/2012/989572.

Cara C, Ruiz C, Oliva JM, Saez F and Castro E (2008) Production of fuel ethanol from steam-explosion pretreated olive tree pruning. *Bioresour. Technol.* 99:1869–76.

Ceballos RM (2018) *Bioethanol and Natural Resources: Substrates*, Chemistry and Engineered Systems. New York: Taylor & Francis Group.

Chakraborty S, Aggarwal V, Mukherjee D and Andras K (2012) Biomass to biofuel: A review on production technology. *Asia-Pac. J. Chem. Eng.* 7:S254–62.

Chandel AK, Chan ES, Rudravaram R, Narasu ML, Rao LV and Ravindra P (2007) Economics and environmental impact of bioethanol production technologies: An Appraisal. *Biotechnol. Mol. Biol. Rev.* 2:14–32.

Chen B, Wan C, Mehmood MA, Chang J-S, Bai F and Zhao X (2017) Manipulating environmental stresses and stress tolerance of microalgae for enhanced production of lipids and value-added products: A review. *Bioresour. Technol.* 244:1198–1206.

Chen BY, Chen SW and Wang HT (2012) Use of different alkaline pretreatments and enzyme models to improve low-cost cellulosic biomass conversion. *Biomass Bioenergy* 39:182–191. doi:10.1016/j.biombioe.2012.01.012

Chen T, Zhang Y, Wang H, Lu W, Zhou Z, Zhang Y and Ren L (2014) Influence of pyrolysis temperature on characteristics and heavy metal adsorptive performance of biochar derived from municipal sewage sludge. *Bioresour.* Technol. 164:47–54.

Cho S, Lee N, Park S, Yu J, Luong TT, Oh Y-K and Lee T (2013) Microalgae cultivation for bioenergy production using wastewaters from a municipal WWTP as nutritional sources. *Bioresour. Technol.* 131:515–20.

Christy MP, Gopinath LR and Divya D (2014) A review on anaerobic decomposition and enhancement of biogas production through enzymes and microorganisms. *Renew. Sustain. Energy Rev.* 34:167–73. doi:10.1016/j.rser.2014.03.010.

Chum H, Faaij A, Moreira J, Berndes G, Dhamija P, Dong H, Gabrielle B, Goss Eng A, Lucht W, Mapako M, Masera Cerutti O, McIntyre T, Minowa T and Pingoud K (2011) Bioenergy. In: Edenhofer O, Pichs-Madruga R, Sokona Y, Seyboth K, Matschoss P, Kadner S, Zwickel T, Eickemeier P, Hansen G, Schlömer S, von Stechow C (Eds.), *IPCC Special Report on Renewable Energy Sources and Climate Change Mitigation.* Cambridge University Press, Cambridge.

Ciferno JP and Marano JJ (2002) *Benchmarking Biomass Gasification Technologies for Fuels*, Chemicals and Hydrogen Production. Washington, DC: US Dep Energy Natl Energy.

Cormier SA, Lomnicki S, Backes W and Dellinger B (2006) Origin and health impacts of emissions of toxic by-products and fine particles from combustion and thermal treatment of hazardous wastes and materials. *Environ Health Perspect.* 114:810–817.

Dafnomilis I, Hoefnagels R, Pratama W, Schott DL, Lodewijks G and Junginger M (2017) Review of solid and liquid biofuel demand and supply in Northwest Europe towards 2030: A comparison of national and regional projections. *Renew. Sustain. Energy Rev.* 78:31–45. doi:10.1016/j.rser.2017.04.108

Dai X, Li X, Zhang D, Chen Y and Dai L (2016) Simultaneous enhancement of methane production and methane content in biogas from waste activated sludge and perennial ryegrass anaerobic co-digestion: The effects of pH and C/N ratio. *Bioresour. Technol.* 216:323–30. doi:10.1016/j.biortech.2016.05.100.

Davis R, Bartling A and Tao L (2020) *Biochemical Conversion of Lignocellulosic Biomass to Hydrocarbon Fuels and Products.* National Renewable Energy Laboratory, State of Technology and Future Research: Golden, CO.

Day JG, Slocombe, SP and Stanley MS (2011) Overcoming biological con straints to enable the exploitation of microalgae for biofuels. *Bioresour. Technol.* 109:245–51.

Dayton DC, Belle-Oudry D and Nordin A (1999) Effect of coal minerals on chlorine and alkali metals released during biomass/coal cofiring. *Energy Fuels* 13:1203–11. doi:10.1021/ef9900841

Deepanraj B, Sivasubramanian V and Jayaraj S (2014) Biogas generation through anaerobic digestion process: An overview. *Res. J. Chem. Environ.* 18:80–93.

Demirbas (2009) Biofuels securing the planet's future energy needs. *Energy Convers. Manag.* 50:2239–49.

Deublein D and Steinhauser A (2008) History and status to date in other countries. In: Deublein D, Steinhauser A (Eds.), *Biogas from Waste and Renewable Resources: An Introduction* (pp. 35–43). Weinheim: Wiley.

Di Bnedetto A (2011) The potential of aquatic biomass for CO2-enhanced fixation and energy production. *Greenhouse Gases Sci. Technol.* 1:58–71.

Dioha IJ, Ikeme CH, Nafi'u T, Soba NI and Yusuf MBS (2013) Effect of carbon to nitrogen ratio on biogas production. *Int. Res. J. Nat. Sci.* 1:1–10.

Doornbusch R and Steenblik R (2007) Biofuels: Is the Cure Worse than the Disease? Paper prepared for the Round Table on Sustainable Development, Organisation for Economic Co-operation and Development (OECD), Paris, 11–12 September, SG/SO/RT(2007)3/REV1. https://www.oecd.org/ dataoecd/9/3/39411732.pdf

Dubey AK, Gupta PK, Garg N and Naithani S (2012) Bioethanol production from waste paper acid pretreated hydrolyzate with xylose fermenting *Pichia stipitis*. Carbo*hydr. Polymer* 88:825–29. doi:10.1016/j.carbpol.2012.01.004.

Duff SJB and Murrayh WD (1996) Bioconversion of forest products industry waste cellulosic to fuel ethanol: A review. *Bioresour. Technol.* 55:1–33. doi:10.1016/0960-8524(95)00122-0.

Edeh I (2020) *Bioethanol Production: An Overview*. IntechOpen. doi:10.5772/intechopen.94895

FAO (2004) Unified Bioenergy Terminology. Rome: FAO.

Fernandez-Fueyo E, Ruiz-Duenas FJ, Jesus Martinez M, Romero A, Hammel KE, Medrano FJ and Martinez AT (2014) Ligninolytic peroxidase genes in the oyster mushroom genome: Heterologous expression, molecular structure, catalytic and stability properties and lignin-degrading ability. *Biotechnol. Biofuels* 7:1–23. doi:10.1186/1754-6834-7-2.

Ferrer I, Ponsá S, Vázquez F and Font X (2008) Increasing biogas production by thermal (70 C) sludge pre-treatment prior to thermophilic anaerobic digestion. *Biochem. Eng. J.* 42:186–92. https://doi. org/10.1016/j.bej.2008.06.020.

Gerardi MH (2003) *The Microbiology of Anaerobic Digesters*. Hoboken: Wiley. doi:10.1002/0471468967.ch14

Getahun T, Gebrehiwot M, Ambelu A, Gerven TV and Brugge BVD (2014) The potential of biogas production from municipal solid waste in a tropical climate. *Environ. Monit. Assess.* 186:4637–46. doi:10.1007/s10661-014-3727-4.

Gkamarazi N (2015) Implementing anaerobic digestion for municipal solid waste treatment: Challenges and prospects. In: Paper presented at the *ICEST*, CEST, Rhodes, Greece, 3–5 September 2015.

Greenwell HC, Laurens LM L, Shields RJ, Lovitt RW and Flynn KJ (2010) Placing micro-algae on the biofuels priority list: A review of the technological challenges. *J. R. Soc. Interface.* 7:703–26.

Haag AL (2007) Algae bloom again. *Nature* 447:520–21.

Habert G, Bouzidi Y, Chen C and Jullien A (2010) Development of a depletion indicator for natural resources used in concrete. *Resour. Conserv.* Recycl. 54:364–376

Hadiyarto A, Soetrisnanto D, Rosyidin I and Fitriana A (2019) Co-digestion of bagasse and waterhyacinth for biogas production with variation of C/N and activated sludge. *J. Phys. Conf.* Ser. 1295:1–6.

Hertel TW, Golub AA, Jones AD, O'Hare M, Plevin RJ and Kammen DM (2010) Effects of US maize ethanol on global land use and greenhouse gas emissions: Estimating market-mediated responses. *Bioscience* 60:223–31. doi:10.1525/bio.2010.60.3.8

Himmel ME, Ding S-Y, Johnson DK, Adney WS, Nimlos MR, Brady JW and Foust TD (2007) Biomass recalcitrance: Engineering plants and enzymes for biofuels production. *Science* 315:804–07.

Hoekman SK, Broch A, Robbins C, Ceniceros E and Natarajan M (2012) Review of biodiesel composition, properties and specifications. *Renew. Sustain. Energy Rev.* 16:143–69 doi:10.1016/j.rser.2011.07.143.

Horan NJ (2018) Introduction. In: Horan N, Yaser A and Wid N (Eds.), *Anaerobic Digestion Processes:* Green Energy and Technology (pp. 1–7). Singapore: Springer.

Hossain SMZ (2019) Biochemical conversion of microalgae biomass into biofuel. *Chem. Eng. Technol. Chem. Eng. Technol.* 42:1–15. doi:10.1002/ceat.201800605.

Hsu C-L, Chang K-S, Lai M-Z, Chang T-C, Chang Y-H and Jang H-D (2011) Pretreatment and hydrolysis of cellulosic agricultural wastes with a cellulase-producing *Streptomyces* for bioethanol production. *Biomass Bioenergy* 35:1878–84.

Hu Q, Sommerfeld M, Jarvis E, Ghirardi M, Posewitz M and Seibert M (2008) Microalgal triacylglycerols as feedstocks for biofuel production: Perspectives and advances. *Plant J.* 54:621–39.

Hwang S, Lee Y and Yang K (2001) Maximization of acetic acid production in partial acido-genesis of swine wastewater. *Biotechnol. Bioeng.* 75:521–29.

Ichsan, Hadiyantob H, Hendrokoc R (2014) Integrated biogas-microalgae from waste waters as the potential biorefinery sources in Indonesia. *Energy Proc.* 47:143–48.

Ishizawa CI, Davis MF, Schell DF and Hohnson DK (2007) Porosity and its effect on the digest-ibility of dilute sulfuric acid pretreated corn stover. *J. Agric. Food Chem.* 55:2575–81.

Jain A and Chaurasia SP (2014) Bioethanol production in membrane bioreactor (MBR) sys-tem: A review. *Int. J. Environ. Res. Dev.* 4:387–94.

James AM, Yuan W, Boyette MD and Wang D (2015) The effect of air flow rate and biomass type on the performance of an updraft biomass gasifier. *Bioresources* 10:3615–24.

Jeffries TW and Jin YS, (2000) Ethanol and thermotolerance in the bioconversion of xylose by yeasts. *Adv. Appl. Microbiol.* 47:221–68.

Jin S and Chen H (2007) Near-infrared analysis of the chemical composition of rice straw. *Ind. Crops Prod.* 26:207–11. doi:10.1016/j.indcrop.2007.03.004

Jingquan L (2002) Optimization of anaerobic digestion of sewage sludge using thermophilic anaerobic pretreatment. Ph.D. thesis, Technical University of Denmark.

Jos B, Hundagi F, Wisudawati RP, Budiyono and Sumardiono S (2018) Study of C/N ratio effect on biogas production of carica solid waste by SS-AD method and LS-AD. *MATEC Web Conf.* 156:1–5. doi:10.1051/matecconf/201815603055.

Joshi B, Bhatt MR, Sharma D, Joshi J, Malla R and Sreerama L (2011) Lignocellulosic etha-nol production: Current practices and recent developments. *Biotechnol. Mol. Biol. Rev.* 6:172–82.

Juodeikiene G, Cernauskas D, Vidmantiene D, Basinskiene L, Bartkiene E, Bakutis B and Baliukoniene V (2014) Combined fermentation for increasing efficiency of bioethanol production from *Fusarium* sp. contaminated barley biomass. *Catal. Today* 223:108–14.

Kaparaju P, Buendia I, Ellegaard L and Angelidakia I (2008) Effects of mixing on methane production during thermophilic anaerobic digestion of manure lab-scale and pilot-scale studies. *Bioresour. Technol.* 99:4919–28.

Karim K, Klasson T, Hoffmann R, Drescher SR, Depaoli DW and Dahhan MHA (2005) Anaerobic digestion of animal waste effect of mixing. *Bioresour. Technol.* 96:1607–12.

Khalid A, Arshad M, Anjum M, Mahmood T and Dawson L (2011) The anaerobic digestion of solid organic waste. *Waste Manag.* 31:1737–44. doi:10.1016/j.wasman.2011.03.021

Khanal SK (2008) *Anaerobic Biotechnology for Bioenergy Production Principles and Applications.* Singapore: Wiley-Blackwell.

Kim D-Y, Lee K, Lee J, Lee Y-H, Han J-I, Park J-Y and Oh Y-K (2017) Acidified flocculation process for harvesting of microalgae: Coagulant reutilization and metal free-microalgae recovery. *Bioresour. Technol.* 239:190–96.

Kim D-Y, Vijayan D, Praveenkumar R, Han J-I, Lee K, Park J-Y, Chang W-S, Lee J-S, Oh Y-K (2016) Cell-wall disruption and lipid/astaxanthin extraction from microalgae: *Chlorella* and *Haematococcus. Bioresour.* Technol. 199:300–10.

Kim I, Seo YH, Kim GY and Han JI (2015) Co-production of bioethanol and biodiesel from corn stover pretreated with nitric acid. *Fuel* 143:285–89. doi:10.1016/j.fuel.2014.11.031

Kobayashi F, Take H, Asada C and Nakamura Y (2004) Methane production from steam-exploded bamboo. *J. Biosci. Bioeng.* 97:426–28.

Kothari R, Pandey AK, Kumar S, Tyagi VV and Tyagi SK (2014) Different aspects of dry anaerobic digestion for bio-energy: An overview. *Renew. Sustain. Energy Rev.* 39:174–95. https://doi. org/10.1016/j.rser.2014.07.011.

Kumagai S, Takahashi Y, Kameda T, Saito Y and Yoshioka T (2021) Quantification of cellulose pyrolyzates via a tube reactor and a pyrolyzer-gas chromatograph/flame ionization detector-based system. *ACS Omega* 6:12022–26. doi:10.1021/acsomega.1c00622

Kumar P, Barrett DM, Delwiche MJ and Stroeve P (2009) Methods for pretreatment of lignocellulosic biomass for efficient hydrolysis and biofuel production. *Indus. Eng. Chem. Res.* 48:3713–29.

Kumar R and Wyman CE (2009) Effects of cellulase and xylanase enzymes on the deconstruction of solids from pretreatment of poplar by leading technologies. *Biotechnol. Prog.* 25:302–14.

Kurian JK, Nair GR, Hussain A and Raghavan GSV (2013) Feedstocks, logistics and pre-treatment processes for sustainable lignocellulosic biorefineries: A comprehensive review, *Renew. Sustain. Energy Rev.* 25:205–19.

Kwietniewska E and Tys J (2014) Process characteristics, inhibition factors and methane yields of anaerobic digestion process, with particular focus on microalgal biomass fermentation. *Renew. Sustain. Energy Rev.* 34:491–500. doi:10.1016/j. rser.2014.03.041.

Kythreotou N, Florides G and Tassou SA (2014) A review of simple to scientifc models for anaerobic digestion. *Renew. Energy* 71:701–14. doi:10.1016/j.renene.2014.05.055.

Lei Y, Xie C, Wang X, Fang Z, Huang Y, Cao S and Liu B (2020) Thermophilic anaerobic digestion of *Arundo donax* cv. *Lvzhou* No. 1 for biogas production: Structure and functional analysis of microbial communities. *BioEnergy Res.* 13:866–77.

Li C, Knierim B, Manisseri C, Arora R, Scheller HV, Auer M, Vogel KP, Simmons BA and Singh S (2010) Comparison of dilute acid and ionic liquid pretreatment of switchgrass: Biomass recalcitrance, delignification and enzymatic saccharification. *Bioresour. Technol.* 101:4900–06. doi:10.1016/j.biortech.2009.10.066.

Liang Y, Siddaramu T, Yesuf J and Sarkany N (2010) Fermentable sugar release from Jatropha seed cakes following lime pretreatment and enzymatic hydrolysis. *Bioresour. Technol.* 101:6417–24.

Liu CG, Xiao Y, Xia XX, Zhao XQ, Peng L, Srinophakun P and Bai FW (2019) Cellulosic ethanol production: Progress, challenges and strategies for solutions. *Biotechnol. Adv.* 37:491–504.

Liu W-J, Jiang H, Yu H-Q (2015) Development of biochar-based functional materials: Toward a sustainable platform carbon material. *Chem. Rev.* 115:12251–85

Liu Y, Wang Y, Liu H and Zhang JA (2015) Enhanced lipid production with undetoxified corncob hydrolysate by *Rhodotorula glutinis* using a high cell density culture strategy. *Bioresour. Technol.* 180:32–39. doi:10.1016/j.biortech.2014.12.093

Lizasoain J, Trulea A, Gittinger J, Kral I, Piringer G, Schedl, A, Nilsen PJ, Potthast A, Gronauer A and Bauer A (2017) Corn stover for biogas production: Effect of steam explosion pretreatment on the gas yields and on the biodegradation kinetics of the primary structural compounds. *Bioresour. Technol.* 244:949–56. doi:10.1016/j.biortech.2017.08.042

Mao C, Feng Y, Wang X and Ren G (2015) Review on research achievements of biogas from anaerobic digestion. *Renew. Sustain. Energy Rev.* 45:540–55. doi:10.1016/j. rser.2015.02.032.

Maurya DPA, Singla A and Negi S (2015) An overview of key pretreatment processes for biological conversion of lignocellulosic biomass to bioethanol. *Biotechnology* 3:1–13.

Miao Z, Grift TE, Hansen AC and Ting KC (2011) Energy requirement for comminution of biomass in relation to particle physical properties. *Ind. Crops Prod.* 33:504–13. doi:10.1016/j.indcrop.2010.12.016

Minty JJ, Foster CE, Liao JC and Lin XN (2013) Design and characterization of synthetic fungal-bacterial consortia for direct production of isobutanol from cellulosic biomass. *Proc. Natl. Acad. Sci. USA* 110:14592–97. doi:10.1073/pnas.1218447110.

Mofijur M, Masjuki H, Kalam M, Atabani A, Shahabuddin M, Palash S and Hazrat M (2013) Effect of biodiesel from various feedstocks on combustion characteristics, engine durability and materials compatibility: A review. *Renew. Sustain. Energy Rev.* 28:441–55.

Montañés R, Pérez M and Solera R (2014) Anaerobic mesophilic co-digestion of sewage sludge and sugar beet pulp lixiviation in batch reactors: Effect of pH control. *Chem. Eng. J.* 255:492–99. doi:10.1016/j.cej.2014.06.074.

Montingelli M, Tedesco S and Olabi A (2015) Biogas production from algal biomass: A review. *Renew. Sust. Energy Rev.* 43:961–72 doi:10.1016/j.biortech.2012.02.111

Müller V (2008) *Bacterial fermentation*. Chichester: John Wiley & Sons Ltd.

Nagy E, Mizsey P, Hancsók J, Boldyryev S and Varbanov P (2015) Analysis of energy saving by combination of distillation and pervaporation for biofuel production. *Chem. Eng. Process. Intensif.* 98:86–94. doi:10.1016/j.cep.2015. 10.010

Nanda S, Mohammad J, Reddy SN, Kozinski JA and Dalai AK (2014) Pathways of lignocellulosic biomass conversion to renewable fuels. *Biomass Conv. Bioref.* 4:157–91.

Narayanaswamy N, Dheeran P, Verma S and Kumar S (2013) Biological pretreatment of lignocellulosic biomass for enzymatic saccharification. In: Fang Z (Eds.), *Pretreatment Techniques for Biofuels and Biorefineries: Green Energy and Technology (pp. 3–34)*. Berlin: Springer. doi:10.1007/978-3-642-32735-3_1

Náthia-Neves G, Berni M, Dragone G, Mussatto SI and Forster-Carneiro T (2018) Anaerobic digestion process: Technological aspects and recent developments. *Int. J Environ. Sci. Technol.* 15:2033–46. doi:10.1007/s13762-018-1682-2.

Nayono SE (2009) *Anaerobic Digestion of Organic Solid Waste for Energy Production*. Karlsruhe: KIT Scientific Publishing.

Neves PV, Pitarelo AP and Ramos LP (2016) Production of cellulosic ethanol from sugarcane bagasse by steam explosion: Effect of extractives content, acid catalysis and different fermentation technologies. *Bioresour. Technol.* 208:184–94. doi:10.1016/j. biortech.2016.02.085

Nigam PS and Singh A (2011) Production of liquid biofuels from renewable resources. *Prog. Energy Combust. Sci.* 37:52–68.

Nijaguna BT (2012) *Biogas Technology*. New Delhi: New Age International Limited Publishers.

Nitsche M and Gbadamosi R (2017) Extractive and azeotropic distillation. In: Nitsche M, Gbadamosi R (Eds.), *Practical Column Design: Guide (pp. 153–64)*. Cham: Springer. doi:10.1007/978-3-319-51688-2_5

Packer M (2009) Algal capture of carbon dioxide, biomass generation as a tool for green house gas mitigation with reference to New Zealand energy strategy and policy. *Energy Policy* 37:3428–37.

Prasad S, Singh A and Joshi HC (2007) Ethanol as an alternative fuel from agricultural, industrial and urban residues. *Resour. Conserv. Recycl.* 50:1–39.

Qiao J, Qiu Y, Yuan X, Shi X, Xu X and Guo R (2013) Molecular characterization of bacterial and archaeal communities in a full-scale anaerobic reactor treating corn straw. *Bioresour. Technol.* 143:512–18.

Rai GD (2010) *Non-Conventional Energy Sources*. New Delhi: Khanna Publishers.

Raja R, Hemaiswarya S, Ashok N, Sridhar S and Rengasamy R (2008) A perspective on bio-technological potential of microalgae. *Crit. Rev. Microbiol.* 34:34–77.

Ramaraj R, Unpaprom Y and Dussadee N (2016) Cultivation of green microalga, *Chlorella vulgaris* for biogas purification. *Int. J. New Technol. Res.* 2:117–22.

Richmond A (2004) *Handbook of Microalgal Culture: Biotechnology and Applied Phycology.* Hoboken: Blackwell Science Ltd.

Rikmann E, Zekker I, Tomingas M, Vabamäe P, Kroon K, Saluste A, Tenno T, Menert A, Loorits L, Rubin SS and Tenno T (2014) Comparison of sulfate-reducing and conventional Anammox upfow anaerobic sludge blanket reactors. *J. Biosci. Bioeng.* 118:426–33.

Roberto V, Zabaniotou AA and Skoulou V (2018) Synergistic effects between lignin and cel-lulose during pyrolysis of agricultural waste. *Energy Fuels* 32:8420–30.

Rodionova MV, Poudyal RS, Tiwari I, Voloshin RA, Zharmukhamedov SK, Nam HG, Zayadan BK, Bruce BD, Hou HJM and Allakhverdiev SI (2016) Biofuel production: Challenges and opportunities. *Int. J Hydrog.* Energy 42:1–12. doi:10.1016/j.ijhydene.2016.11.125.

Rodolfi L, Zittelli GC, Bassi N, Padovani G, Biondi N and Bonini G (2009) Microalgae for oil:strain selection, induction of lipid synthesis and outdoor mass cultivation in a low-cost photobioreactor. *Biotechnol.* Bioeng. 102:100–112.

Roy S (2017) A piece of writing: Future prospect of biofuels. *J Biofuels* 8:49–52. doi:10.5958/0976-4763.2017.00007.1.

Sadhwani N, Adhikari S, Eden MR and Li P (2018) Aspen plus simulation to predict steady state performance of biomass-CO2 gasification in a fluidized bed gasifier. *Biofuels Bioprod. Biorefining* 12:379–89.

Sánchez C (2009) Lignocellulosic residues: Biodegradation and bioconversion by fungi. *Biotechnol. Adv.* 27:185–194.

Sanchez O and Cardona CA (2008) Trends in biotechnological production of fuel ethanol from different feedstocks. *Bioresour. Technol.* 99:5270–95.

Schacht C, Zetzl C and Brunner G (2008) From plant materials to ethanol by means of super-critical fluid technology. *J. Supercrit. Fluids* 46:299–321.

Scott SA, Davery MP, Dennis JS, Horst I, Howe CJ and Lea-Smith D J (2010) Biodiesel from algae:challenges and prospects. *Curr. Opin. Biotechnol.* 21:277–86.

Searchinger T, Heimlich R, Houghton RA, Dong F, Elobeid A, Fabiosa J, Tokgoz S, Hayes D and Yu T-H (2008) Use of U.S. croplands for biofuels increases greenhouse gases through emissions from land-use change. *Science* 319:1238–40.

Senthilguru K, George TS, Vasanthi NS and Kannan KP (2011) Ethanol production from lig-nocellulosic waste. *World J. Sci. Technol.* 1:12–16.

Seo JY, Jeon H-J, Kim JW, Lee J, Oh Y-K, Ahn CW and Lee JW (2018) Simulated sunlight-driven cell lysis of magnetophoretically separated microalgae using ZnFe2O4 octahedrons. *Ind. Eng. Chem. Res.* 57:1655–61.

Serapiglia MJ, Cameron KD, Stipanovic AJ, Abrahamson LP, Volk TA and Smart LB (2013) Yield and woody biomass traits of novel shrub willow hybrids at two contrasting sites. *Bioenerg. Res.* 6:533–46. doi:10.1007/s12155-012-9272-5

Shefali V (2002) Anaerobic digestion of biodegradable organics in municipal solid wastes. M.Sc. thesis, School of Engineering & Applied Science, Columbia University.

Shen DK, Gu S, Luo KH, Wang SR and Fang MX (2010) The pyrolytic degradation of wood-derived lignin from pulping process. *Bioresour. Technol.* 101:6136–46. doi:10.1016/j.biortech.2010.02.078.

Sheth A, Yeboah YD, Godavarty A, Xu Y and Agrawal PK (2004) Catalytic gasification of coal using eutectic salts: Reaction kinetics for hydrogasification using binary and ternary eutectic catalysts. *Fuel.* 82:557–72.

Shi X-S, Dong J-J, Yu J-H, Yin H, Hu S-M, Huang S-X and Yuan X-Z (2017) Effect of hydrau-lic retention time on anaerobic digestion of wheat straw in the semicontinuous continu-ous stirred-tank reactors. *Biomed. Res. Int.* 2017:1–6. doi:10.1155/2017/24578 05.

Shields S and Boopathy R (2011) Ethanol production from lignocellulosic biomass of energy cane. *Int. Biodeterior. Biodegrad.* 65:142–46 doi:10.1016/j.ibiod.2010.10.006

Show KY, Lee DJ and Chang JS (2011) Bioreactor and process design for biohydrogen production. *Bioresour. Technol.* 102:8524–33. doi:10.1016/j.biortech.2011.04.055

Singh S, Cheng G, Sathitsuksanoh N, Wu D, Varanasi P, George A, Balan V, Gao X, Kumar R, Dale BE, Wyman CE and Simmons BA (2015) Comparison of different biomass pretreatment techniques and their impact on chemistry and structure. *Front. Energy Res. Bioenergy Biofuels* 2:1–12. doi:10.3389/fenrg.2014.00062.

Singhania RR, Patel AK, Singh A, Haldar D, Soam S, Chen C-W, Tsai M-L and Dong C-Di (2022) Consolidated bioprocessing of lignocellulosic biomass: Technological advances and challenges. *Bioresour. Technol.* 354:1–9.

Sorigue D, Légeret B, Cuiné S, Morales P, Mirabella B, Guédeney G, Li-Beisson Y, Jetter R, Peltier G and Beisson F (2016) Microalgae synthesize hydrocarbons from long-chain fatty acids via a light-dependent pathway. Plant Physiol. 171:2393–05.

Soudham VP (2015) Biochemical conversion of biomass to biofuels pretreatment-detoxification-hydrolysis-fermentation. Ph.D. thesis, Umeå University, Sweden.

Srichuwong S, Fujiwara M, Wang X, Seyama T, Shiroma R, Arakane M, Mukojima N and Tokuyasu K (2009) Simultaneous saccharification and fermentation (SSF) of very high gravity (VHG) potato mash for the production of ethanol. *Biomass Bioenergy* 33:890–98.

Stavropoulos KP, Kopsahelis A, Zafri C and Kornaros M (2016) Effect of pH on continuous biohydrogen production from end-of-life dairy products (EoL-DPs) via dark fermentation. *Waste Biomass Valoriz.* 7:753–64. doi:10.1007/s12649-016-9548-7.

Sun Y and Cheng J (2002) Hydrolysis of lignocellulosic materials for ethanol production: A review. *Bioresour. Technol.* 83:1–11.

Taherzadeh MJ and Karimi K (2008) Pretreatment of lignocellulosic wastes to improve ethanol and biogas production: A review. *Int. J. Mol. Sci.* 9:1621–51.

Tanjore D and Richard TL (2015) A systems view of lignocellulose hydrolysis. In: Ravindra P, (Ed.), *Advances in Bioprocess Technology* (pp. 387–419). Cham: Springer.

Tse TJ, Wiens DJ and Reaney MJT (2021) Production of bioethanol: A review of factors affecting ethanol yield. *Fermentation* 7:1–18. doi:10.3390/fermentation7040268.

Tursi A (2019) A review on biomass: Importance, chemistry, classification and conversion. *Biofuel Res. J.* 22:962–79.

Vassilev SD Anderson L, Vassilev C and Morgan T (2012) An overview of the organic and inorganic phase composition of biomass. *Fuel* 94:1–33.

Vavilin VA, Rytov SV, Lokshina LY, Pavlostathis SG and Barlaz MA (2003) Distributed model of solid waste anaerobic digestion: Efects of leachate recirculation and pH adjustment. *Biotechnol. Bioengy* 81:66–73. doi:10.1002/bit.10450

Vikram S, Roshan P and Kumar S (2021) Recent modeling approaches to biomass pyrolysis: A review. *Energy Fuels* 35:7406–33. doi:10.1021/acs.energyfuels.1c00251

Viola E, Zimabardi F, Cardinale M, Cardinale G, Braccio G and Gamabacorta E (2008) Processing cereal straws by steam explosion in a pilot plant to enhance digestibility in ruminants. *Bioresour. Technol.* 99:681–89.

Wang F-S, Li C-C, Lin Y-S and Lee W-C (2013) Enhanced ethanol production by continuous fermentation in a two-tank system with cell cycling. *Process Biochem.* 48:1425–28.

Wang GS, Pan XJ, Zhu JY and Gleisner R and Rockwood D (2009) Sulfite pretreatment to overcome recalcitrance of lignocellulose (SPORL) for robust enzymatic saccharification of hardwoods. *Biotechnol. Prog.* 25:1086–109. doi:10.1002/btpr.206

Wang M, Wu M and Huo H (2007) Life-cycle energy and greenhouse gas emission impacts of different corn ethanol plant types. *EnViron. Res. Lett.* 2:1–13.

Ward A, Lewis D and Green F (2014) Anaerobic digestion of algae biomass: A review. *Algal Res.* 5:204–14. doi:10.1016/j.algal.2014.02.001

Ward AJ, Hobbs PJ, Holliman PJ and Jones DL (2008) Optimisation of the anaerobic digestion of agricultural resources. *Biores. Technol.* 99:7928–40. doi:10.1016/j. biortech.2008.02.044.

Wati L, Kumari S and Kundu BS (2007) Paddy straw as substrate for ethanol production. *Indian J. Microbiol.* 47:26–29.

Weimer PJ, Russell JB and Muck RE (2009) Lessons from the cow: What the ruminant animal can teach us about consolidated bioprocessing of cellulosic biomass. *Bioresour. Technol.* 100:5323–31. doi:10.1016/j

Wyman CE, Dale BE, Elander RT, Holtzapple M, Ladisch MR and Lee Y (2005) Coordinated development of leading biomass pretreatment technologies. *Bioresour. Technol.* 96:1959–66.

Xin L, Hong-ying H, Ke G and Ying-xue S (2010) Effects of different nitrogen and phosphorus concentrations on the growth, nutrient uptake and lipid accumulation of a fresh water microalgae Scenedesmus sp. *Bioresour.* Technol. 101:5494–500.

Yang Y and Sha M (2019) *A Beginner's Guide to Bioprocess Modes-Batch*, Fed-B*atch and Continuous Fermentation, Eppendorf Application Note* 408. Eppendorf: Hamburg.

Yu G, Yano S, Inoue H, Inoue S, Endo T and Sawayama S (2010) Pretreatment of rice straw by a hot-compressed water process for enzymatic hydrolysis. *Appl. Biochem.* Biotechnol. 160:539–51.

Zhai N, Zhang T, Yin D, Yang G, Wang X, Ren G and Feng Y (2015) Effect of initial pH on anaerobic co-digestion of kitchen waste and cow manure. *Waste Manag.* 38:126–31. doi:10.1016/j. wasman.2014.12.027.

Zhang X, Yu H, Huang H and Liu Y (2007) Evaluation of biological pretreatment with white rot fungi for the enzymatic hydrolysis of bamboo culms. *Int. Biodeterior. Biodegrad.* 60:159–64.

Zhao X, Cheng K and Liu D (2009) Organosolv pretreatment of lignocellulosic biomass for enzymatic hydrolysis. *Appl. Microbiol. Biotechnol.* 82:815–27.

Zhao X, Zhang L and Liu D (2012) Biomass recalcitrance. Part I: the chemical compositions and physical structures affecting the enzymatic hydrolysis of lignocellulose. *Biofuel Bioprod. Biorefin.* 6:465–82.

Zhu S, Wu Y, Chen Q, Yu Z, Wang C, Jin S, Dinga Y and Wu G (2006) Dissolution of cellulose with ionic liquids and its application: A mini-review. *Green Chem.* 8:325–27.

Zuroff TR, Xiques SB and Curtis WR (2013) Consortia-mediated bioprocessing of cellulose to ethanol with a symbiotic *Clostridium phytofermentans*/yeast coculture. *Biotechnol. Biofuel* 6:1–12. doi:10.1186/1754-6834-6-59.

5 Microbial Approach for Biofuel Production by Biomass Conversion

Khalida Bloch
RK University

Sougata Ghosh
RK University
Kasetsart University

5.1 INTRODUCTION

In modern times, the demand for energy has increased tremendously due to excessive industrial growth and urban development that has occurred globally in the past few decades. Every field related to human activity such as agriculture, textiles, transport, healthcare, and others involves equipment that consumes large amount of energy worldwide. This has led to the higher amount of emission of CO_2 into the environment due to the continuous combustion of the fossil fuels. The gradual depletion of the fossil fuels at an alarming rate has generated a demand to produce alternative fuels (Mahapatra et al., 2021). Hence, biofuel has received attention that is produced from various sources such as bacteria, fungi, algae, plants, and even agricultural wastes (Demirbas et al., 2016). The use of renewable energy acts as an alternative as it has a global environmental impact. Biological materials were utilized earlier as a source of energy (Potters et al., 2010). Biofuels can be biodiesel and/or bioethanol that serve as an alternative to conventional fuel with 18% energy consumption. The biofuel shows 10%–45% oxygen levels and a low amount of sulfur and nitrogen (Demirbas, 2008). On the basis of the feedstocks used, biofuels are classified as first-, second-, third-, and fourth-generation biofuels (Paul et al., 2019). The first-generation biofuels are produced from animal feed, crops, and other food products via fermentation, distillation, and transesterification reaction, while the second-generation biofuels are synthesized from the conversion of lignocellulosic materials and agricultural and forest wastes by different thermophilic and mesophilic microbes. The third-generation biofuels are produced via transesterification reaction from microalgae (Jeswani et al., 2020). The fourth-generation biofuels use genetically engineered microorganisms (Singh et al., 2019). The biofuels are also classified as solid (biochar), liquid (biodiesel, bioethanol, bio-gasoline, bio-oils), and gaseous biofuels (biogas, biohydrogen, and biosyngas) (No, 2019). This chapter emphasizes the production of different types of first-, second-, third-, and fourth-generation biofuels from the conversion of biomass using bacteria, fungi, and algae, that are listed in Table 5.1.

DOI: 10.1201/9781003406501-5

TABLE 5.1

Microorganisms Mediated Biomass Conversion

S. No.	Microorganisms	Type of Biomass	Biofuel Produced	References
1	Marine bacterial consortia: *E. adhaerens* ACe06 KJ754136 and *P. geniculata* ACe07 KJ754137	Fresh seaweed and spent seaweed	Ethanol	Sudhakar et al. (2017)
2	*B. subtilis* KX712301, *B. subtilis* KX712302, *B. subtilis* KX712303, and *P. aeruginosa* KX712304	Nonedible plant oils	Biodiesel	Rana et al. (2019)
3	*B. licheniformis* (MG738312), *B. subtilis* (MG738313), *B. pumilus* (MG738314), and *B. thuringiensis* (MG744606).	Sugarcane molasses	Hydrogen and ethanol	Gabra et al. (2019)
4	*S. fulvissimus* CKS7	Agricultural waste	-	Mihajlovski et al. (2020)
5	*C. acetobutylicum* NCIM 2337	Rice straw	Biobutanol	Rajan et al. (2013)
6	*Perconia* sp. BCC2871	Rice straw	-	Harnpicharnchai et al. (2009)
7	*A. tubingensis* TSIP 9, *T. spathulata* JU4–57, *C. tropicalis* X37, *R. mucilaginosa* G43, and *Y. lipolytica* TISTR 5151	Palm biomass	-	Intasit et al. (2020)
8	*T. reesei*, *A. niger*, and *A. oryzae*.	Sugarcane bagasse	Ethanol	Pirota et al. (2014)
9	*P. ruminantium* C1A	Corn stover	Ethanol	Ranganathan et al. (2017)
10	Extraction from fungal biomass	*M. circinelloides* feedstock	Biodiesel	Vicente et al. (2009)
11	*S. cerevisiae*	Sugar beet pulp and juice	Ethanol	Gumienna et al. (2014)
12	*S. cerevisiae*	Sugarcane molasses	Ethanol	Wu et al. (2020)
13	*S. cerevisiae*	Pretreated pine	Ethanol	Hawkins and Doran-Peterson (2011)
14	*T. reesei*	Wheat biomass	Ethanol	Juodeikiene et al. (2012)
15	*D. hansenii*	Agricultural residue	Ethanol	Menon et al. (2010)
16	*D. bruxellensis* GDB 248	Sugarcane and sweet sorghum bagasse	Ethanol	Reis et al. (2016)
17	*C. kessleri* UTEX2229 and *C. vulgaris* CCAP211/19	Algal biomass	Biodiesel and methane	Caporgno et al. (2015)
18	*Spirogyra* and *Oedogonium*	Algal biomass	Biodiesel and bioethanol	Kumar et al. (2018)
19	*C. vulgaris* and *C.* sp. (GN38)	Algal biomass	Biodiesel	Pan et al. (2017)
20	*Nannochloropsis* PTCC 6016	Algal biomass	Biodiesel	Moazami et al. (2012)

5.2 BACTERIA

In the study carried out by Sudhakar et al. (2017), various industrial spent and fresh biomasses of agarophytes, alginophytes, and seaweed were used for the production of ethanol by biosaccharification mediated by marine bacterial consortia (BC). The seaweeds were collected and washed thoroughly to remove extra materials such as epiphytes. The thallus of the fresh seaweed and the spent biomass collected from the industrial seaweeds were dried for 3 days. The pigments such as phycoerythrin and fucoxanthin were isolated from the red and brown seaweed, respectively. Agar and alginate were also extracted from dried seaweeds. The biosaccharification and hydrolysis processes were carried out using 1% v/v H_2SO_4, while saccharification was carried out using 10% v/v marine BC that was capable to degrade cellulose. Around 2.5% v/v of *Ensifer adhaerens* KJ754136 (ACe06) and *Pseudomonas geniculata* KJ754137 (ACe07) were used along with 5% v/v substrate for the degradation of agar and alginate. *E. adhaerens* KJ754139 AAg01 and AAg02 were inoculated into the spent biomass of agarophyte (*Gracilaria corticata* var. *corticata*), while *Sinomicrobium oceani* KJ754133 (AAl02) and *Sinomicrobium oceani* KJ754140 (AAl04) were added into the spent biomass of alginophyte (*Sargassum ilicifolium*). The presence of sugars in the seaweed and spent seaweed was analyzed. The total carbohydrate content found in *G. corticata* and *Sargassum wightii* was 30.71% ± 4.21% and 39.75% ± 3.25%, respectively. The carbohydrate content was about 8.51% ± 0.28% and 7.25% ± 0.4%, while cellulose content was 0.67% ± 0.03% and 0.6% ± 0.12% for the agar and alginate spent, respectively. The seaweeds were treated using mild acid with bacterial consortia (ABC) and only BC. The yield of the reducing sugar obtained from fresh biomass of red seaweed saccharified using ABC was 13.88 ± 1.31 g/L, while it was 11.85 ± 0.17 g/L saccharified using only BC. The amount of sugar yield in brown seaweeds was 11.44 ± 0.07 and 8.50 ± 0.23 g/L when saccharified using ABC and BC, respectively. The production of ethanol was carried out from seaweeds saccharified using dilute acid treated consortia and only marine consortia. The ethanol obtained from spent biomass collected from alginate industry was 13.0% ± 2.1% w/v when treated with saccharified ABC, while it was 12.1% ± 0.5% w/v from phycocolloid extracted spent biomass using ABC. The maximum yield of ethanol was equivalent to 2.66 ± 0.16 g/L as obtained from the brown seaweed biomass treated using saccharified ABC.

The production of biodiesel was reported via transesterification of nonedible plant oils using lipolytic bacterial strains (Rana et al., 2019). Various lipase-producing bacteria were isolated and evaluated for their ability to produce biodiesel from jatropha, mustard, soybean, and taramira oils. The wastewater containing oils was collected and stored at 4°C. Various parameters like COD, pH, amount of oil, and ghee were tested. The production of lipase was carried out using a submerged fermentation process. The oil toxicity and methanol sensitivity tests were also conducted. The wastewater of the vegetable oil industry was treated using the lipolytic strains. The cell biomass was generated in the Luria Bertani (LB) broth containing peptone, sodium chloride, and yeast extract that was harvested and 1 mL of the same was added into different oils. The reactor was closed to eliminate the evaporation of methanol. The reaction was carried out for 48 h at 37°C under shaking condition (150 rpm). After completion of the incubation, the reaction mixture was kept under static condition for

24 h that resulted in the formation of a yellow colored layer indicating the biodiesel. The pH of the wastewater was neutral with an increased level of COD which in turn helped the isolates to adapt in the presence of a large number of organic compounds. Different bacterial isolates, viz. *Bacillus subtilis* (Q1 KX712301; Q5 KX712302; Q6 KX712303) and *P. aeruginosa* (Q8 KX712304), were obtained. The maximum lipase activity was observed in Q5 at 48 h. The toxicity test for jatropha oil showed no significant inhibition against all isolated strains. The methanol sensitivity test revealed that the strains showed luxuriant growth at 24, 48, and 72 h at 5% concentration. The treatment of wastewater was carried out in a bioreactor where Q1 exhibited the highest COD removal equivalent to 90% within 8 days. The Fourier-transform infrared (FTIR) spectroscope analysis confirmed the synthesis of biodiesel from various oils. The mono-alkyl ester showed peaks at 1,735–1,750 cm which confirmed the presence of ester. The Q1 strain and taramira oil were used to optimize the production of biodiesel. A maximum volumetric yield of 102% was obtained when the reaction took place at 37°C with a 1:9 (oil:methanol) ratio, 30% inoculum, and 6% n-hexane. The reused biomass was able to yield 44% of biodiesel.

In another study, biofuel production was reported using sugarcane molasses (SCM) by Gabra et al. (2019). The production of biofuel involved the most competent N_2-fixing *Bacillus* species. The soil and domestic wastewater samples were collected for the isolation of the bacteria (total of 21). All of the isolates showed nitrogenase activity while only 4 out of 21 were selected that exhibited significant N_2-fixation. The organisms were identified as *Bacillus licheniformis* (MG738312), *B. subtilis* (MG738313), *Bacillus pumilus* (MG738314), and *Bacillus thuringiensis* (MG744606). The *nifH* gene was confirmed in the isolates. The production of hydrogen was carried out by the bacterial isolates in the presence of sucrose as the carbon source. The maximum yield of hydrogen was noted for *B. thuringiensis* (1,400 mL) followed by *B. subtilis*, *B. pumilus*, and *B. licheniformis* that showed production of 1,200 mL, 900 mL, and 800 mL, respectively. The highest H_2 formation rates of 19.44 mL H_2/h, 16.66 mL H_2/h, 12.5 H_2/h, and 11.11 H_2/h were observed for *Bacillus thuringiensis* (MG744606), *B. subtilis* (MG738313), *Bacillus pumilus* (MG738314), and *Bacillus licheniformis* (MG738312), respectively. The ethanol production up to 1.55 and 1.031 g/L from 6% molasses and 0.517 and 0.465 g/L from sucrose were produced by *B. thuringiensis* and *B. subtilis*, respectively. The bacteria were also able to produce acetic acid, butyric acid, and lactic acid. *B. thuringiensis* showed 0.07 g/L of lactic acid production followed by *B. subtilis* with 0.06 g/L, which was increased to 1.1 and 0.55 g/L for acetic acid, respectively, when provided with molasses (6%). The production of butyric acid was 4.23 and 4.07 g/L in the case of *B. thuringiensis* and *B. subtilis*, respectively. Hence, the conversion of biomass using N_2-fixing bacteria plays a major role in the production of the biofuel.

Agricultural waste was used for the production of bioethanol using *Streptomyces fulvissimus* CKS7. The horsetail waste (*Equisetum arvense*) and yellow gentian (*Gentiana lutea*) were obtained from the herbs processing industries. The bacterial strain CKS7 was isolated and identified. The process was carried out using solid-state fermentation (SSF) where the inoculum was prepared in the ISP1 medium. The sunflower meal, soybean meal, wheat, rye, and barley bran were used as substrates. After the extraction of cellulase, amylase, pectinase, and xylanase, the enzyme activity was checked using 3,5-dinitrosalicylic acid (DNSA) method. The maximum enzyme

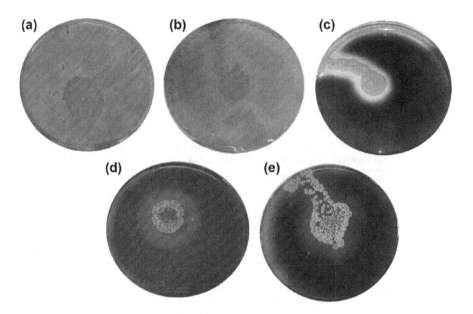

FIGURE 5.1   Hydrolysis of: (a) cellulose-CMC; (b) cellulose-Avicel; (c) starch; (d) pectin; and (e) xylan agar by *S. fulvissimus* CKS7. Reprinted with permission from Mihajlovski et al., 2020. Copyright © 2020 Springer Nature B.V.

activity was reported for the carboxymethyl cellulase, pectinase, amylase, and xylanase with rye bran as the substrate. The maximum formation of sugar was 2.55 mg/L in horsetail after hydrolysis for 72 h. Hydrolysis using enzymes plays an important role in the synthesis of biofuel. The hydrolysis of cellulose by cellulase from the ligno-cellulosic biomass produced reducing sugars like glucose, while xylanase converted cellulose into fermentable sugars as evident from Figure 5.1 (Mihajlovski et al., 2020).

The production of biobutanol was carried out using rice straw in the study conducted by Ranjan et al. (2013). Biobutanol acts as an alternative to gasoline and diesel. The anaerobic fermentation was carried out using *Clostridium acetobutylicum* NCIM 2337. The strain was anaerobically incubated at 37°C. The rice straw hydrolysate was treated using acidified double distilled water containing 1% sulfuric acid (H_2SO_4). The mixture was incubated under stirring at 200 rpm for 24 h at 60°C that was eventually autoclaved followed by cooling and filtration. The batch fermentation was carried out in a vessel (2 L volume) under anaerobic condition. The reaction took place for 12 d at 200 rpm. The hydrolysis of rice straw using acid showed the release of sugars due to shear stress, high pressure, and temperature. This in turn resulted in the loosening of lignin and hemicellulose which converted solid straw into slurry. The total sugar concentration was 13.02 g/L after 6 h, while it was 11.47 g/L in case of the reducing sugars. The sugar present in the hydrolysate was made available to *C. acetobutylicum* NCIM 2337 during fermentation. After 12 d, the amounts of residual total sugars, reducing sugars, and glucose were 15.35, 13.35, and 10.44 g/L, respectively. The production of acetone on 11 d of fermentation was 6.24 g/L while the total amount of biobutanol and ethanol was 13.5 and 0.82 g/L, respectively, after 12 d of the fermentation process.

5.3   FUNGI

In a study performed by Harnpicharnchai et al. (2009), the biomass was converted into carbohydrates using *Periconia* sp. BCC2871, an endophytic fungal strain that produces thermotolerant β-glucosidase, BGL I. The gene responsible for the enzyme was cloned into *Pichia pastoris* KM71 and the resulting recombinant enzyme showed activity at 70°C temperature and pH 5–6 which was considered as the optimum conditions. The high enzyme activity was obtained at high temperature which was maximum (60%) at 70°C after 90 min. The activity was 100% at pH ≥8 at 2 h. The enzyme showed significant activity against the substrates containing carboxymethyl-cellulose. In the presence of glucose, cellobiose, and sucrose, the thermostability of the enzyme was enhanced. The rice straw was used for the hydrolysis and the total reducing sugars was estimated using the DNSA method. The hydrolysis of rice straw using BCC2871 BGL I was efficient indicating its potential to convert agricultural wastes into useful products. The hydrolysis was maximum at pH 6.0 and 60°C that resulted in the release of 70% reduced sugars. A concentration of 132 mg of sugar was released from rice straw/gram.

The production of biodiesel from lignocellulosic biomass using *Aspergillus tubingensis* TSIP 9, *Trichosporonoides spathulata* JU4–57, *Candida tropicalis* X37, *Rhodotorula mucilaginosa* G43, and *Yarrowia lipolytica* TISTR 5151 was investigated by Intasit et al. (2020). The empty fruit bunch (EFB) collected from the palm-oil mill was pretreated with lignocellulosic fungus by SSF. The treated EFB was mixed with fungal spores and fermented at room temperature. The separation of the liquid from the fermented solid was followed by evaluation of the enzyme activity. Various parameters for enzyme hydrolysis, such as time, fungal pretreated EFB (FPEFB), and cellulase, were optimized using the response surface methodology (RSM). The FPEFB showed a high hemicellulose content (23.6% ± 1.9% to 26.1% ± 3.0%). The total sugar concentration was 11–35 g/L at 114–563 mg/g of FPEFB. The production of sugar was increased to 18–20 g/L within 12 h. The highest production of lipids was achieved in the case of *Y. lipolytica* TISTR 5151 with 37.0 ± 0.1 mg/g FPEFB. The lipid yield obtained using SSF was 47.9 ± 1.5 mg/g FPEFB. In case of the fed-batch SSF of nonsterile FPEFB, the yield of lipid was 53.4 ± 0.5 mg/g FPEFB. The bioprocess showed a total lipid yield of 149.3 ± 6.66 mg/g EFB.

In another study, ethanol was produced from the biomass using *Trichoderma reesei, Aspergillus niger*, and *Aspergillus oryzae*. Pirota et al. (2014) investigated the conversion of biomass into ethanol using SSF from steam-exploded sugarcane bagasse (SESB). The whole medium (WM) was comprised of enzymatic cocktails and enzymatic extracts (EE). The conversion of biomass into ethanol involved the fungal strains cultivated under SSF followed by extraction of solid–liquid phase that was further filtered, centrifuged, followed by separation of the solid residue and the EE. The saccharification of SESB was accomplished using the WM of *A. niger* with EE of *T. reesei*, WM of *A. oryzae* with EE of *A. niger*, and EE of *A. niger* with WM of *T. reesei* which showed 66%, 65%, and 64% total reduction of sugars, respectively. The production of ethanol by *Saccharomyces cerevisiae* was investigated from SESB, wheat bran (WB), fungal mycelium, and enzymes. The ethanol produced by *S. cerevisiae* using hydrolyzed material obtained from WM *A. oryzae* + EE *A. niger* was 42.9 g/L, while the WM *A. niger* + EE *T. reesei* and WM *T. reesei* + EE *A. niger*

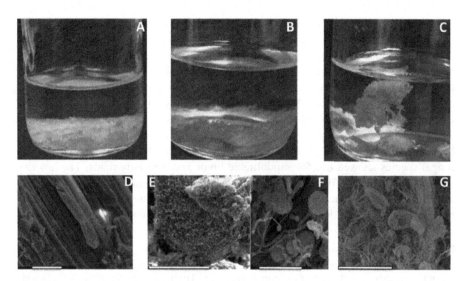

FIGURE 5.2 Culture media with pretreated alkaline corn stover (a), with scanning electron micrograph (SEM) showing the intact structure of its particles prior to fungal inoculation (d, scale bar 50 mM). Growth of strain C1A on alkaline-pretreated corn stover for 2 days resulted in visual growth around corn stover particles (b), with extensive sporangia and rhizoidal colonization (e, scale bar 300 mM) that appears closely associated and penetrating corn stover particles (f, scale bar 30 mM). At the conclusion of the saccharification phase, the loss of corn stover weight and density could be visually ascertained (c), with SEM (g, scale bar 50 mM) showing sporangial and rhizoidal remains, as well as pronounced pitted patterns (arrow) suggesting extensive decay of corn stover particles. Reprinted from Ranganathan et al. 2017.

resulted in 31.2 and 29.5 g/L, respectively. The maximum ethanol production of about 84% was obtained in the WM *A. oryzae* with EE *A. niger*.

The anaerobic fungi *Pecoramyces ruminantium* C1A was used for the production of biofuel from lignocellulosic biomass in the study conducted by Ranganathan et al. (2017). The *Zea mays* (corn stover), *Sorghum bicolor* (sorghum forage), *Saccharum officinarum* var. Ho02, and the mixture of biomass of C3 grass (*Elymus canadensis* L), C4 grass (*Sporobolus compositus*), and *Ambrosia psilostachya* were used as shown in Figure 5.2. The samples were dried overnight at 45°C and sieved. The plant biomass was allowed to incubate for 1 h at 121°C in the presence of 3% sodium hydroxide (NaOH) solution followed by washing with deionized water. The maximum yield of sugars per gram of corn stover was achieved after 48 h. It was observed that during the growth phase, 28.3 mg of hyphal biomass was produced while the amount of the total extracellular proteins was 8.9 mg. Various CIA fermentation products such as acetate, lactate, and formate with concentrations of 79.53, 5.86, and 4.14, respectively, were produced along with 21.29 mg (9.2 mM) ethanol. The gradual release of xylose and glucose in the phase of saccharification was obtained which were 18 mg at 1.65 ± 0.24 µg/mL/h and 55.7 mg at a rate of 6.16 ± 1.1 µg/mL/h, respectively. Around 74.6 mg of glucose and 21.29 mg of ethanol were accumulated in the media. The production of ethanol during the fermentation phase was increased to 43.39 mg at a rate of 78.14 ± 13.50 µg/mL/h.

In another study, the oleaginous microbial species was used as feedstock for the biodiesel production. Vicente et al. (2009) investigated the potential of oleaginous fungi to produce biodiesel. The microbial species showed more than 85% lipid content. The three solvent systems comprising chloroform: methanol, chloroform: methanol: water, and n-hexane was used for the extraction of lipids. The dry biomass of *Mucor circinelloides* MU241 (3.73 ± 0.27 g/L) was used where the maximum lipid extraction was obtained in the chloroform: methanol and chloroform: methanol: water that was equivalent to 19.9 wt.% and 19 wt.%, respectively. The amount of lipid extracted from sunflower, rapeseed, and soybean crops were compared. The fungal species showed 86.15% saponifiable lipids and free fatty acids. The amount of free fatty acids was $31.6\% \pm 1.3\%$ while the strain itself contained $37 \pm 1\%$ of oleic acid and $20\% \pm 1\%$ palmitic acid. The transesterification using microbial oil was carried out using acid catalysts at two different temperatures. The amount of fatty acid methyl esters (FAMEs) was 14.1%–14.6% at 25°C. At 65°C, biodiesel production up to 18.5%, 18.0%, and 17.5% was obtained using sulfuric acid (H_2SO_4), hydrochloric acid (HCl), and BF_3, respectively.

In the study carried out by Gumienna et al. (2014), the raw, concentrated, and thick sugar beet juice and sugar beet pulp were used to produce bioethanol. The raw material was pretreated using pasteurization and sterilization. The process was carried out in the stationary and shaken batch processes. The sugar beet roots and juice were collected from the sugar factory. Around 1.6 mg/L of the reducing sugars and 17.1% of sucrose were present in the roots. The amounts of sucrose in raw, concentrated, and thick juice were 23.9%, 58.0%, and 67.7%, respectively. The amount of reducing sugars present in raw juice was 9.7 mg/L, while it was 53.4 mg/mL in concentrated juice. The thermal treatment of sugar beet pulp was carried out in order to eliminate the contamination. Pasteurization was done for 15 min using a boiling water bath. In another variant, it was sterilized at 121°C for 15 min. The fermentation of the pasteurized sugar beet pulp using *S. cerevisiae* showed 70.5%–77.6% ethanol while sterilized pulp showed 76.1%–83.3% ethanol production. Two different *S. cerevisiae* preparations (Ethanol red and Safdistil C-70) were used for the fermentation process under the stationary phase. The ethanol yield for Safdistil C-70 was 87.7%, which was significantly high as compared to ethanol red.

Enhanced ethanol production from SCM using genetically modified *Saccharomyces cerevisiae* was reported by Wu et al. (2020). Yeast extract peptone dextrose (YEPD) medium was used to cultivate the wild-type strain MF01, ME13, and MC15 isolated from the sugar mill waste. The cultivated cells were harvested followed by DNA extraction. The construction of yeast strains MF01-PHO4 was accomplished by the replacement of ethanol-fermentation-related regulatory gene *PHO4* from MC15 to MF01 via homologous recombination (HR). This also enhanced the time of fermentation and production of ethanol from SCM. The maximum ethanol content (EC) of MF01-PHO4, MF01, and MC15 were 114.71 ± 0.24 mg/L, 108.94 ± 0.71 g/L, and 82.16 ± 0.79 g/L, respectively, at 30°C. Compared to the MF01 strain, the ethanol yield increased by 2.30% and the time of fermentation was decreased by 12.5% in the recombinant strains. The production of ethanol using MF01, MF02, and MF03 from the fermentation of SCM was >106 g/L. The production of ethanol by MF02 and MF03 strains were 114.29 and 106.56 g/L, respectively, at 72 h at 30°C. The genetically engineered MF01-PHO4 showed the highest production of ethanol. The total residual sugars (TRSs) from SCM fermentation were 29.07 ± 0.12 g/L (MF01), 28.68 ± 0.06 g/L (MF01-PHO4), and 44.05 ± 0.06 g/L (MC15) after 56 h at 30°C. Hence, ethanol production was improved by replacing the *PHO4* gene in yeast strains.

The production of ethanol from the biomass of pretreated pine wood (*Pinus taeda*) using the fungal strain *Saccharomyces* AJP50 evolved from XR122N was reported by Hawkins and Doran-Peterson (2011). The daughter strain AJP50 showed more ethanol production as compared to the parent strain in the process of fermentation that used pretreated pine as a substrate (5%, 10%, and 12% w/v) with XR122N (4 g/L dry cell weight (dcw)). Ethanol production was obtained with 0.5 g/L of inoculums while the production rate was enhanced with the increase in the concentration of the inoculum to 1 g/L dcw in 12% w/v solids. The pre-treatment of pine wood was done using SO_2 steam explosion. The increase in the concentration of biomass increased the production of inhibitory substances as well. The presence of high solids with low inoculum generated multiple stresses on the XR122N strain. The directed evolution process of XR122N started with the addition of 2 g/L inoculum into pretreated pine at 17.5% w/v solid loading. The strain AJP40 was added directly to 17.5% solid-loaded pretreated pine when little production of ethanol was observed. After 72 h, the AJP50 strain showed >80% ethanol production which increased to >90% after 120 h.

The efficiency of bioethanol production under the influence of *Trichoderma reesei* β-xylanase was investigated by Juodeikiene et al. (2012). The bioethanol production used *Fusarium* (FM)-contaminated wheat biomass that was stored under moisture. The amount of the fungal toxin trichothecene deoxynivalenol (DON) present in the wheat infected by *Fusarium* spp. was evaluated to be 3.95 mg/kg using an enzyme-linked immunosorbent assay (ELISA). Around 67.5% polymers of wheat were degraded by xylanase from *T. reesei*. The production of monosaccharides and xylo-oligosaccharides from wheat treated using enzymes was dependent on the preparation of xylanase. The xylose (39.9 mg/g) and arabinose (50 mg/g) were released from the wheat. The impact of xylanase on ethanol production was determined by evaluating the combined effect of amylolytic and xylanolytic enzymes in the fermentation process. The saccharification of wheat contaminated with FW using amylolytic enzyme showed low content of soluble dry matter (SDM). The concentration of ethanol present after fermentation of FW was <13.5% with a high SDM of 20%. The use of xylanase along with amylolytic enzyme in FW fermentation resulted in an increase of SDM by 3.7%–35.2% in the wort. The ethanol concentration also increased by 12%–35.3% for FW. The highest ethanol concentration was 4,000 XU/kg in the presence of xylanase.

In the study conducted by Menon et al. (2010), production of ethanol was reported from the agricultural waste. The thermostable xylanase and thermotolerant yeast were used for the production of ethanol from hemicellulose. The oat spelt xylan (OSX) containing xylose, arabinose, and glucose and WB containing hemicellulose, cellulose, starch, klason lignin, and protein were used. The hemicellulose was extracted from WB. The thermostable xylanase was isolated from the alkalothermophilic *Thermomonospora* sp., while the thermotolerant yeast *Debaromyces hansenii* was isolated from rotten grapes. *Thermomonospora* was grown in Reses medium for 96 h at 50°C for xylanase production. After 96 h of fermentation, the broth was collected, centrifuged at 9,600 ×g for 15 min at 4°C. The enzyme was extracted from the supernatant. The enzyme assay was performed followed by immobilization of the cells of *D. hansenii*. The kinetics of the simultaneous saccharification and fermentation (SSF) of OSX and hemicellulose from wheat bran (WBH) using xylanase and *D. hansenii* cells were studied. Treatment of WBH (2.5%) with 150 U/g xylanase and OSX resulted in the saccharification up to 54% and 38%, respectively. The OSX produced 9.1 g/L ethanol in 36 h, while WBH produced 9.5 g/L ethanol in 48 h.

The sugarcane and sweet sorghum bagasse (SSB) were used to produce ethanol using *Dekkera bruxellensis* in a study conducted by Reis et al. (2016). The strain GDB248 was isolated from the ethanol fermentation process. The lignocellulosic biomass of sugarcane bagasse (SCB) and SSB were collected, washed, and dried followed by treatment with H_2O_2 and incubation for 1 h. It was filtered and washed at 70°C to remove the lignin residue. The total sugar released from SCB and SSB after the alkaline treatment was 14.49 ± 0.1 and 18.07 ± 0.4 g/L, respectively, while the efficiency of hydrolysis was 84.28 ± 1.14 and 88.07 ± 1.6, respectively. Around 94% of the sugar was utilized by *D. bruxellensis* after 7 h with 84.4% ethanol production. The amount of glucose was 9.57 and 11.5 g/L, while cellobiose was 0.81 and 1.62 g/L in SCB and SSB, respectively. The quantity of the ethanol obtained was 1.09 and 1.46 g/L from SCB and SSB, respectively, in 1 h.

5.4 ALGAE

Caporgno et al. (2015) investigated the production of biodiesel and methane using freshwater and marine microalgae. The water sample collected from the wastewater plants and natural seawater was used for the cultivation of microalgae. The large particles were removed using centrifugation followed by filtration. Bold Basal Medium (BBM) was used to grow the freshwater microalgae *Chlorella kessleri* UTEX2229 and *Chlorella vulgaris* CCAP211/19, while the marine microalgae *Nannochloropsis oculata* was allowed to grow in the seawater enriched with 25% salinity and 3 mL/L Conway medium. The high algal biomass was produced using a flat-panel airlift photobioreactor (PBR). The algae were cultivated in a batch mode, which was maintained at 25°C via ambient airflow. The pH was maintained as 7.5 and 8 for freshwater and marine microalgae, respectively. Both methane and biodiesel were produced by anaerobic digestion. The microalgal biomass (freeze-dried) was converted to FAME biodiesel via a transesterification reaction. The biomass (dry weight) obtained for *C. kessleri* and *C. vulgaris* were 2.70 ± 0.08 and 2.91 ± 0.02 g/L, respectively. The reduction of nitrogen from the medium was around 96% and 95%, while phosphorus reduction was 99% and 98%, in the case of *C. kessleri* and *C. vulgaris*, respectively. The *N. oculata* showed less uptake efficiency which in turn required optimization of the process for the maximum removal of nutrients. The *C. kessleri* and *C. vulgaris* showed methane production of 346 ± 3 and 415 ± 2 mL_{CH4}/g_{VS}, respectively. The total yield of biodiesel from *C. kessleri* and *C. vulgaris* was 7.4 ± 0.2 and 11.3 ± 0.1 g/100 VS, respectively. The fatty acids in *C. kessleri* were palmitic acid (20.3%), oleic acid (10.4%), linoleic acid (32.2%), and linolenic acid (12.9%). The amount of palmitic, oleic, linoleic, and linolenic acid were 22.2%, 27.2%, 11.6%, and 10.6%, respectively, in *C. vulgaris*. Hence, it can be concluded that due to the presence of high amount of polyunsaturated fatty acids, the microalgae served as a potent source for biodiesel.

The algae collected from the freshwater, identified as *Spirogyra* and *Oedogonium* were used for the biofuel production from 0.650 g dry wt/m^2 of biomass in the study carried out by Kumar et al. (2018). The FAMEs using GC-MS revealed the presence of various mono-saturated fatty acids such as hexadecenoic, oleic, and octadecadi-enoic acids. Around 18.6% of the lipid was recovered from Soxhlet extraction using chloroform: methanol (2:1). The iodine value (38.02 g I$_2$/100 g), saponification value (130.10 mg KOH), specific gravity (0.756 kg), acid value (2.5 mg KOH/g), cetane number (30), high heating value (41.00), and long-chain saturation factor (57 wt.%) of the algal biodiesel were evaluated. The four-stroke engine test of the biodiesel was carried out using two blends of biodiesel: A5B25D70 (algae 5%, butanol 25%, and diesel 70%) and A10B30D60 (algae 10%, butanol 30%, and diesel 60%). The torque values of A5B25D70 and A10B30D60 decreased by 5% and 6%, respectively. The maximum brake power at 2,200 rpm was obtained for the blended fuels. The highest brake-specific fuel consumption (BSFC) was shown by A10B30D60 as compared to A5B25D70. The low heating values of the biodiesel were speculated to be respon-sible for the high BSFC. The BSFC values of A5B25D70 and A10B30D60 increased by 5.3% and 8.7%, respectively. With the addition of butanol, the emission of NOx value was decreased in the blend fuels, whereas less emission of CO was observed in A10B30D60. The single-step hydrolysis of algal biomass using 10% H$_2$SO$_4$ and 10% HCl showed a release of 230.02 ± 03 and 212.24 ± 03 mg/g of fermentable sugars, respectively.

Biofuel was also produced from unbroken and wet microalgal biomass in a study carried out by Pan et al. (2017). *C. vulgaris* and *Chlorococcum* sp. (GN38) were cultivated in 10 L glass reactor at 30°C. The deep eutectic solvent (DES) was mixed with formic, acetic, oxalic, and propanedioic acids at different concentrations rang-ing from 1:1 to 1:3. The lipid was extracted from *Chlorella* sp. and *Chlorococcum* sp. by one- and two-step method. The two-step method was carried out by treating chloride-acetic acid (Ch-Aa) in a 200 mL reactor. Various proportions of DES and algae mud were allowed to react in 20:3, 40:3, 60:3, 80:3, 100:3, and 200:3 at 90°C, 100°C, 110°C, 120°C, and 130°C. The reaction mixture was cooled and 1:1 v/v of hexane/methanol was mixed followed by stirring at 200 rpm for 2 h at 55°C. The lipid reacted with the methanol under the presence of H$_2$SO$_4$ as a catalyst and was converted into biodiesel (FAME). The ratio of DES: methanol (H$_2$SO$_4$): algal bio-mass equivalent to 60:40:3 was used at 110°C and 130°C for *Chlorococcum* sp. and *Chlorella* sp., respectively, for 60 min that was considered as the optimum condition. The total amount of FAME produced by the one-step method was 30% more efficient compared with the two-step method. In total FAME, more than 90% were palmitic, palmitoleic, stearic, oleic, and linoleic acids.

The production of biodiesel on a large scale was accomplished using biomass of *Nannochloropsis* sp. in the study conducted by Moazami et al. (2012). The mass production of microalgae involved indoor raceway ponds with 2,000 L capacity as shown in Figure 5.3. The experiment was conducted for 2 weeks at paddlewheel speeds of 1.4, 2.1, and 2.8 rad/s. The CO$_2$ was continuously injected into the culture at pH 7.5 ± 0.2. The algal cells were cultured for 14 d at 25°C after which the biomass

FIGURE 5.3   Indoor 2000-L open pond for large-scale production of microalgae. Reprinted with permission from Moazami et al. 2012. Copyright © 2012 Elsevier Ltd.

was collected using a vacuum (wet and dry) and passed through a 2 mm nylon mesh. The maximum growth of cells at a paddlewheel speed of 2.1 rad/s was 46 g/L/m². The *Nannochloropsis* PTCC 6016 showed continuous lipid synthesis. The lipid content of 52% was obtained at 2.1 rad/s. The hydrocarbon varied from 70 to 55 mg/L/d cell weight. The high amount of oleic acid (45.4%) in the fraction facilitated the biodiesel production. The strain showed a biomass producing capacity of 130 tons/h/year and biodiesel production of 60,000 L.

5.5   CONCLUSION AND FUTURE PERSPECTIVES

The depletion of the fossil fuels at an alarming rate has raised global concern to meet the energy requirement for the future. Various microbes are employed for production of biofuel as an alternative source of energy. Biogenic energy is an environmentally friendly, rapid, and efficient method, which has attracted attention of the researchers worldwide. Microorganisms are preferred due to the ease of isolation, culturing, and preservation. A large amount of biomass can be generated in a short time. Bacteria, fungi, and algae use numerous enzymes, such as amylase, carboxymethyl cellulase, pectinase, thermotolerant β-glucosidase, and xylanase, for biofuel production. Hence, there is scope for immobilizing these enzymes in a polymeric matrix or on nanoparticles' surface for reuse and recycling purposes. Further, these enzyme immobilization strategies may give more stability to the enzymes. As enzymes are biocatalysts that can show maximum activity only under specific conditions, the reaction parameters such as temperature, pH, enzyme concentration, substrate concentration, and metal ions should be optimized carefully to get maximum biofuel production. Further, strain improvement can be achieved by using recombinant DNA technology that will enable enhancement in biofuel generation. Genetically modified microbes with more linoleic, linolenic, oleic, palmitic, palmitoleic, and stearic acid content may serve as more useful microbes for high biofuel production. However, there are certain challenges like scale-up, downstream processing, purification, commercialization, and acceptance that should be overcome before biofuel can be introduced as a mainstream energy source.

REFERENCES

Caporgno, M.P., Taleb, A., Olkiewicz, M., Font, J., Pruvost, J., Legrand, J., Bengoa, C. (2015). Microalgae cultivation in urban wastewater: Nutrient removal and biomass production for biodiesel and methane. *Algal Res.* 10, 232–239.

Demirbas, A. (2008). Biofuels sources, biofuel policy, biofuel economy and global bio-fuel projections. *Energy Convers. Manag.* 49(8), 2106–2116.

Demirbas, A., Bafail, A., Ahmad, W., Sheikh, M. (2016). Biodiesel production from non-edible plant oils. *Energy Explor. Exploit.* 34(2), 290–318.

Gabra, F.A., Abd-Alla . M.H., Danial A/ W., Abdel-Basset, R., Abdel-Wahab, A.M. (2019). Production of biofuel from sugarcane molasses by diazotrophic *Bacillus* and recycle of spent bacterial biomass as biofertilizer inoculants for oil crops. *Biocatal. Agric. Biotechnol.* 19, 101112.

Gumienna, M., Szambelan, K., Jelen, H., Czarnecki, Z. (2014). Evaluation of ethanol fermentation parameters for bioethanol production from sugar beet pulp and juice. *J Inst. Brew.* 120(4), 543–549.

Harnipicharnchai, P., Champreda, V., Sornlake, W., Eurwilaichitr, L. (2009) A thermotolerant b-glucosidase isolated from an endophytic fungi, *Periconia* sp., with a possible use for biomass conversion to sugars. *Protein Expr. Purif.* 67(2), 61–69.

Hawkins, G.M., Doran-Peterson, J. (2011). A strain of *Saccharomyces cerevisiae* evolved for fermentation of lignocellulosic biomass displays improved growth and fermentative ability in high solids concentrations and in the presence of inhibitory compounds. *Biotechnol. Biofuels.* 4(1), 49.

Intasit, R., Cheirsilp B., Louhasakul, Y., Boonsawang, P. (2020). Consolidated bioprocesses for efficient bioconversion of palm biomass wastes into biodiesel feedstocks by oleaginous fungi and yeasts. *Bioresour. Technol.* 315, 123893.

Jeswani, H.K., Chilvers, A., Azapagic, A. (2020). Environmental sustainablility of biofuels: A review. *Proc. R. Soc. A* 476(2243), 2020035.

Juodeikiene, G., Basinskiene, L., Vidmantiene, D., Makaravicius, T., Bartkiene, E. (2012). Benefits of beta-xylanase for wheat biomass conversion to bioethanol. *J. Sci. Food Agric.* 92(1), 84–91.

Kumar, V., Nanda, M., Joshi, H.C., Singh A., Sharma, S., Verma, M. (2018). Production of biodiesel and bioethanol using algal biomass harvested from freshwater river. *Renew. Energy* 116, 606–612.

Mahapatra, S., Kumar, D., Singh, B., Sachan, P.K. (2021). Biofuels and their sources of production: A review on cleaner sustainable alternative against conventional fuel, in the framework of the food and energy nexus. *Energy Nexus.* 4, 100036.

Menon, V., Prakash, G., Prabhune, A., Rao, M. (2010). Biocatalytic approach for the utilization of hemicellulose for ethanol production from agricultural residue using thermostable xylanase and thermotolerant yeast. *Bioresour. Technol.* 101(14), 5366–5373.

Mihajlovski, K., Buntic, A., Milic, M., Rajilic-Stojanovic, M. Dimitrijevic-Brankovic, S. (2020). From agricultural waste to biofuel: Enzymatic potential of a bacterial isolate *Streptomyces fulvissimus* CKS7 for bioethanol production. *Waste Biomass Valor.* 12, 165–174.

Moazami, N., Ashori, A., Ranjbar, R., Tangestani, M., Eghtesadi, R., Nejad, A.S. (2012). Large-scale biodiesel production using microalgae biomass of *Nannochloropsis*. *Biomass Bioenergy* 39, 449–453.

No, S.Y. (2019). *Production of Liquid Biofuels from Biomass, Application of Liquid Biofuels to Internal Combustion Engines: Green Energy and Technology.* Springer, Singapore.

Pan, Y., Alam, Md A., Wang, Z., Huang, D., Hu, K., Chen, H., Yuan, Z. (2017). One-step production of biodiesel from wet and unbroken microalgae biomass using deep eutectic solvent. *Bioresour. Technol.* 238, 157–163.

Paul, P.E.V., Sangeetha, V., Deepika, R.G. (2019). Emerging trends in the industrial production of chemical products by microorganisms. In: Buddolla V. (Ed.), *Recent Developments in Applied Microbiology and Biochemistry.* Academic Press, New York, pp. 107–125.

Pirota, R.D.P.B., Delabona, P.S., Farimas, C.S. (2014). Simplification of the biomass to ethanol conversion process by using the whole medium of filamentous fungi cultivated under solid-state fermentation. *Bioenergy Res.* 7, 744–752.

Potters, G., Van Goethem . D., Schutte, F. (2010). Promising biofuel resources: Lignocellulose and algae. *Nat. Educ.* 3(9), 14

Rajan, A., Khanna, S., Moholkar, V.S. (2013). Feasibility of rice straw as alternate substrate for biobutanol production. *Appl. Energy* 103, 32–38.

Rana, Q., Laiq Ur Rehman, M., Irfan, M., Ahmed, S., Hasan, F., Shah, A.A., Khan, S., Badshah, M. (2019). Lipolytic bacterial strains mediated transesterification of non-edible plant oils for generation of high-quality biodiesel. *J. Biosci. Bioeng.* 127(5), 609–617.

Ranganathan, A., Smith, O.P., Youssef, N.H., Struchtemeyer, C.G., Atiyeh, H.K., Elshahed, M.S. (2017). Utilizing anaerobic fungi for two-stage sugar extraction and biofuel production from lignocellulosic biomass. *Front. Microbiol.* 8, 635.

Reis, A.L.S., Damilano, E.D., Menezes, R.S.C., de Morais Jr, M.A. (2016). Second-generation ethanol from sugarcane and sweet sorghum bagasses using the yeast *Dekkera bruxellensis. Ind. Crops Prod.* 92, 255–262.

Singh, D., Sharma, D., Soni, S.L., Sharma, S., Sharma, P.K. Jhalani, A. (2019). A review on feedstocks, production processes, and yield for different generations of biodiesel. *Fuel* 262, 116553.

Sudhakar, M.P., Jegatheesan, A., Poonam C., Perumal, K., Arunkumar, K. (2017). Biosaccharification and ethanol production from spent seaweed biomass using marine bacteria and yeast. *Renew. Energy.* 105, 133–139.

Vicente, G., Bautista L.F., Rodrigue, R., Gutierrez, F.J., Sadaba, I., Ruiz-Vazquez, R.M., Toress-Martinez, S., Garre, V. (2009). Biodiesel production from biomass of an oleaginous fungus. *Biochem. Eng. J.* 48(1), 22–27.

Wu, R., Chen, D., Cao, S., Lu, Z., Huang, J., Lu, Q., Chen, Y., Chen, X., Guan, X., Guan, N., Wei, Y., Huang, R. (2020). Enhanced ethanol production from sugarcane molasses by industrially engineered *Saccharomyces cerevisiae* via replacement of the PHO$_4$ gene. *RSC Adv.* 10(4), 2267–2276.

6 Anaerobic Digestion
A Sustainable Biochemical Approach to Convert Biomass to Bioenergy

Gaganpreet Kaur and Nisha Yadav
Sardar Swaran Singh National Institute of Bioenergy
Dr B R Ambedkar National Institute of Technology

Sachin Kumar
Sardar Swaran Singh National Institute of Bioenergy

ABBREVIATIONS

AD	Anaerobic digestion
CH$_4$	Methane
C/N	Carbon/nitrogen
CNG	Compressed natural gas
CO$_2$	Carbon dioxide
GHG	Greenhouse gas
GW	Gigawatt
H$_2$	Hydrogen
H$_2$S	Hydrogen sulphide
HRT	Hydraulic retention time
MAD	Mesophilic anaerobic digestion
MSW	Municipal solid waste
N$_2$	Nitrogen
OLR	Organic loading rate
TAD	Thermophilic anaerobic digestion
VFAs	Volatile fatty acids
VS	Volatile solids

6.1 INTRODUCTION

Increasing population and industrialization lead to increased consumption and demand of energy. The major share of energy demand is fulfilled by conventional energy sources like fossil fuels. But due to their limited source and adverse effect on the environment, fossil fuels are not sustainable as an energy source. Therefore,

DOI: 10.1201/9781003406501-6

there is a need for sustainable and ecofriendly energy sources (Krishnan et al. 2021). Renewable energy sources such as wind, solar energy, biomass/bioenergy, small hydro, etc. can be a better alternative to conventional energy sources. Bioenergy is the most promising type/form of renewable energy, which is generated when the biomass produced through photosynthesis is turned into biofuel (biogas, bioethanol), heat, and electricity (Aryal et al. 2021).

Anaerobic digestion (AD) is one of the techniques, which is used for the generation of biofuel/biogas from organic solid wastes, viz. agricultural residues/wastes, food waste, animal manure, municipal solid waste (MSW), garden waste, etc. In this process, a group of microorganisms break down the organic waste in anaerobic conditions and produce biogas (Ileleji et al. 2015). Biogas is a mixture of gases which is produced in AD and is comprised of methane (CH_4): 50%–75%, carbon dioxide (CO_2): 25%–50%, nitrogen (N_2): 0%–10%, hydrogen sulphide (H_2S): 0%–3%, hydrogen (H_2) 0%–1%, and some other gases in traces (Goswami et al. 2016). This biochemical process of AD is considered to be a sustainable process as it is cost-effective, produces renewable energy, reduces and manages organic wastes, provides employment, environmentally friendly, etc. (Meegoda et al. 2018; Anukam et al. 2019; Khalid et al. 2011). Along with these advantages, AD has some disadvantages as well, such as high retention time, instability and low removal efficiencies of organic compounds, uneasy digestion of lignocellulosic biomass (Park et al. 2005). Different pretreatment methods like thermal, acidic, alkaline, mechanical, etc. and new technologies like thermophilic anaerobic digestion (TAD) can be adopted to get rid of the abovementioned disadvantages of the AD process (Park et al. 2005). Different types of microorganisms play an important role in the process of AD and these can work actively only if the operational conditions are ideal. Operational conditions/factors like pH, C/N ratio, solid loading rate, temperature, hydraulic retention time, particle size, etc., are responsible for the functioning of the AD process and are explained in Section 6.4 (Leung and Wang 2016; Pramanik et al. 2019).

The two main products of the AD process are biogas and digestate. The biogas yield and composition of biogas is different depending upon the composition of the feedstock, where CH_4 is the major combustible component and it ranges between 50% and 75% (Anukam et al. 2019). The CH_4 and CO_2 concentration in the produced biogas totally depends upon the degrading substrate; if sugars, starch, and cellulose are being digested equally then equal amounts of CH_4 and CO_2 is obtained, whereas if fats and proteins are digested in the AD process, then CH_4 concentration would be more (Anukam et al. 2019). Table 6.1 shows the biogas yield and CH_4 concentration of various feedstocks.

Biogas has many applications in households as well as industries. Heating is the most common use of non-upgraded biogas; it can be used in the boilers for heating, cooking, drying processes, heating of buildings, etc. (Fagerström et al. 2018; Abanades et al. 2021). Power generation from renewable resources like biomass, wind energy, solar energy, etc. is also a growing market these days. Biogas uses internal combustion engines or gas turbines to produce electricity, which can help in reducing the dependency on fossil fuels. Over the last decade, the global capacity of biogas-based power has increased significantly, with the capacity of biogas-based power generation expanding from 65 GW in 2010 to 120 GW in 2019, an increase of

TABLE 6.1
Biogas Yield and Methane Concentration of Various Feedstocks

Substrate	Biogas Yield (m³/kg VS)	CH₄ Content (%)	References
Municipal solid waste	0.36	60–65	Vogt et al. (2002)
Paddy Straw	0.34	75.9	Lei et al. (2010)
Vegetable waste	0.4	64	Babaee and Shayegan (2011)
Pig manure	0.43	66.9	Beschkov et al. (2021)
Poultry manure	0.47	57.9	Beschkov (2021)
Napier grass	0.36	57	Deshmukh et al. (2015)
Water hyacinth	0.41	-	Koley et al. (2022)
Cow dung	0.04	-	Singh and Kalamdhad (2021)

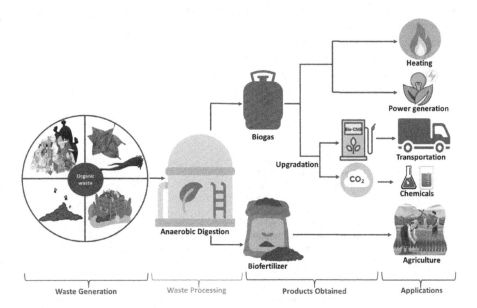

FIGURE 6.1 Anaerobic digestion process: products and their applications.

around 90% (Kabeyi and Olanrewaju 2022). After upgradation, i.e., removal of CO_2 and other impurities, the gas (bioCNG) can be used as transportation fuel, which mitigates greenhouse gas (GHG) emissions (Fagerström et al. 2018). Digestate, the effluent of AD process, can be utilized as an organic fertilizer in gardening and farming. The farmers are recommended to use 5 and 10 tons of the digestate per hectare of irrigated and non-irrigated fields, respectively. It is good for the fertility of the soil, increases the crop yield by 10%–20%, and promotes sustainable agriculture (Kabeyi and Olanrewaju 2022). Figure 6.1 illustrates the products formed in the AD process and its applications.

6.2 STEPS AND BIOCHEMICAL REACTIONS INVOLVED IN AD

AD occurs in four successive phases: hydrolysis, acidogenesis, acetogenesis, and methanogenesis. The essential microbial interactions involved in these steps are described and shown in Figure 6.2 (Meegoda et al. 2018).

6.2.1 HYDROLYSIS

In the initial step, insoluble complex organic molecules that are not accessible to the microorganisms (due to the large size of the molecules) are broken down into soluble organic molecules. During this process, some extracellular enzymes, i.e., amylase, cellulase, protease, xylanase, etc., are secreted by the hydrolytic bacteria which hydrolyse carbohydrates, proteins and fats and oils into simple sugars, amino acids, and fatty acids, respectively (Meegoda et al. 2018; Kothari et al. 2014). Equation 6.1 represents the chemical reaction involved in the hydrolysis (Anukam et al. 2019). As various microorganisms are involved in this process, the hydrolysis rate of different compounds present in the organic wastes is different; hydrolysis of carbohydrates takes lesser time compared to that of proteins and lipids (Leung and Wang 2016). *Clostridium, Acetivibrio, Staphylococcus, Bacteroides, Proteus, Peptococcus, Micrococcus,* etc., are different fermentative bacteria which carry out the hydrolysis of lipids, carbohydrates, proteins, and in organic wastes (Pramanik et al. 2019).

$$C_6H_{10}O_4 + 2H_2O \rightarrow C_6H_{12}O_6 + H_2 \tag{6.1}$$

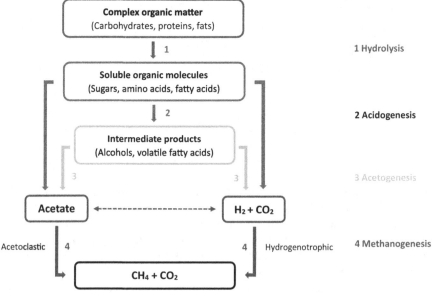

FIGURE 6.2 Schematic of stages of anaerobic digestion process. (Adapted from Meegoda et al. 2018.)

6.2.2 ACIDOGENESIS

In the second stage, i.e., acidogenesis, the products of hydrolysis are consumed by the acidogenic bacteria through their cell wall and are converted into simple molecules like alcohols and volatile fatty acids (VFAs), i.e., propionate, acetate, and butyrate (Meegoda et al. 2018). The concentrations of the products generated during this step vary depending upon the type of microorganisms and their growth parameters such as temperature and pH (Ostrem and Themelis 2004). Microorganisms involved in this stage are acidogenic bacteria such as *Eubacterium limosum, Sarcina, veillonella, Desulfovibrio, Desulfuromonas, Desulfobacter, Selenomonas*, etc. (Pramanik et al. 2019). Acidogenesis occurs at a faster rate because microorganisms involved in this stage has high growth rate with a doubling time of around 30 min comparative to other stages (Meegoda et al. 2018; Kothari et al. 2014). Equations 6.2–6.4 represent the chemical reaction involved in acidogenesis (Anukam et al. 2019).

$$C_6H_{12}O_6 \rightleftharpoons 2CH_3CH_2OH+2CO_2 \qquad (6.2)$$

$$C_6H_{12}O_6+2H2 \rightleftharpoons 2CH_3CH_2COOH+2H_2O \qquad (6.3)$$

$$C_6H_{12}O_6 \rightarrow 3CH_3COOH \qquad (6.4)$$

6.2.3 ACETOGENESIS

The third phase of AD is acetogenesis where acetogenic bacteria utilize the VFAs (formed during acidogenesis), viz. propionic acid, butyric acid, etc., for their growth and produce acetic acid and H_2. The doubling time of acetogens which are involved in this stage such as *Clostridium, Syntrophobacter wolinii, Syntrophomonas wolfeii*, etc., is 1.5–4 days which/that means these are slow-growing bacteria and make the acetogenesis phase slow (Kothari et al. 2014; Pramanik et al. 2019). During the acid-producing stages, the organic material of the feedstock gets converted into organic acids leading to the pH drop of the AD system. This pH drop is favourable for the microorganisms involved in acidogenesis and acetogenesis as these bacteria prefer somewhat acidic environment of pH 4.5–5.5, but it is detrimental to the microorganisms involved in the following step, i.e., methanogenesis (Ostrem and Themelis 2004). Equations 6.5–6.7 represent the chemical reaction involved in acetogenesis (Anukam et al. 2019).

$$CH_3CH_2COO^- +3H_2O \rightleftharpoons CH_3COO^- + H^+ + HCO_3^- +3H_2 \qquad (6.5)$$

$$C_6H_{12}O_6+2H_2O \rightleftharpoons 2CH_3COOH+2CO_2+4H_2 \qquad (6.6)$$

$$CH_3CH_2OH+2H_2O \rightleftharpoons CH_3COO^- +3H_2+H^+ \qquad (6.7)$$

6.2.4 METHANOGENESIS

Methanogenesis is the last step of AD where methanogenic bacteria utilize the products of the first three stages/earlier stages such as CO_2, acetic acid, H_2, etc., and produce CH_4 and CO_2. *Methanobrevibacter ruminantium, M. thermoautotrophicum, M. bryantic, Methanogenium cariaci, M. marinsnigri*, etc., are the methanogens which carry out the process of methanogenesis (Anukam et al. 2019). Methanogenic bacteria grow slowly as compared to other microorganisms involved in the AD process, and their regeneration time is 5–16 days (Meegoda et al. 2018). Methanogens are highly sensitive to oxygen, temperature, and pH; these bacteria grow and work/perform in a neutral to slightly alkaline environment (Ostrem and Themelis 2004; Achinas et al. 2020). Three types of bacteria are involved in methanogenesis: acetoclastic, methylotrophic, and hydrogenotrophic methanogens. Methylotrophic methanogenesis is not a common pathway but if the feedstock contains the methyl group, it can be converted into CH_4 as shown in Equation 6.8 (Leung and Wang 2016). In acetoclastic and hydrogenotrophic methanogenesis, CH_4 is produced by the reduction of acetate and CO_2/H_2 respectively. About 70% of CH_4 is produced by acetoclastic methanogens, whereas the rest 30% is produced by hydrogenotrophic methanogens (Achinas et al. 2020). Equations 6.8–6.11 represent/show the chemical reaction associated with methanogenesis (Anukam et al. 2019; Leung and Wang 2016).

$$3CH_3OH+3H_2 \rightarrow 3CH_4+3H_2O \tag{6.8}$$

$$CH_3COOH \rightarrow CH_4 + CO_2 \tag{6.9}$$

$$CO_2+4H_2 \rightarrow CH_4+2H_2O \tag{6.10}$$

$$2CH_3CH_2OH+CO_2 \rightarrow CH_4+2CH_3COOH \tag{6.11}$$

6.3 TYPES OF ANAEROBIC DIGESTION

The microorganisms are highly dependent on temperature for the AD process. There are commonly two temperature regimes for AD: mesophilic (30°C–40°C) and thermophilic (50°C–60°C) (Zupančič et al. 2012).

6.3.1 MESOPHILIC ANAEROBIC DIGESTION (MAD)

MAD is a biological process that operates at moderate temperatures between 30°C to 40°C, with an optimal range of around 35°C (Kim et al. 2002). This is the most common and commercial AD process and can also handle a wide range of feedstocks (Labatut et al. 2014). MAD has several advantages over other digestion processes. It is less sensitive to temperature and pH fluctuations, making it a more stable process. In addition, the optimum temperature range for MAD is closer to ambient temperatures, reducing the need for energy input to maintain the process temperature (Gebreeyessus and Jenicek 2016; Fernandez et al. 2008; Suryawanshi et al. 2010). However, MAD also has some limitations. The process is relatively slow, with retention time typically ranging from 20 to 30 days, which can limit the processing capacity of the system. Furthermore, the low operating temperature can result in lower biogas production rates compared to TAD. Overall, MAD is a reliable and versatile process that is suitable for various substrates and produces valuable biogas and digestate.

6.3.2 THERMOPHILIC ANAEROBIC DIGESTION

TAD is a digestion process that occurs at higher temperatures, typically between 50°C and 60°C, with the optimal temperature range usually around 55°C (Labatut et al. 2014). During TAD, a consortium of thermophilic microorganisms breaks down complex organic compounds into simpler compounds. At higher temperature, the kinetics and the metabolic reaction rate of microorganisms expedite leading to enhanced substrate utilization. Therefore, the faster substrate conversion decreases the hydraulic retention time (HRT) (Noraini et al. 2017; Gebreeyessus and Jenicek 2016).

Compared to MAD, TAD is more advantageous. The higher operating temperature speeds up the digestion process, reducing retention times and increasing processing capacity (Bravo et al. 2009). TAD can also handle a wider range of feedstocks, including high-strength and complex organic wastes, and produces a higher biogas yield compared to MAD. In addition, TAD can produce a higher-quality digestate that is more stable and has a lower pathogen content.

However, TAD also has some disadvantages. The high operating temperature needs more energy input to maintain the process temperature, which can increase operational costs. TAD is also more sensitive to fluctuations in temperature and pH, which can make the process less stable (Nguyen et al. 2007). Also, the high temperature results in higher free ammonia concentration inside the digester, thus making the process unstable (Lohani and Havukainen 2018). The comparison between MAD and TAD based on various parameters is given in Table 6.2.

TABLE 6.2
Difference between MAD and TAD (Singh et al. 2023)

S. No.	Parameters	MAD	TAD
1.	Biogas production	Low	High
2.	Methane concentration	Low	High
3.	Digester volume	Large	Small
4.	Pretreatment	Yes	No
5.	Hydraulic retention time	High	Low
6.	Biogas yield	Low	High
7.	Temperature maintenance	Less	High

6.4 FACTORS AFFECTING AD

AD depends on various factors for efficient and stable operation. These factors include substrate composition, ORL, pH, Temperature, mixing, microbial population, and the inhibitory substances. The success of AD systems relies on how these factors interact and impact the rate of biogas production and also the quality of the digestate produced. Furthermore, the impact of these factors changes depending on the specific feedstock being used, the system design, and the operating conditions. Therefore, it is crucial to understand the role of each of these factors and their interactions in order to optimize AD performance and maximize the biogas yield and its valuable by-products.

6.4.1 TEMPERATURE

The temperature plays an important role in the AD process, as it greatly affects the kinetics of the process, thermodynamic equilibrium, and the growth rate of microorganisms (Panigrahi and Dubey 2019). There are basically three temperature regimes, viz. psychrophilic, mesophilic, and thermophilic for the AD process (Lohani and Havukainen 2018). The optimal temperature range for AD varies with the feedstock and microorganisms, but generally falls within the mesophilic (20°C–40°C) or thermophilic (45°C–65°C) temperature ranges. At lower temperatures, such as those found in mesophilic conditions, the AD process is slower and less efficient but is generally more stable and less susceptible to fluctuations. In contrast, higher temperatures found in thermophilic conditions can speed up the process and result in higher biogas yields but also increase the risk of process instability and failure. With increase in reaction temperature by 10°C, the biochemical reactions rate can be doubled according to the Arrhenius equation (Panigrahi and Dubey 2019). Temperature also affects the rate of microbial growth and metabolism, with different microorganisms being active at different temperature ranges. For example, mesophilic bacteria are most active between 30°C and 40°C, while thermophilic bacteria are most active between 50°C and 60°C. The activity of different microbial communities also affects the breakdown of different types of organic matter, with some feedstocks being better suited to mesophilic or thermophilic conditions. Therefore, controlling and monitoring the temperature in anaerobic digesters is important to optimize the process and ensure stable and efficient biogas production.

6.4.2 C/N RATIO

The C/N ratio is another important parameter in AD processes which determines the ratio of carbon to N_2-containing compounds in the feedstock, which impacts the microbial population and the rate of decomposition. In general, a C/N ratio of around 20–30 is considered optimal for AD (Panigrahi and Dubey 2019). This range provides enough carbon for energy and growth of the microorganisms, while also providing enough N_2 for protein synthesis. The microorganisms require both carbon and N_2 for their growth and metabolism for the degradation of the organic matter (Sanusi et al. 2018). If the C/N ratio is too high (excessive carbon compared to N_2), the microorganisms may not have enough N_2 to synthesize protein, which can lead to slow digestion rates and high concentration of VFAs in the digester. In case of low C/N ratio (excessive N_2 compared to carbon), the excess N_2 can be released as ammonia, which can inhibit digestion and harm the microorganisms. Therefore, maintaining an optimal C/N ratio is important for efficient and stable AD. This can be achieved by blending different feedstocks with varying C/N ratios or by adding N_2 sources to the reactor as needed. For instance, the co-digestion of carbon-rich substrate can be done with the N_2-rich waste to maintain the optimum C/N ratio (Sawatdeenarunat et al. 2015). An imbalanced C/N ratio can reduce biogas generation due to substrate underutilization and accumulation of inhibitory compounds. These issues can cause process failure, resulting in reduced biogas production and the need for costly reactor cleaning and restart.

6.4.3 Organic Loading Rate

Another important factor involved in AD systems is Organic loading rate (OLR). It basically refers to the volatile solids content that is fed to the anaerobic digester per unit of reactor volume per day. It can be estimated through the following equation (Neves et al. 2018):

$$\text{OLR}[\text{kgVS}(\text{m}^3\text{day})^{-1}] = Q\frac{\text{VS}}{V}$$

where Q is daily flow (kg/day); VS is volatile solids (kg VS (kg)$^{-1}$); and V is volume (m³).

The OLR directly affects the microbial activity, bio-gas yield, and the stability of the AD process. The microorganisms responsible for the utilization of organic matter require a sufficient amount of substrate for their balanced growth and activity (Nkuna et al. 2021). However, if the OLR is too high, the microorganisms may not be able to utilize the substrate efficiently, resulting in substrate underutilization, accumulation of intermediates such as VFAs, and lesser biogas production. If the OLR is too low, the microorganisms may not have enough substrate to maintain their activity and growth, resulting in reduced biogas production. It can be categorized into three types on the basis of solid content: low solid matter (<15%), medium solid content (between 15% and 20%), and high solid content (between 20% and 40%) (Neves et al. 2018). The biogas is directly proportional to the amount of organic matter fed to the digester. However, the OLR needs to be within a specific range to ensure the proper functioning of the digester. So, the factors such as the type and quality of the substrate, reactor design, and microbial population should be determined through careful monitoring and experimentation.

6.4.4 Inhibitory Substances

Inhibitory substances are compounds that can reduce microbial activity and biogas production in AD without necessarily being toxic. These substances include VFAs, organic acids, and ammonia. Inhibitory substances can negatively impact the AD performance by causing process instability, such as acidification, foaming, and reduced biogas production. The high concentrations of VFAs and free ammonia is toxic to hydrolytic and acidogenic microorganisms and retard their growth (Van et al. 2020). The effect of inhibitory substances on AD varies with different factors such as concentration, exposure time, and ionic strength of toxic material. The presence of inhibitory substances affects the microbial community structure, reduces the population of methanogens, and increases the population of acid-forming bacteria. This shift in microbial community can minimize the biogas production and enhance the concentration of intermediate compounds such as VFAs, which can lead to process instability. The VFAs concentration of not more 3,000 mg HAc/L is suitable for normal AD. Some substances in the substrate such as heavy metals, cyanides, antibiotics, etc., can be toxic to the microbial community present in AD (Khanal et al. 2019).

6.4.5 pH

The pH level can significantly affect the activity of microorganisms because different microorganisms require specific pH for their growth and metabolism. The optimal pH range for AD is generally between 6.8 and 7.2. When the pH level is too low or too high, it greatly impacts the activity of microorganisms and the efficiency of the process. Acidogenic and methanogenic bacteria are more pH sensitive (Neves et al. 2018; Labatut and Pronto 2018). If the pH level drops below 6.5, the acidogenic bacteria that are responsible for the first stage of AD can become inhibited, leading to reduced biogas production. At a pH below 6.0, CH_4 production can completely cease, and the process may become unstable. If the pH level is too high (above 7.5), the activity of the methanogenic bacteria responsible for the final stage of AD decreases, resulting in reduced biogas production (Nkuna et al. 2021). Also, high pH levels can also result in the accumulation of ammonia and other toxic substances, which can further inhibit microbial activity and lead to process failure. Therefore, it is essential to monitor and control the pH level in AD systems to ensure optimal conditions for microbial activity and efficient biogas production. This can be done through the addition of buffer solutions, such as sodium bicarbonate or calcium carbonate, to balance the pH level within the optimal range.

6.4.6 Hydraulic Retention Time

HRT is the period of time that the organic matter is retained in an anaerobic digester. It is an important factor in AD as it affects the microbial ecology, the rate and efficiency of biogas production, and the quality of the digested material (Nkuna et al. 2021). The acidification in the digester through VFAs accumulation can occur due to unstable HRT. It can be calculated by using the given equation:

$$HRT = V / Q$$

where V and Q are the volumes of the digester and the feed flow rate in time, respectively.

The optimal HRT depends on different factors, such as the characteristics of the feedstock, the temperature of the digester, and the digester operation (Wainaina et al. 2019). In general, longer HRTs result in higher levels of biogas production, as well as a greater degree of digestion and stabilization of organic matter. This is because longer HRT gives the microorganisms more time for the degradation of the organic matter into biogas and digestate (Mao et al. 2015). However, there is a limit to the benefits of longer HRTs, and beyond a certain point, there may be diminishing returns in terms of biogas production. This is because longer HRTs can also result in lower volumetric biogas production rates, meaning that the amount of biogas produced per unit volume of the digester is reduced. In addition, longer HRTs can increase the risk of process inhibition, which can lead to reduced biogas production and lower-quality digestate.

6.4.7 MIXING

Mixing is another important operational step in AD that affects the process performance. It helps to improve mass transfer by increasing the contact between substrate and microorganisms (Chang et al. 2019). The efficient transfer of nutrients, such as carbon, N_2, phosphorous, and gases, such as CO_2 and CH_4, is crucial for the efficient working of the process. Proper mixing can help to maintain a homogenous mixture, which ensures that all microorganisms have access to the substrate and nutrients (Zhai et al. 2018). The enhanced mass transfer also helps to reduce the concentration of intermediate compounds such as VFAs, which inhibits the activity of methanogenic archaea. Mixing can prevent the stratification of the digester by ensuring the uniform distribution of solids and liquids (Kaparaju et al. 2008). It prevents the solids' accumulation at the bottom of the reactor, which can lead to the formation of dead zones and the accumulation of toxic compounds. Adequate mixing can prevent the formation of scum by ensuring that the substrate is uniformly distributed and the gas bubbles are evenly distributed throughout the digester. The location of the mixer, the strategy of mixing, duration and intensity, are all factors that affect the digester's performance.

6.4.8 TRACE ELEMENTS

Trace elements, also known as micronutrients, are necessary for the balanced growth and activity of microorganisms in AD. They are needed in small amounts but are essential for the efficient and stable operation of AD systems; however, the lack of trace elements can result in decreased microbial activity and lower biogas production (Khanal et al. 2019). Anaerobes also need low concentrations of trace elements in addition to the basic macronutrients such as N_2, phosphorous, carbon, and sulphur. The most studied and essential trace elements are nickel, cobalt, and iron. Trace elements act as co-factors for enzymes involved in various stages of the AD process. Moreover, the supplementation of multiple trace elements can lead to better anaerobic digester performance in comparison to single-element addition (Choong et al. 2016). For example, cobalt is a co-factor for vitamin B_{12}, which is required for the production of CH_4 by methanogenic archaea. They act as co-factors for enzymes, support the microbial growth, and help maintain the proper balance of microbial populations. Ensuring optimal trace element availability is essential for maximizing biogas production and maintaining process stability.

6.4.9 Microbial Population

Microbial populations are responsible for the degradation of complex organic compounds into simpler compounds, such as VFAs and CH_4 through enzymes. The proper understanding of the structure, function, and growth kinetics of the microorganisms is crucial to maintain the process stability (Gagliano et al. 2015). The substrate composition determines the types of microorganisms that can grow and the types of products that are produced. For example, carbohydrates are metabolized by acidogenic bacteria, which produce VFAs, while methanogenic archaea are responsible for the CH_4 production (Shah et al. 2014). The activity of microbial populations affects the AD process stability. A diverse and balanced microbial population is important for maintaining stable process performance, as it can ensure the efficient breakdown of organic matter and prevent the cumulation of inhibitory compounds. Variations in the microbial population can lead to a reduction in process performance, such as reduced biogas production or increased HRT. Therefore, the syntropic population is imperative for mutual interactions among bacteria under stress conditions (Jonge et al. 2020). A diverse microbial population is important for maintaining process stability and ensuring the efficient breakdown of organic matter. A lack of diversity can lead to a reduction in process performance, as certain microorganisms may be inhibited or outcompeted by others. Maintaining a diverse microbial population can be achieved through proper reactor design, substrate management, and microbial inoculation.

6.5 RECENT ADVANCES

The AD technology has gained significant attention in recent years due to its potential to address two critical issues: management of waste and renewable energy production. Recent research in AD process technology has focused on process improvement, enhancing the quality of the biogas produced, and finding new applications for the technology. Nowadays, the TAD is getting more attention due to its various merits such as fast rate of substrate utilization, low HRT, and high biogas production rate in comparison to MAD. Nevertheless, TAD technology needs more in-depth knowledge to reach commercialization because of some drawbacks such as VFAs inhibition, ammonia accumulation, and high energy demand. Various studies have targeted the wastes to overcome the VFAs concentration in TAD. One such study showed that the sawdust-derived biochar can reduce the VFAs' concentration and elevate microbial activities to enhance CH_4 production in TAD (Wang et al. 2019). The co-digestion of different feedstocks, supplementation of conductive materials, the addition of nanoparticles, and improved digester designs are some of the recent methods focused to enhance the AD process feasibility and CH_4 production (Tiwari et al. 2021; Baniamerian et al. 2021; Karki et al. 2021). Researchers are exploring the microbial communities involved in AD and how they can be manipulated to improve the process instabilities. New methods for digestate treatment have been explored to remove contaminants and produce high-quality fertilizer. Moreover, sensor-based technologies and data analytics tools to improve process monitoring and control are coming to the forefront. For instance, researchers have been investigating the use of artificial intelligence algorithms to predict biogas production and optimize process parameters and the overall economic viability of the process (Cruz et al. 2022; Offie et al. 2023). Recently, emerging bioelectrochemical techniques such as microbial electrolysis cells have been integrated with the AD to mitigate VFAs inhibition and enhance microbial interactions (Huang et al. 2020). However, the work is limited to lab-scale studies till now. Further research could be necessary for large-scale implementation.

6.6 FUTURE PROSPECTS

The future prospects of AD look promising, and there are several reasons. With the rising renewable energy sources' demand, there is a growing interest in the development of technologies that can harness renewable energy from organic waste. AD offers a sustainable and cost-effective solution for converting organic waste into renewable energy. As the population continues to grow, there is a corresponding increase in the generation of waste materials. AD can help to manage this waste by converting it into valuable products and mitigating GHG emissions. The use of AD technology in the agricultural sector can provide a renewable energy source such as cooking, electricity, etc., for farms. In addition, the digestate, the leftover of the AD process, can be used as a nutrient-rich fertilizer. Many governments around the globe are offering incentives to promote the renewable energy technologies such as AD. These incentives can include tax credits, grants, and other financial incentives, which can encourage the development of AD systems. Advances in technology have made AD more efficient and cost-effective. This includes the use of advanced control systems, improved reactor designs, and better waste pretreatment technologies. Therefore, the plenty of waste available in India could be better utilized in AD in an ecofriendly and sustainable manner. Furthermore, biorefinery concept could be explored where multiple products such as chemicals, solid and liquid fuels, polymers, etc., can be produced by utilizing various feedstocks. The anaerobic digester is a key component of an anaerobic biorefinery where raw materials are biotransformed into a variety of intermediate and high-end products. The use of biorefinery appears to be the best technique because it produces new materials and products while also producing electricity and protecting the environment.

ACKNOWLEDGEMENT

Authors Ms. Gaganpreet Kaur and Ms. Nisha Yadav are grateful to SSS-NIBE Bioenergy Promotion Fellowship.

REFERENCES

Abanades, S., Abbaspour, H., Ahmadi, A., Das, B., Ehyaei, M.A., Esmaeilion, F., El Haj Assad, M., Hajilounezhad, T., Jamali, D.H., Hmida, A. and Ozgoli, H.A., 2021. A critical review of biogas production and usage with legislations framework across the globe. *International Journal of Environmental Science and Technology*, 19, 3377–3400.

Achinas, S., Achinas, V. and Euverink, G.J.W., 2020. Microbiology and biochemistry of anaerobic digesters: An overview. *Bioreactors*, 17–26. doi:10.1016/B978-0-12-821264-6.00002-4

Ali Shah, F., Mahmood, Q., Maroof Shah, M., Pervez, A. and Ahmad Asad, S., 2014. Microbial ecology of anaerobic digesters: The key players of anaerobiosis. *The Scientific World Journal*. doi:10.1155/2017/3852369

Anukam, A., Mohammadi, A., Naqvi, M. and Granström, K., 2019. A review of the chemistry of anaerobic digestion: Methods of accelerating and optimizing process efficiency. *Processes*, 7(8), 504. doi:10.3390/pr7080504

Aryal, N., Ottosen, L.D.M., Kofoed, M.V.W. and Pant, D. (Eds.), 2021. *Emerging Technologies and Biological Systems for Biogas Upgrading*. Acadamic Press, London.

Babaee, A. and Shayegan, J., 2011, November. Effect of organic loading rates (OLR) on production of methane from anaerobic digestion of vegetables waste. In: *World Renewable Energy Congress-Sweden*, 8–13 May, 2011, Linköping; Sweden (No. 057, 411–417). Linköping University Electronic Press, Linköping. doi:10.3384/ecp11057411

Beschkov, V., 2021. *Biogas Production: Evaluation and Possible Applications*. IntechOpen. doi:10.5772/intechopen.101544

Chang, C.C., Kuo-Dahab, C., Chapman, T. and Mei, Y., 2019. Anaerobic digestion, mixing, environmental fate, and transport. *Water Environment Research*, 91(10), 1210–1222.

Choong, Y.Y., Norli, I., Abdullah, A.Z. and Yhaya, M.F., 2016. Impacts of trace element supplementation on the performance of anaerobic digestion process: A critical review. *Bioresource Technology*, 209, 369–379.

De Jonge, N., Davidsson, Å., la Cour Jansen, J. and Nielsen, J.L., 2020. Characterisation of microbial communities for improved management of anaerobic digestion of food waste. *Waste Management*, 117, 124–135.

Deshmukh, A., Nagarnaik, P.B. and Daryapurkar, R.A., 2015, November. Assessment of biogas generation potential of Napier Grass. In: *2015 7th International Conference on Emerging Trends in Engineering & Technology (ICETET)* (pp. 68–71). IEEE. doi:10.1109/ICETET.2015.35

Donoso-Bravo, A., Retamal, C., Carballa, M., Ruiz-Filippi, G. and Chamy, R., 2009. Influence of temperature on the hydrolysis, acidogenesis and methanogenesis in mesophilic anaerobic digestion: Parameter identification and modeling application. *Water Science and Technology*, 60(1), 9–17.

Fagerström, A., Al Seadi, T., Rasi, S. and Briseid, T. 2018. The role of anaerobic digestion and biogas in the circular economy. In: Murphy J.D. (Ed.), *IEA Bioenergy Task 37, 2018:8*. IEA, Paris

Fernández, J., Pérez, M. and Romero, L.I., 2008. Effect of substrate concentration on dry mesophilic anaerobic digestion of organic fraction of municipal solid waste (OFMSW). *Bioresource Technology*, 99(14), 6075–6080.

Gagliano, M.C., Braguglia, C.M., Gallipoli, A., Gianico, A. and Rossetti, S., 2015. Microbial diversity in innovative mesophilic/thermophilic temperature-phased anaerobic digestion of sludge. *Environmental Science and Pollution Research*, 22, 7339–7348.

Gebreeyessus, G.D. and Jenicek, P., 2016. Thermophilic versus mesophilic anaerobic digestion of sewage sludge: A comparative review. *Bioengineering*, 3(2), 15.

Goswami, R., Chattopadhyay, P., Shome, A., Banerjee, S.N., Chakraborty, A.K., Mathew, A.K. and Chaudhury, S., 2016. An overview of physico-chemical mechanisms of biogas production by microbial communities: A step towards sustainable waste management. *3 Biotech*, 6, 1–12. doi:10.1007/s13205-016-0395-9

Ileleji, K.E., Martin, C. and Jones, D., 2015. Basics of energy production through anaerobic digestion of livestock manure. In: *Bioenergy* (pp. 287–295). Academic Press, London. doi:10.1016/B978-0-12-407909-0.00017-1

Kabeyi, M.J.B. and Olanrewaju, O.A., 2022. Biogas production and applications in the sustainable energy transition. *Journal of Energy*, doi:10.1155/2022/8750221

Kaparaju, P., Buendia, I., Ellegaard, L. and Angelidakia, I., 2008. Effects of mixing on methane production during thermophilic anaerobic digestion of manure: Lab-scale and pilot-scale studies. *Bioresource Technology*, 99(11), 4919–4928.

Khalid, A., Arshad, M., Anjum, M., Mahmood, T. and Dawson, L., 2011. The anaerobic digestion of solid organic waste. *Waste Management*, 31(8), 1737–1744. doi:10.1016/j.wasman.2011.03.021

Khanal, S.K., Nindhia, T.G.T. and Nitayavardhana, S., 2019. Biogas from wastes: Processes and applications. In: *Sustainable Resource Recovery and Zero Waste Approaches* (pp. 165–174). Elsevier, Amsterdam. doi:10.1016/B978-0-444-64200-4.00011-6

Kim, M., Ahn, Y.H. and Speece, R.E., 2002. Comparative process stability and efficiency of anaerobic digestion; mesophilic vs. thermophilic. *Water Research*, 36(17), 4369–4385.

Koley, A., Mukhopadhyay, P., Show, B.K., Ghosh, A., Balachandran, S. and Santiniketan, V.B., 2022. OP30: Biogas production potentiality of Water Hyacinth, Pistia and Duckweed: A comparative analysis *National Symposium: "Recent Trends in Sustainable Technology - Techno - Commercial Developments"*, 978-93-5636-245-1.

Kothari, R., Pandey, A.K., Kumar, S., Tyagi, V.V. and Tyagi, S.K., 2014. Different aspects of dry anaerobic digestion for bio-energy: An overview. *Renewable and Sustainable Energy Reviews*, 39, 174–195. doi:10.1016/j.rser.2014.07.011

Krishnan, S.K., Kandasamy, S. and Subbiah, K., 2021. Fabrication of microbial fuel cells with nanoelectrodes for enhanced bioenergy production. In: *Nanomaterials* (pp. 677–687). Academic Press, London. doi:10.1016/B978-0-12-822401-4.00003-9

Labatut, R.A. and Pronto, J.L., 2018. Sustainable waste-to-energy technologies: Anaerobic digestion. In: *Sustainable Food Waste-to-Energy Systems* (pp. 47–67). Academic Press, London.

Labatut, R.A., Angenent, L.T. and Scott, N.R., 2014. Conventional mesophilic vs. thermophilic anaerobic digestion: A trade-off between performance and stability? *Water Research*, 53, 249–258.

Lei, Z., Chen, J., Zhang, Z. and Sugiura, N., 2010. Methane production from rice straw with acclimated anaerobic sludge: Effect of phosphate supplementation. *Bioresource Technology*, 101(12), 4343–4348. doi:10.1016/j.biortech.2010.01.083

Leung, D.Y. and Wang, J., 2016. An overview on biogas generation from anaerobic digestion of food waste. *International Journal of Green Energy*, 13(2), 119–131. doi:10.1080/15435075.2014.909355

Lohani, S.P. and Havukainen, J., 2018. Anaerobic digestion: Factors affecting anaerobic digestion process. In: *Waste Bioremediation* (pp. 343–359). doi:10.1007/978-981-10-7413-4_18

Mao, C., Feng, Y., Wang, X. and Ren, G., 2015. Review on research achievements of biogas from anaerobic digestion. *Renewable And Sustainable Energy Reviews*, 45, 540–555.

Meegoda, J.N., Li, B., Patel, K. and Wang, L.B., 2018. A review of the processes, parameters, and optimization of anaerobic digestion. *International Journal of Environmental Research and Public Health*, 15(10), 2224. doi:10.3390/ijerph15102224

Náthia-Neves, G., Berni, M., Dragone, G., Mussatto, S.I. and Forster-Carneiro, T., 2018. Anaerobic digestion process: Technological aspects and recent developments. *International Journal of Environmental Science and Technology*, 15, 2033–2046.

Nguyen, P.H.L., Kuruparan, P. and Visvanathan, C., 2007. Anaerobic digestion of municipal solid waste as a treatment prior to landfill. *Bioresource Technology*, 98(2), 380–387.

Nkuna, R., Roopnarain, A., Rashama, C. and Adeleke, R., 2022. Insights into organic loading rates of anaerobic digestion for biogas production: A review. *Critical Reviews in Biotechnology*, 42(4), 487–507.

Noraini, M., Sanusi, S., Elham, O.S.J., Sukor, M.Z. and Halim, K., 2017. Factors affecting production of biogas from organic solid waste via anaerobic digestion process: A review. *Solid State Science and Technology*, 25(1), 29–39.

Ostrem, K. and Themelis, N.J., 2004. Greening waste: Anaerobic digestion for treating the organic fraction of municipal solid wastes. M.S. thesis, Earth Engineering Center Columbia University.

Panigrahi, S. and Dubey, B.K., 2019. A critical review on operating parameters and strategies to improve the biogas yield from anaerobic digestion of organic fraction of municipal solid waste. *Renewable Energy*, 143, 779–797.

Park, C., Lee, C., Kim, S., Chen, Y. and Chase, H.A., 2005. Upgrading of anaerobic digestion by incorporating two different hydrolysis processes. *Journal of Bioscience and
Bioengineering*, 100(2), 164–167. doi:10.1263/jbb.100.164

Pramanik, S.K., Suja, F.B., Zain, S.M. and Pramanik, B.K., 2019. The anaerobic digestion
process of biogas production from food waste: Prospects and constraints. *Bioresource
Technology Reports*, 8, 100310. doi:10.1016/j.biteb.2019.100310

Sanusi, S.N.A., Elham, O.S.J., Sukor, M.Z. and Noraini, M., Factors affecting production of
biogas from organic solid waste via anaerobic digestion process: A review. *Advanced
Organic Waste Management*. doi:10.1016/B978-0-323-85792-5.00020-4

Sawatdeenarunat, C., Surendra, K.C., Takara, D., Oechsner, H. and Khanal, S.K., 2015.
Anaerobic digestion of lignocellulosic biomass: Challenges and opportunities.
Bioresource Technology, 178, 178–186.

Singh, P. and Kalamdhad, A.S., 2021. A comprehensive assessment of state-wise biogas potential and its utilization in India. *Biomass Conversion and Biorefinery*, 1, 1–23.

Singh, R., Hans, M., Kumar, S. and Yadav, Y.K., 2023. Thermophilic anaerobic digestion:
An advancement towards enhanced biogas production from lignocellulosic biomass.
Sustainability, 15(3), 1859.

Suryawanshi, P.C., Chaudhari, A.B. and Kothari, R.M., 2010. Mesophilic anaerobic digestion:
First option for waste treatment in tropical regions. *Critical Reviews in Biotechnology*,
30(4), 259–282.

Van, D.P., Fujiwara, T., Tho, B.L., Toan, P.P.S. and Minh, G.H., 2020. A review of anaerobic
digestion systems for biodegradable waste: Configurations, operating parameters, and
current trends. *Environmental Engineering Research*, 25(1), 1–17.

Vogt, G.M., Liu, H.W., Kennedy, K.J., Vogt, H.S. and Holbein, B.E., 2002. Super blue box
recycling (SUBBOR) enhanced two-stage anaerobic digestion process for recycling
municipal solid waste: Laboratory pilot studies. *Bioresource Technology*, 85(3), 291–
299. doi:10.1016/s0960-8524(02)00114-1

Wainaina, S., Lukitawesa, Kumar Awasthi, M. and Taherzadeh, M.J., 2019. Bioengineering of
anaerobic digestion for volatile fatty acids, hydrogen or methane production: A critical
review. *Bioengineered*, 10(1), 437–458.

Zhai, X., Kariyama, I.D. and Wu, B., 2018. Investigation of the effect of intermittent minimal
mixing intensity on methane production during anaerobic digestion of dairy manure.
Computers and Electronics in Agriculture, 155, 121–129.

Zupančič, G.D. and Grilc, V., 2012. Anaerobic treatment and biogas production from organic
waste. *Management of Organic Waste*, 2. DOI: 10.5772/32756

7 Applications of Biomass-Derived Materials for Energy Production

Gurkanwal Kaur, Jaspreet Kaur, and
Monica Sachdeva Taggar
Punjab Agricultural University

7.1 INTRODUCTION

The unprecedented increase in the human population all over the world has led to ever rising energy-intensive demands. Worldwide energy consumption is reported to escalate by 1.2% annually, and is expected to touch 18.9 billion tons (of oil equivalent) mark by the year 2040 (Koyama, 2017). Presently, the limited supply of non-replenishable resources of energy such as coal, crude oil, and natural gas are diminishing continuously, thereby necessitating the need to develop alternative sources of energy (Quah, 2019). Moreover, an extensive exploitation of the fossil fuels due to rapid expansion of industrialization has severely affected the environment, attributable to emission of greenhouse gases (CO_2, SO_2, and NOx) accompanied by escalating levels of toxic compounds (polycyclic hydrocarbons, mercury, etc.) (Kumar et al., 2020). In contemplation of securing global energy security and to efficiently combat with the gruesome environmental challenges, development of renewable, sustainable, and economical energy alternatives has emerged as the pressing priority in recent times (Hajilary et al., 2020). Focusing on these aspects, 17 sustainable development goals for the year 2030, listed by the United Nations, emphasized the use of affordable and replenishable sources of energy (Guney, 2019).

Among various renewable energy production sources, biomass is the most abundantly available bioresource where 2×10^{11} tons of biomass (dry carbon basis) was produced globally in 2016 (Liu et al., 2019). The plant-derived organic matter produced via the process of photosynthesis involving the interconversion of CO_2 and water into carbohydrates in the presence of sunlight is referred to as biomass which further undergoes interconversions via natural/artificial processing (Li et al., 2016). A considerable deal of attention has been received by functional carbonaceous materials derived from various types of biomasses, namely, agricultural wastes, industrial wastes, sewage sludge/municipal waste, animal excreta, algal waste, etc. (Yang et al., 2019). Several biomass conversion technologies, including torrefaction, pyrolysis,

DOI: 10.1201/9781003406501-7

127

carbonization, combustion, etc., have been explored depending upon the appositeness of the product (Kumar et al., 2020; Krerkkaiwan et al., 2015). Based on the chemical composition of the selected biomass, various kinds of carbon-rich products, such as biochar, bio-oil (liquid biofuel), and syngas, can be produced for the purpose of energy generation (Chen et al., 2020). Biomass is believed to be 'carbon-neutral' since it has the potential to be used directly or converted to other forms, where the amount of CO_2 released balances the amount of CO_2 captured from the surroundings during the time of its growth (Antar et al., 2021).

Biochar is yielded by thermochemical decomposition of lignocellulosic agro-residues and other organic materials available abundantly, and possesses high porosity, surface area, biocompatibility, and functionality making it amenable to surface modifications (Bhatia et al., 2021). Removal of cyanotoxins, improving water quality and nutrient retention capacity of soil, synthesis of electrodes in supercapacitors, and acting as catalyst in fuel cells are some of the important implications of biochar production (Gaurav et al., 2021; Ding et al., 2020). Similarly, liquid biofuel, also referred to as bio-oil, is being produced from a variety of feedstocks using advanced technologies involving low temperature pyrolytic reactions (Lehmann et al., 2006). Contrary to the fossil fuels, bio-oils are considered as cleaner fuels since the latter release significantly reduced levels of sulphur and nitrogen oxides (Xiu & Shahbazi, 2012). Syngas, a gaseous fuel, is constituted mainly of carbon monoxide and hydrogen, and its production involves pyrolysis and gasification of different feedstocks such as wood, bamboo, food waste, pine sawdust, refuse oils, sewage sludge, etc. (Zheng et al., 2016; Campoy et al., 2014). High carbon content in syngas is ascribed to applicability as an alternative to natural gas in various industries (Guruviah et al., 2019). Different properties of the source biomass account for various environmental, catalytic, and electrical applications of the resultant carbonaceous materials. These biomasses acquired materials have an immense potential for producing renewable energy that would help to establish sustainable developmental routes for generations to come and make the planet Earth an energy-secure planet.

This chapter focuses on a detailed insight into the applications of biomass-derived products, such as biosyngas, biochar, and bio-oil, for the purpose of energy generation. It also discusses potential bioresources that could be converted into energy-rich carbonaceous materials by employing suitable production techniques. Further, factors affecting the production of the aforementioned materials have also been discussed briefly which directly influence the applicability of the same.

7.2   VARIOUS BIOMASS RESOURCES

The paradigm shifts in the global energy market from the accustomed and non-replenishable sources such as oil, coal, and natural gas to carbon-neutral biomass-derived bio-oils, biosyngas, biochar, and other such replenishable resources is the need of the hour. Bioenergy or biomass-derived energy is the largest form of replenishable and sustainable form of energy available today. It is derived from various bioenergy crops, lignocellulosic agricultural and forest residues, algal biomass, municipal solid waste/sludge, etc. A brief outline of the extensively prevalent sources of biomass feedstock for bioenergy production is highlighted in Figure 7.1.

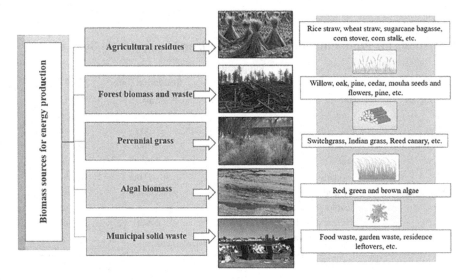

FIGURE 7.1   Different biomass sources available for renewable energy production.

The plant-derived bioresources have certain biochemical constituents in their structure, which mainly include a framework of cellulosic and hemicellulosic components conjoined with a network of lignin polymeric chains. Cellulose is a homopolymer formed by linear chains of glucose moieties interconnected by β-1,4-glycosidic bonds (Heinze, 2016). Hemicellulose is a heterogeneous carbohydrate polymer comprising xylans, mannans, glucomannans, and xyloglucans, and their relative compositions vary in different types of agricultural residues and other biomass materials (Scheller, 2010). Lignin, however, is a phenolic polymer made up of precursors of p-coumaroyl alcohol, coniferyl alcohol, and sinapyl alcohol (Xiang & Li, 2018). Higher amount of cellulose is necessary to produce volatile compounds which are further important for the oxidation process during combustion of biomass for bioenergy generation (Cao et al., 2019). The hemicellulosic components are connected to the cellulose fibrils that provide strength and crystallinity to the lignocellulosic biomass structure via hemicellulose-lignin crosslinking (Agrawal et al., 2017). Lower lignin and hemicellulose content in the targeted bioresource is desirable for energy production since lignin undertakes the role of a chemical binding agent that makes it tough to dissociate cellulose from hemicellulose, thereby preventing the generation of amorphous and hydrolysable cellulose (Antar et al., 2021). Cellulose subjected to acid hydrolysis yields rod-like cellulose nanocrystals, which when disintegrated with mechanical shear force leads to the formation of cellulose nanofibrils/microfibrils (Yang et al., 2019). The hydroxyl groups present in the chemical structure of cellulose allow further modification of the same, enhancing its mechanical stability, biocompatibility, and biodegradability (Trache et al., 2017). Considering the variability in the cellulosic, hemicellulosic, and lignin content of different feedstocks, different synthetic strategies are adopted to produce biomass-based carbonaceous materials for energy production, which have been discussed in the subsequent sections.

TABLE 7.1

Methods for Production of Biochar, Bio-oil, and Syngas from Biomass

Method	Type	Temperature (°C)	Residence Time	Final Product
Pyrolysis	Slow	300–700	Few min-days	Biochar
				Syngas
	Fast	500–1,000	5–30 s	Bio-oil
Torrefaction	Steam	<260	10 min	Biochar
	Wet	180–260	5–240 min	Biochar
	Oxidative	200–300	30–60 min	Biochar
Carbonization	Hydrothermal carbonization	220–240	Several hours	Hydrochar
	Molten state carbonization	250–350	1–6 h	Porous carbon

7.3 METHODS FOR PRODUCTION OF BIOCHAR, BIO-OIL, AND SYNGAS

There are different kinds of methods that can be employed to produce biomass-derived energy resources as suggested in Table 7.1. These methods are mainly categorized into three broad categories, namely, pyrolysis, torrefaction, and carbonization. Biomass conversion into carbonaceous materials like biochar, bio-oil, and syngas is determined by various parameters that include reaction temperature and pressure, type of feedstock, biomass moisture content, etc. These methods have been briefly discussed in the following sections in order to understand the product generation from the abundantly available biomass.

7.3.1 PYROLYSIS

Involving thermal degradation of biomass between a temperature range of 250°C–900°C, pyrolysis is carried out in an oxygen-deficient environment for the transfiguration of organic material in the biomass into solid (biochar), liquid (bio-oil), and gaseous (carbon dioxide, carbon monoxide, syngas, and hydrogen) products (Ismail et al., 2023). The word "pyrolysis" has been originally derived from Greek word "pyro", which means fire; and lysis, which in relevance to the process, means breakdown into fundamental structural parts. The process of pyrolysis involves multiple steps, viz. depolymerization, fragmentation, and crosslinking of biomass constituents (lignin, cellulose, and hemicellulose) at defined temperatures (Ansari et al., 2019). A process is termed as pyrolysis under certain conditions when the substrate is in a degradation state and the nature of bond cleavage is thermal (Devi et al., 2021). Pyrolysis process is mainly of two types depending upon the residence time, heating rate, temperature, and pressure, which are differentiated as: (i) slow pyrolysis which involves the production of heat and biochar and (ii) fast pyrolysis which involves the production of biochar and bio-oils (Zaman et al., 2017).

Fast pyrolysis process involves the liquefication process that transforms the solid residue into liquid bio-oil. This process generally operates at high temperatures (300–1,000°C) with low residence time (5–30 s) (Sakhiya et al., 2020). It has been suggested that the fume residence time during fast pyrolysis should be kept in hot zone in order to achieve a good quality of bio-oil (Wang et al., 2014). On the contrary, slow pyrolysis is generally operated at lower temperatures (300°C–550°C) with longer residence time, which is varied between several minutes to days. Prospecting towards the major pyrolysis application, biochar is a solid product of slow pyrolysis with an approximate yield of 35% of the dry biomass weight (Daful & Chandraratne, 2018). The main advantage of biochar is that it has the potential to be used as a modifying agent to ameliorate the quality of soil (Arni, 2018). There are several reactors, viz. wagon reactors, paddle kin, agitated sand rotating kilns, and bubbling fluidized bed, which are available for biochar production. The main parameters that determine the yield and efficiency of biochar during pyrolysis process are the type of biomass, its quality, and the reaction temperature range (Wei et al., 2019). During pyrolysis, the rise in temperature beyond a certain point results in substantially lower biochar and higher syngas production yield (Yaashikaa et al., 2020). During biomass pyrolysis, some carcinogenic and phytotoxic compounds may be generated under certain conditions like extreme pyrolysis parameters, unsuitable feedstock type, etc. (Ndriangu et al., 2019). Comprehensively, the results of pyrolysis process involve loss of nitrogen and sulphur content of biomass, alteration of heavy metals into less toxic forms, elimination of pathogenic substances, etc. (Paz-Ferreiro et al., 2018; Maguire & Agblevor, 2010).

7.3.2 Torrefaction

Torrefaction is a heat-intensive pre-treatment process that is mainly used for the production of biosolids (biochar). It is relatively a milder form of pyrolysis since it generally utilizes a moderate temperature range (200–300°C) during product development (Waheed et al., 2022). The biomass is degraded to remove oxygen, moisture content, and carbon dioxide by using inert atmospheric conditions (Yu et al., 2017). The torrefaction process involves transformation in chemical and physical nature of biomass, reduction in moisture content, and increase in the intrinsic energy content of the derivatised product (Tumuluru et al., 2021). Factors including atmospheric pressure, temperature, heating rate, residence time, biomass moisture content, biomass flexibility, and particle size critically affect the torrefaction process. It has been suggested that the moisture content of the product obtained after this thermal treatment should be preferably less than 10% (Waheed et al., 2022). Evaporation of moisture, thermal degradation of hydrogen, and oxygen-enriched components of organic compounds accompanied by the release of volatile substances have been reported during the torrefaction process (Beckman et al., 2012). At the end of this process, a solid product is formed that comprises a significantly high energy and substantially low moisture content (Tumuluru et al., 2021). Torrefaction is broadly categorized into three types: (i) steam torrefaction, (ii) wet torrefaction, and (iii) and oxidative torrefaction. Steam torrefaction process utilizes steam for the treatment of biomass at a temperature of <260°C where the residence time is about 10 min. During wet torrefaction (also known as hydrothermal carbonization, since water treatment of biomass is involved), a temperature range of 180°C–260°C is coordinated with a residence time of 5–240 min. In oxidizing torrefaction process, the selected biomass is treated with certain oxidizing agents to produce heat energy along with the carbonized product (Yaashikaa et al., 2020).

7.3.3 HYDROTHERMAL CARBONIZATION

Hydrothermal carbonization, as the name suggests, is a thermal process that degrades the biomass by using water. Hydrothermal carbonization process was discovered by Bergius in 1913 and this process has been found to be similar to the process of transformation of cellulose into coal. Formerly, it was referred to as artificial coalification, which was later described as hydrothermal carbonization, moist torrefaction, and/or subcritical water treatment (Jien et al., 2019). This process is normally carried out in a closed reactor at a temperature above 180°C (Cheng and Li, 2018). Hydrochar is the primary solid product formed amidst the process described, which is certainly different from the char generated during pyrolysis (Cao et al., 2017). Wet biomass including animal waste, sewage sludge, and compost is mainly used for hydrochar production. Hydrothermal carbonization entails an intricate reaction process in which various chemical transformations involving dehydration, retro-aldol condensation as well as isomerization are performed sequentially (Titirici et al., 2012). This process involves heating of wet biomass at high temperatures ranging between 220°C and 240°C and atmospheric pressure adjusted between 2 and 10 MPa for several hours (Gabhane et al., 2020). The yield of hydrochar generally varies between 30% and 80% and is susceptible to the nature of biomass and overall process conditions (Cheng et al., 2017). Several carbonaceous compounds corresponding to distinctive morphological and chemical properties are produced during hydrothermal carbonization. This process has several advantages over torrefaction and pyrolysis treatments including substantial reduction in O/C ratio, higher calorific values, enhanced hydrophobicity, and better grinding ability (Gao et al., 2013). Hydrochar can be used to make activated charcoal, adsorbent and solid fuel; improve the soil quality; store hydrogen; and treat wastewater (Kumar et al., 2023).

7.3.4 MOLTEN SALT CARBONIZATION (MSC)

MSC process involves the use of molten salts for the generation of porous carbon from biomass by breaking the native cellulose and lignin components present in the biomaterial (Lu et al., 2016). The process involves three simple steps: (i) liquefaction of salts to its melting point, (ii) immersion of biomass into molten salts followed by carbonization (>400°C temperature and inert atmosphere), and (iii) washing of salts with acid (HCl) and distilled water. This process presents two main advantages: (i) prevention in the generation of harmful by-products and (ii) easy removal and recycling of salts (Diaz et al., 2021). Various kinds of salts used in the MSC process include carbonate, nitrite, alkali, hydroxide, alkaline earth metal cation, and phosphate anions. The low cost and abundant availability of these salts make them versatile to use. The main advantages of using these salts are a wide range of working temperature, suitable thermal stability, high capacity of heat, low viscosity, low vapour pressure, and greater solubility for impurities in the MSC. The thermal instability generated due to ionic nature of molten salts can degrade the feedstock into gas, liquid, and solid carbon material (Egun et al., 2022). The process of MSC process is primarily of two main types: mixing-type MSC and immersion-type MSC. These two processes can be distinguished by the polymerization or depolymerization reaction catalysed during biomass conversion. The mixing method starts before the melting point of the salt as compared to immersion process, which ultimately leads to the solubilization of carbon source into molten salts. The MSC process offers several advantages like (i) high yield of carbon, (ii) formation of pores between carbonaceous matter and emitted reactive gases, and (iii) it can be flexibly used for various carbon sources (Diaz et al., 2021).

7.4 APPLICATIONS OF BIOMASS-DERIVED MATERIALS

7.4.1 BIOCHAR

As discussed above, pyrolysis of the feedstock mainly yields a carbonaceous end product referred to as biochar, which holds back the chemical energy of the native feedstock (Nanda et al., 2016). Coal has been substituted by biochar in power plants heeding to the environmental concerns that have risen because of fossil fuel combustion. In addition, the need for mining and the associated health hazards have been reduced remarkably because of the production and use of biochar (Hajilary et al., 2020). Biochar, also being an excellent source of carbon sequestration, can potentially sequester 0.50 billion tons of CO_2 per annum, given that 340 tonnes of biochar is produced per year (Kumar et al., 2020). The C-C bonds of biochar constitute of a high energy content, thereby resulting in its large calorific values (28–32 MJ/kg), which are reportedly higher than that of natural coal (24 MJ/kg) (Abdullah et al., 2010). The biochar produced by pyrolysis and hydrochars derived from hydrothermal carbonization have gained significant amount of attention for their application in reducing soil contaminant content, consequently improving soil fertility. Their peculiar properties such as resistance to microbial degradation, high surface area, good stability, high cation exchange capacity, and remarkable porosity facilitate an efficient sorption of polyaromatic compounds, organic and inorganic contaminants, heavy metals, and various other toxic substances from soil and sewage (Yargicoglu et al., 2015). Biochar/hydrochar when doped with suitable activation agents are capable of adsorbing various contaminants in the industrial waste water. Hydrochar produced from orange peels was activated using H_3PO_4 and then employed to adsorb pharmaceutical organic contaminants (flurbiprofen, salicylic acid, and diclofenac sodium) from waste water (Fernandez et al., 2015). Biochar has been asserted to be used as a catalyst for degradation of plastic waste where alumina, silica, and aluminium hydroxides in biochar bind with the plastic structure and facilitate its degradation and subsequent energy generation (Krerkkaiwan et al., 2015).

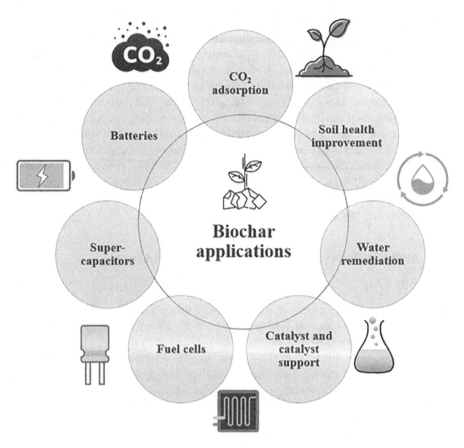

FIGURE 7.2 Various applications of biochar for environment detoxification and energy production.

Various other applications of biochar for energy production are highlighted in Figure 7.2. In recent times, the role of biochar as a cathodic catalyst has been explored in microbial fuel cells that basically transform different chemical compounds in wastewater into electricity by exploiting micro-organism metabolism (Janicek et al., 2015). The tendency of biochar to effectively interact with molten carbonate electrolyte enhances the diffusivity of biochar during the reaction and initiates electron transport. The feasibility to induce electrical conductivity makes biochar applicable in the fuel cells. Microbial fuel cell is a relatively new technology developed for electricity generation by utilizing the metabolic activity of micro-organisms in order to catalyse redox reactions and harvest direct current from the same (Cao et al., 2007). Doping of the highly porous biochar with certain heteroatoms or nanocomposites makes it competent for catalysing the oxygen reduction reactions (Chen et al., 2014).

Heteroatoms of sulphur, phosphorus, boron, nitrogen, iron, nickel, and cobalt have been doped to enhance the catalytic preferability of biochar-based materials in various zinc-air batteries and membrane fuel cells (Wang et al., 2017; Gao et al., 2015). Nitrogen-doping of biochar has been reported to enhance its specific capacity which facilitates lithium and sodium-ion batteries with an interface for charge transfer. The ability of biochar to capture sulphur and polysulphides in certain battery models improves the battery stability by inhibiting active material expansion (Luo et al., 2018). Biochar can also be tuned for the purpose of energy storage in the form of supercapacitors by modifying its porosity, conductivity, and surface functional groups. Doping of biochar with metal oxides resulted in enhanced energy density and improved capacitance in the supercapacitors supplemented with quick charge/discharge capabilities (Gao et al., 2011). The use of biochar has also been reported for biohydrogen production where methods like water splitting (Yang et al., 2021), steam reforming (Harun et al., 2020), and anaerobic digestion (Sharma & Melkania, 2017) have been explored. The use of biochar as an electrocatalyst is gaining attention due its good electrical conductivity and considerably low cost. These carbon-based materials are functionalized with various transition metal compounds that facilitate the structure with more active sites for faster electron transfer (Bhatia et al., 2021). Advancements are being made to improve these technologies for promoting energy security and sustainability. Some of the recent applications of biochar-based materials for energy storage and conversion are represented in Table 7.2.

In addition to the applications discussed above, biochar also acts as a catalyst for various esterification and transesterification reactions to produce bio-oils, including biodiesel (Foroutan et al., 2021), which is discussed in the following section.

TABLE 7.2

Applications of Various Biochar-Based Materials for Energy Generation, Storage, and Conversion

Biomass Feedstock	Treatment Conditions	Application	References
Watermelon peels	Water splitting of nanocomposite doped with $CoCl_2$ and activated by KOH (600°C–800°C)	Electrocatalytic hydrogen production	Yang et al. (2021)
Sargassum oligocystum	Heterogenous biochar supplemented with CaO and K_2CO_3	Biodiesel from waste edible oil	Foroutan et al (2021)
Deoiled *Azolla* biomass	Pyrolysis (600°C, 3 h) KOH activation of biochar at 1:4 ratio (KOH: biochar)	Bioelectrode (anode)	Hemalatha et al. (2020)
Watermelon	Acid pre-treatment followed by carbonization (700°C)	Bioelectrode (cathode)	Zhong et al. (2019)
Corn stalks	Pyrolysis (300°C, 3 h) with $CaCl_2$	Lithium-ion batteries	Li et al. (2018)
Municipal solid waste	Anaerobic digestion using co-culture of *Enterobacter aerogenes* and *E. coli*	Hydrogen production	Sharma and Melkania (2017)
Spruce whitewood	Fast pyrolysis (600°C) KOH activation of biochar at 3.5:1 ratio (KOH: biochar)	Supercapacitor	Dehkhoda et al. (2016)
Pine sawdust and pinewood	Carbonization (1,000°C, 1 h) in biomass gasifier	Fuel cell	Huggins et al. (2014)

7.4.2 BIO-OILS

The liquid fuels, also referred to as bio-oils, are a mixture of constituents of varying molecular sizes resulting from disintegration of lignin, cellulose, and hemicellulose. They are generally composed of alcohols, acids, aldehydes, esters, guaiacols, ketones, furans, lignin-derived phenols and sugars (Baloch et al., 2018). Bio-oils are produced via pyrolytic and hydrothermal reactions from biomass where treatment conditions such as heating rate, temperatures, and vapour residence time are of great importance. Fast pyrolysis is reported to yield superior quality bio-oil which has a viscosity at par with the conventionally used diesel (Kumar et al., 2020). A sharp increase in the world's biofuel production has been observed in less than two decades, where 95.4 Mt of oil equivalent of biofuel was produced in 2018 as compared to 9.2 Mt of oil equivalent in 2000 (Wang et al., 2019). First-generation biofuels, including bioethanol, are mainly derived from food crops such as corn, beet, sugarcane, oilseed rape, soybean oil, etc., based on their sugar, starch, and oil content. However, the most commercially amenable production of biofuels is attributed to the lignocellulosic materials that are composed of cellulose, hemicellulose, and lignin in their biochemical structure. The fuels produced from these feedstocks are called second-generation biofuels (Binod et al., 2019). The third generation of biofuels, which is a relatively advanced field, are derived from algae and marine plants/seaweeds. Since algal growth is prevalent in wastewater, sewage, saltwater, and other areas which are considered inappropriate for crop production, the third-generation bio-oils have a tremendous potential to replace fossil fuels (Panahi et al., 2019). The most discussed strategies for biofuel production from biomass feedstocks are thermochemical (pyrolysis, hydrothermal liquefaction, super critical water oxidation) and biochemical (microbe-assisted) conversion along with transesterification (Lee et al., 2019). Heating values of bio-oils produced from different biomasses using one of the most efficient conversion processes, i.e., hydrothermal liquefaction process, are shown in Figure 7.3. The quality and yield of bio-oils are highly influenced by the organic composition of the feedstock used (Minowa et al., 1995). Plant or algal oils, when subjected to transesterification with alcohols, result in biodiesel formation (Foroutan et al., 2021). The presence of C14–C20 long carbon chain fatty acids in

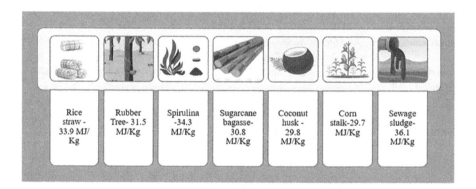

FIGURE 7.3 Heating values of bio-oils produced from different kinds of biomasses (obtained using hydrothermal liquefaction process).

the chemical structure of biodiesel results in high energy density, which makes it a desirable source of renewable energy for the existing engines (Bhatia et al., 2017). Biochar-mediated catalytic transesterification processes for biodiesel production have been reported. The porosity and hydrophobicity of biochar facilitated efficient active site binding of the reactant and removal of undesired products from the reaction mixture, respectively (Bhatia et al., 2021a). Cost effective and high-quality production of bio-oils in a biorefinery-based approach is still a developing avenue that aims to replace fossil fuels for sustainable development and waste utilization.

7.4.3 BIOSYNGAS

The synthesis gas or syngas earlier produced from petroleum, residual oils, and natural gas is now being replaced by biosyngas that is derived from biomass. This transition is considered to have a positive impact on the carbon footprint, since a reduction in the amount of greenhouse gas emissions (methane, nitrous oxide, carbon dioxide, sulphur hexafluoride, hydrofluorocarbons, etc.) is observed (dos Santos et al., 2020). Biosyngas is chemically analogous to syngas (constituting of carbon monoxide and hydrogen with traces of methane, carbon dioxide, and water); therefore, it can be used for similar purposes (Boerrigter et al., 2005). It is a combustible gas that can be used in the generation of electrical energy in fuel cells and turbines (Wender, 1996). An abundant amount of syngas is produced by the process of gasification where partial biomass oxidation occurs in a temperature range of 700°C–900°C (Cha et al., 2016). Heating biochar at higher temperatures also results in the formation of this gaseous product (Goyal et al., 2008). Biosyngas is reportedly used to produce alkane, alcohol, and hydrogen, and is therefore considered as a high energy source (Rostrup-Nielsen, 2001). The alcohol produced gasification is a potential fuel for transportation as well as a feed for fuel cells (Kumar et al., 2020). High carbon content in syngas signifies its applicability to be used as a substitute of natural gas to be used in automobiles and industries. This fuel gas is quite versatile since it can be used in boilers, engines, turbines, distributed in pipelines, and blended with other gas fuels. Upgraded technologies have been employed to convert syngas into other biofuels and high-value-added chemicals besides the production of hydrogen (Bermudez & Fidalgo, 2016). Different kinds of syngas-derived products are illustrated in Figure 7.4.

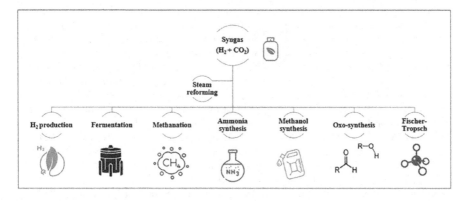

FIGURE 7.4 Generation of several products derived from syngas.

7.5 CONCLUSION

Energy and environmental impacts are the most crucial issues that are influencing industrial development in the present times. Biomass-derived products such as biochar, bio-oil, and biosyngas have a wide variety of applications such as physisorption of toxic compounds and heavy metals from soil and water, carbon sequestration, removal of greenhouse gases, along with energy production, conversion, and storage. The carbon-enriched biomass by-products have an immense potential in the synthesis of polymeric composites, which can be modified and used for environmental and industrial purposes. Functionalization of biomass-derived carbon enables it to be employed as a catalyst and catalyst support, supercapacitors, etc., which has facilitated an economical and more efficient way of energy generation. Using the biodegradable feedstock for its cellulosic, lignin, and hemicellulosic content by modifying the inherent properties using various physical and chemical conversions is a green route towards tackling the issues of sustainability and energy security. Replacing the expensive carbon-based materials, such as activated carbon, with relatively cheaper biochar and hydrochar should be the target of the researchers. So far, pyrolysis is the most feasible method for bulk biomass conversion into useful carbonaceous materials in terms of large-scale production. However, hydrothermal, ionothermal, and MSC methods show immense potential that need to be optimized for commercialization. The current outlook displays the need to improve production of biomass-derived materials to ensure a competitive position in the oil as well as gas market. Understanding the economic viability, extended lifetime, and selectivity of these carbonaceous products is required for their standardized production.

REFERENCES

Abdullah, H., Mediaswanti, K. A., & Wu, H. (2010). Biochar as a fuel: 2. Significant differences in fuel quality and ash properties of biochars from various biomass components of Mallee trees. *Energy & Fuels, 24,* 1972–1979. doi:10.1021/ef901435f

Agarwal, B., Ahluwalia, V., Pandey, A., Sangwan, R. S., & Elumalai, S. (2017). Sustainable production of chemicals and energy fuel precursors from lignocellulosic fractions. In A. K. Agarwal, R. A. Agarwal, T. Gupta, & B. R. Gurjar (Eds.), *Biofuels: Technology, Challenges and Prospects* (pp. 7–33). Singapore: Springer Publishers. doi:10.1007/978-981-10-3791-7_2

Ansari, K. B., Arora, J. S., Chew, J. W., Dauenhauer, P. J., & Mushrif, S. H. (2019). Fast pyrolysis of cellulose, hemicellulose, and lignin: Effect of operating temperature on bio-oil yield and composition and insights into the intrinsic pyrolysis chemistry. *Industrial & Engineering Chemistry Research, 58,* 15838–15852. doi:10.1021/acs.iecr.9b00920

Antar, M., Lyu, D., Nazari, M., Shah, A., Zhou, X., & Smith, D. L. (2021). Biomass for a sustainable bioeconomy: An overview of world biomass production and utilization. *Renewable and Sustainable Energy Reviews, 139,* 110691. doi:10.1016/j.rser.2020.110691

Arni, S. A. (2018). Comparison of slow and fast pyrolysis for converting biomass into fuel. *Renewable Energy, 124,* 197–201. doi:10.1016/j.renene.2017.04.060

Baloch, H. A., Nizamuddin, S., Siddiqui, M. T. H., Riaz, S., Jatoi, A. S., Dumbre, D. K., Mubarak, N. M., Srinivasan, M. P., & Griffin, G. J. (2018). Recent advances in production and upgrading of bio-oil from biomass: A critical overview. *Journal of Environmental Chemical Engineering, 6,* 5101–5118. doi:10.1016/j.jece.2018.07.050

Beckman, J., Hertel, T., Taheripour, F., & Tyner, W. (2012). Structural change in the biofuels era. *European Review of Agricultural Economics, 39,* 137–156. doi:10.1093/erae/jbr041

Bermudez, J. M., & Fidalgo, B. (2016). Production of bio-syngas and bio-hydrogen via gasification. In R. Luque, C. S. K. Lin, K. Wilson, & J. Clark (Eds.), *Handbook of Biofuels Production* (2nd ed., pp. 431–494). Cambridge: Woodhead Publishing. doi:10.1016/B978-0-08-100455-5.00015-1

Bhatia, S. K., Bhatia, R. K., & Yang, Y. H. (2017). An overview of microdiesel: A sustainable future source of renewable energy. *Renewable and Sustainable Energy Reviews*, *79*, 1078–1090. doi:10.1016/j.rser.2017.05.138

Bhatia, S. K., Bhatia, R. K., Jeon, J. M., Pugazhendhi, A., Awasthi, M. K., Kumar, D., Kumar, G., & Yang, Y. H. (2021a). An overview on advancements in biobased transesterification methods for biodiesel production: Oil resources, extraction, biocatalysts, and process intensification technologies. *Fuel*, *285*, 119117. doi:10.1016/j.fuel.2020.119117

Bhatia, S. K., Palai, A. K., Kumar, A., Bhatia, R. K., Patel, A. K., Thakur, V. K., & Yang, Y H. (2021b). Trends in renewable energy production employing biomass-based biochar. *Bioresource Technology*, *340*, 125644. doi:10.1016/j.biortech.2021.125644

Binod, P., Gnansounou, E., Sindhu, R., & Pandey, A. (2019). Enzymes for second generation biofuels: Recent developments and future perspectives. *Bioresources Technology Reports*, *5*, 317–325. doi:10.1016/j.biteb.2018.06.005

Boerrigter, H., & Rauch, R. (2005). Syngas production and utilisation. In *Handbook Biomass Gasification*. BTG. Enschede: Biomass technology group.

Campoy, M., Gómez-Barea, A., Ollero, P., & Nilsson, S. (2014). Gasification of wastes in a pilot fluidized bed gasifier. *Fuel Processing Technology*, *121*, 63–69. doi:10.1016/j.fuproc.2013.12.019

Cantrell, K. B., Hunt, P. G., Uchimiya, M., Novak, J. M., & Ro, K. S. (2012). Impact of pyrolysis temperature and manure source on physicochemical characteristics of biochar. *Bioresource Technology*, *107,* 419–428. doi:10.1016/j.biortech.2011.11.084

Cao, D., Sun, Y., & Wang, G. (2007). Direct carbon fuel cell: Fundamentals and recent developments. *Journal of Power Sources*, *167*, 250–257. doi:10.1016/j.jpowsour.2007.02.034

Cao, W., Li, J., Martí-Rosselló, T., & Zhang, X. (2019). Experimental study on the ignition characteristics of cellulose, hemicellulose, lignin and their mixtures. *Journal of the Energy Institute*, *92*, 1303–1312. doi:10.1016/j.joei.2018.10.004

Cao, X., Sun, S., & Sun, R. (2017). Application of biochar-based catalysts in biomass upgrading: A review. *RSC Advances, 7*, 48793–48805. doi:10.1039/C7RA09307A

Cha, J. S., Park, S. H., Jung, S. C., Ryu, C., Jeon, J. K., Shin, M. C., & Park, Y. K. (2016). Production and utilization of biochar: A review. *Journal of Industrial and Engineering Chemistry*, *40*, 1–15. doi:10.1016/j.jiec.2016.06.002

Chen, P., Wang, L. K., Wang, G., Gao, M. R., Ge, J., Yuan, W. J., Shen, Y. H., Xie, A. J., & Yu, S. H. (2014). Nitrogen-doped nanoporous carbon nanosheets derived from plant biomass: An efficient catalyst for oxygen reduction reaction. *Energy & Environmental Science*, *7*, 4095–4103. doi:10.1039/C4EE02531H

Chen, Q., Tan, X., Liu, Y., Liu, S., Li, M., Gu, Y., Zhang, P., Ye, S., Yang, Z., & Yang, Y. (2020). Biomass-derived porous graphitic carbon materials for energy and environmental applications. *Journal of Materials Chemistry A*, *8*, 5773–5811. doi:10.1039/C9TA11618D

Cheng, B. H., Zeng, R. J., & Jiang, H. (2017). Recent developments of post-modification of biochar for electrochemical energy storage. *Bioresources Technology, 246*, 224–233. doi:10.1016/j.biortech.2017.07.060.

Cheng, F., & Li, X. (2018). Preparation and application of biochar-based catalysts for biofuel production. *Catalysts*, *8*, 346. doi:10.3390/catal8090346

Daful, A. G., & Chandraratne, M. R. (2018). Biochar production from biomass waste-derived material. In *Reference Module in Materials Science and Materials Engineering* (pp. 11249–11258). Amsterdam: Elsevier. doi:10.1016/B978-0-12-803581-8.11249-4

Dehkhoda, A. M., Gyenge, E., & Ellis, N. (2016). A novel method to tailor the porous structure of KOH-activated biochar and its application in capacitive deionization and energy storage. *Biomass and Bioenergy*, *87*, 107–121. doi:10.1016/j.biombioe.2016.02.023

Devi, M., Rawat, S., & Sharma, S. (2021). A comprehensive review of the pyrolysis process: From carbon nanomaterial synthesis to waste treatment, *Oxford Open Materials Science, 1*, itab014. doi:10.1093/oxfmat/itab014

Díez, N., Fuertes, A. B., & Sevilla, M. (2021). Molten salt strategies towards carbon materials for energy storage and conversion. *Energy Storage Materials, 38,* 50–69. doi:10.1016/j.ensm.2021.02.048

Ding, Y., Wang, T., Dong, D., & Zhang, Y. (2020). Using biochar and coal as the electrode material for supercapacitor applications. *Frontiers in Energy Research, 7,* 159. doi:10.3389/fenrg.2019.00159

dos Santos, R. G., & Alencar, A. C. (2020). Biomass-derived syngas production via gasification process and its catalytic conversion into fuels by Fischer Tropsch synthesis: A review. *International Journal of Hydrogen Energy, 45,* 18114–18132. doi:10.1016/j.ijhydene.2019.07.133

Egun, I. L., He, H., Hu, D., & Chen, G. Z. (2022). Molten Salt carbonization and activation of biomass to functional biocarbon. *Advanced Sustainable System, 6,* 2200294. doi:10.1002/adsu.202200294

Fernandez, M. E., Ledesma, B., Román, S., Bonelli, P. R., & Cukierman, A. L. (2015). Development and characterization of activated hydrochars from orange peels as potential adsorbents for emerging organic contaminants. *Bioresource Technology, 183,* 221–228. doi:10.1016/j.biortech.2015.02.035.

Foroutan, R., Mohammadi, R., Razeghi, J., & Ramavandi, B. (2021). Biodiesel production from edible oils using algal biochar/CaO/K2CO3 as a heterogeneous and recyclable catalyst. *Renewable Energy, 168,* 1207–1216. doi:10.1016/j.renene.2020.12.094

Gabhane, J. W., Bhange, V. P., Patil, P. D. Bankar, S. T., & Kumar, S. (2020). Recent trends in biochar production methods and its application as a soil health conditioner: A review. *SN Applied Sciences, 2,* 1307. doi:10.1007/s42452-020-3121-5

Gao, S., Liu, H., Geng, K., & Wei, X. (2015). Honeysuckles-derived porous nitrogen, sulfur, dual-doped carbon as high-performance metal-free oxygen electroreduction catalyst. *Nano Energy, 12,* 785–793. doi:10.1016/j.nanoen.2015.02.004

Gao, Y., Wang, X., Wang, J., Li, X., Cheng, J., Yang, H., & Chen, H. (2013). Effect of residence time on chemical and structural properties of hydrochar obtained by hydrothermal carbonization of water hyacinth. *Energy, 58,* 376–383. doi:10.1016/j.energy.2013.06.023

Gao, Z., Wang, J., Li, Z., Yang, W., Wang, B., Hou, M., He, Y., Liu, Q., Mann, T., Yang, P., Zhang, M., & Liu, L. (2011). Graphene nanosheet/Ni^{2+}/Al^{3+} layered double-hydroxide composite as a novel electrode for a supercapacitor. *Chemistry of Materials, 23,* 3509–3516. doi:10.1021/cm200975x

Goyal, H. B., Seal, D., & Saxena, R. C. (2008). Bio-fuels from thermochemical conversion of renewable resources: A review. *Renewable and Sustainable Energy Reviews, 12,* 504–517. doi:10.1016/j.rser.2006.07.014

Güney, T. (2019). Renewable energy, non-renewable energy, and sustainable development. *International Journal of Sustainable Development & World Ecology, 26,* 389–397. doi: 10.1080/13504509.2019.1595214

Gurav, R., Bhatia, S. K., Choi, T. R., Choi, Y. K., Kim, H. J., Song, H. S., Lee, S. M., Park, S. L., Lee, H. S., Koh, J., Jeon, J. M., Yoon, J. J., & Yang, Y. H. (2021). Application of macroalgal biomass derived biochar and bioelectrochemical system with Shewanella for the adsorptive removal and biodegradation of toxic azo dye. *Chemosphere, 264,* 128539. doi:10.1016/j.chemosphere.2020.128539

Guruviah, K. D., Sivasankaran, C., & Bharathiraja, B. (2019). Thermochemical conversion: Bio-oil and syngas production. In A. A. Rastegari, A. N. Yadav, & A. Gupta (Eds.), *Prospects of Renewable Bioprocessing in Future Energy Systems* (pp. 251–267). Cham: Springer. doi:10.1007/978-3-030-14463-0_9

Hajilary, N., Rezakazemi, M., & Shahi, A. (2020). CO2 emission reduction by zero flaring startup in gas refinery. *Materials Science for Energy Technologies, 3*, 218–224. doi:10.1016/j.mset.2019.10.013

Harun, K., Adhikari, S., & Jahromi, H. (2020). Hydrogen production via thermocatalytic decomposition of methane using carbon-based catalysts. *RSC Advances, 10*, 40882–40893. doi:10.1039/D0RA07440C

Heinze, T. (2016). Cellulose: Structure and properties. In O. J. Rojas (Ed.), *Cellulose Chemistry and Properties: Fibers, Nanocelluloses and Advanced Materials* (pp. 1–52). Cham: Springer. doi:10.1007/978-3-319-26015-0

Hemalatha, M., Sravan, J. S., Min, B., & Mohan, S. V. (2020). Concomitant use of Azolla derived bioelectrode as anode and hydrolysate as substrate for microbial fuel cell and electro-fermentation applications. *Science of The Total Environment, 707*, 135851. doi:10.1016/j.scitotenv.2019.135851

Huggins, T., Wang, H., Kearns, J., Jenkins, P., & Ren, Z. J. (2014). Biochar as a sustainable electrode material for electricity production in microbial fuel cells. *Bioresource Technology, 157*, 114–119. doi:10.1016/j.biortech.2014.01.058

Ismail, I. S., Othman, M. F. H., Rashidi, N. A., & Yusup, S. (2023). Recent progress on production technologies of food waste-based biochar and its fabrication method as electrode materials in energy storage application. *Biomass Conversion and Biorefinery, 7*, 1–17. doi:10.1007/s13399-023-03763-3

Janicek, A., Gao, N., Fan, Y., & Liu, H. (2015). High performance activated carbon/carbon cloth cathodes for microbial fuel cells. *Fuel Cells, 15*, 855–861. doi:10.1002/fuce.201500120

Jien, S. H. (2019). Physical characteristics of biochars and their effects on soil physical properties. In Y. S. Ok, D. C. W. Tsang, & J. M. Novak (Eds.), *Biochar from Biomass and Waste* (pp. 21–35). Amsterdam: Elsevier. doi:10.1016/C2016-0-01974-5

Koyama, K. (2017). The role and future of fossil fuel. *IEEJ Energy Journal,* Special Issue, 80–83.

Krerkkaiwan, S., Mueangta, S., Thammarat, P., Jaisat, L., & Kuchonthara, P. (2015). Catalytic biomass-derived tar decomposition using char from the co-pyrolysis of coal and giant leucaena wood biomass. *Energy & Fuels, 29*, 3119–3126. doi:10.1021/ef502792x

Kumar, A., Bhattacharya, T., Hasnain, S. M., Nayak, A. K., & Hasnain, M. S. (2020). Applications of biomass-derived materials for energy production, conversion, and storage. *Materials Science for Energy Technologies, 3*, 905–920. doi:10.1016/j.mset.2020.10.012

Kumar, S., Soomro, S. A., Harijan, K., Uqaili, M. A., Kumar, L. (2023). Advancements of biochar-based catalyst for improved production of biodiesel: A comprehensive review. *Energies, 16*, 644. https:// doi.org/10.3390/en16020644

Lee, S. Y., Sankaran, R., Chew, K. W., Tan, C. H., Krishnamoorthy, R., Chu, D. T., & Show, P. L. (2019). Waste to bioenergy: A review on the recent conversion technologies. *BMC Energy, 1*, 1–22. doi:10.1186/s42500-019-0004-7

Lehmann, J., Gaunt, J., & Rondon, M. (2006). Bio-char sequestration in terrestrial ecosystems: A review. *Mitigation and Adaptation Strategies for Global Change, 11*, 403–427. doi:10.1007/s11027-005-9006-5

Li, H., Yuan, D., Tang, C., Wang, S., Sun, J., Li, Z., Tang, T., Wang, F., Gong, H., & He, C. (2016). Lignin-derived interconnected hierarchical porous carbon monolith with large areal/volumetric capacitances for supercapacitor. *Carbon, 100*, 151–157. doi:10.1016/j.carbon.2015.12.075

Li, Y., Li, C., Qi, H., Yu, K., & Li, X. (2018). Formation mechanism and characterization of porous biomass carbon for excellent performance lithium-ion batteries. *RSC Advances, 8*, 12666–12671. doi:10.1039/C8RA02002G

Liu, W. J., Jiang, H., & Yu, H. Q. (2019). Emerging applications of biochar-based materials for energy storage and conversion. *Energy & Environmental Science, 12,* 1751–1779. doi:10.1039/C9EE00206E

Lu, B., Hu, L., Yin, H., Xiao, W., & Wang, D. (2016). One-step molten salt carbonization (MSC) of firwood biomass for capacitive carbon. *RSC Advances, 6,* 106485–106490. doi:10.1039/C6RA22191B

Luo, S., Sun, W., Ke, J., Wang, Y., Liu, S., Hong, X., Li, Y., Chen, Y., Xie, W., & Zheng, C. (2018). A 3D conductive network of porous carbon nanoparticles interconnected with carbon nanotubes as the sulfur host for long cycle life lithium-sulfur batteries. *Nanoscale, 10,* 22601–22611.doi:10.1039/C8NR06109B

Maguire, R. O., & Agblevor, F. A. (2010). *Biochar in Agricultural Systems.* Blacksburg, VA: Laboratory Procedures, Virginia Tech Soil Testing Laboratory, Virginia Cooperative Extension. https://pubs.ext.vt.edu/452/452-881/452-881_pdf.pdf.

Minowa, T., Murakami, M., Dote, Y., Ogi, T., & Yokoyama, S. Y. (1995). Oil production from garbage by thermochemical liquefaction. *Biomass and Bioenergy, 8,* 117–120. doi:10.1 016/0961-9534(95)00017-2

Nanda, S., Dalai, A. K., Berruti, F., & Kozinski, J. A. (2016). Biochar as an exceptional biore-source for energy, agronomy, carbon sequestration, activated carbon and specialty mate-rials. *Waste and Biomass Valorization, 7,* 201–235. doi:10.1007/s12649-015-9459-z

Ndriangu S. M., Li,u Y., Xu, K., & Song, S. (2019). Risk evaluation of pyrolyzed biocharfrom multiple wastes. *Journal of Chemistry,* 4506314. doi:10.1155/2019/4506314

Panahi, H. K. S., Dehhaghi, M., Aghbashlo, M., Karimi, K., & Tabatabaei, M. (2019). Shifting fuel feedstock from oil wells to sea: Iran outlook and potential for biofuel production from brown macroalgae (*ochrophyta; phaeophyceae*). *Renewable and Sustainable Energy Reviews, 112,* 626–642. doi:10.1016/j.rser.2019.06.023

Paz-Ferreiro, J., Nieto, A., Méndez, A., Askeland, M. P. J., & Gascó, G. (2018). Biochar from biosolids pyrolysis: A review. *International Journal of Environmental Research and Public Health, 15,* 956. doi:10.3390/ijerph15050956

Quah, R. V., Tan, Y. H., Mubarak, N. M., Khalid, M., Abdullah, E. C., & Nolasco-Hipolito, C. (2019). An overview of biodiesel production using recyclable biomass and non-biomass derived magnetic catalysts. *Journal of Environmental Chemical Engineering, 7,* 103219. doi:10.1016/j.jece.2019.103219

Rostrup-Nielsen, J. R. (2001). Conversion of hydrocarbons and alcohols for fuel cells. *Physical Chemistry Chemical Physics, 3,* 283–288. doi:10.1039/B004660O

Sakhiya, A. K., Anand, A., & Kaushal, P. (2020). Production, activation, and applications of biochar in recent times. *Biochar, 2,* 253–285. doi:10.1007/s42773-020-00047-1

Scheller, H. V., & Ulvskov, P. (2010). Hemicelluloses. *Annual Review of Plant Biology, 61,* 263–289. doi:10.1146/annurev-arplant-042809-112315

Sharma, P., & Melkania, U. (2017). Biochar-enhanced hydrogen production from organic fraction of municipal solid waste using co-culture of *Enterobacter aerogenes* and *E. coli. International Journal of Hydrogen Energy, 42,* 18865–18874. doi:10.1016/j.ijhydene.2017.06.171

Titirici, M. M., Thomas, A., & Antonietti, M. (2007). Back in the black: Hydrothermal carbon-ization of plant material as an efficient chemical process to treat the CO_2 problem? *New Journal of Chemistry, 31,* 787. doi:10.1039/b616045j

Titirici, M. M., White, R. J., Falco, C., Sevilla, M. (2012). Black perspectives for a green future: Hydrothermal carbons for environment protection and energy storage. *Energy & Environmental Science, 5,* 6796–6822. doi:10.1039/C2EE21166A

Trache, D., Hussin, M. H., Haafiz, M. M., & Thakur, V. K. (2017). Recent progress in cel-lulose nanocrystals: Sources and production. *Nanoscale, 9,* 1763–1786. doi:10.1039/C6NR09494E

Tumuluru, J. S., Ghiasi, B., Soelberg, N. R., & Sokhansanj, S. (2021). Biomass torrefaction process, product properties, reactor types, and moving bed reactor design concepts. *Frontiers in Energy Research, 9*, 728140. doi:10.3389/fenrg.2021.728140

Waheed, A., Naqvi, S. R., & Ali, I. (2022). Co-torrefaction progress of biomass residue/waste obtained for high-value bio-solid products. *Energies, 15*, 8297. doi:10.3390/en15218297

Wang, Y., Yin, R., & Liu, R. (2014). Characterization of biochar from fast pyrolysis and its effect on chemical properties of the tea garden soil. *Journal of Analytical and Applied Pyrolysis, 110*, 375–381. doi:10.1016/j.jaap.2014.10.006

Wang, H., Yin, F. X., Chen, B. H., He, X. B., Lv, P. L., Ye, C. Y., & Liu, D. J. (2017). ZIF-67 incorporated with carbon derived from pomelo peels: A highly efficient bifunctional catalyst for oxygen reduction/evolution reactions. *Applied Catalysis B: Environmental, 205*, 55–67. doi:10.1016/j.apcatb.2016.12.016

Wang, T. (2019). *Global Biofuel Production by Select Country 2018*. https://www.statista.com/statistics/274168/biofuel-production-inleading-countries-in-oil-equivalent/ (Accessed 4/4/2023).

Wei, J., Tu, C., Yuan, G., Liu, Y., Bi, D., Xiao, L., Lu, J. B. K. G., Theng, B. K. G., Wang, H., Zhang, L., & Zhang, X. (2019). Assessing the effect of pyrolysis temperature on the molecular properties and copper sorption capacity of a halophyte biochar. *Environmental Pollution, 251*, 56–65. doi:10.1016/j.envpol.2019.04.128

Wender, I. (1996). Reactions of synthesis gas. *Fuel Processing Technology, 48*, 189–297. doi:10.1016/S0378-3820(96)01048-X

Xiang, Y., & Li, Z. (2018). Lignin-based functional nanocomposites. In X. J. Loh, D. Kai, & Z. Li (Eds.), *Functional Materials from Lignin* (pp. 49–80). doi:10.1142/q0153

Xiu, S., & Shahbazi, A. (2012). Bio-oil production and upgrading research: A review. *Renewable and Sustainable Energy Reviews, 16*, 4406–4414. doi:10.1016/j.rser.2012.04.028

Yaashikaa, P. R., Kumar, P. S., Varjani, S., Saravanan, A. (2020). A critical review on the biochar production techniques, characterization, stability and applications for circular bioeconomy. *Biotechnology Reports, 28*, e00570. doi:10.1016/j.btre.2020.e00570

Yang, D. P., Li, Z., Liu, M., Zhang, X., Chen, Y., Xue, H., Ye, E., & Luque, R. (2019). Biomass-derived carbonaceous materials: Recent progress in synthetic approaches, advantages, and applications. *ACS Sustainable Chemistry & Engineering, 7*, 4564–4585. doi:10.1021/acssuschemeng.8b06030

Yang, Z., Yang, R., Dong, G., Xiang, M., Hui, J., Ou, J., & Qin, H. (2021). Biochar nanocomposite derived from watermelon peels for electrocatalytic hydrogen production. *ACS Omega, 6*(3), 2066–2073. doi:10.1021/acsomega.0c05018

Yargicoglu, E. N., Sadasivam, B. Y., Reddy, K. R., & Spokas, K. (2015). Physical and chemical characterization of waste wood derived biochars. *Waste Management, 36*, 256–268. doi:10.1016/j.wasman.2014.10.029

Yu, K. L., Lau, B. F., Show, P. L., Ong, H. C., Ling, T. C., Chen, W. H., Ng, E. P., & Chang, J. S. (2017). Recent developments on algal biochar production and characterization. *Bioresource Technology, 246*, 2–11. doi:10.1016/j.biortech.2017.08.009

Zaman, C. W., Pal, K., Yehye, W. A., Sagadevan, S., Shah, S. T., Adebisi, G. A., Marliana, E., Rafique R. F., & Johan, R. B. (2017). A sustainable way to generate energy from waste. *Pyrolysis*. doi:10:10.5772/intechopen.69036

Zheng, X., Chen, C., Ying, Z., & Wang, B. (2016). Experimental study on gasification performance of bamboo and PE from municipal solid waste in a bench-scale fixed bed reactor. *Energy Conversion and Management, 117*, 393–399. doi:10.1016/j.enconman.2016.03.044

Zhong, K., Li, M., Yang, Y., Zhang, H., Zhang, B., Tang, J., Yan, J., Su, M., & Yang, Z. (2019). Nitrogen-doped biochar derived from watermelon rind as oxygen reducti on catalyst in air cathode microbial fuel cells. *Applied Energy, 242*, 516–525. doi:10.1016/j.apenergy.2019.03.050

8 Recent Advances and Challenges in Biomass Research

Ankush Sharma and Shiksha Arora
Indian Institute of Maize Research, ICAR

Loveleen Kaur Sarao
Punjab Agricultural University

8.1 INTRODUCTION

Energy is a fundamental prerequisite for the growth of practically every part of a culture in the world. Ecosystems, life itself, and human civilizations all depend on energy (Guo et al., 2015; Yuan et al., 2008). The use of conventional energy sources, however, might result in a number of issues. First off, conventional energy (i.e. fossil fuel) is nonrenewable and its overuse will result in a significant energy problem, which is currently a major issue for the entire globe. As carbon dioxide and other greenhouse gases (GHGs) rise as a result of using traditional fossil fuels, this can also be a polluting source that hastens global warming. Third, the nitrogen oxides released as a result of the combustion of fossil fuels affect human health (Cowie et al., 2018; Jåstad et al., 2020). Unfortunately, 80% of the world's energy comes from fossil fuels, and in the next 20 years, that percentage will rise to more than 50%. As a result, bio-energy, a potent renewable fuel that can replace fossil fuels, has grown over the past few decades, particularly in North America and Europe, with the goals of addressing the rise in global population, ensuring energy security, and reducing global warming (Alexander et al., 2017).

Biomass-based energy generation is one of the major focus areas of renewable energy programs in India. The strength of India's biomass resources mostly lies in the agricultural sector. A large quantity of crop residue biomass is generated in India. Biomass resources are relatively uniformly available in India compared to other renewable sources. Realizing the potential of bio-energy generation, the Ministry of New and Renewable Energy (MNRE), India, has initiated several biomass programs, with encouraging degree of success (Dash & Bingfa, 2018).

Biomass consists mainly of carbon dioxide and lignin, which are generated by the photosynthesis process, thus helping in easy absorption of solar energy from plant crops. Biofuels are used to define the biomass that is produced in a more comfortable type for domestic applications such as biogas. It is widely used for domestic usage, but it is also included for the production of crude oil and strong fuel-like

DOI: 10.1201/9781003406501-8

pellets of timber. Although many kinds of biomass can be transformed directly into heat or other forms of energy, certain kinds of biomass are only transformed into an advanced organic fuel in a more effective manner. These materials may have several desirable characteristics such as improved retention, ease of operation, improved functionality or higher energy density (Rajmohan et al., 2021).

A variety of biomasses have been tested for bio-energy generation including woody biomass and loose biomass such as rice husk, cashew nut shell, areca nut, and sugarcane residue. Village-level decentralized biomass power generation of kilowatt scale has also been commissioned in the country. Scientists around the world have paid much attention to the balance between bio-energy production and environmental protection by considering multiple approaches, including the best management practices (BMPs) (Kang et al., 2014).

Today, biogas technology for renewable bio-energy production continues to be of great interest among developing nations (particularly in rural geographies), but interest is also growing in many developed countries. The USA and Western Europe have placed an emphasis on large-scale commercial facilities to utilize an assortment of substrates which include municipal waste, agricultural residues, animal manure, and food waste, among others. The approach, however, requires large capital investments and is much too costly for decentralized communities by orders of magnitude. In recent years, there have been some research on the implementation of household (small-to-medium scale) digesters for remote rural regions, but in addition to all of the technical challenges faced by the USA and Europe, a number of other considerations (i.e. limitation to cheap construction materials, difficulties in the transportation of resources, and simplicity of operation for people with little or no formal education) have slowed research and development (Dung et al., 2014; Sawatdeenarunat et al., 2016).

Based on the feedstocks, biofuels are categorized as first-generation, second-generation, third-generation, and fourth-generation biofuels (see Figure 8.1). Biofuels derived from crop plants such as jatropha, almond, barley, camelina, coconut, copra, fish oil, groundnut, laurel, oat, poppy seed, okra seed, rice bran, sesame, sunflower, sorghum,

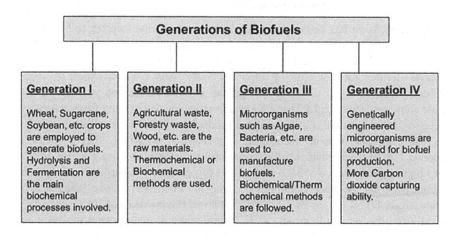

FIGURE 8.1 Different biofuel generations.

wheat soybean, rapeseed, and karanja are termed as first-generation feedstocks. Biofuel production from such feedstocks faces criticism due to the food versus fuel dilemma. The value of food waste has not been paid much attention due to the lack of application policies and suitable management laws. Materials are also used to produce biofuels, known as second-generation feedstocks. However, these feedstocks do not have a stable supply to fulfill future energy needs. Alternatively, microorganisms, called third-generation feedstocks, can be used. A wide variety of microorganisms have been identified which serve as a sink for CO_2 and produce biofuel. Genetically engineered microorganisms and algae constitute fourth-generation feedstock material. Optimization of processes for effective and efficient biomass conversion for all generations of biofuels and for true integration in a biorefinery will require appropriate separation technologies. Compared to other separation technologies, the low energy consumption, greater separation efficiency, reduced number of processing steps, and high quality of the final product are the main attractions of membrane separation processes in biorefining and bio-energy production (He et al., 2012).

Biomass consists mainly of carbon dioxide and lignin, which are generated by the photosynthesis process, thus helping in easy absorption of solar energy from plant crops. Biofuels are used to define the biomass that is produced in a more comfortable type for domestic applications such as biogas. It is widely used for domestic usage, but it is also included for the production of crude oil and strong fuel-like pellets of timber. Although many kinds of biomass can be transformed directly into heat or other forms of energy, certain kinds of biomass are only transformed into an advanced organic fuel in a more effective manner. These materials may have several desirable characteristics such as improved retention, ease of operation, improved functionality, or higher energy density (Chen et al., 2020).

8.2 ADVANCES IN BIOMASS ENERGY RESEARCH

8.2.1 GROWTH IN THE PRODUCTION OF ENERGY USING FIRST- AND SECOND-GENERATION FEEDSTOCKS

Traditionally, yeast has been used in anaerobic fermentation to produce bioethanol. However, using commercially available New Aule Alcohol yeast and New Aule Baker's yeast, Jayus et al. (2016) reported the impact of aeration during sugarcane molasses fermentation on ethanol output. In comparison to alcohol yeast (0.4 ggL), Baker's yeast (0.7 ggL) demonstrated greater ethanol production per unit of substrate consumed (yp/s, ggL). Using *Saccharomyces* species isolated from grapes and molasses, Muruaga et al. (2016) reported producing ethanol from sugarcane molasses. Using molasses with an initial sugar concentration of 250 g/L, they were able to produce 128.7 g/L of ethanol, which translates to an yp/s of 0.6 g/g/L. Because of their high sugar contents—53.0% and 60.0%, respectively—sugar beet molasses and viscous juice are the other viable raw materials for ethanol synthesis. Using *S. cerevisiae* (strain DTN) in free and immobilized form, Razmovski and Vučurović (2012) investigated the very high gravity (VHG) fermentation of sugar beet molasses and viscous juice. With an initial sugar concentration of 300 g/L, the maximal ethanol concentrations produced by the immobilized yeast during VHG fermentation were 83.2 and 132.4 g/L from thick juice and sugar beet molasses, respectively. Together with beet molasses, the by-products (raw juice, thin juice, and thick juice) produced during the processing of sugar beets were used as feedstock for the manufacturing of bioethanol. According to reports, there was little variation in the volume of ethanol produced (v/v), but there was a large variation in the length of the fermentation process. Whereas molasses' optimal fermentation time was 50 h, that of intermediates was 36 h. In addition, it was stated that aqueous sugars recovered from 1 kg of sugar beet can provide 0.07 kg of ethanol. Another factor that increased ethanol yields was the use of immobilized yeast in various modes of fermentation, such as batch fermentation inside a bioreactor and continuous fermentation in a fluidized-bed device (Muktham et al., 2016).

The following are significant recent advancements and advances in biomass-derived energy research that are discussed.

8.2.2 Evolution in Feedstocks

The resulting ethanol yield in the manufacturing of ethanol from maize grain depends on the type and grade of corn grain used. The efficacy of starch saccharification is lower in maize samples with greater starch content, according to research on 258 corn cultivars used to produce bioethanol. The production of ethanol is impacted by various factors such as grain quality, corn quality, including kernel composition, endosperm hardness, planting site, and the presence of mycotoxins, had an impact on. Also, the abundance of free sugar in maize kernels may reduce the need for enzymes during saccharification, leading to increased ethanol yields. The synthesis of ethanol from maize of the high sugary corn genotype, high stay green (HSG), and its parent field corn lines, penetration frictional component (PFC), demonstrated that HSG corn required 1.5 kg/tonne of dry corn for enzyme requirements, but PFC corn required 2 kg/tonne. So, it is clear that corn grain starch concentration is not the only variable impacting ethanol productivity. Another important cereal for grain distilleries and ethanol production is wheat, which took the position of barley 30 years ago. The starch content of flour is increased by dry milling wheat to remove the bran from the grain, which raises the ethanol titer. Whereas Thomas et al. (1993) reported a maximum ethanol titer of 23.8% (v/v) under VHG conditions, Sosulski and Sosulski (1994) reported ethanol production utilizing wheat flour from dry milling with a maximum ethanol percentage of 15%–5.89% (v/v) at 344–367 L/tonne of wheat flour. Furthermore, according to a recent study, using 1 MJ of bioethanol made from wheat instead of 1 MJ of gasoline can cut emissions of GHGs by 42.5%–61.2%. Another method for producing bioethanol from cassava flour at the lab and pilot scale level was reported and used simultaneous saccharification and fermentation (SSF) with 315 g/L of slurry concentration (Muktham et al., 2016).

8.2.3 Harnessing Dairy Sector in Biomass Energy Production

8.2.3.1 Hydrogen Production via Dairy Biomass

A promising method is the production of biological hydrogen from waste utilizing dark fermentation because it doesn't produce any pollutants while running, uses less energy, and can employ organic pollutants as sustainable energy feedstocks (Marone et al., 2015). A diversity of acceptable substrates and commensurate high biohydrogen generation are two benefits of fermentative biohydrogen production. The ability to combine biomass usage with biohydrogen synthesis while simultaneously treating waste substrates is a crucial benefit of fermentative biohydrogen generation. The implications of dairy manure (DM) preparation to optimize biohydrogen output have not been thoroughly investigated. For instance, Concetti et al. demonstrated that a batch fermentation procedure using sterilized pretreatment at a mesophilic temperature of 37°C enhanced biohydrogen output from 10 to 38 mL H_2/g VS of substrate (Concetti et al., 2013). According to Yokoyama et al., 2007, batch reactors running at a pH of 8.2 result in a maximum biohydrogen production of 0.7 mL H_2/gL VS of substrate from non-pretreated cow slurry. Also, this method utilized the slurry's native microflora at a mesophilic setting of 37°C (Yokoyama et al., 2007). The greatest biohydrogen output from sterilized buffalo manure was only 6.6 mL H_2/gL VS of substrate under the same working conditions. However, employing acid-pretreated DM of 50 g VSL1 of substrate in batch reactors at pH 7, 37°C, a maximum biohydrogen production yield of 18 m H_2/gL VS of substrate was achieved (Zhu et al., 2021)

There have been reports of biohydrogen generation yields of 160–170 L H_2/kgL TOC of cheese whey at pH 6.0–6.5. The process is rigorously controlled by a number of variables, including the inoculum, reactor type, substrate/feed composition, organic loading rate, residence time, reaction temperature, pH, and pretreatment methods used. Because pure strains like *Escherichia coli* and *Clostridium saccharoperbutyacetonicum* provided 2.7 mol H_2 mol/L lactose (158 L H_2 kg/L COD lactose), it is intriguing to note that a mixed consortium of both dairy sludge bacteria and pure variant can be utilized for the best dark fermentation yield. The best substrate for fermentation is raw feed, or untreated dairy whey, because all of the nutrients in the dairy waste will be utilized by the microbial species (Li et al., 2021). With yield amounts of 3 mol H_2 mol/L lactose and acetate and iso-butyrate concentrations surpassing 5 g/L, co-fermenting dairy bovine dung with cheese whey with plenty of protein and ash accessible from both the waste streams appears to be most beneficial (Awasthi et al., 2022).

8.2.3.2 Bio-Oil from Dairy Manure

Due to its abundant availability, cheap cost, and rich lignocellulosic composition, DM is a significant second-generation biomass resource that can also be used to make bio-oil. For turning DM into bio-oil, hydrothermal liquefaction (HTL) has been viewed as a potential approach. Midgett et al. (2012) analyzed the effectiveness of applying HTL technology to convert various readily accessible bio-waste and biomass resources to bio-oil and studied the increasing efficacy of three catalysts on the production of bio-oil from DM. The oil produced from DM was separated, and the final products included post-processed water (PPW), water-soluble fraction (WSF), acetone-soluble fraction (ASF), and char at various temperatures. These findings suggested that DM might be recovered with a high energy yield (70.8%). Crucially, this strategy makes it possible to convert garbage with low energy content into high energy bio-oil (Midgett et al., 2012). This research also demonstrated that HTL does not require a significant oil concentration for efficient conversion (Zhu et al., 2021).

8.2.3.3 Bioethanol Production from Dairy Manure

Vancov et al. also looked into the ethanol synthesis from feedlot DM. Compost was incubated using 2.5% H_2SO_4 at 121°C, over 90 min, then CTec 2 was used to saccharify it (Novozyme). After 6 h of fermentation, *Saccharomyces cerevisiae* successfully fermented manure hydrolysates to yield 7.3 g/L ethanol (Vancov et al., 2015). By using the manufacture of bioethanol from manure as an example, the life cycle assessment (LCA) demonstrates how the bioethanol manufacturing process reduces the need for manure disposal and the use of leftover raw materials. This practical method might balance the environmental impact of the bioethanol production process. Importantly, lowering the cost of nitrogen and biomass supplies will keep the cost of ethanol production low. After pretreating manure, diverse end-products (such as citrate, isocitrate, succinate, and malate) may additionally be created from fermentable sugars based on the Embden Meyerhof Parnas (EMP) pathway. *Z. mobilis* may also produce ethanol, butanol, and isopropanol based on the Entner–Doudoroff (ED) pathway. To date, many strains have been tested and modified for the production of desired products (such biofuels and chemical building blocks) from various feedstocks (Zhu et al., 2021).

8.2.4 IMPROVEMENTS IN THE USE OF MICROBES IN THE GENERATION OF BIOMASS ENERGY

8.2.4.1 Microbial Fuel Cells for Waste Management

Due to its simultaneous ability to produce bioelectricity and treat wastewater, microbial fuel cells (MFCs) are regarded as a flexible and extremely promising technology. Due to the quick metabolism of dissolved sugars, lactic acid, and proteins, algal biomass and lactose-tolerant species of bacteria are both ideal candidates for the generation of bioelectricity from dairy effluent with dramatic, simultaneous reductions in Chemical Oxygen Demand (COD) levels in the wastewater and manure (Li et al., 2021; Sivakumar, 2021).

By passing across the proton exchange membrane of the MFC, the electrons produced by the ETC mechanism of bacterial respiration finish the redox reaction at the cathode end of the cell. Research has shown that using a cathode material that is appropriate, like graphite in mesh, high-power density materials like felt, metal-coated (Pt-coated Ti metal), activated carbon, etc., have been produced in large quantities (Awasthi et al., 2022).

The absence of reliable voltage generation has been a major MFC flaw. Using organic waste in a batch approach does not result in stable microbial bioelectricity. One explanation is that batch mode operation causes the bacteria to quickly use organic substrate at the start of the process, which causes significant voltage generation initially and a progressive decline in electricity with time (Jain et al., 2022; Sarsaiya et al., 2019). A fed-batch method must be used because the batch process is unstable. A maximum of 621.13 w/m² was also attained in several trials with increased substrate loading and an ideal pH range of 6.7–7.0. While working on dairy wastewater in a single chamber anodic MFC with a residence time of 15 days, Shewanella algae (MTCC-10608) demonstrated a removal efficiency of COD of 92.21% and current density of 143.3 mA/m² at peak 286 h of incubation (Awasthi et al., 2022).

The use of agricultural waste and its effluents to generate bioelectricity has been demonstrated and can be viewed as an alternate site of renewable energy technology, supported by a variety of microbial communities that are typically found in the anodic solution of most MFCs. Since these types of microbial populations confer characteristics of electrogens for effective electron transfer, especially to support redox reactions, this ultimately characterizes the aforementioned agricultural waste as a source of bioelectricity. Wheat straw effluent, rice straw hydrolysate (without pretreatment), maize stover, and the use of glucose as a substrate are among the agricultural wastes that contribute most to the energy production in MFCs. However, rice straw was shown to have the highest percentage rate of COD removal in MFCs. With a high energy recovery and a high percentage of COD reduction, the processing of mustard tuber wastewater in the dual MFC has shown positive results (Pandit et al., 2021).

Similar to how using vegetable waste extract gives U-shaped MFCs a higher power output than dual-chambered MFCs, overall some compelling findings were made regarding the production of electricity using effluent from slaughterhouses and animal carcasses. Cattle dung and manure wash effluent in particular were thought to be ideal substrates with a successful high proportion of COD elimination for the production of bioelectricity (Pandit et al., 2021).

8.2.4.2　Biodiesel Production from Microalgae

Numerous studies have discussed the advantages of using microalgae as opposed to other raw materials for the production of biodiesel. Practically speaking, they are easily cultivable, develop with little to no attention, utilize water that people cannot use, and readily assimilate nutrients. Up to 250 times as much oil can be produced by microorganisms per acre as soybeans. In general, making biodiesel from microalgae is likely the only option to produce enough automotive fuel to meet the current demand for gasoline. Microalgae produce 7-to-31 times more oil than palm oil (Rahpeyma & Raheb, 2019). An experimental study showed that, in comparison to the usage of fossil fuels, a microalgal-biodiesel-based energy method reduced GHG emissions by 45.77%. By using the effects of acids, alkalis, catalysts, and nanocatalysts in thermochemical and hydrothermal processes, biodiesel was created from the cellular lipid content of microalgae with alcohol generating fatty acid esters. *Chlamydomonas reinhardtii*, a type of microalgae, produced 15 2% w/w dry-biomass biodiesel. Around 45.6% of total lipid and 85.8% of FAME (fatty acid methyl ester) were produced by *Chlorella protothecoides* (Wang et al., 2016). Other microalgae species, such as *Spirulina maxima*, *Spirulina* sp., and *Tetraselmis elliptica*, respectively, produced 86.10%, 99.32%, and 37% biodiesel. Around 95.1% of *Chlorella pyrenoidosa*'s lipids were converted to biodiesel by a transesterification process in the presence of graphene oxide, a solid acid catalyst (nanomaterial). Due to the influence of calcinated CaO, Euglena sanguine produced a lipid content of over 98%. A good amount of biodiesel can be produced by *Nannochlorum* sp., *N. oceanica*, *H. pluvialis*, *Anabaena*, *Chlorococcum* sp., and *B. sudetica* (Hossain & Mahlia, 2019).

8.2.4.3　Generation of Biomass Using Microalgae

It is crucial to have knowledge of how microorganisms behave while turning some elements into other beneficial systems. In addition, it means that a variety of materials can be employed to create practical products and by-products, enabling their responsible and controlled use. Algae are the third-generation biofuels among all the microorganisms. Algae range in size from tiny single-celled creatures to multicellular bacteria, some of which display an unusually complex morphology. Similar to crops, algae primarily require sunlight, carbon dioxide, and water for the creation of biomass (Packer, 2009). Microalgae are capable of surviving heterotrophically, and lipid droplets are frequently used to transport external carbon sources for the cells. Algae can effectively absorb atmospheric carbon dioxide and can be cultivated in greenhouses or other controlled environments. They are made up of components that can be employed as raw materials in the production of biodiesel. With physiological and biochemical adaptations for low-light conditions, such as densely packed thylakoids and sizable light-harvesting antenna complexes, microalgae may efficiently collect light energy. With more intense light, this could overwhelm the process of photosynthesis and cause the generation of reactive oxygen and harmful stress on cells (Rajmohan et al., 2021).

8.2.5 DEVELOPMENTS IN BIO-ENERGY CROPS

Bio-energy crops can be grown for two distinct markets: liquid transport fuels and power generation (electricity, heat, and combined heat and power) (Karp & Shield, 2008). Many options exist for bio-energy crops to protect the environment. Due to their perpetual nature, they are resilient to pests and diseases. The phenotypic, architectural, biochemical, and physiological properties of bio-energy plans have improved, which are favorable traits in the production of biofuels. Furthermore, bio-energy crop cultivars grow more quickly than other crops because they can withstand biotic and abiotic challenges. Bio-energy crops also require fewer biological, chemical, or physical pretreatments, which lowers the cost of processing biomass. Plant varieties with desirable morphological, phenotypic, and biochemical characteristics have been created with the help of traditional breeding techniques and genetic modification (Baenziger et al., 2006; Lee, 1998). Such initiatives primarily aim to increase agricultural quality and productivity. Moreover, food crops can be genetically altered to produce more starch and a higher C:N ratio in order to modify them for bio-energy production. By the production of cellulases and cellulosomes, such alteration could change the lignin manufacturing pathway for improved preprocessing. By recognizing natural differences and using genetic modification to create transgenic plants, bio-energy crop characteristics can be improved. Bio-energy crops that have undergone genetic modification are more tolerant of unfavorable conditions and have higher growth rates and caloric values. The significant degree of similarity identified among the grass or *Poplar* spp. genomes may make it easier to transfer gene function from these species to grass species that are more genetically resistant, such as switchgrass, miscanthus, and short rotation coppice. Due to its ease of propagation, rapid growth in short growing coppice cycles, and minimal fertilizer needs, willow has been deemed a promising biomass crop. Willow plants must be routinely kept clear from pests and illnesses to produce more fruit. Genetic engineering can increase the yields of willow without considerably raising the need for fertilizers and water (Yadav et al., 2019).

TABLE 8.1

Pretreatment Techniques Followed for Bio-Energy Production

Physical Pretreatment	Chemical Pretreatment	Biological Pretreatment	Physicochemical Pretreatment
Microwaving	Acid hydrolysis	Fungi	Irradiation
Milling	Alkaline hydrolysis	Bacteria	Steam explosion
Extrusion	Organic solvent	Enzymes	Plasma
Sonification	Ozonolysis	Termite treatment	Wet oxidation

The information currently available about the European Union (EU) model for the bioethanol from all parts of the sweet sorghum crop suggests an output/input ratio of 1.7–7.3 (depending on the method selected for the by-products exploitation), a significant reduction in GHGs in accordance with the RES Directive, and a low water footprint. The reduction of the transport sector's contribution to GHG emissions is the primary environmental benefit of using bioethanol in place of gasoline and/or as bio-ETBE (ethyl tert-butyl ether) in place of fossil brake fluid. This claim is assumed to be true for bioethanol made from sweet sorghum because, in this instance, the savings estimated using the approach suggested by the EU is 71%, one of the most morally upright figures among the feasible ones (Pathak, 2014).

8.2.6   Advances in Biomass Processing Techniques

Though there are many different techniques followed for pretreatment of biomass raw materials (shown in Table 8.1.), here we are looking at some of the latest techniques that are developed for effective manufacturing of bio-energy which includes the recent pretreatment methods and the new groundbreaking processing techniques and methods that are being embraced to achieve the aim of bio-energy manufacturing smoothly, precisely, and quickly.

8.2.6.1   Deep Eutectic Solvent (DES)

In order to create aromatic chemicals from lignocellulosic biomass, a "lignin-first" approach has recently been developed. DESs have a few benefits for the subsequent treatment, such as minimal toxicity and biocompatibility. Also, the hardly volatile components of DESs render them simple to recycle and reuse without affecting the effectiveness of the treatment. Recent studies have turned their attention to DES and demonstrated various innovative DES-based processes as well as DES pretreatment features (Ning et al., 2021).

8.2.6.2   Ionic Liquids

Ionic liquids (ILs) are thought of as green solvents because of their special solvation characteristics. ILs have low toxicity, good thermal stability, and low vapor pressure requirements. To supply pure cellulose for further hydrolysis, these ILs selectively eliminate the lignin and hemicellulose portion of the feedstock. With a large biomass input, the IL pretreatment process could be run more effectively in continuous mode (Brandt et al., 2011). Using the IL treatment, various studies have been able to remove 15%–92% of the lignin from biomass (Arora et al., 2010; Brandt et al., 2011). An IL combination of 1-butyl-3-methylimidazolium hydrogen sulfate [C_4Cl im], [H_2SO_4], 1-butyl-3-methylimidazolium chloride [C_4Cl im]Cl, 1-butyl-3-methylimidazolium acetate [C_4Cl im]MeCO$_2$], and 1,3-dimethylimidazolium methyl sulfate [C_4Cl im]. Biomass materials such as miscanthus, pine, willow, maple wood, switchgrass, and oak were pretreated using water. In order to employ the cellulose-enriched fraction for enzymatic hydrolysis, the lignin component was removed (15%–92%). After IL and enzyme hydrolysis, 25% of the hemicellulose and 90% of the glucose in the biomass were liberated. The pretreatment of rice straw, corncobs, barley straw, and wheat bran used typical acid and alkali techniques. For acid-pretreated biomass, the yield of reducing sugars ranged from 34% to 49%, while for alkali-treated biomass, it ranged from 46% to 65%. In a different investigation, pretreated maize stover with ILs produced larger yields of glucose and xylose than pretreated corn stover with AFEX (Bhatia et al., 2019).

8.2.6.3 Supercritical Fluids–Based Pretreatments

The peculiar characteristics of supercritical fluids (SCFs), which resemble both gases and liquids, include diffusivity, viscosities similar to those of gases, and densities similar to those of liquids. SCFs have a lower solvation power than other fluids and are more susceptible to minute variations in temperature and pressure. Due to its lower critical temperature (31°C), pressure (73.8 bar), and solubility (7.118 cal/cm^3) 0.5, CO_2 is classified as supercritical CO_2 (scCO_2). Compared to many other compounds, such as ammonia (132.3°C, 112.8 bar), methanol (240°C, 79.6 bar), and water, carbon dioxide does have a lower critical temperature and pressure (374.2°C, 221.2 bar). Enzymatic hydrolysis followed by subcritical CO_2 pretreatment of eucalyptus chips (at 180°C and 50 bar for 80 min) results in peak glucan conversions of 92% (Zhang & Wu, 2015). Wheat straw was exposed to pressure variations (80–120 bar) for 30 min at 190°C, and the outcome was a greater yield of glucose. In a different investigation, scCO_2 was applied to rice straw for 15–45 min at 10–30 MPa and 40°C–110°C. Rice straw that had been pretreated produced glucose yields of 32% as opposed to untreated rice straw's glucose yield of 27% (Bhatia et al., 2019).

8.2.6.4 Ultrasonic Pretreatment

Because of their intense energy and penetrability, ultrasonic vibrations have the ability to displace the lignocellulose crystal structure. According to the study, normally increasing ultrasonic frequency and time could lead to greater pretreatment efficiency; however, the pretreatment effectiveness will not rise any further when the frequency is larger than 100 kHz. Ultrasonic pretreatment is often carried out at 20–80 kHz for 20–150 min, which might reduce grass and bagasse's degree of polymerization and cellulose crystallinity by 1.5%–29.2%. Together with chemical preparation, ultrasonic pretreatment is frequently used (Zhao et al., 2022).

8.2.6.5   Microwave Radiation Pretreatment

A novel form of heat pretreatment known as microwave radiation has the potential to destabilize the lignocellulose structure and liberate the intracellular cellulose. Some suggest that the walnut shells were processed with 550–600 W of microwave power for 25 min, yielding a maximal hemicellulose conversion efficiency of 96.4% and a xylose yield of 12.40 g/L (Zhao et al., 2022).

8.2.6.6   Alkali Pretreatment

The alkaline pretreatment's high pH is able to dissolve more lignin and some hemicellulose, reduce cellulose crystallinity, and promote the subsequent enzymatic hydrolysis and residual solids' fermentation. For lignocellulose with high lignin content, an alkaline pretreatment is typically utilized. Common pretreatment agents include NaOH, KOH, NH_3–H_2O, $Ca(OH)_2$, etc. The normal operating parameters for alkaline pretreatment involve an alkali concentration range of 0%–2% at 50°C–100°C for a number of hours, or a concentration range of 2%–7% under 100–200°C for a brief contact period (10–90 min). Moreover, the alkali pretreatment exhibits adaptable performance at even low temperatures and can be used for a wider range of temperatures (Zhao et al., 2022).

8.2.6.7   Termite Pretreatment

One of the most organic lignocellulose-decomposition creatures on the globe is the termite. There are several different lignocellulose-decomposing bacteria that can release extremely potent lignocellulose-degrading enzymes, according to research on the gut microflora of termites. Moreover, the termites may drastically reduce the size of wood fibers and disassemble the structure of poplar wood at 27°C for 45 days, with a final lignin degradation of 60% (Zhao et al., 2022).

8.2.6.8   Transesterification

For the purpose of generating biodiesel, biomass such as microalgal oils have been transesterified using both heterogeneous and homogeneous catalysis. Relatively homogenous alkaline catalysis remains the most often used technique for producing biodiesel because it catalyzes the process at low temperatures and atmospheric pressure and can generate a significant conversion yield quickly. Alkaline catalysts, such as sodium hydroxide (NaOH) and potassium hydroxide (KOH), are widely used; however, due to the high concentration of free fatty acids in microalgal oils, alkaline catalysts end up causing the free fatty acids to produce soap and are therefore unsuitable for the production of microalgal biodiesel. Acid catalysts are used to get over the restriction of a high free fatty acid content as long as the content of free fatty acids is larger than 1%. The most popular acid catalysts are sulfuric acid (H_2SO_4) and hydrochloric acid (HCl). They need higher temperatures and slower response times than alkaline catalysts. At the beginning, some research used an acid catalyst to esterify free fatty acids into esters. The oils go through a further transesterification phase using an alkaline catalyst after the free fatty acid concentration has been reduced to less than 1%. Despite the remarkable conversion yields homogeneous catalysts can attain, catalyst loss still happens after the reaction. Due to their advantages with regard to recovery and reuse, heterogeneous catalysts are recognized to make a substantial contribution to the future in this area (Anekwe et al., 2022).

8.2.6.9   Boost in the Production of Beta-Glucosidase Enzyme

Including both the creation and breakdown of the glycosidic bond, beta-glucosidase is a dual-character enzyme that has a tremendous amount of promise for use in industry. The manufacture of cellulosic biofuel involves the conversion of lignocellulosic biomass into sugar, followed by the fermentation process, which produces the biofuel. Cellobiose and cello-oligosaccharide are ultimately converted into a monomeric unit of glucose by the important enzyme known as BGL. The rate-limiting step in the technique for producing biofuels, however, is the inadequate BGL production (Srivastava et al., 2019).

Due to its dual nature, which includes both the formation and breakdown of the glycosidic linkages, beta-glucosidase has a tremendous amount of promise from an industrial standpoint. The conversion of lignocellulosic biomass into sugar is the first step in the manufacturing of cellulosic biofuels, which is then followed by the fermentation process that produces the biofuel. The primary enzyme, BGL, is responsible for converting cellobiose and cello-oligosaccharide into a monomeric form of glucose. However, due to the insufficient BGL generation, it becomes a rate-limiting step in the technology used to produce biofuel (Srivastava et al., 2019)

8.2.6.10   Biorefining through Synthetic Biology and System Biology

The significant advancement of synthetic and systems biology technology in recent years has given researchers studying lignocellulose biorefinery new insights and resources. In order to find cellulase-genes, metagenomic, transcriptome, and metagenomics technologies have recently been created that bypass microbial pure-breeding and directly read the genome, transcriptome, and proteome data of the microbial community in the original environment. The investigation of new cellulase gene extraction, heterologous protein expression, purification, and degradation methods was later made possible thanks to research on the level of enzyme expression and its mechanism. A metagene library was created from the complete DNA and total RNA that were directly extracted from termite intestines by a variety of research organizations, including the US Department of Energy. Sequencing techniques make it clear that there are many different genes connected to cellulose and hemicellulose hydrolysis, which furthers understanding of the diversity of cellulase (Rai et al., 2022).

For better understanding of the several processing techniques, refer to Figure 8.2.

8.2.7   OTHER APPROACHES FOR ENHANCED BIOFUEL PRODUCTION

The production of biofuel might be enhanced by altering the cell walls' makeup. Through the use of contemporary molecular biology approaches, synthetic and systemic biology, microbial community-based approaches, metabolic engineering techniques, and nanotechnology-based approaches to increase plant cell wall digestibility, biofuel production from lignocellulosic biomass may be improved (Rai et al., 2022).

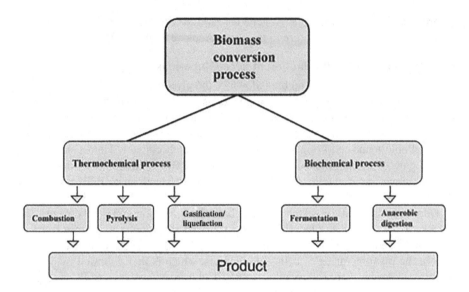

FIGURE 8.2   Processing techniques of biomass raw materials.

8.2.8 POLICIES AND CERTIFICATIONS

Specific regulations are implied to help the participants and make the process stable and long-lasting. Jaiv Indhan Vatavaran-anukool Fasal awashesh Nivaran Yojana is one such program in India. This program guarantees that entrepreneurs engaged in bio-energy production using LCB from the agricultural sector and other renewable feedstocks will get financial and fiscal support. India's cabinet proposed the National Biofuel Policy (NBP, 2018) with a target date of 2030.

The MNRE is another program devoted to installing and commercializing power and electricity generated from biomass. It provides farmers in rural regions with adequate training by holding several training workshops and seminars to increase their expertise and information in this field, as well as free installation, financial help, and a 5-year maintenance warranty on the bio-energy plant (Ministry of Drinking Water & Sanitation, 2018; Neu et al, 2002). The Government of India has set a goal of supplying 175 GW of renewable energy-based power to rural areas by 2022 with the purpose of concurrently reducing GHG emissions by 30%–35% till 2030 (Arora et al., 2023).

The New National Biogas and Organic Manure Program (NNBOMP), the program on energy from urban, industrial, and agriculture waste/residues, and the new National Policy on Biofuels were all introduced in 2018 in the area of bio-energy. In order to solve upcoming energy concerns, these projects aim to modernize the current policies and programs. The Indian government has also started a number of initiatives to increase biogas generation in the nation. In order to make it easier to set up off-grid biogas production units, the government recently announced two central programs: the NNBOMP, and the Biogas Power Generation and Thermal Energy Application Program (BPGTP) (Duarah et al., 2022).

The National Policy on Biofuels (2018) was anticipated to increase consumer affordability while reducing production costs. Realizing the ambition, goals, and strategy for creating biofuels involves a variety of institutional interventions as well as the technical framework, cash incentives, pricing, and pricing policies (Duarah et al., 2022).

8.3 CHALLENGES IN BIO-ENERGY PRODUCTION

8.3.1 PROCESSING OF RAW MATERIALS

The presence of lignin macromolecules makes the cellulose and hemicellulose microfibrils accessible for dissolution into the readily convertible glucan chains, which is why researchers are working hard to find a means to reduce the resistance against processing. Several lignin-containing biomasses have already been studied, but many more need to be explored in order to discover a solution for the problem of the commercial production of bio-energy being constrained. Pretreatment is an additional step that can be used to process biomass more effectively. This step is necessary in order to depolymerize the macromolecule and release the sugar molecules for use in the production of biomaterials and bio-energy in all lignocellulosic agricultural-based and woody biomass. Pretreatments are frequently combined; for instance, microwaves are employed to improve the hydrolysis of the biomass (Joshi and Mehmood, 2011; Prasad et al., 2007). The feedstock's prominence and consistency have a significant impact on the application of pretreatment's competence and proficiency. The storage of the feedstock presents another difficulty in the supply chain (Babin, 2021). Due to the moisture in the feedstock, water molecules increase the volume of the woody feedstocks, which increases the supply cost. Truckloads of biomass are consequently needed for distribution (Arora et al., 2023). The chemically processed biomasses produce fictitious lignin made of HMF and furfural that adheres to the cellulose surface and impairs the effectiveness of anaerobic digestion. Under chemical pretreatment, recovery of pretreatment compounds is challenging. The price of the required enzymes is a problem that largely affects biological pretreatment. This is so that they can be used in their highly purified forms. A better strategy would involve isolating microorganisms and encouraging them to create lignocellulosic enzymes. Despite the method's accomplishments, it is still regarded as the slowest type of pretreatment, and the issue of bacteria depleting the sugars produced in the bioreactor continues to be a problem (Khan & Ahring, 2021). Little sugars can now be converted into bioproducts and biofuels as a result. In addition to other problems associated with their use, such as water, solvent, and energy consumption, acid hydrotropes result in undesirable residual sugar contamination. As high concentrations of the solvents are required, it is advised that future reactors be built with the specific acidity potentials of the reaction mixture of this solvent in mind. Since they are made from petroleum, ILs are a hallmark of energy-intensive recovery procedures and are not sustainable, which raises questions about their cost and environmental toxicity (Ifeanyi–Nze & Omiyale, 2022).

8.3.2 LACK OF DRIVE AMONG LANDOWNERS AND FARMERS

Processing facilities need a consistent and reasonably priced supply of the raw material to maintain the production of biofuels and bio-energy. This would be dependent on the landowner or farmer from the separate urban and rural areas. These vendors must be encouraged and given support in order to assist the government in achieving national energy independence Although this field is still emerging, landowners and farmers should anticipate uncertainty and unfamiliarity. So, it is necessary to inform and educate individuals about the significance of the currently emerging market (Ahorsu et al., 2018).

8.3.3 BIODIVERSITY AT RISK

Forests are a significant source of woody raw material; thus, it is important to manage them and keep them safe from biotic and abiotic hazards like wildfire. By outperforming the disturbance-causing administrations and natural stand dynamics, forest management and sustainability work hand in hand with ecological improvement. Forests offer wood and lignocellulosic biomass in addition to biodegradable materials, waste, and by-products that are used in the production of bio-energy. This forest biomass extraction would have negative consequences on the natural ecology and the habitat of other living things (Arora et al., 2023). The expansion of first-generation biofuels has been linked to environmental deterioration, including biodiversity losses from clearing biodiverse areas, because these fuels compete with agriculture land. Moreover, the production of biofuels can amplify other pressures that harm biodiversity directly or indirectly, such as CO_2 emissions from production and land-use systems, pollution emissions, and water depletion (Correa et al., 2017).

8.3.4 FEATURES OF BIOMASS

Due to its carbohydrate structure rather than the hydrocarbon structure of traditional fossil fuels, biomass is a highly oxygenated fuel. Because of its lower energy value, less predictable behavior of highly O_2 carbohydrates, and increased volatility, inorganic vapors, smoking, and soot formation, biomass with a high oxygen content (especially among agrarian and herbaceous biomass, wood and woody biomass, and algae) is not advantageous (Vassilev et al., 2015).

8.3.5 SUPPLY CONCERNS OF BIO-ENERGY

The dangers of biological production are some fundamental supply-side issues that the bio-energy sector, along with the food and forestry sectors, has to deal with. While these risks are typically connected to the availability of residues, primarily from the forestry and wood industry, for bioheat and biopower, they are typically connected to crop production for current first-generation biofuels. El Nino, drought, other weather-related effects, fire, and pests (including insects, plant diseases, and vertebrate species) can significantly restrict the supply of biomass feedstocks. These factors also have an impact on the production of food and fiber. Both specialized energy crops and residues fall under this description. Uncertainties regarding the consequences of climate change, in that rising temperatures and changing rainfall patterns could significantly alter the appropriateness of certain regions of the world for the production of particular crops, are among the longer-term supply-side issues. Moreover, soil and water deterioration, for instance, as a result of poor irrigation techniques or excessive crop waste removal, can negatively affect cultivation output and, in the worst case scenario, render further production unprofitable (Bauen et al., 2011).

8.3.6 TECHNOLOGICAL HAZARDS AND OBSTACLES

Technically speaking, solid biomass feedstock is challenging to handle and store since it often has a low bulk density and is available in an array of shapes and types. One of the major technical obstacles that still needs to be solved, especially for gasification units that run under pressure, is the reliability of the feeding systems into the boiler or reactor. Except for those used in heating applications, commercially available methods often have poor small-scale economics. This is a specific issue since it is challenging to supply large plants with mostly lignocellulosic feedstocks because of inadequate resource availability, distribution, density, and logistics (Bauen et al., 2011). Using a range of various technologies can improve the effectiveness with which forest biomass can be turned into energy as well as the quality of the products that can be produced from it. Nevertheless, these technologies are still in the R&D&P (research, development, and prototyping) stage. As many of these advances have substantial processing costs, they have not yet been widely used. To cut costs and enable full commercialization, more development effort is required (Manikandan et al., 2023).

The research in bio-energy production is constantly evolving and several new methods and technological advancements are being made. Yet, there are some hurdles that have to be crossed for boosting the production of bio-energy and reducing our dependence on fossil fuels. A brief description of the advances and challenges in bio-energy research is shown in Table 8.2.

TABLE 8.2
Advances and Challenges in Bio-Energy Production

Advancements to bio-energy production	Challenges to bio-energy production
New and better processing techniques have been developed	Processing constraints are still prevalent
Several new feedstock materials are being employed for	Poses the threat of loss in biodiversity
New bio-energy crops are being discovered to be utilized in bio-energy production	Farmers and landowners are still hesitant to be involved
Several molecular techniques are paving the way for better yield in lesser time	Weather phenomena or natural calamities can affect the supply of feedstock materials

8.4   CONCLUSION

Although the term "bio-energy" has only recently become fashionable, plant-based energy has been around since the 1980s, when concerns over oil availability and prices led to the use of plant materials for heat and power. It has long been pursued (Karp & Shield, 2008). In the global energy consumption and the fight against climate change, bio-energy has garnered a lot of attention and taken on a significant role. The potential of bio-energy will be enormous in the near future, despite the fact that it only makes up 14% of the world's energy consumption at the moment (World Energy Resources 2016, https://www.world energy.org/publications/2016/world energy resources–2016/). In addition, the production of sustainable bio-energy can effectively lower the risk of energy poverty and promote economic development, particularly in developing nations. As a result, governments all over the world are working to both encourage the production of bio-energy and find the best laws or policies to control its growth (Wu et al., 2018). Primary energy from renewable sources like bio-energy doesn't produce carbon dioxide when used sustainably (Fischer & Schrattenholzer, 2001). Currently, hydrogen is more expensive than other traditional energy sources. An alternative method of producing hydrogen will be through microbial processes. Utilizing waste to produce hydrogen energy could lower production costs, increasing availability, and lowering the price of hydrogen gas (Chong et al., 2009). An increasing amount of non-edible oil is produced using an algae-based biodiesel, which does not compete with soil food production. Algal biodiesel production per hectare is higher than that of traditional oil refineries (Rajmohan et al., 2021).

The production of ethanol and biodiesel from bio-energy crop feedstocks (cellulose or sugar, starch plants) can also significantly contribute to the development of the rural economy, increase energy efficiency, and make productive use of lands with environmental damage. Drought-resistant and carbon-sequestering bio-energy crops exist (Yadav et al., 2021).

The creation of novel methods for biomass conversion, mixing, process monitoring, and process control, as well as significant improvements in the performance of currently used technologies, are necessary for the development of additional biofuel facilities. However, reducing the cost of synthesizing biofuels is the main issue (Anekwe et al., 2022). Not only is it urgently necessary to increase biomass production but also to improve the industrial processes used to produce biofuel, including biodiesel. The development of future energy crops has opened up new vistas, paving the way for the substitution of fossil fuels with renewable energy. This is made possible by advancements in omics and other cutting-edge disciplines (Khan et al., 2021). The study of biomass discovery and its transformation into value-added biomaterials has come a long way. Many methods for enhancing bio-oil have been researched. The management of process parameter settings and the use of catalysts are some of these. Biomass selection, biomass pretreatments (chemical and thermal), and biomass blending co-pyrolysis (biomass–biomass, biomass–plastics, biomass–coal, and biomass–shale) are other examples (Mohammed et al., 2023).

Commercialization of biomass conversion technologies for energy, liquid biofuel, and chemicals faces some difficulties because of the physical structure and chemical makeup of biomass. Low economic returns, ineffective conversion, and uncertainty regarding environmental effects are the main problems (Pang, 2019).

REFERENCES

Ahorsu, R., Medina, F., & Constantí, M. (2018). Significance and challenges of biomass as a suitable feedstock for bio-energy and biochemical production: A review. *Energies, 11*(12). doi:10.3390/en11123366

Alexander, P., Brown, C., Arneth, A., Finnigan, J., Moran, D., & Rounsevell, M. D. A. (2017). Losses, inefficiencies and waste in the global food system. *Agricultural Systems, 153,* 190–200. doi:10.1016/j.agsy.2017.01.014

Anekwe, I. M. S., Armah, E. K., & Tetteh, E. K. (2022). Bio-energy production: Emerging technologies. In: M. Samer (Ed.), *Biomass, Biorefineries and Bioeconomy* (Vol. 225). doi:10.5772/intechopen.102692

Arora, R., Manisseri, C., Li, C., Ong, M. D., Scheller, H. V., Vogel, K., Simmons, B. A., & Singh, S. (2010). Monitoring and analyzing process streams towards understanding ionic liquid pretreatment of switchgrass (*Panicum virgatum L.*). *Bioenergy Research, 3*(2), 134–145. doi:10.1007/s12155-010-9087-1

Arora, S., Sarao, L. K., & Singh, A. (2023). Bio-energy from cellulose of woody biomass. In: N. Srivastava, B. Verma, & P. Mishra (Eds.), *Agroindustrial Waste for Green Fuel Application. Clean Energy Production Technologies.* Springer, Singapore. doi:10.1007/978-981-19-6230-1_4

Baenziger, P. S., Russell, W. K., Graef, G. L., & Campbell, B. T. (2006). Improving lives: 50 Years of crop breeding, genetics, and cytology (C-1). *Crop Science, 46*(5), 2230–2244. doi:10.2135/cropsci2005.11.0404gas

Bauen, A., Berndes, G., Junginger, M., Londo, M., Vuille, F., Ball, R., Bole, T., Chudziak, C., Faaij, A., & Mozaffarian, H. (2011). Bioenergy: A sustainable and reliable energy source: A review of status and prospects. *Agricultural and Biosystems Engineering Technical Reports and White Papers 16.* https://lib.dr.iastate.edu/abe_eng_reports/16

Bhatia, S. K., Sadashiv Jagtap, S., Ashok Bedekar, A., Kant Bhatia, R., Kumar Patel, A., Pant, D., Rajesh Banu, J., Rao, C. V, Kim, Y.-G., & Yang, Y.-H. (2019). Recent developments in pretreatment technologies on lignocellulosic biomass: Effect of key parameters, technological improvements, and challenges. *Bioresource Technology. 300.* doi:10.1016/j.biortech.2019.122724.

Brandt, A., Ray, M. J., To, T. Q., Leak, D. J., Murphy, R. J., & Welton, T. (2011). Ionic liquid pretreatment of lignocellulosic biomass with ionic liquid-water mixtures. *Green Chemistry, 13*(9), 2489–2499. doi:10.1039/c1gc15374a

Chong, M. L., Sabaratnam, V., Shirai, Y., & Hassan, M. A. (2009). Biohydrogen production from biomass and industrial wastes by dark fermentation. *International Journal of Hydrogen Energy, 34*(8), 3277–3287. doi:10.1016/j.ijhydene.2009.02.010

Correa, D. F., Beyer, H. L., Possingham, H. P., Thomas-Hall, S. R., & Schenk, P. M. (2017). Biodiversity impacts of bio-energy production: Microalgae vs. first generation biofuels. *Renewable and Sustainable Energy Reviews, 74,* 1131–1146. doi:10.1016/j.rser.2017.02.068

Cowie, A. L., Brandão, M., & Soimakallio, S. (2018). Quantifying the climate effects of forest-based bio-energy. In: *Managing Global Warming: An Interface of Technology and Human Issues* (pp. 399–418). doi:10.1016/B978-0-12-814104-5.00013-2

Dash, S. K., & Lingfa, P. (2018). An overview of biodiesel production and its utilization in diesel engines. *IOP Conference Series: Materials Science and Engineering, 377*(1). doi:10.1088/1757-899X/377/1/012006

Duarah, P., Haldar, D., Patel, A. K., Dong, C. Di, Singhania, R. R., & Purkait, M. K. (2022). A review on global perspectives of sustainable development in bio-energy generation. In *Bioresource Technology, 348.* doi:10.1016/j.biortech.2022.126791

Dung, T. N. B., Sen, B., Chen, C. C., Kumar, G., & Lin, C. Y. (2014). Food waste to bio-energy via anaerobic processes. *Energy Procedia, 61,* 307–312. doi:10.1016/j.egypro.2014.11.1113

Guo, M., Song, W., & Buhain, J. (2015). Bio-energy and biofuels: History, status, and per-spective. *Renewable and Sustainable Energy Reviews, 42*, 712–725. doi:10.1016/j.rser.2014.10.013

Gold, S., & Seuring, S. (2011). Supply chain and logistics issues of bio-energy production. *Journal of Cleaner Production, 19*(1), 32–42. doi:10.1016/j.jclepro.2010.08.009

He, Y., Bagley, D. M., Leung, K. T., Liss, S. N., & Liao, B. Q. (2012). Recent advances in mem-brane technologies for biorefining and bio-energy production. *Biotechnology Advances, 30*(4), 817–858. doi:10.1016/j.biotechadv.2012.01.015

Hossain, N., & Mahlia, T. M. I. (2019). Progress in physicochemical parameters of microalgae cultivation for biofuel production. *Critical Reviews in Biotechnology, 39*(6), pp. 835–859. doi:10.1080/07388551.2019.1624945

Ifeanyi-Nze, F. O., & Omiyale, C. O. (2022). Insights into the recent advances in the pre-treatment of biomass for sustainable bio-energy and bio-products synthesis: Challenges and future directions. *European Journal of Sustainable Development Research, 7*(1), em0209. doi:10.29333/ejosdr/12722

Iyodo Mohammed, H., Garba, K., Isa Ahmed, S., & Garba Abubakar, L. (2023). Recent advances on strategies for upgrading biomass pyrolysis vapor to value-added bio-oils for bio-energy and chemicals. *Sustainable Energy Technologies and Assessments, 55*. doi:10.1016/j.seta.2022.102984

Jain, A., Sarsaiya, S., Kumar Awasthi, M., Singh, R., Rajput, R., Mishra, U. C., Chen, J., & Shi, J. (2022). Bioenergy and bioproducts from bio-waste and its associated modern circular economy: Current research trends, challenges, and future outlooks. *Fuel, 307*. doi:10.1016/j.fuel.2021.121859

Jåstad, E. O., Bolkesjø, T. F., Trømborg, E., & Rørstad, P. K. (2020). The role of woody bio-mass for reduction of fossil GHG emissions in the future North European energy sector. *Applied Energy, 274*, 115360. doi:10.1016/j.apenergy.2020.115360

Jayus, Nurhayati, Mayzuhroh, A., Arindhani, S., & Caroenchai, C. (2016). Studies on bio-ethanol production of commercial baker's and alcohol yeast under aerated culture using sugarcane molasses as the media. *Agriculture and Agricultural Science Procedia, 9*, 493–499. doi:10.1016/j.aaspro.2016.02.168

Joshi O, Mehmood SR. (2011). Factors affecting nonindustrial private forest landowners' willingness to supply woody biomass for bio-energy. *Biomass and Bio-Energy, 35*(1), 186–192. doi:10.1016/j.biombioe.2010.08.016

Kang, Q., Appels, L., Tan, T., & Dewil, R. (2014). Bioethanol from lignocellulosic bio-mass: Current findings determine research priorities. *Scientific World Journal*. 298153. doi:10.1155/2014/298153

Karp, A., & Shield, I. (2008). bio-energy from plants and the sustainable yield challenge. *New Phytologist, 179*(1), 15–32. doi:10.1111/j.1469-8137.2008.02432.x

Khan, M. S., Mustafa, G., Joyia, A., & Mirza, S. A. (2021). Sugarcane as future bio-energy crop: Potential genetic and genomic approaches. *Sugarcane—Biotechnology for Biofuels*. doi:10.5772/intechopen.97581.

Kumar Awasthi, M., Paul, A., Kumar, V., Sar, T., Kumar, D., Sarsaiya, S., Liu, H., Zhang, Z., Binod, P., Sindhu, R., Kumar, V., & Taherzadeh, M. J. (2022). Recent trends and devel-opments on integrated biochemical conversion process for valorization of dairy waste to value added bioproducts: A review. *Bioresource Technology, 344*. doi:10.1016/j.biortech.2021.126193

Lee, M. (1998). Genome projects and gene pools plant breeding? *Proceedings of Natural Acadamic of Science. 95*, 2001–2004.

Li, Y., Qi, C., Zhang, Y., Li, Y., Wang, Y., Li, G., & Luo, W. (2021). Anaerobic digestion of agricultural wastes from liquid to solid state: Performance and environ-economic com-parison. *Bioresource Technology, 332*. doi:10.1016/j.biortech.2021.125080

Manikandan, S., Vickram, S., Sirohi, R., Subbaiya, R., Krishnan, R. Y., Karmegam, N., Sumathijones, C., Rajagopal, R., Chang, S. W., Ravindran, B., & Awasthi, M. K. (2023). Critical review of biochemical pathways to transformation of waste and biomass into bio-energy. *Bioresource Technology*, *372*. doi:10.1016/j.biortech.2023.128679

Marone, A., Varrone, C., Fiocchetti, F., Giussani, B., Izzo, G., Mentuccia, L., Rosa, S., & Signorini, A. (2015). Optimization of substrate composition for biohydrogen production from buffalo slurry co-fermented with cheese whey and crude glycerol, using microbial mixed culture. *International Journal of Hydrogen Energy*, *40*(1), 209–218. doi:10.1016/j.ijhydene.2014.11.008

Midgett, J. S., Stevens, B. E., Dassey, A. J., Spivey, J. J., & Theegala, C. S. (2012). Assessing feedstocks and catalysts for production of bio-oils from hydrothermal liquefaction. *Waste and Biomass Valorization*, *3*(3), 259–268. doi:10.1007/s12649-012-9129-3

Muktham, R., K. Bhargava, S., Bankupalli, S., & S. Ball, A. (2016). A review on 1st and 2nd; generation bioethanol production-recent progress. *Journal of Sustainable Bio-Energy Systems*, *6*(3), 72–92. doi:10.4236/jsbs.2016.63008

Muruaga, M. L., Carvalho, K. G., Domínguez, J. M., de Souza Oliveira, R. P., & Perotti, N. (2016). Isolation and characterization of Saccharomyces species for bioethanol production from sugarcane molasses: Studies of scale up in bioreactor. *Renewable Energy*, *85*, 649–656. doi:10.1016/j.renene.2015.07.008

Ning, P., Yang, G., Hu, L., Sun, J., Shi, L., Zhou, Y., Wang, Z., & Yang, J. (2021). Recent advances in the valorization of plant biomass. *Biotechnology for Biofuels*, *14*(1). doi:10.1186/s13068-021-01949-3

Packer, M. (2009). Algal capture of carbon dioxide; biomass generation as a tool for greenhouse gas mitigation with reference to New Zealand energy strategy and policy. *Energy Policy*, *37*(9), 3428–3437. doi:10.1016/j.enpol.2008.12.025

Pandit, S., Savla, N., Sonawane, J. M., Sani, A. M., Gupta, P. K., Mathuriya, A. S., Rai, A. K., Jadhav, D. A., Jung, S. P., & Prasad, R. (2021). Agricultural waste and wastewater as feedstock for bioelectricity generation using microbial fuel cells: Recent advances. *Fermentation*, *7*(3). doi:10.3390/fermentation7030169

Pang, S. (2019). Advances in thermochemical conversion of woody biomass to energy, fuels and chemicals. *Biotechnology Advances*, *37*(4), 589–597. doi:10.1016/j.biotechadv.2018.11.004

Pathak, V. V. (2014). Assessment of solid waste management and energy recovery from waste materials in Lucknow Zoo: A case study bioprocessing of aquatic biomass for energy generation and climate change mitigation. *Recent Advancements and Challenges View Project Algal Technology*, *3*, 14–22. https://www.researchgate.net/publication/263275333

Rai, A. K., Al Makishah, N. H., Wen, Z., Gupta, G., Pandit, S., & Prasad, R. (2022). Recent Developments in Lignocellulosic biofuels, a renewable source of bio-energy. *Fermentation*, *8*(4). doi:10.3390/fermentation8040161

Rahpeyma, S. S., & Raheb, J. (2019). Microalgae Biodiesel as a Valuable Alternative to Fossil Fuels. *Bioenergy Research*, *12*(4), 958–965. doi:10.1007/s12155-019-10033-6

Rajmohan, K. S., Ramya, C., & Varjani, S. (2021). Trends and advances in bio-energy production and sustainable solid waste management. *Energy and Environment*, *32*(6), 1059–1085. doi:10.1177/0958305X19882415

Razmovski, R., & Vučurović, V. (2012). Bioethanol production from sugar beet molasses and thick juice using Saccharomyces cerevisiae immobilized on maize stem ground tissue. *Fuel*, *92*(1), 1–8. doi:10.1016/j.fuel.2011.07.046

Sarsaiya, S., Jain, A., Kumar Awasthi, S., Duan, Y., Kumar Awasthi, M., & Shi, J. (2019). Microbial dynamics for lignocellulosic waste bioconversion and its importance with modern circular economy, challenges and future perspectives. *Bioresource Technology*, *291*. doi:10.1016/j.biortech.2019.121905

Sawatdeenarunat, C., Nguyen, D., Surendra, K. C., Shrestha, S., Rajendran, K., Oechsner, H., Xie, L., & Khanal, S. K. (2016). Anaerobic biorefinery: Current status, challenges, and opportunities. *Bioresource Technology*, *215*, 304–313. doi:10.1016/j.biortech.2016.03.074

Sivakumar, D. (2021). Wastewater treatment and bioelectricity production in microbial fuel cell: Salt bridge configurations. *International Journal of Environmental Science and Technology*, *18*(6), 1379–1394. doi:10.1007/s13762-020-02864-0

Sosulski, K., & Sosulski, F. (1994). Wheat as a feedstock for fuel ethanol. *Applied Biochemistry and Biotechnology*, *45*, 169–180. doi:10.1007/BF02941796

Srivastava, N., Rathour, R., Jha, S., Pandey, K., Srivastava, M., Thakur, V. K., Sengar, R. S., Gupta, V. K., Mazumder, P. B., Khan, A. F., & Mishra, P. K. (2019). Microbial beta glucosidase enzymes: Recent advances in biomass conversation for biofuels application. *Biomolecules*, *9*(6). doi:10.3390/biom9060220

Thomas, K. C., Hynes, S. H., Jones, A. M., & Ingledew, W. M. (1993). Production of fuel alcohol from wheat by VHG technology effect of sugar concentration and fermentation temperature. *Applied Biochemistry and Biotechnology*, *2*(1), 211–226.

Unther Fischer, G., & Schrattenholzer, L. (2001). Global bio-energy potentials through 2050. *Biomass and Bio-Energy*, *20*, 151–159. doi:10.1016/S0961-9534(00)00074-X

Vancov, T., Schneider, R. C. S., Palmer, J., McIntosh, S., & Stuetz, R. (2015). Potential use of feedlot cattle manure for bioethanol production. *Bioresource Technology*, *183*, 120–128. doi:10.1016/j.biortech.2015.02.027

Vassilev, S. V., Vassileva, C. G., & Vassilev, V. S. (2015). Advantages and disadvantages of composition and properties of biomass in comparison with coal: An overview. *Fuel*, *158*, 330–350. doi:10.1016/j.fuel.2015.05.050

Wang, S., Zhu, J., Dai, L., Zhao, X., Liu, D., & Du, W. (2016). A novel process on lipid extraction from microalgae for biodiesel production. *Energy*, *115*, 963–968. doi:10.1016/j.energy.2016.09.078

Wu, Y., Zhao, F., Liu, S., Wang, L., Qiu, L., Alexandrov, G., & Jothiprakash, V. (2018). bio-energy production and environmental impacts. *Geoscience Letters*, *5*(1). doi:10.1186/s40562-018-0114-y

Yadav, G., Shanmugam, S., Sivaramakrishnan, R., Kumar, D., Mathimani, T., Brindhadevi, K., Pugazhendhi, A., & Rajendran, K. (2021). Mechanism and challenges behind algae as a wastewater treatment choice for bio-energy production and beyond. *Fuel*, *285*. doi:10.1016/j.fuel.2020.119093

Yadav, P., Priyanka, P., Kumar, D., Yadav, A., & Yadav, K. (2019). *Bioenergy Crops: Recent Advances and Future Outlook* (pp. 315–335). doi:10.1007/978-3-030-14463-0_12

Yokoyama, H., Waki, M., Moriya, N., Yasuda, T., Tanaka, Y., & Haga, K. (2007). Effect of fermentation temperature on hydrogen production from cow waste slurry by using anaerobic microflora within the slurry. *Applied Microbiology and Biotechnology*, *74*(2), 474–483. doi:10.1007/s00253-006-0647-4

Yuan, J. S., Tiller, K. H., Al-Ahmad, H., Stewart, N. R., & Stewart, C. N. (2008). Plants to power: Bioenergy to fuel the future. *Trends in Plant Science*, *13*(8), 421–429. doi:10.1016/j.tplants.2008.06.001

Zhang, H., & Wu, S. (2015). Pretreatment of eucalyptus using subcritical CO_2 for sugar production. *Journal of Chemical Technology and Biotechnology*, *90*(9), 1640–1645. doi:10.1002/jctb.4470

Zhao, L., Sun, Z. F., Zhang, C. C., Nan, J., Ren, N. Q., Lee, D. J., & Chen, C. (2022). Advances in pretreatment of lignocellulosic biomass for bio-energy production: Challenges and perspectives. *Bioresource Technology*, *343*. doi:10.1016/j.biortech.2021.126123

Zhu, Q. L., Wu, B., Pisutpaisal, N., Wang, Y. W., Ma, K. Dong, Dai, L. C., Qin, H., Tan, F. R., Maeda, T., Xu, Y. sheng, Hu, G. Q., & He, M. X. (2021). Bio-energy from dairy manure: Technologies, challenges and opportunities. *Science of the Total Environment*, *790*. doi:10.1016/j.scitotenv.2021.148199

9 Omics Technology Approaches for the Generation of Biofuels

Jyoti Sarwan, Mithila V. Nair, Smile Sharma,
Nazim Uddin, and K. Jagadeesh Chandra Bose
Chandigarh University

9.1 INTRODUCTION TO BIOFUELS AND BIOENERGY

Biofuels are fuels that are produced from biomass, which is organic material derived from plants and animals. They are considered renewable because they can be produced from crops that can be grown year after year, unlike fossil fuels, which are limited in supply and can take millions of years to form. Biofuels and bioenergy are renewable sources of energy derived from biological materials, such as crops, wood, and organic waste. They offer an alternative to non-renewable sources of energy, such as fossil fuels, which contribute to environmental degradation, climate change, and energy security concerns. Biofuels are fuels that are made from biomass, which can include agricultural crops, forestry residues, and municipal waste.

There are several examples of biofuels such as biodiesels, bioethanol, and biogas. For processing of ethanol, the main components required are corn, sugarcane, ligno-cellulosic biomass, and other cellulosic biomass that have large quantity of sugar and starch. However, to produce the biodiesel, there is a need for cooking oils, recycled cooking oils, and animal fats. In a similar way, the production of the biogas requires the decomposition of livestock waste, food scraps, etc.

Bioenergy is power generated from biomass using a variety of techniques, including combustion, gasification, or fermentation. Energy can be produced from biomass in the form of heat, electricity, or other forms. For instance, biogas can be used to produce electricity in power plants, while wood pellets can be burnt to heat houses or businesses. Biofuels and bioenergy present a possible substitute for non-renewable energy sources because they are sustainable, renewable, and can be generated locally, thereby reducing reliance on foreign oil. The development of biofuels could have negative environmental effects, including deforestation, altered land use, water scarcity, and competition for resources with food crops. Therefore, it is essential to carefully regulate the growth of biofuels and bioenergy to ensure that their advantages are maximised while minimising any potential drawbacks.

9.2   WHAT IS BIOMASS CONVERSION?

Biomass conversion is the process of converting organic matter into a biofuel. This can be done through a few methods, including fermentation, gasification, pyrolysis, and anaerobic digestion. Each method has its own advantages and disadvantages, but all of them aim to produce a fuel that can be used in place of fossil fuels. The most common type of biomass conversion is fermentation. This involves using microorganisms to break down the carbohydrates in biomass into simpler molecules like alcohols or acids. These products can then be used as fuel or further processed into other chemicals. Fermentation is a relatively simple process and doesn't require high temperatures or pressures, making it well suited for small-scale operations. However, it generally has low yields and can produce undesirable side products that need to be removed before the final fuel product can be used.

Gasification is another common method of biomass conversion. In this process, biomass is heated in the absence of oxygen to produce a mixture of gases like carbon monoxide, hydrogen, and methane. These gases can be combusted to generate electricity or further processed into other fuels like methanol or diesel. Gasification is generally more efficient than fermentation and can be scaled up to large industrial operations. However, it requires expensive equipment and high temperatures, making it less suitable for small-scale operations. Pyrolysis is like gasification but uses lower temperatures and shorter heating times. This produces a liquid fuel.

9.3   THE PROCESS OF CONVERTING BIOMASS
INTO BIOFUELS

The process of converting biomass into biofuels is a two-step process. The first step is to convert the biomass into a sugar-rich syrup, which is then fermented into alcohol. The second step is to convert the alcohol into fuel. There are several different ways to convert biomass into sugar-rich syrup. One way is to use enzymes to break down the cellulose in the biomass into glucose [1,2]. The glucose can then be fermented into alcohol. Another way to convert biomass into sugar-rich syrup is to gasifier the biomass, which breaks down the cellulose into gases that can be converted into glucose. Once the sugar-rich syrup has been fermented into alcohol, it can be converted into fuel using one of two processes. The first process is called catalytic cracking, which uses a catalyst to break down the molecules in the alcohol and produce gasoline, diesel, and other fuels. The second process is called pyrolysis, which involves heating the alcohol without the presence of oxygen, so that it decomposes into gases that can be used as fuel.

9.4   THE BENEFITS OF BIOMASS CONVERSION

The process of transforming organic matter into a useful type of energy is known as biomass conversion. The most typical method of converting biomass is combustion, which involves burning organic material to create heat and/or power. However, there are also alternative processes, such as anaerobic digestion, pyrolysis, and gasification, which can turn biomass into fuel. The conversion of biomass has many advantages. The fact that it offers a renewable energy source is arguably its biggest advantage. Biomass is a renewable resource that can be renewed through sustainable land management techniques, in contrast to fossil fuels, which are finite resources that will eventually run out. By lowering greenhouse gas emissions, biomass conversion can also aid in lessening the consequences of climate change. Biomass energy sources have the potential to be carbon-neutral or even carbon-negative when correctly handled. The ability of biomass conversion to support job creation and economic growth in rural regions is another important advantage. Because biomass is often produced and processed locally, it presents an opportunity for farmers and small companies. In addition, many biomass conversion methods use little to no imported resources, which helps local economies even more. The conversion of biomass also has a variety of positive environmental effects. By lowering emissions from combustion processes, it can, for instance, support sustainable land management strategies, lessen reliance on fossil fuels, and enhance air quality.

9.5 CLASSIFICATION OF BIOFUELS

There are two main types of biofuels: First-generation and second-generation biofuels are the two primary categories of biofuels. Sugar, starch, or vegetable oil obtained from crops like corn, sugarcane, or soybeans is used to make first-generation biofuels. These fuels are already used in many regions of the world and are reasonably simple to generate. First-generation biofuels' effects on land use, food prices, and overall sustainability are all subjects of concern, though. Vegetable oil, corn, sugarcane, and other edible crops are frequently used to produce first-generation biofuels. Second-generation biofuels are produced from non-food crops or waste materials such as municipal solid waste, agricultural waste, wood chips, and non-edible plant materials like switchgrass. They do not compete with food crops for land use, and they utilise waste items that would otherwise be thrown away, making them generally more sustainable than first-generation biofuels. Third-generation biofuels are also available; they are produced using algae or other aquatic plants. Algae or other microorganisms are used in the production of third-generation biofuels. Because these fuels can be cultivated on non-arable ground and don't need freshwater resources, they have the potential to be even more sustainable than second-generation biofuels. Transportation (using ethanol combined with petrol, biodiesel, and renewable diesel, for example), heating, and electricity generation are only a few uses for biofuels. Their use must be carefully controlled to ensure their sustainability and lessen their impact on food prices and land use. They offer the potential to reduce greenhouse gas emissions and reliance on fossil fuels. In addition, there is another category of biofuels, each with unique benefits and drawbacks. These are the several kinds of biofuels:

a. **Bioethanol**: Ethanol is a type of alcohol that can be produced from crops such as corn, sugarcane, and wheat. Crops including corn, sugarcane, and wheat can be used to make ethanol, a form of alcohol. It is frequently used as a fuel additive to lower emissions in automobiles with petrol engines. In cars made to run on ethanol, ethanol can also be used as a solo fuel.

b. **Biodiesel**: Biodiesel is a renewable diesel fuel that is made from vegetable oils, animal fats, or recycled restaurant grease. Using recovered restaurant grease, animal fats, or plant oils, biodiesel is a renewable diesel fuel. It doesn't require as much adjustment as regular diesel and can be utilised in most diesel engines.

c. **Biogas**: Produced from organic waste materials like food scraps, manure, and agricultural by-products, biogas is a renewable gas. It can be used to produce heat, power, and transportation fuel.

d. **Bio-jet fuel**: Made from vegetable oils, animal fats, or other biomass feedstocks, this form of fuel is a renewable substitute for conventional jet fuel. To cut greenhouse gas emissions, it is currently employed in commercial aircraft.

e. **Pyrolysis oil**: Produced by heating biomass without oxygen, pyrolysis oil is a liquid biofuel. For engines, turbines, and boilers, it might serve as fuel.

f. **Algal biofuels**: Algae can be produced in ponds, tanks, or other enclosed systems and is the source of algal biofuels. Algae can be cultivated in non-potable water and in an area that is not suited for growing food crops, and they can yield a lot of biomasses per unit of land. Biofuels have a long history of use in energy, power, and transportation, much as the use of algae. The concept of creating methane gas from algae was first proposed in the 1950s, and it gained popularity during the 1970s energy crisis. The Aquatic Species Programme (ASP), supported by the US Department of Energy, made investments between 1980 and 1996 with the goal of producing oil from microalgae, and in the middle of the 20th century, commercialisation for their fatty acid and lipid contents began. Algal biomass can also be converted utilising a range of conversion methods, primarily three types of processes: biochemical, thermochemical, and chemical, into various third-generation biofuels and other by-products.

The most recent method for creating algae-based biofuels is growing the algae, harvesting it, drying it, extracting the oil, and trans esterifying it to create biodiesel. These cultivation methods are influenced by the kind, quality, and market value of the biofuel that will be produced [3]. Therefore, the procedures employed to harvest algae for the purpose of producing biofuel are crucial. Since microalgae are so small, they are normally produced by immersion in water that has been examined for important elements including temperature, pH level, CO_2 level, nutrients, amount of light reaching the water body, and photoperiod. Direct and conventional transesterification are the two types available. Without any pre-treatments, direct or in situ transesterification and oil extraction take place simultaneously. It produces more biodiesel than the standard variety and

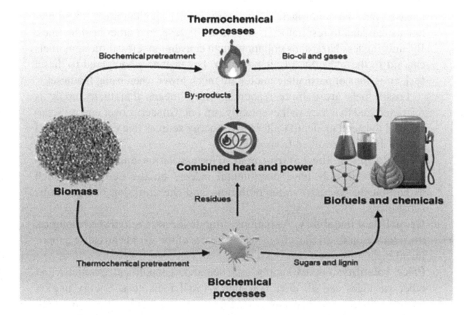

FIGURE 9.1   Figure representing the biomass conversion into biofuel [4].

is crucial for future development. The typical method, however, requires two steps since a mechanical process must come before the lipid extraction. This method is more time- and energy-consuming because of the pre-treatment activities (Figure 9.1).

9.6 ADVANTAGES OF BIOFUELS WHEN COMPARED TO TRADITIONAL FOSSIL FUEL

The choice of biofuel is influenced by things including the availability of feedstocks, the cost of production, and its suitability for different uses. Each form of biofuel has specific benefits and drawbacks. There are a few issues with using natural fuels like coal, oil, and natural gas, which has prompted the creation and use of alternate energy sources like biofuels. The following are some of the main issues with conventional fossil fuels:

a. **Climatic Change**: Fossil fuels are the main source of greenhouse gas emissions, which are a factor for climate change on a worldwide scale. Fossil fuel combustion produces greenhouse gases in the atmosphere, including carbon dioxide, which traps heat and causes temperatures to rise. Other effects include a rise in the sea level, altered precipitation patterns, and more frequent and severe weather events.

b. **Air pollution**: When fossil fuels are used, pollutants such as nitrogen oxides, sulphur dioxide, particulate matter, and volatile organic compounds are released into the atmosphere. Particularly in metropolitan areas, these pollutants can lead to respiratory issues, heart disease, and other health issues. By emitting less hazardous pollutants than conventional fossil fuels, biofuels can aid in the reduction of air pollution. For instance, compared to diesel fuel, emissions of particulate matter are much lower when using biodiesel.

Fossil fuels are a finite resource, which means that their supply is restricted and that they will eventually run out. Concerns over energy security and the need to identify alternative energy sources that can be produced sustainably have resulted from this.

c. **Degradation of the environment**: The extraction and transportation of fossil fuels may have detrimental effects on the ecosystem, such as habitat destruction, water pollution, and the emission of hazardous substances.

d. **Geopolitical instability**: As nations struggle for access to these resources, the reliance on fossil fuels from specific parts of the world can cause geopolitical instability.

e. **Price volatility**: Global supply and demand, geopolitical conflicts, and other variables can all alter how much fossil fuels cost. Energy market uncertainty and volatility may result from this. As a result of the issues with conventional fossil fuels, interest in alternative energy sources like biofuels has surged. These fuels provide a more sustainable and renewable energy choice. The choice of biofuel is influenced by several variables, including the cost of production, the availability of feedstocks, and the suitability for different uses.

f. **Renewable and sustainable**: Plants, which can be cultivated and harvested year after year, are used to make biofuels. In other words, if the biomass is cultivated and managed appropriately, they can be produced sustainably as a renewable source of energy.

g. **Lowers greenhouse gas emissions**: When compared to fossil fuels, biofuels have the potential to lower greenhouse gas emissions. The carbon dioxide that is emitted when biofuels are burnt is balanced by the carbon dioxide that plants collect during photosynthesis, resulting in a net-zero or even a net-negative emission.

h. **Energy security**: By lowering reliance on foreign oil and diversifying the energy mix, biofuels can boost energy security. Because they may be produced domestically, biofuels can lessen the need to import crude oil from other nations.

i. **Job creation**: The production and consumption of biofuels can lead to the development of positions in the distribution, processing, and farming industries. This can benefit rural areas economically and support them.

j. **Better rural development**: By generating jobs, raising incomes, and promoting economic expansion in rural areas, the production of biofuels can boost rural development. Biofuels have renewable nature, the potential for reducing greenhouse gas emissions, and the potential to create jobs and stimulate economic growth. However, it is important to ensure that biofuels are produced sustainably and do not have unintended negative impacts on food prices, land use, or biodiversity.

9.7 BIOENERGY

Bioenergy can replace traditional fossil fuels by lowering greenhouse gas emissions and improving energy security. Bioenergy is a renewable and sustainable energy source that can help reduce dependence on finite fossil fuels. By reducing greenhouse gas emissions and boosting energy security, bioenergy can take the place of conventional fossil fuels. Bioenergy is a sustainable and renewable energy source that can lessen reliance on depletable fossil resources. In addition, indigenous production of bioenergy can lessen dependency on imported energy sources and increase energy security. Bioenergy is probably going to play a bigger role in the world's energy mix as technology for producing it develops and costs come down. Because biofuels are a subset of bioenergy, the two are related. Biomass, or organic material like plants, trees, and agricultural waste, is a sustainable energy source that is used to create bioenergy. Bioenergy can be utilised to produce fuel, heat, or power. A special kind of bioenergy called biofuels is created from biomass and utilised to replace conventional fossil fuels.

9.8 BIOENERGY'S POTENTIAL

In several ways, bioenergy has the potential to replace fossil fuels. Diesel and petrol can be swapped out for liquid biofuels like ethanol and biodiesel. Corn, sugarcane, soybeans, algae, and other feedstocks can all be used to make these biofuels. These biofuels are becoming more cost-competitive with fossil fuels as production technology advances. Utilising biomass to produce energy or heat is known as biopower. Anaerobic digestion, gasification, and direct combustion of biomass are all options for doing this. Various feedstocks, such as agricultural and forestry wastes as well as energy crops like switchgrass and willow, can be used to create biopower. Anaerobic digestion of organic matter, such as animal manure, food waste, and sewage, can produce biogas, a renewable natural gas. Natural gas can be replaced with biogas in the production of power and heat. Chemicals made from renewable biomass feedstocks, as opposed to fossil fuels, are known as biochemicals. Biochemicals include, for instance, bioplastics, biodegradable cleaners, and bio-based solvents. Typically, biofuels are used to replace fossil fuels like petrol, diesel, and others. They can be utilised for several purposes, including power production and transportation. Because they release fewer greenhouse emissions when burnt, biofuels are a cleaner substitute for fossil fuels. In addition, indigenous production of biofuels can lessen the need for imported oil. Biofuels are one sort of bioenergy; hence, in general, the two are related. While biofuels are a particular kind of bioenergy that can be useful to act as a replacement for conventional fossil fuels, bioenergy is a renewable and sustainable energy source that can help to reduce greenhouse gas emissions and increase energy security.

9.9 INDIA'S BIOFUEL RESEARCH STATUS

To lessen its reliance on fossil fuels and lessen the effects of climate change, India had been actively pursuing research and development in the field of biofuels. Several recent research projects and advancements in India are linked to biofuels. To boost the usage of biofuels in the nation and develop a sustainable biofuel economy, the Indian government introduced the National Biofuel Policy in 2018. By 2030, the strategy aims to blend 20% of biofuels into petrol and diesel. The development of ethanol, a biofuel generated from sugarcane, maize, and other agricultural waste, has been a priority for India. The government has established production goals of 10% and 20% ethanol-blended petrol (EBP) by 2022 and 2025, respectively. Several ethanol plants have been built across the nation to accomplish this. India is also looking into the creation of biodiesel, a fuel generated from organic ingredients like vegetable oils and animal fats. The Indian Institute of Technology, Delhi, has invented a method to turn wasted cooking oil into biodiesel. Algal fuel production is being studied as a potential source of biofuels. Several Indian research institutions are developing the technologies needed to generate ethanol from algae. India is looking into waste-to-energy systems to make biofuels. A technique to create biofuel from municipal solid trash has been developed by the Indian Institute of Technology, Madras. Overall, the necessity for India to lessen its reliance on imported fossil fuels, increase energy security, and cut greenhouse gas emissions is what is driving the country's focus on biofuels (Figure 9.2).

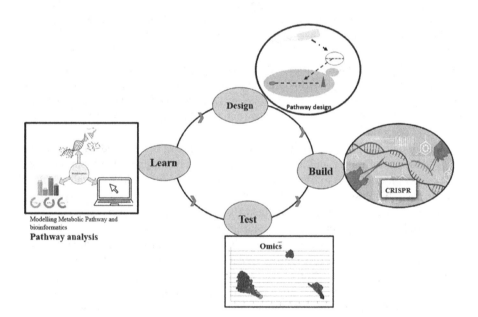

FIGURE 9.2   The major basic ideology of omics technology for product formation [5].

9.10   GLOBAL RESEARCH ON BIOFUELS

Advanced biofuels generated from non-food crops or waste materials are the subject of current research. Compared to fossil fuels, these biofuels have the potential to considerably lower greenhouse gas emissions. Several nations have established goals to expand the production and use of advanced biofuels, including the United States, Brazil, and the European Union. Algae are being investigated as a potential source of biofuels. Some species of algae have a high oil content that can be used to make biofuels, and they can be grown in saltwater or wastewater. Large-scale fuel production from algae-based biofuels is possible without displacing food crops on arable ground. Researchers are also investigating the production of biofuels using waste-to-energy systems. By converting municipal solid trash, agricultural waste, and forestry by-products into biofuels, waste and greenhouse gas emissions can be decreased. Genetic engineering is being used to develop crops that are optimised for biofuel production. For example, researchers are developing varieties of sugarcane and switchgrass that have higher yields and are more resistant to pests and diseases. Governments around the world are providing policy support for biofuels to encourage their production and use. This includes incentives such as tax credits and mandates for the use of biofuels in transportation fuels. In general, global research on biofuels is focused on developing sustainable, low-carbon alternatives to fossil fuels. Advances in technology and policy support are driving the growth of the biofuels industry, with the aim of reducing greenhouse gas emissions and improving energy security.

9.11   OMICS TECHNOLOGIES

Omics technologies refer to the various high-throughput methods used to study large-scale biological data. These technologies are based on analysing the genome, transcriptome, proteome, metabolome, and other "-omes" of an organism or sample.

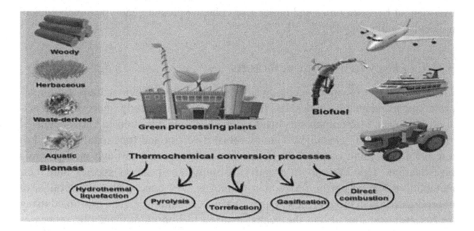

FIGURE 9.3   Thermo-chemical conversion of lignocellulosic biomass for production of biofuel [4].

a. **Genomics**: This is the study of the complete DNA sequence of an organism or sample. Genomics can provide information about gene expression, genetic variation, and evolutionary relationships.

b. **Transcriptomics**: This is the study of the complete set of RNA molecules (transcripts) produced by a cell or organism. Transcriptomics can provide insights into gene expression, alternative splicing, and RNA editing.

c. **Proteomics**: This is the study of the complete set of proteins produced by a cell or organism. Proteomics can provide information about protein function, post-translational modifications, and protein–protein interactions.

d. **Metabolomics**: This is the study of the complete set of small molecules (metabolites) produced by a cell or organism. Metabolomics can provide information about metabolic pathways, cellular metabolism, and disease biomarkers.

e. **Epigenomics**: This is the study of the modifications to DNA and histones that affect gene expression without changing the DNA sequence. Epigenomics can provide insights into gene regulation, cell differentiation, and disease states.

f. **Glycomics**: This is the study of the complete set of carbohydrates (glycans) produced by a cell or organism. Glycomics can provide information about cellular interactions, immune response, and disease biomarkers.

g. **Lipidomics**: This is the study of the complete set of lipids produced by a cell or organism. Lipidomics can provide insights into cellular metabolism, membrane structure, and disease biomarkers.

These omics technologies are often used in combination to provide a more complete picture of biological systems. For example, transcriptomics and proteomics can be used together to study gene expression and protein function, while metabolomics and lipidomics can be used together to study cellular metabolism (Figure 9.4). An overview of between omes and omics from phenotype and genotype metabolites in molecular technology [5].

9.11.1 Omics Technology in Biofuels

The development of omics technology, which includes genomes, transcriptomics, proteomics, and metabolomics, has the potential to significantly improve the generation of biogas. Microorganisms involved in anaerobic digestion can be recognised and described using genomics. The microbial diversity and functional potential of microbial communities in biogas reactors can be revealed through metagenomic investigation. This can aid in identifying important bacteria species and their metabolic pathways in the production of biogas. The gene expression profiles of microorganisms in biogas reactors under various conditions can be examined using transcriptomics. Identifying the genes involved in important metabolic processes like methanogenesis and acetogenesis as well as optimising reactor settings for optimal methane generation might both benefit from this. The proteins that bacteria in

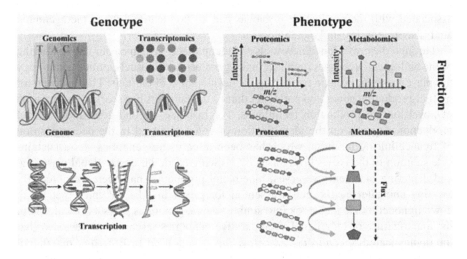

FIGURE 9.4 An overview between omes and omics from phenotype and genotype metabolites in molecular technology [5].

biogas reactors express under various conditions can be identified using proteomics. The identification of important enzymes engaged in metabolic pathways and the improvement of reactor parameters for optimal methane generation can both benefit from this. It is possible to analyse the metabolic processes involved in the formation of biogas using metabolomics and to pinpoint the important metabolites. This can aid in maximising reactor parameters for maximal biogas generation, such as substrate concentration and pH. Omics technology can aid in the optimisation of reactor settings for maximum biogas production and can offer insightful information regarding the microbial communities and metabolic pathways involved in the production of biogas. By identifying important microbial communities and metabolic pathways involved in the anaerobic digestion process, omics technology, which includes genomics, proteomics, and metabolomics, has the potential to improve biogas generation. For instance, metagenomic analysis has been utilised to identify the microbial communities involved in the production of biogas from various feedstocks, and the results have shown that the kind of feedstock used had an impact on the microbial community structure [6]. The expression of essential enzymes involved in the breakdown of complex organic molecules was found to be elevated during the anaerobic digestion process, according to research using proteomics to identify the proteins involved in the metabolic pathways of biogas production. Key metabolites associated with biogas production have been identified using metabolomics, as have metabolic pathways that can be targeted for genetic engineering to increase biogas output. Through the identification of important microbial communities and metabolic pathways involved in the anaerobic digestion process, the omics technology offers the potential to enhance biogas generation. The microbial populations engaged in the production of biogas, the proteins participating in the metabolic pathways of the production of biogas, and the important metabolites associated with the production of biogas can all be identified using metagenomic analysis, proteomics, and metabolomics.

Through the provision of new tools and techniques to enhance our understanding of the biological processes involved in the generation of biofuels, omics technology has played a crucial role in furthering the development of biofuels. The discovery of biofuel-producing genes and pathways thanks to genomics has sparked the creation of novel biofuel production techniques. For instance, second-generation biofuel production depends on the identification of genes involved in the decomposition of lignocellulosic biomass, which has been done using genomics. As a sustainable solution to the issues brought on by growing crude oil costs, global warming, and depleting petroleum reserves, bioethanol production has been promoted. By locating and altering the genes essential for producing the enzymes that break down lignocellulosic feedstock into simple sugars, genomics plays a critical role in the manufacture of bioethanol. Comparative genomics research has revealed that the domesticated *Saccharomyces cerevisiae* strains used in Brazilian bioethanol production are all related to the yeasts used to make cachaça. The quickest and most effective approach to generate bioethanol, according to some researchers, is to genetically modify a strain of yeast that produces the fuel. Creating bioethanol from sugarcane is a significant large-scale technique that can produce biofuel effectively and affordably. It has been demonstrated that manufacturing bioethanol fuel from sugarcane is economically feasible. By using bioethanol as an alternative fuel, you can fight vehicle pollution and cut greenhouse gas emissions. Using genetically altered *Zymomonas mobilis* will speed up the manufacture of bioethanol from leftover fruits and vegetables. In conclusion, genomics is critical in locating and changing the genes that create the enzymes needed to convert lignocellulosic feedstock into simple sugars, which is necessary for the generation of bioethanol. It has been demonstrated that it is economically feasible to produce bioethanol from sugarcane, and using bioethanol as an alternative fuel can help reduce greenhouse gas emissions and fight vehicular pollution (Figure 9.5). Status of omics in different industrial products [5].

Transcriptomics in biofuels generation: Transcriptomics has been used to study gene expression patterns in organisms used for biofuel production, providing insights into the metabolic pathways involved in biofuel production. Transcriptomics has been used to identify genes that are upregulated during lipid accumulation in algae, which can be used to produce biofuels. Proteomics has been used to identify the proteins involved in biofuel production, providing insights into the metabolic pathways and regulatory mechanisms involved in biofuel production. Proteomics has been used to identify enzymes involved in lignocellulosic biomass degradation and lipid accumulation in algae, which can be targeted for improvement in biofuel production. Metabolomics has been used to study the metabolic pathways involved in biofuel production, providing insights into the metabolic fluxes and regulatory mechanisms involved in biofuel production. Metabolomics has been used to identify metabolic pathways involved in lignocellulosic biomass degradation, lipid accumulation in algae, and fermentation pathways in bacteria, which can be targeted for improvement in biofuel production. Systems biology has been used to integrate omics data and develop models of biological systems involved in biofuel production, providing a

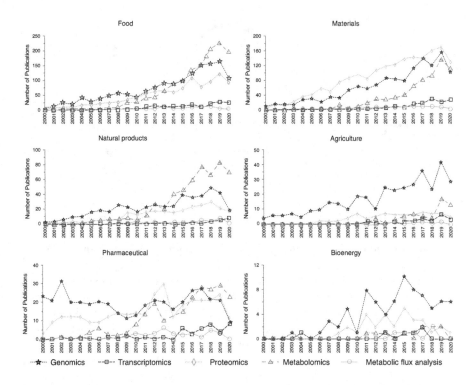

FIGURE 9.5   Status of omics in different industrial products [5].

holistic view of the metabolic pathways involved in biofuel production. Systems biology has been used to develop models of the metabolic pathways involved in lignocellulosic biomass degradation, lipid accumulation in algae, and fermentation pathways in bacteria, which can be used to improve biofuel production processes. Ultimately, omics technology has provided new insights and tools to improve the understanding of the biological processes involved in biofuel production, leading to the development of new biofuel production methods and improved efficiency of biofuel production processes (Figure 9.6). An overview from biomass to omics and further product formation [5].

a. **Metabolomics in biofuels generation**: An effective tool for increasing the production of bioethanol is metabolomics. The metabolic characteristics of a *Clostridium thermocellum* wild-type (WT) strain and an ethanol-tolerant strain grown without (ET0) or with (ET3) 3% (v/v) exogenous ethanol were examined by Zhu et al. [7] using systematic metabolomics. According to the research, the ethanol-tolerant strain exhibited a larger concentration of several metabolites, like organic acids and amino acids, which may have something to do with its increased ethanol tolerance [8]. Like this, Seong et al. [9] used metabolomic analysis to characterise an alcohol yeast of the *Issatchenkia orientalis* MTY1 strain and its various tolerance against high temperature and acidity environments. The study

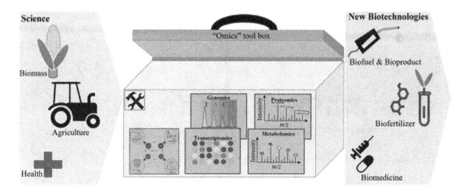

FIGURE 9.6   An overview from biomass to omics and further product formation [5].

discovered that *I. orientalis* MTY1 has greater growth ranges than standard yeast of Saccharomyces cerevisiae at pH 2–8 and 30°C–45°C. These findings show how metabolic pathways and genes that may serve as possible targets for the creation of genetically modified microbes to produce bioethanol can be identified using metabolomics. The study of all the metabolites in a biological system is known as metabolomics. By identifying and characterising the metabolites involved in the metabolic pathways of bioethanol production, it plays a significant part in the generation of bioethanol. By identifying important metabolites associated with Saccharomyces cerevisiae's quick adaptation to several inhibitors of furfural, acetic acid, and phenol, metabolomics can aid in the optimisation of bioethanol production processes. Users can rapidly explore metabolomics data and quickly find intriguing and reusable metabolomics datasets thanks to the metabolome change system, which was created by the EMBL-EBI-coordinated Coordination of Standards in Metabolomics (COSMOS) project. Comprehensive metabolite profiling and thermodynamic analysis can reveal the limitations in xylose-fermenting Saccharomyces cerevisiae. In conclusion, the identification and characterisation of the metabolites involved in the metabolic pathways of bioethanol synthesis are key steps in the development of metabolomics, which can aid in the optimisation of bioethanol production processes. The metabolome Xchange technology enables users to quickly find intriguing and reusable metabolomics datasets and rapidly query metabolomics data.

b. **Proteomics in the production of biofuels**: By identifying and characterising the proteins involved in the metabolic pathways of bioethanol generation, proteomics plays a significant role in the production of bioethanol. Bicolor sorghum *L. Moench* is a high-biomass feedstock that is quickly becoming available for the manufacture of bioethanol and lignocellulosic biomass, and conventional proteomics methods have been utilised to describe the sorghum proteome. *Trichoderma reesei* is frequently employed in the biotechnology sector, primarily in the manufacture of biofuels. In

2008, the genome of this microbe was sequenced, bringing up new research opportunities. The optimisation of bioethanol production methods can benefit from the identification of the proteins involved in the metabolic pathways of bioethanol generation. The identification and characterisation of the proteins involved in the metabolic pathways of the synthesis of bioethanol are significant tasks for proteomics, which can aid in the optimisation of the processes for the manufacture of bioethanol.

c. **Mutagenic studies in the production of biofuels**: The production of bioethanol has been enhanced through the application of mutagenic studies. For instance, *Penicillium echinulatum*'s new genetic variant was discovered using mutagenesis, selection, micro fermentation, and microanalysis techniques [10]. A study also investigated using a novel marine yeast strain called *Saccharomyces cerevisiae AZ65* and seawater-based media to lessen the water footprint of bioethanol production [11]. By locating essential metabolites associated with Saccharomyces cerevisiae's quick adaptation to several inhibitors of furfural, acetic acid, and phenol, bioethanol production processes can be optimised [12]. A kinetic model was built for the overall reduced sugar consumption and the production of bioethanol [12], and it can lower product costs to employ cheap and abundant maize stover in yeast fermentation. In conclusion, mutagenesis studies have been used to enhance the production of bioethanol, and optimisation of the bioethanol production processes can be accomplished by identifying important metabolites associated with *Saccharomyces cerevisiae*'s quick response to a variety of inhibitors. Costs of production can also be decreased by using a cheap and plentiful feedstock like maize stover. Plant cell walls are composed of cellulose, a complex carbohydrate. Given that it can be converted into simple sugars and then fermented to create bioethanol, it has the potential to be a source of renewable energy. Cellulose is particularly helpful in biofuels for several reasons, including the fact that it is plentiful and accessible in agricultural waste, forestry leftovers, and energy crops. It can lessen competition for food crops and land use if it is used as a feedstock for biofuel production. Because it is a renewable resource and has a smaller carbon impact than fossil fuels, cellulose is a sustainable feedstock for biofuels. Because cellulose is more plentiful and doesn't need as much energy or water to grow as traditional feedstocks like maize and sugarcane, making biofuels from it should be less expensive. Energy security can be improved and dependence on foreign oil can be decreased by using cellulose as a feedstock for biofuels. Because the carbon emitted during combustion is balanced by the carbon absorbed during plant growth, biofuels made from cellulose are regarded as carbon-neutral. Overall, using cellulose as a feedstock for biofuels has the potential to offer a reliable, affordable, and plentiful source of renewable energy, with advantages for the economy and the environment.

The process of making bioethanol from cellulose requires numerous phases. The pre-treatment of the biomass is the first stage in the process of turning the cellulose and hemicellulose into simple sugars that are simple to ferment. Chemical treatment, steam explosion, and enzymatic hydrolysis are examples of pre-treatment techniques. In the second phase, yeast or bacteria are used to ferment the simple carbohydrates into ethanol. To acquire the finished product, the generated ethanol is next refined through distillation and dehydration. Compared to more conventional feedstocks like corn and sugarcane, using cellulose as a feedstock to produce bioethanol has several benefits. In forestry by-products, energy crops, and agricultural waste, cellulose is plentiful and easily accessible. Its usage as a feedstock to produce bioethanol may help lessen competition for agricultural land and food crops. However, there are still several technical and financial difficulties with the manufacture of bioethanol from cellulose. The yield of simple sugars is frequently insufficient to make the pre-treatment phase economically viable due to its high cost and energy requirements. Research is still being done to increase the viability of producing bioethanol from this feedstock by pre-treating and fermenting cellulose in more efficient and affordable ways.

Cellulase enzymes with higher binding specificity for cellulose have been created via site-directed mutagenesis, which can decrease the efficiency of cellulases' unproductive binding to lignin and improve the efficiency of the enzymatic hydrolysis of biomass [13]. The ratios and synergetic effects of each cellulase are also simultaneously adjusted, and these factors are crucial for efficient cellulose degradation [14]. The synergistic action of the cellulolytic enzymes endoglucanase, cellobiohydrolases, and glucosidase is necessary for the effective breakdown of cellulosic biomass [15]. In *Sporotrichum* (*Chrysosporium*) thermophile, cellobiose dehydrogenases have been discovered, and they may be responsible for the non-hydrolytic (initial) attack on cellulose [16]. In conclusion, cellulases with improved cellulose binding specificity have been engineered via site-directed mutagenesis, which can improve the effectiveness of enzymatic hydrolysis of biomass. For efficient cellulose degradation, the ratios and synergistic effects of each cellulase are adjusted at the same time at different expression levels. In this organism, cellobiose dehydrogenases have been discovered, and they may be responsible for the non-hydrolytic (first) attack on cellulose.

d. **Enzyme engineering in the production of biofuels**: Enzyme engineering can help to lower the price of cellulases and increase their effectiveness in the synthesis of bioethanol. In addition, cellulase synthesis during solid-state fermentation can be enhanced by optimising process variables such as temperature, pH, and substrate concentration [17,18]. Cell wall–degrading enzymes can assist to lower the cost of producing bioethanol, and plant molecular farming has been created as an efficient way to do this [19]. In conclusion, some techniques that can be utilised to enhance cellulase production and lower the cost of bioethanol production include enzyme engineering, process parameter optimisation, and plant molecular farming.

e. **Genomics in the production of biofuels**: Genomics plays a significant role in the production of algal biofuels by identifying the critical elements involved in lipid productivity and initiating hypothesis-driven strain-improvement efforts [20]. In the past, there weren't enough algal genome sequences accessible, and attempts to use bioprospecting methods to find the best-producing wild species had mainly failed [21]. The capacity to pinpoint metabolic pathways and genes that may serve as potential targets in the creation of genetically modified microalgae with enhanced lipid production capability has recently advanced due to advancements in algal genomics [22]. A unique system for algal biofuel production has been created that integrates algal biomass production, wastewater treatment, and hydrothermal liquefaction of biomass to bio-crude oil [23]. Low-lipid content algae have been suggested as a source of biofuel. In conclusion, genomics is essential for pinpointing the critical factors influencing lipid productivity and for launching hypothesis-driven strain-improvement initiatives in the production of algal biofuels [24,25]. The capacity to identify metabolic pathways and genes that are possible targets in the development of genetically altered microalgae with increased lipid production abilities has recently advanced thanks to advancements in algal genomics. A unique system for algal biofuel production has been created that integrates algal biomass production, wastewater treatment, and hydrothermal liquefaction of biomass to bio-crude oil. The utilisation of low-lipid content algae for biofuel production has been proposed.

9.11.2 OMICS TECHNOLOGY IN ALGAE-BASED BIOFUELS

Algal biofuels hold out the possibility of becoming more environmentally friendly than biofuels made from terrestrial oil crops and bioethanol made from maize and sugarcane [26]. Algae can be grown in brackish, saline, and wastewater with little competition, as well as on unproductive soil that is unfit for agriculture [26]. The most effective low-cost method for producing algal-derived biofuels is to use gravity settling, which may be improved by flocculation, without the use of chemical flocculants. Algal biofuels would primarily be based on the algal cell's high lipid content, making them a suitable feedstock for high-energy-density transportation fuels including biodiesel, green diesel, green jet fuel, and green petrol. Following the recovery and refinement of the algal oil, the downstream conversion to biodiesel or green diesel is well understood. Algal wastewater treatment and hydrothermal liquefaction have been integrated in a unique method for the generation of algal biofuel [27]. Traditional biofuels can be replaced with algal biofuels, which are a promising and ecological alternative. By combining hydrothermal liquefaction with algal wastewater treatment, biofuel output can be increased while waste is reduced. It has become easier to use and to identify the metabolic pathways targeted for the genes coded for genetically modified microalgae thanks to recent studies and their successful contribution to the combination of genomics in association with omics technology. Cyanobacteria have many applications in industries and are actively contributing to the same. Cyanobacteria have successfully produced a wide range of bioproducts, including biowaste, in industrial settings. Other products that can be obtained from cyanobacteria include high-value bioactive components, recombinant proteins, biofuels, isopropanol, hydrogen, and biofuels. Omics technology is the technology that has made it simple to create scientific methodology; it is based on bioeconomic and has many industrial applications.

a. **Algae-based biofuels and proteomics**: By identifying and characterising the proteins involved in the metabolic pathways of lipid biosynthesis, proteomics plays a significant role in the development of algae-based biofuel. To clarify the essential elements involved in lipid productivity and to launch hypothesis-driven strain-improvement techniques, transcriptomics and proteomics investigations have been used. Proteomics can be used to identify the important enzymes involved in lipid biosynthesis and to maximise the generation of biofuels from algae [28]. The identification of the proteins involved in an algae's reaction to environmental challenges, such as nutrient shortage and high light levels, which can lower lipid output, is another application of proteomics [29]. The identification and characterisation of the proteins involved in the metabolic pathways of lipid biosynthesis in algae is vital for optimising the generation of biofuels. This is where proteomics comes in. Aside from helping to discover the proteins involved in the response of algae to environmental challenges, transcriptomics and proteomics investigations can also be utilised to launch hypothesis-driven strain-improvement methods.

b. **Algae-based biofuels and metabolomics:** By identifying and characterising the metabolites involved in the metabolic pathways of lipid biosynthesis, metabolomics plays a significant role in the development of algal biofuels. By identifying important metabolites associated with lipid productivity and the metabolic pathways that can be targeted for genetic engineering to boost lipid productivity, metabolomics can help to maximise the generation of biofuels. The identification of the metabolites involved in how algae react to environmental challenges such as nutrient shortage and high light levels, which can lower lipid output, is another benefit of metabolomics. In conclusion, the identification and characterisation of the metabolites involved in the metabolic pathways of lipid biosynthesis in the creation of algae biofuels are significant tasks for metabolomics. By identifying important metabolites associated with lipid productivity and the metabolic pathways that can be targeted for genetic engineering to increase lipid productivity, metabolomics can help to maximise the generation of biofuels.

Through the identification of crucial elements involved in lipid productivity and the launch of hypothesis-driven strain-improvement initiatives, omics technology, which includes genomics, proteomics, and metabolomics, has the potential to revolutionise the production of algal biofuels [30]. For instance, genomics can be utilised to pinpoint possible targets for metabolic pathways and genes in the creation of genetically modified microalgae with enhanced lipid production capacity [31]. In order to maximise the generation of biofuels, proteomics can be utilised to identify the proteins involved in the metabolic pathways of lipid biosynthesis [32]. Key metabolites associated with lipid productivity can be found using metabolomics, as can metabolic pathways that can be targeted for genetic engineering to increase lipid productivity [33]. The generation of algal strains with higher lipid productivity and other desirable features for the production of biofuels can result from the fusion of omics technology with other methodologies, such as metabolic engineering and synthetic biology [34]. As a result, omics technology has the potential to dramatically increase the production of algal biofuels by pinpointing the critical factors influencing lipid productivity and enabling the creation of genetically modified algae strains with enhanced lipid productivity and other desirable characteristics.

Cooking oil waste has the potential to be used as a feedstock for the generation of biodiesel. According to a study, waste cooking oil may be transformed into biodiesel utilising a two-step process that starts with base- and acid-catalysed transesterification. Cooking oil waste disposal can have a negative environmental impact, but it can also lessen our reliance on fossil fuels if waste cooking oil is used as a feedstock for the creation of biodiesel [35]. In addition, since waste cooking oil is a cheap feedstock, using it as a source for the creation of biodiesel can assist to lower the cost of making it [36]. The quality of the used waste cooking oil as a feedstock, however, can have an impact on the quality of the biodiesel generated, and the presence of impurities in the used waste cooking oil can have an impact on the process' effectiveness in producing biodiesel [37]. In summary, employed as a feedstock for biodiesel production, spent cooking oil has the potential to reduce the environmental impact of used cooking oil disposal, reduce reliance on fossil fuels, and lower the cost of biodiesel production. The efficiency of the biodiesel production process, however, must take into account the quality of the used waste cooking oil as a feedstock. Bacteria are essential for the production of biofuels, particularly in the transformation of organic matter into biogas, biohydrogen, and bioethanol. Biogas, a combination of methane and carbon dioxide, is created using anaerobic microorganisms. In the absence of oxygen, these bacteria decompose organic waste, such as food scraps, animal dung, and agricultural wastes, to create biogas. Biohydrogen can be produced by photosynthetic bacteria like *Rhodobacter sphaeroides* through photosynthesis. These bacteria use sunlight to convert water and organic materials into hydrogen gas. Through a process known as homofermentative fermentation, lactic acid bacteria can create bioethanol. Using sugars like glucose as fuel, these bacteria generate ethanol and carbon dioxide. Agricultural waste and forestry waste are examples of lignocellulosic biomass that can be broken down into sugars by cellulolytic bacteria such as *Clostridium thermocellum* and *Caldicellulosiruptor bescii*. These sugars can then be fermented to create bioethanol. Gas fermentation is a method used by acetogenic bacteria, like *Clostridium ljungdahlii*, to make biofuels like ethanol and butanol. As a by-product, these bacteria create biofuels using carbon dioxide or monoxide as their carbon source.

9.12   BACTERIA ASSOCIATED WITH THE PRODUCTION OF BIOFUELS

Bacteria have the potential to be used for biofuel generation due to their ability to produce a variety of biofuels, including ethanol, butanol, and hydrogen. For example, *Clostridium acetobutylicum* is a bacterium that can produce butanol through the acetone-butanol-ethanol (ABE) fermentation process In addition, *Escherichia coli* has been engineered to produce ethanol from glucose through the Entner–Doudoroff pathway. Bacteria can also be used to produce hydrogen through the dark fermentation process, which involves the breakdown of organic matter by bacteria in the absence of light. The use of bacteria for biofuel generation can be advantageous due to their ability to utilise a wide range of feedstocks, including waste biomass and lignocellulosic materials. However, the efficiency of bacterial biofuel production can be affected by factors such as substrate availability, product toxicity, and the need for expensive downstream processing. In conclusion, bacteria have the potential to be used for biofuel generation due to their ability to produce a variety of biofuels and to utilise a wide range of feedstocks. However, more investigation is required to improve bacterial biofuel production and to overcome the difficulties in producing bacterial biofuel.

A technology that has promise for generating energy from organic waste or harvested plant material is bacterial-based biogas production. A variety of microorganisms, including bacteria from the genus *Clostridium*, create biogas in anaerobic reactors. The possibility of pathogenic *Clostridium* species developing during the fermentation process is a worry, though. Disturbances like feeding rates, which can affect the effectiveness of biogas production, can have an impact on the relative richness of microbial populations. According to research on the bacterial and archaeal communities identified in anaerobic digestion reactors used to produce biogas, the type of feedstock used had an impact on the microbial community structure. Therefore, bacteria-based biogas production is a viable method for generating energy from harvested plant material or organic waste. However, the possible proliferation of pathogenic Clostridium species during the fermentation process is a cause for concern, and disturbances like feeding rates can have an impact on how effectively biogas is produced. The kind of feedstock utilised has an impact on the composition of the microbial population in anaerobic digestion reactors used to produce biogas.

9.13   ARTIFICIAL BIOLOGY METHODS

Advanced biofuels are being created by engineering microbes with synthetic biology techniques. These methods can be utilised to create new biofuels that cannot be produced by natural organisms and to enhance metabolic pathways for enhanced biofuel production. To absorb and use carbon dioxide emissions, carbon capture systems are being linked with the production of biofuels. It is possible to produce biofuels that are carbon-neutral or even carbon-negative by absorbing carbon dioxide from the atmosphere or industrial sources. Researchers are looking into novel ways to turn garbage into energy, like making biofuels from municipal solid waste or used cooking oil. These trash-to-energy devices can contribute to waste reduction and offer a sustainable energy source. To replace conventional fossil fuels with sustainable alternatives, algae-based biofuels are currently being developed. Algae can provide significant amounts of oil that can be turned into biofuels and can be cultivated using wastewater or saltwater, which reduces the demand for freshwater resources. A variety of bio-based products, including biofuels, chemicals, and materials, are being produced in biorefineries. Biorefineries can maximise resource utilisation and reduce waste by integrating several bio-based technologies. These recent events demonstrate how crucial a role biofuel could play in the shift to a low-carbon economy. Agricultural and forestry leftovers, which are examples of lignocellulosic biomass, have the potential to be used as a feedstock for the manufacture of biofuels. The generation of biofuels from lignocellulosic biomass can help reduce reliance on fossil fuels and greenhouse gas emissions. However, because of its complex structure and resistance to degradation, lignocellulosic biomass is more challenging to convert into biofuels than conventional feedstocks, such as maize and sugarcane. To address these issues, several strategies have been developed, such as using fungal-bacterial communities to transform raw cellulolytic feedstocks into biofuels. In addition, to gain a better understanding of the metabolic processes involved in converting the copious pentose sugars found in hemicellulose into biofuels, comparative genomics of xylose-fermenting fungi has been explored. Lignocellulosic biomass has the potential to be used as a feedstock for the production of biofuels, which can lessen reliance on fossil fuels and greenhouse gas emissions. Plant pathogenic fungus is also being investigated as a largely unexplored resource for novel hydrolytic enzymes for biomass conversion.

9.14 ADDITIONAL MOLECULAR METHODS FOR BIOFUELS RESEARCH

The creation of biofuels has benefited greatly from the use of molecular biology technologies. Using the potent technique of genetic engineering, scientists can change the genetic make-up of microorganisms to enhance their capacity to produce biofuels. For instance, scientists can insert genes that help microbes break down cellulose more quickly or yield more ethanol. The study of all the RNA molecules generated by a cell or organism is known as transcriptomics. Researchers can determine which genes in microbes are active during the creation of biofuel by employing transcriptomics. Utilising this knowledge, metabolic pathways can be modified for increased biofuel production. The study of all the tiny molecules that a cell or organism produces is known as metabolomics. Researchers can determine which metabolic pathways in microbes are active during the production of biofuel by using metabolomics. Utilising this knowledge, metabolic pathways can be modified for increased biofuel production. The study of all the proteins that a cell or organism produces is known as proteomics. Researchers can determine which proteins are essential for microorganisms to produce biofuel by using proteomics. Utilising this knowledge, metabolic pathways can be modified for increased biofuel production.

9.15   CRISPR-CAS TECHNOLOGY FOR THE PRODUCTION OF BIOFUEL

The use of CRISPR-Cas technology in the creation of biofuels has the potential to be very beneficial. To increase the effectiveness and productivity of microorganisms utilised in the generation of biofuels, such as yeast and bacteria, CRISPR-Cas can be used to precisely modify their genomes. For instance, CRISPR-Cas can be used to remove genes that prevent the synthesis of biofuels or introduce genes that boost biofuel yield. In microorganisms employed in the generation of biofuel, CRISPR-Cas can also be utilised to control gene expression. Researchers can optimise metabolic pathways for enhanced biofuel production by turning genes on or off by utilising CRISPR-Cas to target genes or gene regulatory areas. To create new kinds of biofuels, CRISPR-Cas can be utilised to modify the complete metabolic processes of microorganisms. Researchers can construct novel metabolic pathways that produce biofuels with enhanced characteristics or that aren't present in nature by utilising CRISPR-Cas to add or delete genes in existing metabolic pathways. Microbial communities can be engineered using CRISPR-Cas to produce biofuels. Researchers can modify the structure of microbial communities for increased biofuel production by focusing on microbes or genes in the communities. CRISPR-Cas: As was already noted, CRISPR-Cas is a potent genetic engineering technique that can be used to increase the productivity and effectiveness of microorganisms employed in the generation of biofuels. The development of new and more effective microbes for the production of biofuels as well as the improvement of our understanding of the metabolic pathways involved have both benefited from the use of molecular biology technologies. These instruments will become more crucial as they grow, contributing to the creation of sustainable and affordable biofuels.

By permitting accurate and effective genetic modification of the microorganisms utilised in biofuel production, CRISPR-Cas technology has the potential to revolutionise the production of biofuels [38]. It has been reported that the cellulolytic bacterium Clostridium has developed type I-B and type II CRISPR/Cas genome-editing systems. This technology can be used to engineer microorganisms to produce biofuels from a variety of feedstocks, including lignocellulosic biomass and waste materials. By optimising metabolic pathways and reducing the production of unwanted by-products, CRISPR-Cas technology can also be used to increase the efficiency of biofuel production. It can also cut down on the time and expense involved with conventional genetic engineering methods. As a result of permitting accurate and effective genetic modification of the microorganisms employed in the manufacture of biofuel, CRISPR-Cas technology has the potential to dramatically enhance the production of biofuel. A noteworthy advancement in this area is the creation of type I-B and type II CRISPR/Cas genome-editing systems in the cellulolytic bacterium, Clostridium. The outcome demonstrates the revolutionary nature of CRISPR-based genome editing as a tool for genetic engineering and the benefits it offers for cellular metabolic engineering. To produce biofuel, the study examines the current developments and obstacles in CRISPR-mediated genome editing in unconventional yeasts. The cellulolytic *Clostridium thermocellum* has developed both type I-B and type II CRISPR/Cas genome-editing systems [39]. To enhance the ability to engineer C, the work characterises native Type I-B and heterologous Type II Clustered Regularly Interspaced Short Palindromic Repeat (CRISPR)/cas (CRISPR associated) systems. *Clostridium thermocellum* to produce biofuel: The findings demonstrate that cyanobacteria have been effectively edited in their genomes using CRISPR systems like CRISPR-Cas9 and CRISPR-Cpf1. To develop the tools and create genetic strategies based on CRISPR-Cpf1, the study used the cyanobacterium Anabaena PCC 7120 as a model strain. This allowed the realisation of genetic studies that had previously been challenging.

9.16 CONCLUSION

In conclusion, genome-editing techniques, especially those based on CRISPR, have enormous potential to enhance the production of biofuels by permitting precise and effective genetic engineering of the microbes that produce it. Some notable advancements in this area include the utilisation of CRISPR-Cpf1 in cyanobacteria and the development of type I-B and type II CRISPR/Cas genome-editing systems in *Clostridium thermocellum*. Transcriptomics, the study of an organism's entire set of RNA transcripts, has been used to pinpoint critical pathways and genes for the synthesis of biofuels [40–42]. To develop next-generation biofuels from the microalgae *Dunaliella tertiolecta*, Rismani-Yazdi et al. [40] used transcriptome sequencing and annotation to discover pathways and genes of interest. The investigation of triacylglycerol biosynthetic routes in an unsequenced microalga by Guarnieri et al. [41] could hasten the commercialisation of algae-derived biofuels. To determine the metabolic processes causing high-rate lipid production in these oleaginous species, Tanaka et al. [42] examined the genome and transcriptome of the oleaginous diatom *Fistulifera solaris JPCC DA0580*. In addition, switchgrass, a developing biofuel crop, was employed by Xie et al. [43] to identify stress-responsive microRNAs using high-throughput deep sequencing, which made it easier to choose gene targets to increase biomass and biofuel yield. Since this can speed up the commercialisation of biofuels and increase biomass and biofuel yield, transcriptomics has been utilised to pinpoint important pathways and genes for biofuel production.

By identifying crucial elements in biofuel production and fostering the creation of genetically modified microorganisms with enhanced biofuel production capacities, omics technology has the potential to revolutionise the biofuel industry. By reducing the creation of undesirable by-products and optimising metabolic pathways, omics technology can also increase the effectiveness of biofuel production. Recent advances in omics technology, such as the use of CRISPR-Cas for genome editing and transcriptomics for identifying pathways and genes crucial for the generation of biofuels, have showed significant promise in enhancing the production of biofuels. In conclusion, omics technology has the potential to dramatically enhance biofuel production by pinpointing crucial elements involved in the process and fostering the creation of genetically modified microbes with enhanced biofuel production skills [44–48].

The production of biofuels has a bright future since it is seen as an important means of lowering carbon emissions and advancing renewable energy sources. The development of new biofuel production technologies, such as genetic engineering, will probably improve efficiency and lower production costs. The availability of feedstock for biofuels may expand because of advances in agriculture and biotechnology, which will help make these fuels more affordable. Production facilities will probably expand as demand for biofuels rises, resulting in increased economies of scale and lower production costs. As consumers, governments, and businesses increasingly use biofuels, more money may be spent on research and development, which could raise biofuel efficiency and lessen their environmental impact. To build a more sustainable and dependable energy system, biofuels can be used with other renewable energy sources like solar and wind. Overall, the production of biofuel appears to have a bright future because it could offer a sustainable alternative to current fossil fuels. To make it a viable and competitive energy source, though, further research, development, and investment will be needed. Biofuels can play a significant role by replacing the expensive and outdated energy resources because of the depletion of non-renewable energy sources. Since many years ago, people have begun using biofuels in the United States, Europe, and Asia. There are some methods that the auto industry will use to build engines that run effectively on biofuel. We have tried to gather a significant amount of information in this review that describes the procedures and techniques utilised to create biofuels from biowaste. Every year, plants lose their leaves during specific seasons. Agricultural wastes are another major source of biowaste, which can be recycled in the most effective way possible. Conventional energy sources are not only in low supply but conventional energy sources are also contributing to environmental pollution through CO_2 emissions. Therefore, it is highly advised to use biofuels like bioethanol, biodiesel, biochar, and bio-oils.

REFERENCES

[1] Bose K, J. C., & Sarwan, J. (2023). Multi-enzymatic degradation potential against wastes by the novel isolate of Bacillus. *Biomass Convers. Biorefin.*, 1–14.

[2] Sarwan, J., & Bose, J. C. (2021). Importance of microbial cellulases and their industrial applications. *Ann. Romanian Soc. Cell Biol.*, 3568–3575.

[3] Zabed, H. M., Akter, S., Yun, J., Zhang, G., Awad, F. N., Qi, X., & Sahu, J. N. (2019). Recent advances in biological pretreatment of microalgae and lignocellulosic biomass for biofuel production. *Renew. Sust. Energ. Rev.*, 105, 105–128.

[4] Osman, A.I., Mehta, N., Elgarahy, A.M. *et al.* Conversion of biomass to biofuels and life cycle assessment: A review. *Environ. Chem. Lett.* 19, 4075–4118 (2021). doi:10.1007/s10311-021-01273-0

[5] Amer, B., & Baidoo, E. E. (2021). Omics-driven biotechnology for industrial applications. *Fron. Bioeng. Biotechnol.*, 9, 613307. doi:10.3389/fbioe.2021.613307

[6] Puchajda, B., & Oleszkiewicz, J. (2008). Impact of sludge thickening on energy recovery from anaerobic digestion. *Water Sci. Technol.*, 3(57), 395–401. doi:10.2166/wst.2008.021

[7] Zhu, X., Cui, J., Feng, Y., Fa, Y., Zhang, J., & Cui, Q. (2013). Metabolic adaption of Ethanol-Tolerant *Clostridium thermocellum*. *PLoS One*, 8(7), e70631. https://doi.org/10.1371/journal.pone.0070631.

[8] Amorim, H. V., Lopes, M. L., de Castro Oliveira, J. V., Buckeridge, M. S., & Goldman, G. H. (2011). Scientific challenges of bioethanol production in Brazil. *Appl. Microbiol. Biotechnol.*, 5(91), 1267–1275. doi:10.1007/s00253-011-3437-6

[9] Seong, Y. J., Lee, H.J., Lee, J. E., Kim, S., Lee, D.Y., Kim, K.H., & Park, Y.C. (2017). Physiological and metabolomic analysis of Issatchenkia orientalis MTY1 with multiple tolerance for cellulosic bioethanol production. *Biotechnol J.*, 12(11). doi: 10.1002/biot.201700110.

[10] Dillon, A. J. P., Bettio, M., Pozzan, F. G., Andrighetti, T., & Camassola, M. (2011). A new *Penicillium echinulatum* strain with faster cellulase secretion obtained using hydrogen peroxide mutagenesis and screening with 2-deoxyglucose. *J. Appl. Microbiol.*, 1(111), 48–53. doi:10.1111/j.1365-2672.2011.05026.x

[11] Zaky, A. S., Greetham, D., Tucker, G. A., & Du, C. (2018). The establishment of a marine focused biorefinery for bioethanol production using seawater and a novel marine yeast strain. *Sci. Rep.*, 1(8). doi:10.1038/s41598-018-30660-x

[12] Tian, S. Q., & Chen, Z. C. (2016). Dynamic analysis of bioethanol production from corn stover and immobilized yeast. *Bioresources*, 3(11). doi:10.15376/biores.11.3.6040-6049

[13] Strobel, K. L., Pfeiffer, K. A., Blanch, H. W., & Clark, D. S. (2016). Engineering Cel7A carbohydrate binding module and linker for reduced lignin inhibition. *Biotechnol. Bioeng.*, 6(113), 1369–1374. doi:10.1002/bit.25889

[14] Yamada, R., Taniguchi, N., Tanaka, T., Ogino, C., Fukuda, H., & Kondo, A (2010). Cocktail δ-integration: A novel method to construct cellulolytic enzyme expression ratio-optimized yeast strains. *Microb. Cell Fact.*, 1(9). doi:10.1186/1475-2859-9-32

[15] Yamada, R., Nakatani, Y., Ogino, C., & Kondo, A. (2013). Efficient direct ethanol production from cellulose by cellulase- and cellodextrin transporter-co-expressing Saccharomyces cerevisiae. *AMB Expr.*, 1(3). doi:10.1186/2191-0855-3-34

[16] Canevascini, G., Borer, P., & Dreyer, J. L. (1991). Cellobiose dehydrogenases of *Sporotrichum* (Chrysosporium) *thermophile*. *Eur. J. Biochem.*, 1(198), 43–52. doi:10.1111/j.1432-1033.1991.tb15984.x

[17] Singhania, R. R., Sukumaran, R. K., & Pandey, A. (2007). Improved cellulase production by *Trichoderma reesei* RUT C30 under SSF through process optimization. *Appl. Biochem. Biotechnol.*, 1(142), 60–70. doi:10.1007/s12010-007-0019-2

[18] Mekala, N. K., Singhania, R. R., Sukumaran, R. K., & Pandey, A. (2008). Cellulase production under solid-state fermentation by *Trichoderma reesei* RUT C30: Statistical optimization of process parameters. *Appl. Biochem. Biotechnol.*, 2–3(151), 122–131. doi:10.1007/s12010-008-8156-9

[19] Jung, S., Lee, D. S., Kim, Y. O., Joshi, C. P., & Bae, H. J. (2013). Improved recombinant cellulase expression in chloroplast of tobacco through promoter engineering and 5′ amplification promoting sequence. *Plant Mol. Biol.*, 4–5(83), 317–328. doi:10.1007/s11103-013-0088-2

[20] Guarnieri, M. T., Nag, A., Smolinski, S. L., Darzins, A., Seibert, M., & Pienkos, P. T. (2011). Examination of triacylglycerol biosynthetic pathways via De Novo transcriptomic and proteomic analyses in an unsequenced microalga. *PLoS ONE*, 10(6), e25851. doi:10.1371/journal.pone.0025851

[21] Unkefer, C. J., Sayre, R. T., Magnuson, J. K., Anderson, D. B., Baxter, I., Blaby, I. K., & Olivares, J. A. (2017). Review of the algal biology program within the national alliance for advanced biofuels and bioproducts. *Algal Res.*, (22), 187–215. doi:10.1016/j.algal.2016.06.002

[22] Misra, N., Panda, P. K., & Parida, B. K. (2013). Agrigenomics for microalgal biofuel production: An overview of various bioinformatics resources and recent studies to link OMICS to bioenergy and bioeconomy. *OMICS: J. Integra. Biol.*, 11(17), 537–549. doi:10.1089/omi.2013.0025

[23] Zhou, Y., Schideman, L., Zhang, Y., Yu, G., Wang, Z., & Pham, M. (2011). Resolving bottlenecks in current algal wastewater treatment paradigms: A synergistic combination of low-lipid algal wastewater treatment and hydrothermal liquefaction for large-scale biofuel production. *Proc. Water Environ. Fed.*, 6(2011), 347–361. doi:10.2175/193864711802837084

[24] Hannon, M., Gimpel, J., Tran, M., Rasala, B., & Mayfield, S. (2010). Biofuels from algae: Challenges and potential. *Biofuels*, 5(1), 763–784. doi:10.4155/bfs.10.44

[25] Saad, M. G., Dosoky, N. S., Zoromba, M. S., & Shafik, H. M. (2019). Algal biofuels: Current status and key challenges. *Energies*, 10(12), 1920. doi:10.3390/en12101920

[26] Pienkos, P. T., & Darzins, A. L. (2009). The promise and challenges of microalgal-derived biofuels. *Biofuels Bioprod. Bioref.*, 4(3), 431–440. doi:10.1002/bbb.159

[27] Zhou, Y., Schideman, L., Yu, G., & Zhang, Y. (2013). A synergistic combination of algal wastewater treatment and hydrothermal biofuel production maximized by nutrient and carbon recycling. *Energy Environ. Sci.*, 12(6), 3765. doi:10.1039/c3ee24241b

[28] Wang, B., Wang, J., Zhang, W., & Meldrum, D. R. (2012). Application of synthetic biology in cyanobacteria and algae. *Front. Microbiol.*, (3). doi:10.3389/fmicb.2012.00344

[29] Salami, R., Kordi, M., Bolouri, P., Delangiz, N., & Asgari Lajayer, B. (2021). Algae-based biorefinery as a sustainable renewable resource. *Circ. Econ. Sust.*, 4(1), 1349–1365. doi:10.1007/s43615-021-00088-z

[30] Carere, C. R., Sparling, R., Cicek, N., & Levin, D. B. (2008). Third generation biofuels via direct cellulose fermentation. *IJMS*, 7(9), 1342–1360. doi:10.3390/ijms9071342

[31] Wang, H., Laughinghouse IV, H. D., Anderson, M. A., Chen, F., Willliams, E., Place, A. R., ... & Hill, R. T. (2012). Novel bacterial isolate from permian groundwater, capable of aggregating potential biofuel-producing microalga *Nannochloropsis oceanica* IMET1. *Appl. Environ. Microbiol.*, 5(78), 1445–1453. doi:10.1128/aem.06474-11

[32] Wang, H., Hill, R. T., Zheng, T., Hu, X., & Wang, B. (2014). Effects of bacterial communities on biofuel-producing microalgae: Stimulation, inhibition and harvesting. *Crit. Rev. Biotechnol.*, 2(36), 341–352. doi:10.3109/07388551.2014.961402

[33] Farrokh, P., Sheikhpour, M., Kasaeian, A., Asadi, H., & Bavandi, R. (2019). Cyanobacteria as an eco-friendly resource for biofuel production: A critical review. *Biotechnol. Prog.*, 5(35). doi:10.1002/btpr.2835

[34] Pei, G., Sun, T., Chen, S., Chen, L., & Zhang, W. (2017). Systematic and functional identification of small non-coding RNAs associated with exogenous biofuel stress in cyanobacterium *Synechocystis* sp. PCC 6803. *Biotechnol. Biofuels*, 1(10). doi:10.1186/s13068-017-0743-y

[35] Farese, R. V., & Walther, T. C. (2009). Lipid droplets finally get a little R-E-S-P-E-C-T. *Cell*, 5(139), 855–860. doi:10.1016/j.cell.2009.11.005

[36] Letcher, P. M., Lopez, S., Schmieder, R., Lee, P. A., Behnke, C., Powell, M. J., & McBride, R. C. (2013). Characterization of *Amoeboaphelidium protococcarum*, an algal parasite new to the cryptomycota isolated from an outdoor algal pond used for the production of biofuel. *PLoS ONE*, 2(8), e56232. doi:10.1371/journal.pone.0056232

[37] Tu, Q., Lu, M., Thiansathit, W., & Keener, T. C. (2016). Review of water consumption and water conservation technologies in the algal biofuel process. *Water Environ. Res.*, 1(88), 21–28. doi:10.2175/106143015x14362865227517

[38] Walker, J. E., Lanahan, A. A., Zheng, T., Toruno, C., Lynd, L. R., Cameron, J. C., & Eckert, C. A. (2020). Development of both type I-B and type II CRISPR/Cas genome editing systems in the cellulolytic bacterium *Clostridium thermocellum. Metabol. Eng. Commun.*, 10, e00116. doi:10.1016/j.mec.2019.e00116

[39] Sharon, A. A., Nikam, A. N., Jaleel, A., Tamhane, V. A., & Rao, S. P. (2018). Method for label-free quantitative proteomics for *Sorghum bicolor* L. Moench. *Tropic. Plant Biol.*, 1–2(11), 78–91. doi:10.1007/s12042-018-9202-6

[40] Rismani-Yazdi, H., Haznedaroglu, B. Z., Bibby, K., & Peccia, J. (2011). Transcriptome sequencing and annotation of the microalgae *Dunaliella tertiolecta*: Pathway description and gene discovery for production of next-generation biofuels. *BMC Genomics*, 1(12). doi:10.1186/1471-2164-12-148

[41] Guarnieri, M. T., Nag, A., Smolinski, S. L., Darzins, A., Seibert, M., & Pienkos, P. T. (2011). Examination of triacylglycerol biosynthetic pathways via De Novo transcriptomic and proteomic analyses in an unsequenced microalga. *PLoS ONE*, 10(6), e25851. doi:10.1371/journal.pone.0025851

[42] Tanaka, T., Maeda, Y., Veluchamy, A., Tanaka, M., Abida, H., Maréchal, E., ... & Fujibuchi, W. (2015). Oil accumulation by the oleaginous diatom *Fistulifera solaris* as revealed by the genome and transcriptome. *Plant Cell*, 1(27), 162–176. doi:10.1105/tpc.114.135194

[43] Xie, F., Stewart Jr, C. N., Taki, F. A., He, Q., Liu, H., & Zhang, B. (2013). High-throughput deep sequencing shows that microRNAs play important roles in switchgrass responses to drought and salinity stress. *Plant Biotechnol. J.*, 3(12), 354–366. doi:10.1111/pbi.12142

[44] Gupta, V. K., Steindorff, A. S., de Paula, R. G., Silva-Rocha, R., Mach-Aigner, A. R., Mach, R. L., & Silva, R. N. (2016). The post-genomic era of *Trichoderma reesei*: What's next? *Trends Biotechnol.*, 12(34), 970–982. doi:10.1016/j.tibtech.2016.06.003

[45] Kaspar, H., Dettmer, K., Gronwald, W., & Oefner, P. J. (2008). Automated GC-MS analysis of free amino acids in biological fluids. *J. Chromatogr. B*, 2(870), 222–232. doi:10.1016/j.jchromb.2008.06.018

[46] Cook, C. E., Bergman, M. T., Finn, R. D., Cochrane, G., Birney, E., & Apweiler, R. (2016). The European Bioinformatics Institute in 2016: Data growth and integration. *Nucleic Acids Res.*, D1(44), D20–D26. doi:10.1093/nar/gkv1352

[47] Wang, X., Li, B. Z., Ding, M. Z., Zhang, W. W., & Yuan, Y. J. (2013). Metabolomic analysis reveals key metabolites related to the rapid adaptation of *Saccharomyce cerevisiae* to multiple inhibitors of furfural, acetic acid, and phenol. *OMICS: J. Integrat. Biol.*, 3(17), 150–159. doi:10.1089/omi.2012.0093

[48] Klimacek, M., Krahulec, S., Sauer, U., & Nidetzky, B. (2010). Limitations in xylose-fermenting Saccharomyces cerevisiae, made evident through comprehensive metabolite profiling and thermodynamic analysis. *Appl. Environ. Microbiol.*, 22(76), 7566–7574. doi:10.1128/aem.01787-10

10 Multidisciplinary Approaches for Biomass Energy Production

Yamini Tripathi
Helmholtz Centre for Infection Research

Priya Agarwal
Indian Institute of Science Education and
Research Thiruvananthapuram (IISER TVM)

Manisha Bisht
L.S.M Govt. P.G-College

Manoj K. Pal
Graphic Era (Deemed to be University)

10.1 INTRODUCTION

Interdisciplinary approaches play a crucial role in the development of sustainable biofuel production processes. The integration of various fields, including biology, chemistry, engineering, and environmental science, is necessary to overcome the challenges associated with biofuel production, such as feedstock availability, conversion efficiency, and environmental impact. Through interdisciplinary collaborations, novel technologies have been developed that utilize different types of biomass to produce biofuels with high energy content and low emissions (Godbole et al., 2021). Furthermore, interdisciplinary approaches have helped identify strategies to improve the economic viability of biofuels, such as co-product generation and biorefinery concepts. It is evident that interdisciplinary approaches will continue to be essential for the advancement of biofuel production, as the field moves towards more sustainable and efficient processes. A number of methods and approaches are used including physical, chemical, and biochemical (Hajilary et al., 2019; Kumar et al., 2009). The selection of a biofuel production method largely depends on the type of raw material utilized. Consequently, biofuels are classified into various types, namely, first-generation, second-generation, third-generation, and fourth-generation biofuels (Jeswani et al., 2020).

In the physical pretreatment of biomass, various techniques, such as grinding, chipping, shredding, milling, compression/expansion, agitation, and pulse-electric

DOI: 10.1201/9781003406501-10

field (PEF) treatment are commonly used (Aslanzadeh et al., 2014; Chakravarty and Mandavgane, 2022).

However, physical pretreatment methods have certain limitations, such as their inability to remove lignin, which restricts the accessibility of enzymes to cellulose (Berlin et al., 2006; Mooney et al., 1999). Moreover, these methods require high energy consumption, and their large-scale implementation may not be cost-effective, thereby posing significant environmental and safety concerns (Agbor et al., 2011). Chemical pretreatment is mainly used to improve the efficiency of enzymatic hydrolysis. Chemical methods use various types of acids, bases, oxidizing agents, and solvents to alter the biomass structure and composition (Arhin et al., 2023). However, they also have some limitations, including high costs, corrosiveness, and potential by-product formation (Melero et al., 2023). Therefore, the selection of appropriate chemical pretreatment methods should be based on a thorough understanding of their effects on biomass and the downstream processes (Grippi et al., 2020).

Application of enzymatic methods has gained significant attention as an alternative to traditional chemical methods for biofuel production (Rocha-Gaso et al., 2009). Enzymes are highly specific bio-catalysts that can convert complex biomass into simple sugars, which can be used as feedstock for biofuel production (Barelli et al., 2021). In the enzymatic method, complex biopolymers such as cellulose and hemicellulose are converted into simple sugars such as glucose and xylose. Hydrolysis, fermentation, and transesterification are some commonly used enzymatic processes (Rao et al., 2020). Enzymatic methods have several advantages over traditional chemical methods, including high selectivity, mild reaction conditions, and low environmental impact. However, enzymatic methods also have several challenges, including high enzyme costs, enzyme stability, and enzyme inhibition (Onumaegbu et al., 2018). Addressing these challenges is critical to realizing the full potential of enzymatic methods in biofuel production.

In addition, progress in genetic engineering, computational biology (e.g. NGS), multi-omic databases, and genome-scale metabolic reconstruction have enabled the rapid identification of new targets or pathways for strain development (Godbole et al., 2021). This reduces the need for extensive experimental work, leading to time and cost savings. By integrating multiple interdisciplinary approaches, it is possible to enhance the yield of biofuel with reduction in the cost and time. Various new approaches have garnered much attention over the decade and this chapter will look into some of the most promising approaches for biomass conversion.

10.2 BIOMASS CONVERSION

Biomass is defined as biological mass that is used as a renewable energy source. Various types of biomasses are used in biofuel production, and each required unique pretreatment. Some commonly used biomasses are as follows:

- **Wood and wood waste**: This includes forest residues, sawdust, wood chips, and bark.
- **Agricultural residues**: This includes crop residues like straw, corn stover, and sugarcane bagasse.

- **Energy crops**: These are crops that are grown specifically for energy production, such as switchgrass, miscanthus, and fast-growing trees like willow.
- **Municipal solid waste**: This includes organic waste from households and businesses, such as food waste and yard trimmings.
- **Algae**: Algae can be grown in ponds or tanks and can be used to produce biofuels like biodiesel and bioethanol.
- **Animal waste**: This includes manure from livestock, which can be converted into biogas through anaerobic digestion.
- **Industrial waste**: This includes waste from industries like paper and pulp mills, which can be used to produce biofuels.

Each type of biomass has its own advantages and disadvantages, and the suitability of each type depends on factors like availability, cost, and the type of biofuel being produced.

Biomass is converted to energy through different processes, including direct combustion or heat generation. The following are the approaches for converting biomass into energy and other useful products:

1. **Thermal conversion**: This involves the direct burning of biomass to produce heat energy, which can be used for various purposes such as heating and power generation.
2. **Biological conversion**: This approach involves using microorganisms such as yeast, bacteria, or algae to produce energy in the form of biofuels, such as ethanol or biodiesel.
3. **Chemical conversion**: This involves the use of chemical processes to convert biomass into energy or other useful products. One example is the production of bio-oils through the fast pyrolysis of biomass.
4. **Gasification**: This involves heating biomass in the absence of oxygen to produce a mixture of gases, including carbon monoxide and hydrogen, which can be further processed to produce biofuels or other chemicals.
5. **Fermentation**: This involves using microorganisms to convert biomass into biofuels, such as ethanol or butanol, through a process of anaerobic digestion.
6. **Thermochemical conversion**: This involves the use of heat and chemical reactions to convert biomass into energy or other useful products. One example is the production of biochar, a type of charcoal that can be used as a soil amendment or a fuel source (Figure 10.1).

10.2.1 PHYSICAL APPROACHES TO BIOMASS CONVERSION

Physical methods are used to produce high-value products with the help of lignocellulosic materials. Physical approaches to biomass conversion use physical methods to modify the structure of biomass without chemical or biological processes, leading to an increase in the efficiency of energy extraction. Some common physical approaches to biomass conversion are as follows.

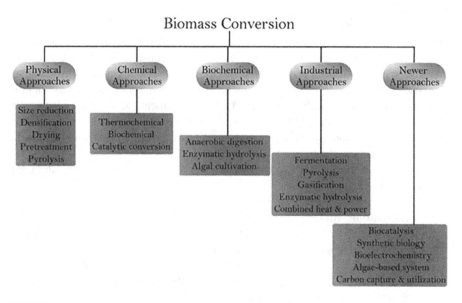

FIGURE 10.1   Various types of biomass conversion processes.

10.2.1.1   Size Reduction

Size reduction is an important step in the production of biofuels from biomass, as it increases the surface area of the biomass and makes it easier to convert into biofuels. The size reduction process involves breaking down the biomass into smaller particles, which can be achieved through various mechanical methods (Garba, 2021), such as:

a. **Chipping and shredding**: Biomass is passed through chippers and shredders to break it down into small pieces. This is commonly used for wood and agricultural residues.

b. **Grinding and milling**: Biomass is ground or milled using a hammer mill or other milling equipment to create a finer, more consistent particle size. This is commonly used for corn stover and other agricultural residues.

c. **Pelletizing:** Biomass is compressed into pellets using a pellet mill. Pelletizing is commonly used for wood and agricultural residues, and produces a uniform, dense fuel that is easier to handle and transport.

d. **Pyrolysis:** Biomass is heated in the absence of oxygen to produce a charcoal-like substance called biochar. The biochar can be used as a solid fuel or processed further into liquid biofuels (Jiang et al., 2019).

The size reduction process is important in biofuel production, as it increases the surface area of the biomass, making it easier to convert into biofuels. This improves the efficiency of the conversion process and reduces the energy required to produce biofuels from biomass.

10.2.1.2 Densification

Densification is the process of increasing the bulk density of biomass by compressing it into a more compact form, which makes it easier to transport and store (Manickam et al., 2006). Densification can also improve the efficiency of biomass conversion, as it increases the energy density of the biomass, making it more suitable for use in bio-energy applications (Tumuluru et al., 2011). After densification, the materials usually have lower moisture content. Densification can improve the efficiency and economics of biomass conversion, by reducing transportation and storage costs, and increasing the energy density of the biomass.

There are several methods of densification in biomass conversion (Vaish et al., 2022), including:

a. **Pelletization**: Biomass is compressed into small, cylindrical pellets using a pellet mill. Pellets are typically between 6 and 8 mm in diameter and 10–30 mm in length, and have a high energy density.

b. **Briquetting**: Biomass is compressed into larger, irregularly shaped briquettes using a briquetting machine. Briquettes can be made from a variety of biomass feedstocks, including wood, agricultural residues, and waste materials.

c. **Torrefaction**: Biomass is heated in the absence of oxygen to remove moisture and volatile organic compounds, creating a more stable, energy-dense material. Torrefaction can be used to improve the properties of biomass for use in combustion or gasification.

d. **Pyrolysis**: Biomass is heated in the absence of oxygen to produce a bio-oil, syngas, and biochar. Biochar can be used as a solid fuel, and has a high energy density due to its low moisture content.

Physical approaches to biomass conversion have the advantage of being simple, efficient, and inexpensive, and they have the potential to provide a significant contribution to meeting the growing demand for renewable energy. Each of the abovementioned approaches has its own advantages and disadvantages, and the best approach for a particular application will depend on factors such as the availability of resources, the desired end product, and the environmental impact of the process. The quality and quantity of biomass feedstock, availability, end-product selection, process economics, and environmental concerns also affect the process. The development of new technologies and the increasing demand for renewable energy and bio-based products are driving research in the field of biomass conversion.

10.2.2 CHEMICAL APPROACHES TO BIOMASS CONVERSION

The chemical conversion aims at converting the entire biomass into gases, which are then synthesized into the desired chemicals or used directly. There are several chemical approaches to biomass conversion, including the following.

10.2.2.1 Thermochemical Conversion

This approach involves the use of heat and chemical reactions to convert biomass into energy. Examples include combustion, gasification, and pyrolysis. Fischer–Tropsch synthesis of syngas into liquid transport fuels is an example of thermochemical conversion (Lappas and Heracleous, 2016). Thermochemical conversion can be divided into:

a. **Combustion**: Combustion involves high-temperature exothermic oxidation in oxygen-rich ambience to hot flue gas.
b. **Carbonization**: The carbon content of organic materials is increased through thermochemical decomposition.
c. **Torrefication**: The biomass is heated slowly, but to a lower temperature range of 200–300°C without or little contact with oxygen (Basu, 2018a).
d. **Pyrolysis**: It involves rapid heating in the total absence of oxygen (Basu, 2018b).
e. **Gasification**: It involves chemical reactions in an oxygen-deficient environment producing product gases with heating values (Basu, 2018c).
f. **Liquefaction**: Here, the large molecules of solid feedstock are decomposed into liquids having smaller molecules. This occurs in the presence of a catalyst and at a still lower temperature (Elliott et al., 2015).

10.2.3 BIOCHEMICAL CONVERSION

Biochemical conversion of biomass into biofuel is a process of using microorganisms such as bacteria, fungi, or enzymes to convert organic matter derived from plant or animal biomass into biofuels such as ethanol, butanol, and other biofuels (Okolie et al., 2021). The process involves breaking down the complex organic molecules present in biomass into simpler compounds that can be used as a source of energy. There are several methods of biochemical conversion, including fermentation, enzymatic hydrolysis, and anaerobic digestion (Bundhoo and Mohee, 2018). The biochemical conversion technologies refer to the conversion of biomass into corresponding products through certain physical, chemical, and biological pretreatments. Pretreatments in the biochemical conversion technologies of biomass aim to help reach ideal conversion effects, not to produce final products, which is the essential difference between the aforementioned physical and chemical conversion of biomass. Biomass can be turned into different products, such as hydrogen, biogas, ethanol, acetone, butanol, organic acids (pyruvate, lactate, oxalic acid, levulinic acid, citric acid), 2,3-butanediol, 1,4-butanediol, isobutanol, xylitol, mannitol, and xanthan gum by selecting different microorganisms in the process of biochemical conversion (Cherubini, 2010). Biochemical conversion is a promising method for producing biofuels as it offers several advantages over other methods. It is a renewable source of energy that can reduce greenhouse gas emissions and dependence on fossil fuels. In addition, it can utilize a wide range of feedstocks, including agricultural waste, forest residues, and energy crops. Advantages and limitations of some commonly used techniques and approaches are listed in Table 10.1.

TABLE 10.1

Some Commonly Used Biomass Conversion Approaches and Methods

Technique	Advantages	Disadvantages
Combustion	• Low capital cost • Widely available • High conversion efficiency • Can be used to generate heat and electricity	• High CO_2 emissions • Requires large amounts of water for cooling • High levels of ash and pollutants are produced
Gasification	• High conversion efficiency • Low CO_2 emissions • Low levels of ash and pollutants produced • Can be used to generate heat, electricity, and biofuels	• High capital cost • Complex technology • Requires specialized skills to operate
Anaerobic digestion	• Low CO_2 emissions • High conversion efficiency • Low levels of ash and pollutants produced • Can be used to generate biogas, which can be used for heating, electricity generation, and transportation of fuel	• High capital cost • Requires careful management to prevent contamination and maintain stable conditions • Low energy density of biogas compared to other fuels
Pyrolysis	• High conversion efficiency • Can produce a range of bio-based products, including biochar, bio-oil, and syngas • Low levels of ash and pollutants produced	• High capital cost • Complex technology • Requires specialized skills to operate • Low energy density of bio-oils compared to other fuels
Fermentation	• Low CO_2 emissions • High conversion efficiency • Can produce a range of bio-based products, including bioethanol, biobutanol, and biohydrogen	• High capital cost • Complex technology • Requires specialized skills to operate • Can be vulnerable to contamination and process upsets

10.2.3.1 Fermentation

This is a process that involves the use of microorganisms, such as yeast or bacteria, to convert sugars and other carbohydrates found in biomass into alcohols, such as ethanol, or organic acids, such as lactic acid (Kucharska et al., 2020). A pure culture of selected yeast strains can be used to make bioethanol, which is a clean fuel for transportation (Oh and Jin, 2020). The fermentation processes can be classified into two main types, namely, aerobic and anaerobic, based on the presence or absence of oxygen.

10.2.3.2   Anaerobic Digestion

Anaerobic digestion is a process that uses microorganisms to break down organic matter in the absence of oxygen in a closed container, producing biogas as a by-product (Bozan et al., 2017). This process can be used to convert biomass into biofuel, particularly in the form of methane, which can be used for electricity generation and as a transportation fuel (Klassen et al., 2017). The methane produced in anaerobic digestion can be used as a fuel for electricity generation or transportation, either directly or after further processing (Ambaye et al., 2021). The remaining organic matter, known as digestate, can be used as a fertilizer or soil amendment. The advantages of this method are that it can be applied on a wide range of substrates, even those with high moisture contents and impurities.

10.2.3.3   Enzymatic Hydrolysis

This is a process in which enzymes are used to break down the complex carbohydrates, such as cellulose and hemicellulose, into simple sugars that can then be fermented into various products such as ethanol (Sheng et al., 2021). Enzymatic hydrolysis offers several advantages over other methods of biomass conversion. It can utilize a wide range of feedstocks, including agricultural waste, forestry residues, and energy crops. In addition, it produces high-quality biofuels that can be used in transportation and can reduce greenhouse gas emissions and dependence on fossil fuels. However, the technology can be expensive to implement, and the enzymes used in the process can be costly (Osman et al., 2021). In addition, the efficiency of enzymatic hydrolysis can be affected by factors such as the composition and structure of the biomass and the conditions under which the enzymes are used (Robak and Balcerek, 2018). Ongoing research is focused on improving the efficiency and reducing the cost of enzymatic hydrolysis to make it a more viable method for the production of biofuels from biomass.

10.2.3.4   Algae Cultivation

Algae can be grown using renewable energy sources and then harvested and processed to produce biofuels and other value-added products. Biofuel developed from algal biomass is considered as third-generation biofuel (Kaloudas et al., 2021). The ability of microalgae to efficiently convert sunlight, CO_2, and water into a variety of products (carbohydrate, protein, lipid, etc.) makes them a suitable feedstock for sustainable biofuel production. Furthermore, many genetic manipulations were laid down to increase the biofuel yield, which are collectively considered as a fourth-generation biofuel (Godbole et al., 2021). Currently, the energy balance ratio is not considered economically feasible due to high energy input, time, and cost incurred during the transformation of microalgae biomass into biofuel (Kumar et al., 2020).

Biochemical conversion of biomass offers several advantages, including the use of renewable feedstocks, the production of biodegradable products, and the reduction of greenhouse gas emissions compared to traditional fossil fuel–based processes.

10.2.4 Newer Approaches to Biomass Conversion

10.2.4.1 Advanced Thermal Technologies

Advanced thermal technologies in biomass conversion refer to the use of high-temperature processes to convert biomass into energy, fuels, and chemicals. Examples include gasification, pyrolysis, and supercritical fluid extraction (Wen et al., 2009). Supercritical fluid enhances enzymatic hydrolysis (Li et al., 2022), and it can be used as a pretreatment technique. They can also be directly involved in liquid fuel and char formation from lignin transformation and carbohydrate hydrolysis (Escobar et al., 2020).

10.2.4.2 Synthetic Biology

Synthetic biology involves the design and construction of new biological systems or the modification of existing biological systems for specific applications (Madhavan et al., 2017). In biofuels, synthetic biology has been used to improve the efficiency and productivity of biofuel production, as well as to create new types of biofuels. Synthetic biology is being used to create new types of biofuels, such as bio-based jet fuel and diesel.

Here are some synthetic biological approaches that are being used in biofuels:

a. **Engineering microorganisms**: Synthetic biology and metabolic engineering techniques are being used to engineer microorganisms, such as bacteria, yeast, and algae, to produce biofuels more efficiently. This involves modifying the genetic makeup of the microorganisms to optimize their metabolic pathways for biofuel production. It has been widely used in *Escherichia coli*, *S. cerevisiae*, *Zymomonas mobilis*, and *Chlamydomonas reinhardtii* to enhance ethanol production (Chubukov et al., 2018; Radakovits et al., 2010). Synthetic biology is expanding its reach to non-traditional yeasts, such as *Hansenula polymorpha*, *Kluyveromyces lactis*, *Pichia pastoris*, and *Yarrowia lipolytica*, as part of its ongoing development (Madhavan et al., 2017).

b. *Developing new feedstock*: Synthetic biology is being used to develop new types of feedstock for biofuel production, such as algae and non-food crops (Myburg et al., 2019). For example, scientists have engineered a variety of switchgrass that can be broken down more easily into sugars for biofuel production (Fu et al., 2011). They have also developed a type of sorghum that has higher yields and can be grown in drought-prone areas (Nenciu et al., 2022).

   A number of additional approaches are being employed in genetic manipulations of algae and weed plants to convert them into desired biomass, for example, metagenomics and bioinformatics.

c. **Bioelectrochemistry**: Bioelectrochemistry is the study of the interactions between biological systems and electrical phenomena. In the context of biofuels, bioelectrochemistry can be used to produce electricity from renewable sources such as biomass or to convert electricity into fuel. One such application is microbial fuel cells (MFCs) (Saba et al., 2017). They work by placing bacteria (electrochemically active bacteria (EAB)) in an anode chamber where they break down organic matter to produce electrons and protons. The electrons are then transported to a cathode chamber through an external circuit, generating electricity in the process. MFCs have the potential to be used for wastewater treatment and electricity production simultaneously, making them a promising technology for sustainable bio-fuel production (Jaiswal et al., 2020). In addition, researchers are working on improving MFCs by developing new materials for the anode and cath-ode, optimizing the design of the reactor, and enhancing the activity of the microorganisms used (Hoang et al., 2022).

d. **Carbon capture and utilization**: This involves the capture of carbon diox-ide from industrial processes and the use of this carbon as a feedstock for the production of chemicals and fuels from biomass.

The advantage of using industrial carbon dioxide for biomass produc-tion is that it provides a way to capture and utilize carbon dioxide emis-sions from industrial processes, thereby reducing greenhouse gas emissions (Yang et al., 2021). In addition, the biomass produced can serve as a sustain-able alternative to fossil fuels or other non-renewable resources.

10.3 CONCLUSION

Interdisciplinary biomass conversion approaches involve combining knowledge and techniques from fields such as chemistry, biology, engineering, and materials science to create innovative and integrated solutions for biomass conversion. The choice of biomass conversion technology will depend on the specific goals and conditions. Interdisciplinary biomass conversion approaches are becoming increasingly impor-tant due to the need for sustainable and renewable energy sources. Biomass can be converted into a range of products, including biofuels, chemicals, and materials, and can help reduce reliance on fossil fuels and mitigate climate change. Biomass can be converted into a range of products, including biofuels, chemicals, and materials, and can help reduce reliance on fossil fuels and mitigate climate change. However, continuous efforts are going for the optimization of processes such as biomass pro-duction, harvesting, pretreatment, and conversion. The future of interdisciplinary biomass conversion methods relies on different factors such as capital cost, conver-sion efficiency, and environmental impact.

REFERENCES

Agbor, V.B., Cicek, N., Sparling, R., Berlin, A., Levin, D.B., 2011. Biomass pretreatment: Fundamentals toward application. *Biotechnol. Adv.* doi:10.1016/j.biotechadv.2011.05.005

Ambaye, T.G., Vaccari, M., Bonilla-Petriciolet, A., Prasad, S., van Hullebusch, E.D., Rtimi, S., 2021. Emerging technologies for biofuel production: A critical review on recent progress, challenges and perspectives. *J. Environ. Manag.* doi:10.1016/j.jenvman.2021.112627

Arhin, S.G., Cesaro, A., Di Capua, F., Esposito, G., 2023. Recent progress and challenges in biotechnological valorization of lignocellulosic materials: Towards sustainable biofu-els and platform chemicals synthesis. *Sci. Total Environ.* 857, 159333. doi:10.1016/j. scitotenv.2022.159333

Aslanzadeh, S., Ishola, M.M., Richards, T., Taherzadeh, M.J., 2014. An overview of existing individual unit operations. In: Qureshi, N., Hodge, D.B., Vertès, A.A. (Eds.), *Biorefineries: Integrated Biochemical Processes for Liquid Biofuels*. Elsevier, Amsterdam, pp. 3–36. doi:10.1016/B978-0-444-59498-3.00001-4

Barelli, L., Bidini, G., Pelosi, D., Sisani, E., 2021. Enzymatic biofuel cells: A review on flow designs. *Energies* doi:10.3390/en14040910

Basu, P., 2018a. Torrefaction. In: *Biomass Gasification, Pyrolysis and Torrefaction: Practical Design and Theory*. Academic Press, London, pp. 93–154. doi:10.1016/B978-0-12-812992-0.00004-2

Basu, P., 2018b. Pyrolysis. In: *Biomass Gasification, Pyrolysis and Torrefaction: Practical Design and Theory*. Academic Press, London, pp. 155–187. doi:10.1016/B978-0-12-812992-0.00005-4

Basu, P., 2018c. Gasification theory. In: *Biomass Gasification, Pyrolysis and Torrefaction: Practical Design and Theory*. Academic Press, London, pp. 211–262. doi:10.1016/B978-0-12-812992-0.00007-8

Berlin, A., Balakshin, M., Gilkes, N., Kadla, J., Maximenko, V., Kubo, S., Saddler, J., 2006. Inhibition of cellulase, xylanase and β-glucosidase activities by softwood lignin preparations. *J. Biotechnol.* 125, 198–209. doi:10.1134/S0965545X06020155

Bozan, M., Akyol, Ç., Ince, O., Aydin, S., Ince, B., 2017. Application of next-generation sequencing methods for microbial monitoring of anaerobic digestion of lignocellulosic biomass. *Appl. Microbiol. Biotechnol.* doi:10.1007/s00253-017-8438-7

Bundhoo, Z.M.A., Mohee, R., 2018. Ultrasound-assisted biological conversion of biomass and waste materials to biofuels: A review. *Ultrason. Sonochem.* doi:10.1016/j.ultsonch.2017.07.025

Chakravarty, I., Mandavgane, S.A., 2022. Anaerobic digestion of agrowastes: End-of-life step of a biorefinery. In: Gurunathan, B., Sahadevan, R., Akmar, Z. (Eds.), *Biofuels and Bioenergy: Opportunities and Challenges*. Elsevier, Amsterdam, pp. 233–251. doi:10.1016/B978-0-323-85269-2.00005-8

Cherubini, F., 2010. The biorefinery concept: Using biomass instead of oil for producing energy and chemicals. *Energy Convers. Manag.* 51, 1412–1421. doi:10.1016/j.enconman.2010.01.015

Chubukov, V., Mukhopadhyay, A., Petzold, C.J., Keasling, J.D., Martín, H.G., 2018. Synthetic and systems biology for microbial production of commodity chemicals. *NPJ Syst. Biol. Appl.* 2, 16009. doi:10.1038/npjsba.2016.9

Elliott, D.C., Biller, P., Ross, A.B., Schmidt, A.J., Jones, S.B., 2015. Hydrothermal liquefaction of biomass: Developments from batch to continuous process. *Bioresour. Technol.* 178, 147–156. doi:10.1016/j.biortech.2014.09.132

Escobar, E.L.N., da Silva, T.A., Pirich, C.L., Corazza, M.L., Pereira Ramos, L., 2020. Supercritical fluids: A promising technique for biomass pretreatment and fractionation. *Front. Bioeng. Biotechnol.* doi:10.3389/fbioe.2020.00252

Fu, C., Mielenz, J.R., Xiao, X., Ge, Y., Hamilton, C.Y., Rodriguez, M., Chen, F., Foston, M., Ragauskas, A., Bouton, J., Dixon, R.A., Wang, Z.Y., 2011. Genetic manipulation of lignin reduces recalcitrance and improves ethanol production from switchgrass. *Proc. Natl. Acad. Sci. USA* 108, 3803–3808. doi:10.1073/pnas.1100310108

Garba, A., 2021. Biomass conversion technologies for bioenergy generation: An introduction. In: Basso, T.P., Basso, T.O., Basso, L.C. (Eds.), *Biotechnological Applications of Biomass*. IntechOpen, Rijeka. doi:10.5772/intechopen.93669

Godbole, V., Pal, M.K., Gautam, P., 2021. A critical perspective on the scope of interdisciplinary approaches used in fourth-generation biofuel production. *Algal Res.* 58, 102436. doi:10.1016/j.algal.2021.102436

Grippi, D., Clemente, R., Bernal, M.P., 2020. Chemical and bioenergetic characterization of biofuels from plant biomass: Perspectives for southern europe. *Appl. Sci.* doi:10.3390/app10103571

Hajilary, N., Rezakazemi, M., Shirazian, S., 2019. Biofuel types and membrane separation. *Environ. Chem. Lett.* 17, 1–18. doi:10.1007/s10311-018-0777-9

Hoang, A.T., Nižetić, S., Ng, K.H., Papadopoulos, A.M., Le, A.T., Kumar, S., Hadiyanto, H., Pham, V.V., 2022. Microbial fuel cells for bioelectricity production from waste as sustainable prospect of future energy sector. *Chemosphere* 287. doi:10.1016/j.chemosphere.2021.132285

Jaiswal, K.K., Kumar, V., Vlaskin, M.S., Sharma, N., Rautela, I., Nanda, M., Arora, N., Singh, A., Chauhan, P.K., 2020. Microalgae fuel cell for wastewater treatment: Recent advances and challenges. *J. Water Process. Eng.* 38. doi:10.1016/j.jwpe.2020.101549

Jeswani, H.K., Chilvers, A., Azapagic, A., 2020. Environmental sustainability of biofuels: A review. *Proc. R. Soc. A Math. Phys. Eng. Sci.* 476, 20200351. doi:10.1098/rspa.2020.0351

Jiang, S.F., Sheng, G.P., Jiang, H., 2019. Advances in the characterization methods of biomass pyrolysis products. *ACS Sustain. Chem. Eng.* 7, 12639–12655. doi:10.1021/acssuschemeng.9b00868

Kaloudas, D., Pavlova, N., Penchovsky, R., 2021. Lignocellulose, algal biomass, biofuels and biohydrogen: A review. *Environ. Chem. Lett.* doi:10.1007/s10311-021-01213-y

Klassen, V., Blifernez-Klassen, O., Wibberg, D., Winkler, A., Kalinowski, J., Posten, C., Kruse, O., 2017. Highly efficient methane generation from untreated microalgae biomass. *Biotechnol. Biofuels* 10, 186. doi:10.1186/s13068-017-0871-4

Kucharska, K., Słupek, E., Cieśliński, H., Kamiński, M., 2020. Advantageous conditions of saccharification of lignocellulosic biomass for biofuels generation via fermentation processes. *Chem. Pap.* 74, 1199–1209. doi:10.1007/s11696-019-00960-1

Kumar, A., Jones, D.D., Hanna, M.A., 2009. Thermochemical biomass gasification: A review of the current status of the technology. *Energies* doi:10.3390/en20300556

Kumar, M., Sun, Y., Rathour, R., Pandey, A., Thakur, I.S., Tsang, D.C.W., 2020. Algae as potential feedstock for the production of biofuels and value-added products: Opportunities and challenges. *Sci. Total Environ.* doi:10.1016/j.scitotenv.2020.137116

Lappas, A., Heracleous, E., 2016. Production of biofuels via Fischer-Tropsch synthesis: Biomass-to-liquids. In: Luque, R., Lin, C.S.K., Wilson, K., Clark, J. (Eds.), *Handbook of Biofuels Production: Processes and Technologies* (2nd edn, pp. 549–593). Elsevier, Amsterdam. doi:10.1016/B978-0-08-100455-5.00018-7

Li, X., Shi, Y., Kong, W., Wei, J., Song, W., Wang, S., 2022. Improving enzymatic hydrolysis of lignocellulosic biomass by bio-coordinated physicochemical pretreatment-A review. *Energy Rep.* 8, 696–709. doi:10.1016/j.egyr.2021.12.015

Madhavan, A., Jose, A.A., Binod, P., Sindhu, R., Sukumaran, R.K., Pandey, A., Castro, G.E., 2017. Synthetic biology and metabolic engineering approaches and its impact on non-conventional yeast and biofuel production. *Front. Energy Res.* doi:10.3389/fenrg.2017.00008

Manickam, I.N., Ravindran, D., Subramanian, P., 2006. Biomass densification methods and mechanism. *Cogener. Distrib. Gener. J.* 21, 33–45. doi:10.1080/15453660609509098

Melero, J.A., Morales, G., Paniagua, M., López-Aguado, C., 2023. Chemical routes for the conversion of cellulosic platform molecules into high-energy-density biofuels. In: Luque, R., Lin, C.S.K., Wilson, K., Du, C. (Eds.), *Handbook of Biofuels Production* (3rd edn.). Woodhead Publishing, Sawaston, pp. 361–397. doi:10.1016/b978-0-323-91193-1.00004-4

Mooney, C.A., Mansfield, S.D., Beatson, R.P., Saddler, J.N., 1999. The effect of fiber characteristics on hydrolysis and cellulase accessibility to softwood substrates. *Enzyme Microb. Technol.* 25, 644–650. doi:10.1016/S0141-0229 (99)00098-8

Myburg, A.A., Hussey, S.G., Wang, J.P., Street, N.R., Mizrachi, E., 2019. Systems and synthetic biology of forest trees: A bioengineering paradigm for woody biomass feedstocks. *Front. Plant Sci.* doi:10.3389/fpls.2019.00775

Nenciu, F., Paraschiv, M., Kuncser, R., Stan, C., Cocarta, D., Vladut, V.N., 2022. High-grade chemicals and biofuels produced from marginal lands using an integrated approach of alcoholic fermentation and pyrolysis of sweet sorghum biomass residues. *Sustain.* 14. doi:10.3390/su14010402

Oh, E.J., Jin, Y.S., 2020. Engineering of Saccharomyces cerevisiae for efficient fermentation of cellulose. *FEMS Yeast Res.* doi:10.1093/femsyr/foz089

Okolie, J.A., Mukherjee, A., Nanda, S., Dalai, A.K., Kozinski, J.A., 2021. Next-generation biofuels and platform biochemicals from lignocellulosic biomass. *Int. J. Energy Res.* doi:10.1002/er.6697

Onumaegbu, C., Mooney, J., Alaswad, A., Olabi, A.G., 2018. Pre-treatment methods for production of biofuel from microalgae biomass. *Renew. Sustain. Energy Rev.* doi:10.1016/j. rser.2018.04.015

Osman, A.I., Mehta, N., Elgarahy, A.M., Al-Hinai, A., Al-Muhtaseb, A.H., Rooney, D.W., 2021. Conversion of biomass to biofuels and life cycle assessment: A review. *Environ. Chem. Lett.* doi:10.1007/s10311-021-01273-0

Radakovits, R., Jinkerson, R.E., Darzins, A., Posewitz, M.C., 2010. Genetic engineering of algae for enhanced biofuel production. *Eukaryot. Cell.* doi:10.1128/EC.00364-09

Rao, A., Sathiavelu, A., Mythili, S., 2020. Mini review on nanoimmobilization of lipase and cellulase for biofuel production. *Biofuels* 11, 191–200. doi:10.1080/17597269.2017.13 48187

Robak, K., Balcerek, M., 2018. Review of second generation bioethanol production from residual biomass. *Food Technol. Biotechnol.* doi:10.17113/ftb.56.02.18.5428

Rocha-Gaso, M.I., March-Iborra, C., Montoya-Baides, Á., Arnau-Vives, A., 2009. Surface generated acoustic wave biosensors for the detection of pathogens: A review. *Sensors* 9, 5740–5769. doi:10.3390/s90705740

Saba, B., Christy, A.D., Yu, Z., Co, A.C., 2017. Sustainable power generation from bacterio-algal microbial fuel cells (MFCs): An overview. *Renew. Sustain. Energy Rev.* doi:10.1016/j.rser.2017.01.115

Sheng, Y., Lam, S.S., Wu, Y., Ge, S., Wu, J., Cai, L., Huang, Z., Le, Q. Van, Sonne, C., Xia, C., 2021. Enzymatic conversion of pretreated lignocellulosic biomass: A review on influence of structural changes of lignin. *Bioresour. Technol.* doi:10.1016/j.biortech.2020.124631

Tumuluru, J.S., Wright, C.T., Hess, J.R., Kenney, K.L., 2011. A review of biomass densification systems to develop uniform feedstock commodities for bioenergy application. *Biofuels Bioprod. Biorefining* 5, 683–707. doi:10.1002/bbb.324

Vaish, S., Sharma, N.K., Kaur, G., 2022. A review on various types of densification/briquetting technologies of biomass residues. *IOP Conf. Ser. Mater. Sci. Eng.* 1228, 012019. doi:10 .1088/1757-899x/1228/1/012019

Wen, D., Jiang, H., Zhang, K., 2009. Supercritical fluids technology for clean biofuel production. *Prog. Nat. Sci.* 19, 273–284. doi:10.1016/j.pnsc.2008.09.001

Yang, F., Meerman, J.C., Faaij, A.P.C., 2021. Carbon capture and biomass in industry: A techno-economic analysis and comparison of negative emission options. *Renew. Sustain. Energy Rev.* 144, 111028. doi:10.1016/j.rser.2021.111028

11 Lignocellulosic Biomass Pretreatment for Enhanced Bioenergy Recovery

Anita Saini
Maharaja Agrasen University

Simran Jot Kaur
Sri Guru Granth Sahib World University

11.1 INTRODUCTION

Lignocellulosic biomass (LCB) mainly pertains to the cell walls of plants that are synthesized by the plants through the process of photosynthesis. The LCB is diverse and widely available, making it a sustainable feedstock for producing various value-added products (Zheng et al., 2022). Examples of LCB include agricultural waste, such as corn stover, wheat straw, and sugarcane bagasse, forest waste, such as sawdust and wood chips, agro-industrial waste, domestic waste, etc. Energy crops, for instance, switchgrass and *Miscanthus*, are also being used as sources of LCB. The LCB is a valuable resource for different biorefinery products. It also offers the benefits of employability and economic progress in different geographical regions (Zhang et al., 2018).

LCB or lignocellulose is a composite of polymers of sugars (cellulose and hemicellulose, the holocellulose), and phenolic (lignin) compounds that make up the essential part of plant cell walls (Chen and Chen, 2014) (Figure 11.1). The biomass from different sources varies in its lignocellulosic composition (Table 11.1). The glucose molecules arranged linearly, constitute the cellulose structure, which acts as the primary supportive element of the cell wall. It is a rigid and fibrous material, having crystalline and amorphous regions that provide strength to the structure. Hemicellulose is another biopolymer that is heterogeneous in its composition and usually made of C5, C6 sugars and sugar acids that are arranged in a branched structure in the biomolecule. It is more amorphous as compared to cellulose and is hydrolyzed readily upon treatment with chemicals and hydrolytic enzymes. Lignin is a heteropolymer of three different lignol units, i.e., p-coumaryl alcohol, coniferyl alcohol, and sinapyl alcohol, interconnected through a variety of C-C, ester (R-COOR'), and ether (R-O-R') bonds. The lignin functions as an adhesive substance that cements cellulose to the

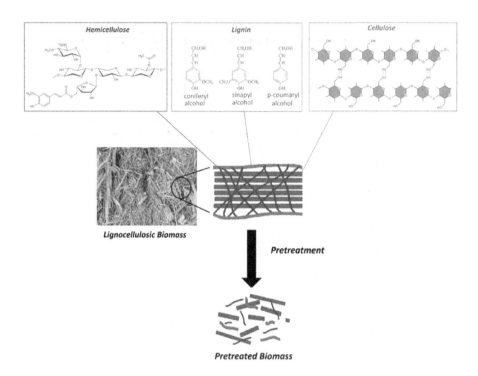

FIGURE 11.1   Lignocellulosic biomass: structure and effect of pretreatment.

TABLE 11.1
Chemical Components of Several Lignocellulosic Biomass

Lignocellulosic Biomass		Cellulose (%)	Hemicellulose (%)	Lignin (%)	References
Wood (hardwood & softwood)	Eucalyptus wood	57.3	16.8	25.9	Ngernyen (2007)
	Pine wood chips	32.09	14.22	31.15	Cotana et al. (2014)
	Poplar	49	23	27	Rigo et al. (2019)
	Spruce	44	23.88	...	Swiatek et al. (2020)
	Beech wood	36.33	34.74	...	Swiatek et al. (2021)
Grass	Napier grass	35.7	26.9	5.2	Rocha-Meneses et al. (2019)
	Switchgrass	31.8	25	31.2	Bonfiglio et al. (2021)
Agricultural waste	Wheat straw	35–39	23–30	16–24	Isikgor and Becer (2015)
	Barley straw	36–43	24–33	6.3–9.8	Isikgor and Becer (2015)
					(*Continued*)

TABLE 11.1 (*Continued*)

Chemical Components of Several Lignocellulosic Biomass

Lignocellulosic Biomass		Cellulose (%)	Hemicellulose (%)	Lignin (%)	References
	Rice straw	29.2–34.7	23–25.9	17–19	Isikgor and Becer (2015)
	Sorghum straw	32–35	24–27	15–21	Isikgor and Becer (2015)
	Sugarcane bagasse	38–50	22–28	19–33	Mkhize et al. (2016)
Agro-industrial waste	Banana peel waste	9.9	41.38	8.9	Kabenge et al. (2018)
	Pineapple waste	26.8	9.55	17.7	Nordin et al. (2020)
Others	Sawdust	40.1	11.56	24.2	Nam et al. (2020)
	Rubberwood chip	49.5	12.1	20.2	Nam et al. (2021)

hemicellulose. It accounts for the stiffness of the cell wall and the resistance of the plant cell walls to deterioration and degradation. LCB has an intricate three-dimensional structure, in which cellulose is enveloped by a dense framework of hemicellulose and lignin (Smith, 2019). The hydrogen, acetyl, or ester bonds play a role in cross-linking the cellulose component to the hemicellulose. The lignin is often joined to the hemicellulose through ester bonds and to the cellulose through ester, hydrogen, or ether linkages, as well as C-C bonds. The strong covalent bonds, such as ether and carbon–carbon bonds, provide robustness to the lignocellulose structure. Due to the complex network of strong and multiple weak bonds between lignocellulose constituents, the breakdown of LCB requires pretreatment prior to its conversion to biorefinery products (Zoghlami and Paes, 2019). The pretreatment makes alterations in the physical architecture and chemical composition of the biomass. The breakdown of bonds loosens up the intricate interior of the biomass so that it becomes more accessible to hydrolytic enzymes or chemicals, thereby increasing the productivity of downstream processes (Figure 11.2). This chapter discusses various pretreatment strategies that can be employed to deconstruct the lignocellulose structure so as to increase the yield of different types of bioenergy from the biomass.

11.2 PRETREATMENT METHODS

Various pretreatment methods known to pretreat biomass can broadly be categorized as physical, chemical, and biological methods (Kumar and Sharma, 2017) (Figure 11.2). The physical pretreatment aims at biomass deconstruction through physical alteration in the biomass structure that increases the exposed surface and penetrability of biomass, making the interior of the structure more accessible and amenable to subsequent processing. The chemical methods make use of acidic or alkaline chemicals or organic solvents that may break down or modify the lignin and

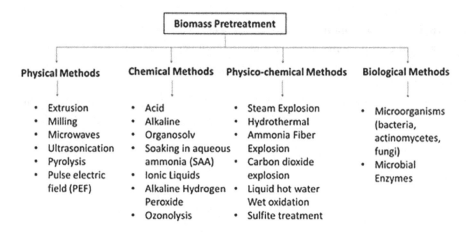

FIGURE 11.2   Common methods to pretreat lignocellulosic biomass.

hemicellulose and may also modify the biomass surface so as to make it more responsive to enzymatic hydrolysis. The biological methods make use of microorganisms
(fungi as well as bacteria) or their enzymes that degrade or modify the lignin and
hemicellulose polymers of the biomass. Also, various combinations of different pretreatment methods may be employed to accomplish more efficient hydrolysis of the
biomass structure. These methods vary in their action mechanism, efficiency, benefits, and limitations (Table 11.2). The suitability of a method depends on the biomass
characteristics and the type of bioconversion process. For example, biomass with a
higher lignin content needs a method specific to delignification, i.e., alkali pretreatment. Several methods are known to have certain limitations, such as cost-ineffective-
ness, generation of products interfering in downstream processes along the bioenergy
production pathway, loss of useful component(s) to different extents, etc. In general,
the pretreatment method for a bioconversion process is selected by taking several
points into consideration, i.e., (i) the method should involve the use of the minimum
amount of catalyst, if any, (ii) the catalyst should be inexpensive, recyclable, and
noncorrosive, (iii) the method should lead to minimum loss of the holocellulose (cellulose + hemicellulose) fraction of biomass, (iv) the method should generate as few
inhibitory products as possible, (v) the method should not be very energy-intensive,
and (vi) the method should fractionate biomass components efficiently and generate
relatively pure streams of co-products for the production of high-value biorefinery
products (Kumar and Sharma, 2017).

11.3   PHYSICAL PRETREATMENT METHODS

Physical methods for biomass pretreatment may include mechanical methods, soni-
cation, irradiation, etc. Mechanical methods aim at biomass size reduction, which is
linked to other changes in biomass physical characteristics, such as an exposure of the
biomass exterior surface and reduced orderness and chain length in cellulose (Kumar
and Sharma, 2017). The size reduction can be achieved through different mechanisms

TABLE 11.2

Comparison of Major Types of Biomass Pretreatment Methods

Physical Methods		Chemical Methods		Biological Methods	
• Increase biomass surface area and porosity • Increase energy densities • Reduce cellulose crystallinity		• Increase cellulose accessibility • Alter lignin structure • Hydrolyze hemicellulose into sugar fractions		• Lignin degradation (complete or partial) • Hemicellulose degradation	
Advantages	**Disadvantages**	**Advantages**	**Disadvantages**	**Advantages**	**Disadvantages**
Simple and Convenient	Often requires additional steps of pretreatment	Relatively high removal efficiency for lignin and hemicellulose	High cost of chemicals (requires recovery and recycle of chemicals)	Environment friendly (doesn't involve use of chemicals)	Slow process (requires long residence time)
Applicable to high biomass loads					Requires large space
No use of chemicals	High cost of power		Often requires harsh conditions	Operated at mild conditions	
			Problem of equipment corrosion	Suitable for biomass with high as well as low moisture levels	
			Formation of inhibitory chemicals	No inhibitors formation	
			Environmental pollution by chemicals	No need of sophisticated instrument	

such as cutting, breaking, compression, shearing, tearing, etc. (Arce and Kratky, 2022). Commonly, it is achieved through processes such as milling, chipping, shredding, extrusion, etc. Other physical methods may involve ultrasonication, microwave irradiation, ozonolysis, pyrolysis, etc. (Moodley and Trois, 2021). Generally, size reduction precedes other pretreatment methods as it makes biomass more responsive to other pretreatment agents due to the accessibility of a more interactive surface area on the physically altered biomass. The most prominent limitation of the physical pretreatment is the high energy demands of the method (Moodley and Trois, 2021). However, this limitation is often overcome by combining physical pretreatment with other methods.

11.4 MECHANICAL METHODS: MILLING, CHIPPING, SHREDDING, AND MECHANICAL EXTRUSION

The bulky biomass can be converted into smaller particles of different sizes using milling, grinding, or chipping processes. The biomass size can be decreased by 10–30 mm by chipping. Milling and grinding, however, are more efficient at reducing the size to nearly 0.2 mm (Kumar and Sharma, 2017). Milling involves grinding the biomass with the help of a mechanical mill or grinder. The mills can be of different types, such as hammer mills, disc mills, ball mills, vibratory mills, etc. (Kim et al., 2016). The process can be grouped into two types, i.e., dry and wet milling, depending on the moisture level (dry and wet) of the biomass. Vibro-ball milling is useful at the industrial level and can be used for both dry and wet biomass. Chipping involves the slicing of biomass material using special cutting machines referred to as chippers. Shredding is primarily based on tearing forces on the biomass. The choice of the method for mechanical size reduction depends on biomass characteristics as well as its subsequent processing. Guo et al. (2012) used mechanical shredding at 50°C–70°C for the biomass pretreatment, followed by digestion of the biomass. The study showed that mechanical pretreatment resulted in the rupturing of biomass structure, which led to a 23%–59% reduction in the time required for biomass digestion. Heller et al. (2023) have observed the positive effect of ball milling on the anaerobic digestion of horse manure. The pretreatment improved the specific methane yield by >37%, i.e., up to 243 L of methane per kg of volatile solids (VS). Rizal et al. (2018) have documented the high efficiency of wet disc-, hammer-, and ball-milling processes for glucose recovery from oil palm biomass. Monlau et al. (2019) have also inspected the influence of vibro-ball milling on the physico-chemical composition and enzymatic breakdown of solid digestates (dry and moist) obtained from an agricultural waste-based biogas plant. They observed an improvement in the particle size as well as the crystallinity index (27%–75%) through dry milling of the biomass. Contrarily, wet milling did not show much impact on the biomass characteristics.

In the extrusion process, the material is passed through a specific-sized cross-section in a dye (Zheng and Rehmann, 2014). The material experiences a compressive force, and a shearing effect is produced, which deconstructs the biomass structure. A special device, the extruder, is often used for this method (Figure 11.3). Extruders cause high shearing, faster heat transference, and thorough mixing in a relatively shorter duration. Extruders may be of different types, e.g., single- and twin-screw extruders (Zheng and Rehmann, 2014; Konan et al., 2022) (Figure 11.3). In a twin-screw extruder, two screws are aligned side by side inside a cylindrical barrel, which has a dye attached to its end. In a single-screw extruder, the inside of the barrel is fitted with a single screw. When the motor of the machine drives screws in rotatory motion around their axis, mechanical force is exerted on the biomass, contributing to its mechanical disruption. Twin-screw extruders may be made to revolve in similar (co-rotation) or opposed (counter-rotation) directions. The screw configurations also vary and depend on differences in position, pitches, size, staggering angles, and spaces within the extruder. The process can be run in batch, fed-batch, or continuous modes (Konan et al., 2022). The increased specific area of the extruded biomass is useful for subsequent processing of the biomass. This method can be coupled with other methods of biomass pretreatment and is also suitable for high biomass loadings (Lamsal et al., 2010; Konan et al., 2022). Lamsal et al. (2010) assessed the outcome of grinding and thermo-mechanical extrusion treatment on wheat bran and found that the extrusion process yielded more reducing sugars compared to that from the grinding of the bran. The best results were obtained at an extrusion screw speed of 7 Hz at 150°C and at a speed of 3.7 Hz at the highest temperature of 110°C. When they used a

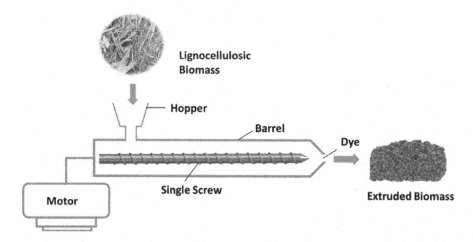

FIGURE 11.3 (a) Scheme of mechanical extrusion of biomass using single-screw extruder and (b) twin-screw barrel (used in twin-screw extruder).

solvent blend (NaOH, urea, and thiourea) and $CaCl_2$ along with extrusion treatment, they achieved 60%–73% conversion efficiency after removal of solvents by thorough washing. In another study, Papepinto and Genon (2016) studied the effect of extrusion on anaerobic digestion of nine maize silage samples and one mixed sample of silage maize and manure. The extrusion resulted in a 0%–15% increase in the yield of biogas and methane from the tested samples.

11.5 PYROLYSIS

Pyrolysis pretreatment involves disintegration of the biomass structure at higher temperatures of around 400°C–650°C (Zadeh et al., 2020). No oxidizing agent, such as oxygen, is used during this treatment process. The process yields diverse types of products, such as solid biochar, liquid bio-oil, and syngas (containing CO_2, CO, H_2, and CH_4). Biochar is a porous and carbon-rich solid that can be used as fuel and for other applications, including the catalysis of biomass for fuel production processes (Waqas et al., 2018). It can be upgraded to charcoal for cooking. Bio-oil is rich in polyaromatic hydrocarbons (PAH) and phenol (an oxygenated aromatic compound), which can be improved into high-quality hydrocarbon liquid fuels (biofuel) for use in transport and power generation. The syngas can be used for generating power (Zadeh et al., 2020). The rate of the process can be varied, which is subject to the requirements of the final product. Biochar production is favored at a slow pyrolysis rate, whereas bi-oil production is enhanced when the pyrolysis heating rate is faster (Zadeh et al., 2020). A study by Varma and Mondal (2017) reported sugarcane bagasse biomass' pyrolysis in semi-batch mode and obtained the highest 45.23 wt.% of bio-oil at 500°C from biomass particles ranging in size from 0.5 to 0.6 mm at a pyrolysis rate of 50°C/min while maintaining the flow of nitrogen at a rate of 100 cm^3 per min. Rao et al. (2022) performed pyrolysis of *Ficus religiosa* wood and bark. The optimized conditions of temperature (450°C), biomass particle size (1 mm), and sweep gas flow rate

(2 m^3 per hour) yielded 47.5 wt.% of bio-oil. Selvarajoo et al. (2022) produced biochar from citrus peels using a slow pyrolysis process and obtained biochar with a higher heating value and carbon content at 500°C.

11.6 IRRADIATION

During this treatment, the lignocellulose is irradiated with high-energy radiations such as γ-rays, electron beams, microwaves, and sound waves. Different radiations use different mechanisms to deconstruct the LCB structure (Figure 11.4). The selection of the radiation type is done based on multiple aspects, such as biomass characteristics, downstream processing of biomass, and economic considerations. The interaction of biomass with high-energy radiation modifies the physical structure of the biomass, which increases the exposed surface, decreases cellulose chain length and crystallinity, etc. Further, the hemicellulose and lignin portions may undergo partial removal. Gamma (γ) rays are very low-wavelength (high-energy) radiations that can penetrate deeper parts of the biomass. When γ radiation bombards the biomass at a very high speed, the high energy carried by it leads to the ionization of biomass molecules. This alters the biomass structure and composition through polymer disintegration and cross-linking reactions. Exposure to high doses of gamma rays causes depolymerization or delignification of the biomass components. During electron beam irradiations, the kinetic energy of accelerated and highly charged electrons causes multiple alterations in biomass using mechanisms similar to those of other ionizing rays (Henniges et al., 2013). In the ultrasonic method, biomass disintegration is done through exposure of biomass suspension to ultrahigh-frequency sound waves (100 kHz to 1 MHz). Microwave irradiation makes use of electromagnetic waves in the frequency range of 300 MHz–300 GHz that produce thermal effects. Microwave heating is considered good as it results in simultaneous and uniform heating of all parts of the biomass, thus avoiding overheating of a specific part. Also, the pulsed electric field may be used to pretreat

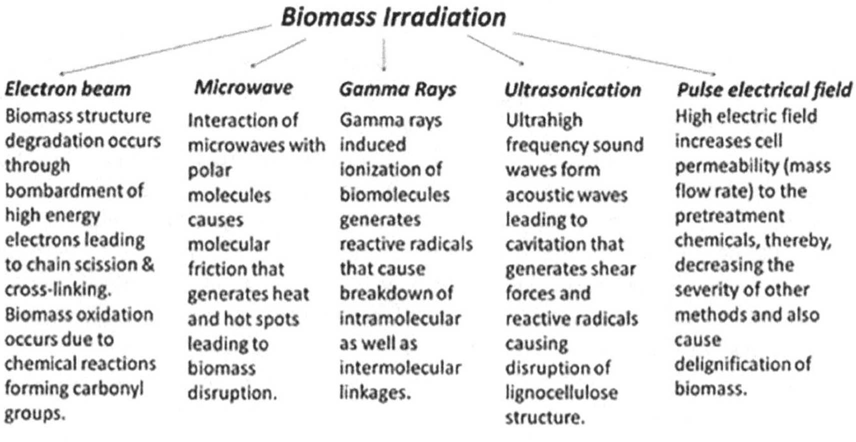

Biomass Irradiation

Electron beam	Microwave	Gamma Rays	Ultrasonication	Pulse electrical field
Biomass structure degradation occurs through bombardment of high energy electrons leading to chain scission & cross-linking. Biomass oxidation occurs due to chemical reactions forming carbonyl groups.	Interaction of microwaves with polar molecules causes molecular friction that generates heat and hot spots leading to biomass disruption.	Gamma rays induced ionization of biomolecules generates reactive radicals that cause breakdown of intramolecular as well as intermolecular linkages.	Ultrahigh frequency sound waves form acoustic waves leading to cavitation that generates shear forces and reactive radicals causing disruption of lignocellulose structure.	High electric field increases cell permeability (mass flow rate) to the pretreatment chemicals, thereby, decreasing the severity of other methods and also cause delignification of biomass.

FIGURE 11.4   Mechanism of action of different types of radiations in biomass deconstruction.

biomass. The method involves exposure of the biomass to surplus transmembrane voltage across the plant cell's membrane, which brings about the electroporation of the membrane. Increased permeability of porous membranes facilitates the flow of solvents, etc. used in other pretreatment methods without requiring harsh conditions for their diffusion (Golberg et al., 2016).

The efficacy of irradiation depends on the frequency of radiation, exposure time, lignocellulose composition, and medium characteristics that may interfere with the movement of radiation. The method offers the advantages of controlled and selective degradation of biomass constituents, without involving the use of corrosive and expensive chemicals requiring recycling. No fermentation inhibitors are formed in this method, and no cooling or neutralization of pretreated biomass is required. Studies have reported that other pretreatment methods become more effective when combined with irradiation. Bak (2014) pretreated rice straw with electron beam irradiation and found that water-soaked biomass exposed to an electron beam (80 kGy dose, at 1 MeV dose and 0.12 mA), resulted in the highest of the theoretical maximum possible hydrolysis yield (70.4%), which was larger than the untreated (29.5%) and unsoaked, irradiated (52.1%) biomass. Roy et al. (2020) have discussed the effect of green solvents on LCB. They reported that deep eutectic solvents are cost-effective, recyclable, and greener solvents. Furthermore, their efficiency can be improved multiplefold with the assistance of microwaves and sound waves. Su et al. (2020) observed the outcome of gamma irradiation on the *Miscanthus* biomass structure. The irradiation reduced the cellulose crystallinity, biomass particle size, and DP. The treated biomass had a less stable structure and a lower tolerance to heat. The cellulolytic hydrolysis of the treated biomass could produce 557.58 mg reducing sugars per gram of biomass, which is higher than that obtained from untreated biomass. Szwarc and Szwarc (2020) have used a pulsed electric field for biogas production from maize silage. They observed slight changes in the lignocellulosic composition of biomass and achieved a production of 751.97 mL of biogas per gram of VS, which was 14% greater than for the control. Chang et al. (2021) assessed the effect of microwaves on corncob biomass used for producing biohydrogen. The method used a combination of microwave irradiation and rhamnolipid surfactant supplementation. They achieved production of 27.34 mL of hydrogen per gram of total solids (TS) under optimized conditions, which amounted to an 80.94% increase in the hydrogen yield and 90.44% increase in substrate conversion efficiency compared to the use of untreated biomass.

11.7 CHEMICAL PRETREATMENT METHODS

11.7.1 ACID PRETREATMENT

In this method, acids, either in concentrated or diluted form, are used to break open the lignocellulosic cell wall structure, which may eventually improve the amenability of the biomass to enzymatic saccharification (de Jong and Gosselink, 2014; Amin et al., 2017). Sulfuric acid (H_2SO_4) and hydrochloric acid (HCl) are known to be among the most frequently employed acids. However, phosphoric acid, nitric acid, and organic acids are also used (Baruah et al., 2018). During the concentrated acid pretreatment, 30%–70% acid concentrations are used at milder temperatures (usually not more than 100°C). The acids destabilize the van der Waals interactions and hydrolyze the hydrogen and covalent bonds in the lignocellulosic constituents. It often results in the solubilization of hemicellulose, with a reduction in cellulose to a lesser extent. Among different hemicelluloses, the xylan component is most susceptible and is solubilized primarily (Amin et al., 2017). Lignin is not solubilized during acid pretreatment. However, significant modifications in the lignin structure occur so that a more opened-up structure is obtained, which increases the accessibility of the biomass interior to the cellulolytic enzymes (Amin et al., 2017). The main issues with the use of concentrated acids are the toxicity and corrosiveness caused by most acids, which contribute to the operation and maintenance costs of the process. Also, alkali is required in large amounts to neutralize the acidic biomass before its subsequent use (Nazari et al., 2018).

Dilute acid pretreatment, involving acid concentrations less than 4% (by weight), can hydrolyze the hemicellulose at significantly high reaction rates. However, the acid hydrolysis during bioethanol production from cellulose may pose problems, i.e., the sugars constituting the cellulose and hemicellulose, including glucose, xylose, galactose, mannose, etc., are dehydrated during this pretreatment at a higher temperature and result in the formation of compounds such as furfural and hydroxymethyl furfural (HMF) that are inhibitory to the fermentation microorganisms. The parameters such as treatment time, acid concentration, temperature, etc., need to be optimized depending on the biomass as well as the type of the desired end product. The hydrolysate obtained from the barley straw pretreated at 110°C–190°C for 2–20 min, in the presence of 0.1%–2% sulfuric acid, when used for the production of furfural, resulted in 48.5% of theoretical furfural production under optimized conditions (Kim et al., 2015). Deshavath et al. (2017) employed a dilute acid pretreatment method for the sorghum biomass to dissolve hemicellulose. The pretreatment resulted in the production of the minimum amount of fermentation inhibitors. Subsequently, 78.6% and 82.8% of the theoretical ethanol yield were achieved from the fermentation of nondetoxified and detoxified hydrolyzate, respectively. In a study by Jiang et al. (2016), two-step acid hydrolysis of sugarcane bagasse and *Jatropha* hulls was performed to obtain sugars. The subsequent process of fermentation in a 3 L bioreactor at 35°C for 24 h resulted in the production of 2.15, 2.06, and 1.95 mol of hydrogen per mol of total reducing sugar, from glucose, bagasse hydrolysate, and *Jatropha* hull hydrolysate, respectively.

11.8 ALKALI PRETREATMENT

The alkali pretreatment method has been a preferred method in the LCB biorefinery. The method is applicable for pretreatment as well as the fractionation of lignocellulose into lignin and carbohydrates. The chemicals used in this pretreatment method are highly selective for lignin solubilization (Kim et al., 2016). The method involves the use of noncorrosive chemicals such as hydroxides of Na, K, Ca, or ammonium and anhydrous ammonia, with NaOH and lime being most common among all because of their effectiveness and other advantages (Kulshreshtha, 2022). This treatment uses moderate temperatures; consequently, negligible sugar is degraded and inhibitory compounds are generated (Kim et al., 2016). During alkali pretreatment, the ester linkages connecting lignin to the hemicellulose are disrupted, causing the lignin and hemicellulose portions to dissolve. Several groups (acetyl) and substitutions (uronic acid) present in the hemicellulose are removed. The cellulose swells; its polymer length is reduced, and its structure becomes more amorphous. All these factors increase the digestibility of cellulose (Chen et al., 2013; Baruah et al., 2018). Ghasemian et al. (2016) have reported enhanced production of biogas and H_2 from KOH (alkali)-pretreated stalks and bolls of cotton plants. The highest amount of VS obtained from treated stalks was 246 mL/g and from treated bolls was 219 mL/g. The pretreated bolls produced 17 mL/g of hydrogen. Kim (2018) produced bioethanol from empty palm fruit bunch fiber, using alkali-thermal pretreatment. The treatment resulted in 55.4%–56.9% delignification. The fed-batch fermentation of pretreated and saccharified hydrolyzate for 20 h produced 33.8 g/L ethanol (1.57 g/L/h productivity). Gil-Montenegro et al. (2019) carried out alkaline pretreatment of Brewer's Spent Grain for the extraction of arabinoxylans to obtain xylose for its fermentation to xylitol. They achieved 58% of the final yield in their study simulated at a Colombian biorefinery.

11.9   OZONOLYSIS

Ozonolysis is categorized among the most promising methods known for biomass pretreatment. The method is used primarily for lignin disintegration and shows minimal effects on the holocellulosic portions. During ozonolysis, hemicelluloses and lignin parts may get solubilized when the biomass–water mixture is exposed to a specific temperature for a certain time. Ozone works as a strong oxidizing agent, which cleaves the C=C double bonds in lignin (component rich in C=C bonds). A mild effect is seen on the hemicellulose component. The efficacy of the treatment is governed by diverse factors, such as ozone concentration, flow rate, source and type of biomass, moisture levels, particle size, pH conditions, etc. (Travaini et al., 2016). Therefore, optimization of conditions is often done for different types of biomass. The method is advantageous not only in its efficacy for lignin removal; lesser amounts of products inhibitory to subsequent steps of enzymatic saccharification or fermentation are generated, and the treatment can be performed at milder conditions of temperature and pressure. The method, however, is not adequate for moisture-rich biomass and is expensive, requiring large amounts of ozone. In a work by Travaini et al. (2013), ozonolysis of bagasse at room temperature resulted in the selective removal of lignin, enriching the biomass with carbohydrates that yielded high amounts of fermentable sugars upon enzymatic hydrolysis of pretreated biomass. Also, no furfural or hydroxymethylfurfural inhibitors were generated in the process. Panneerselvam et al. (2013) evaluated the effect of ozonolysis on energy grasses, i.e., *Saccharum arundinaceum*, *S. ravennae*, *Miscanthus giganteus*, and *M. sinensis 'Gracillimus'*. The treatment resulted in the effective removal of up to 59.9% of lignin, without degradation of cellulose. However, the inhibitory action of lignin degradation products was also observed during the process. Domanski et al. (2017) performed ozonolysis of rye straw for anaerobic digestion to produce biogas. Under optimized conditions, i.e., a 15 g biomass load treated for 60 min with a low dose of ozone (100 g ozone per m^3) caused a rise in methane production, i.e., 291.71

dm^3 methane/kg VS. In additional research work by Patil et al. (2021), they used ozonolysis combined with thermal pretreatment for the rice straw followed by its anaerobic digestion, and the process yielded 374 mL of methane/gram of VS. The pretreatment enhanced the rate of biomass hydrolysis by up to 26%.

11.10 OXIDATIVE PRETREATMENT

In the oxidative pretreatment method, the aqueous biomass is treated with an oxidizing agent. Hydrogen peroxide (H_2O_2), peracetic acid ($C_2H_4O_3$), and chlorine dioxide are the most commonly used chemicals for this treatment. The method is also used in combination with alkaline treatments such as alkaline-H_2O_2 treatment or alkaline-peracetic acid pretreatment. H_2O_2 is used widely. The method effectively removes hemicellulose and lignin portions, thereby increasing the cellulosic content of the biomass (Maurya et al., 2015). The delignification is primarily caused by the release of OH and O^{-2} (hydroxyl and superoxide, respectively) radicals. The peracetic acid causes the formation of the hydronium ion, H_3O^+, which results in multiple changes in lignin due to interaction with lignin electron sites (Han et al., 2020). Oxidative pretreatment may alter the lignin structure and biomass surface such that nonspecific adsorption of enzymes is minimized. This is attributed to the formation of OH• and O^{-2} as a result of reactions between H_2O_2 and the hydroperoxide anion (HOO⁻) that cause delignification. The chlorine dioxide exerts oxidative action, causing the shortening of lignin chains, the opening of aromatic rings, and the oxidation of side groups. In a study by Han et al. (2020), mild oxidative pretreatment of sugarcane bagasse using alkaline H_2O_2, two-step alkaline $C_2H_4O_3$, and sodium chlorite resulted in biomass delignification, disruption, and deformation. The results showed an enhancement in glucan conversion rate during the subsequent step of enzymatic saccharification of pretreated biomass by more than 90%.

11.11 ORGANOSOLV PRETREATMENT

Organosolv pretreatment encompasses the treatment of biomass with an aqueous solution of organic solvents under optimal conditions of temperature and pressure. Methyl alcohol, ethyl alcohol, acetone, and ethylene glycol are preferred solvents for this method. The method is useful in recovering a nearly pure lignin by-product, which is extracted out of the biomass, and a cellulose-rich biomass is obtained (Meng et al., 2020). A catalyst may or may not be added during this treatment. In the absence of a catalyst, the process requires higher temperatures and a larger concentration of solvent. This is called auto-catalyzed organosolv pretreatment. The acetic acid generated during the process acts as a catalyst for treatment. Acid-catalyzed organosolv pretreatment, using mineral acids or strong organic acids, solubilizes the lignin and hemicelluloses under milder operating conditions. Usually, solvents with low boiling points are preferred because their recovery is relatively easier. Otherwise, solvents may interfere with enzymatic hydrolysis or fermentation processes. The recovery of solvents also makes it a cost-effective method. Meng et al. (2020) studied what impact approaches such as solvent-enhanced lignocellulosic fractionation (CELF) and γ-valerolactone pretreatment have on woody biomass. The CELF pretreatment removed lignin without affecting cellulose significantly. Ethanol changed the DP of the cellulose polymer, whereas γ-valerolactone caused destruction of the crystalline cellulose. All types of increased biomass surface area are exposed to callulolytic enzymes. Viola et al. (2021) used a biphasic system containing n-butanol or 2-methyltetrahydrofuran and an aqueous solution of oxalic acid to pretreat wheat straw and eucalyptus residues. The optimization process was carried out to achieve the highest recovery of cellulose and hemicellulose components, as well as glucose sugar. Butanol resulted in glucose recovery of nearly 98% of the theoretical yield, while this value was 67% for 2-methyltetrahydrofuran. However, 2-methyltetrahydrofuran showed the highest level of delignification with approximately 90% efficiency in the case of wheat straw.

11.12   IONIC LIQUIDS (ILS)

ILs are relatively recently explored chemicals categorized as green solvents for lignocellulosic pretreatment. ILs refer to the organic salts of combinations of anions and cations. ILs possess certain useful characteristics such as high thermal and chemical stability, nontoxicity, low vapor pressure, economic viability, noncorrosiveness, and recyclable nature (Den et al., 2018; Usmani et al., 2020). Examples include 1-ethyl-3-methylimidazolium acetate, 1-butyl-3-methylimidazolium chloride, 1-allyl-3-methylimidazolium chloride, 1-butyl-3-methylimidazolium acetate, 1-butyl-3-methylimidazolium hydrogen sulfate, etc. (Usmani et al., 2020). The liquids can dissolve cellulosic portions of the biomass under milder conditions (Den et al., 2018). The atoms of hydroxyl groups (O and H) in cellulose interact with ILs and form a pair of electron donors and acceptors. Hydrogen can accept electrons, and cellulose can donate electrons. In the ILs, anions and cations behave as electron donors and acceptors, respectively. Thus, mixing cellulose with ILs results in the formation of a composite of electron donors and acceptors. Interactions between cellulose molecules and IL anions disrupt linkages within cellulose and cause the dissolution of this component of lignocellulose (Den et al., 2018; Usmani et al., 2020). Lignin is also dissolved by ILs as a result of the breakdown of the hydrogen bonds and linkages between the nucleus and chains of lignin components (Zhang et al., 2017). Padrino et al. (2018) used 1-ethyl-3-methylimidazolium acetate to pretreat barley straw and investigated its effect on the outcome of anaerobic digestion of pretreated biomass. They observed high biochemical methane potential under different mesophilic and thermophilic scales of temperature during the digestion process. The results revealed an increase of up to 28% and 80% in biochemical methane potential during thermophilic and mesophilic anaerobic digestion for 35 days compared to the native biomass. Ziaei-Rad et al. (2021) have also documented the role of [TEA] [HSO$_4$] IL pretreatment in delignification (80%) of the wheat straw, which enhanced the yield of saccharification by 87.19% and subsequent ethanol fermentation by 84.34% of the theoretical yield (compared to 10.76% yield in untreated biomass).

11.13   PHYSICOCHEMICAL PRETREATMENT METHODS

Physicochemical pretreatment methods engage a blended approach, which optimally influences the physical components as well as the chemical bonds and linkages within the biomass structure (Zhao et al., 2012). These techniques operate in an intensive manner and disintegrate the biomass structure by employing explosive processes (steam or carbon dioxide explosion, or ammonia fiber explosion, AFEX) and other treatments such as liquid hot water, wet oxidation (WO), and SPORL treatment (Taylor et al., 2019).

11.14   STEAM EXPLOSION (SE)

SE is one of the most beneficial, competent, and environmentally friendly techniques for pretreating lignocellulosic masses. This is also known as *autolysis*. In SE, pressures as high as 0.5–4.8 MPa and temperatures ranging from 160°C to 260°C are utilized in such a way that biomass is exposed to saturated steam for a specific duration, and then pressure is declined and brought to atmospheric pressure or even lower. This results in the explosive unwinding of the biomass into component fibers and fiber bundles. A possible mechanism behind this is the major variations in the chemical structure of biomass, which are facilitated by a sudden disruption affecting the hydrogen bonds in the lignocellulosic matrix, ultimately leading to the hydrolysis of strong bonds (Taylor et al., 2019). The exploded material often retains more water, has a larger porosity and surface area, and shows a lesser bulk density. LCB is often pre-soaked before pretreatment so as to enhance penetration efficiency and swelling. During the pre-soaking, free moisture imparts buffering effects, thus decreasing the heat transfer but increasing the energy demand. Moreover, the entrapped moisture increases the pretreatment efficiency by softening the fibers (Sui and Chen, 2016). Parameters such as the amount of moisture, duration of treatment, chip size, and temperature impact the SE process.

Datar et al. (2007) employed the SE method for generating biogas from hemicellulose and obtained high yields of hydrogen. Vivekanand et al. (2013) utilized steam for pretreating birch samples at varying temperatures of 170°C–230°C for 10 min and found 220°C to be the optimal temperature for getting maximum enzymatic hydrolysis yield. Lizasoain et al. (2017) used SE technology to produce biogas using corn stover as a feedstock, and the treatment at 160°C for 2 min resulted in an increase in the methane yield of 22%. Álvarez et al. (2021) found that pretreatment of barley straw by SE at 180°C for 30 min led to the recovery of 90% of the cellulose (in the solid fraction) and 82% of the hemicellulose (in the liquid fraction). Following various approaches, the highest glucose yield (92 g/L) was obtained by utilizing a solid load of 20% and enzyme dosages of 30 FPU/g, respectively. The fermentation process resulted in a maximum 50 g/L yield of bioethanol. Aski et al. (2019) have reported an enhanced production rate (24%) of biogas from rice straw pretreated by SE. A significant reduction from 18.6% to 13.0% was noticed in the lignin content of the rice straw using a 10 min cycle at 10 bar, while the methane yield was enhanced by 113%, 104%, and 147% at a 10 min cycle at 10 bar, a 15 min cycle at 15 bar in the presence of 35% moisture, and a 10 min cycle at 15 bar with 70% moisture, respectively. Xia et al. (2020) pretreated industrial vinegar residue by SE for the extraction of hemicelluloses (glucans, xylans, and arabinans) to obtain sugars (glucose, xylose, and arabinose) for their fermentation to butanol. They reported a butanol titer of 7.98 g/L produced by a superior strain of *Clostridium acetobutylicum* Tust-00, exhibiting high inhibitor tolerance to inhibitors, by the fermentation of the sugars. Furthermore, to improve productivity and overall process economics, the SE has been utilized in different blends with additives and pretreatment methods. The use of acid- and alkali-based catalysts is one of the major modifications in the process (Bensah and Mensah, 2018).

11.15   ACID-CATALYZED STEAM EXPLOSION (ACSE)

This is an acid-based SE in which acid plays a major role, such that the SE is undertaken only after treating with the respective acids. In this process, the biomass is drenched with SO_2 or CO_2 or soaked with dilute acid for 0.5–25 h at low atmospheric pressures at the temperature (5°C–100°C) required for the treatment. This process renders substrates more reactive upon dissolution of the hemicelluloses into their monomer units, while cellulosic hydrolysis by the enzymes makes the process more promising for the disorganization of the cellulose bundles (Bensah and Mensah, 2018). However, it has been noticed that SO_2 impregnates biomass substrates in a suitable and efficient way, but for the removal of hemicellulose, the process utilizes stringent conditions compared to dilute acids (Panagiotopoulos et al., 2013). SO_2 and CO_2-based SEs cause the emergence of pores of variable dimensions in the surface layers of the cell wall of pretreated biomass. But in SO_2-based applications, this effect is more pronounced due to its greater cumulative stringency (Coralles et al., 2012). Although, in contrast to SO_2, CO_2 has a lower solubility, it is still preferred and found to be superior to SO_2 due to its greater availability, lower toxicity, and noncorrosive nature. However, major disadvantages of the SO_2-catalyzed treatments are the toxic nature of SO_2 and the undesirable release of degradation products. Moreover, expensive reactors are required for the acid-based SE that can withstand corrosion (Sipos et al., 2010). Guerrero et al. (2017) utilized acid-impregnated SE on banana stems and reported high glucose levels (91% yield) at optimal conditions of 170°C, 2.2% H_2SO_4 (v/v), and a 5-min treatment time.

11.16   ALKALI-CATALYZED STEAM EXPLOSION

This variant of SE has garnered less attention in comparison to acid-based SE. Alkaline-based SE enhances the enzymatic degradability of the biomass by improving the delignification rate of the substrate. The method, however, comes with its own set of merits and demerits. In terms of its advantages, since no additional or external catalysts are required, it is one of the most economical methods for agricultural wastes and hard woods. Owing to the use of higher-energy steam and minimal amounts of water, pretreatment of the biomass can be done at high solid loading, which ultimately reduces the capital expenditure, thus making the overall process affordable. In addition, lesser sugar dilution is observed in the pretreated liquor, and the requirement for waste stream recovery is reduced or excluded. Despite these significant benefits, there are several drawbacks associated with this method. The generation of furan-derived compounds, weak acids, or phenolic inhibitors severely influences the downstream steps involved in biomass conversion to bioenergy (Alvira et al., 2010). The enhanced degradation of cellulose and hemicellulose can be attributed to the stringent conditions applied during this treatment. Its effect has been found to be less pronounced on softwood and unexploded materials. Further, pressure and high temperature requirements during pretreatment influence other factors such as heat recovery, handling of materials, reactor operation, and energy management (Ibbett et al., 2011). Park et al. (2012) utilized alkaline-based SE on Eucalyptus and reported a promising glucose recovery of 66.55% due to the higher enzymatic digestibility of the biomass relative to the uncatalyzed SE.

11.17 AMMONIA FIBER EXPLOSION

This pretreatment method is a suitable alternative to SE. In this method, heated liquid ammonia (NH_3) is utilized for the fibrous explosion of the lignocellulose. It's a solvent-dependent process in which a mixture of moist biomass in an approximate 1:1 ratio with the liquid (anhydrous) ammonia is subjected to pressures of 17.2–20.6 bar at temperatures ranging from 60°C to 200°C for a time interval of 5–30 min, followed by an immediate release of the pressure. Usually, less than 2 kg of ammonia is added per gram of dry biomass in the reactor (Mood et al., 2013). The severe pressure and temperature conditions, along with liquid ammonia, cause the surface area to expand, reduce the crystallinity of the cellulose, and disintegrate or modify the lignin structure. AFEX pretreatment can be optimized by altering four variables such as water loading, ammonia loading, temperature, and blowdown pressure (El-Naggar et al., 2014). This pretreatment creates a nano-scale meshwork of interconnected channels in the biomass structure by cleaving the lignin-carbohydrate ester linkages through ammonolytic and hydrolytic reactions. These reactions induce major structural and chemical alterations in the biomass. This facilitates enhanced enzymatic access to the cellulose, mainly by partially removing the extractives and their subsequent deposition on the biomass exterior (Chundawat et al., 2013). The partial solubilization of lignin aids in increasing biomass porosity (Mathew et al., 2016). This method is more beneficial for herbaceous residues, corn stover, and rice straw than for woody biomass with a lower lignin content (Brodeur et al., 2011).

As no liquid stream is produced in this method, it is a dry-to-dry pretreatment, making it more suitable and cost-effective as compared to many other methods (Li et al., 2010). AFEX decreases the requirements for nutrient supplementation, post-pretreatment washing, and stream separation and generates useful intermediates that can be used to synthesize various value-added bioproducts. Moderate temperatures and low energy requirements make the AFEX process economically viable. Even fractionation of sizeable solids up to the size of 5 cm can be carried out by AFEX while obtaining better yields (Alvira et al., 2010). Furthermore, the negligible formation of inhibitors in the process makes it superior to several other pretreatment methods. Although this method proves to be an effective pretreatment method, significant operating expenses are required initially due to the use of ammonia reagent. So, it can be recycled and recovered, due to its volatile nature, to curtail the overall operating cost and reduce environmental damage (Bensah and Mensah, 2018).

Alizadeh et al. (2005) reported that AFEX carried out for the switchgrass at optimized treatment conditions of near 100°C reactor temperature, 1:1 ratio of ammonia/kg of dry matter, 80% moisture level, and 5 min of treatment duration caused 93% conversion of glucan. Subsequently, the fermentation process produced 0.2 g of ethanol per gram of dry biomass, a yield about 2.5 times higher than that obtained from the native biomass. Li et al. (2010) carried out the AFEX pretreatment of forage and sweet sorghum bagasse for the extraction of glucans and xylans to obtain glucose and xylose for their fermentation to ethanol by *Saccharomyces cerevisiae*. They achieved an ethanol yield of 30.9 g/L from the forage sorghum bagasse and 42.3 g/L from the sweet sorghum bagasse. Lau et al. (2010) pretreated empty palm fruit bunch fiber (EPFBF) using AFEX and followed its enzymatic saccharification. The pretreatment reduced the particle size of the biomass, which generated 90% of the maximum theoretical yield of the sugars in 72 h of hydrolysis time. Emam and Seyed (2013) revealed that AFEX-treated wheat straw, at optimal treatment conditions of an ammonia-to-wheat straw ratio of 1:1, 95°C temperature, 70% moisture content, and a residence time of 5 min, and generated an 89% sugar yield upon enzymatic saccharification of the pretreated straw. This yield was appreciably greater than that obtained from the hydrolysis of untreated wheat straw (26%). The released sugars can be further fermented to produce effective biofuels.

Zhang et al. (2022) utilized AFEX along with NaOH for the delignification of *P. sinese*, oak, and camphor wood and further subjected the biomass to enzymatic hydrolysis for biohydrogen production by dark fermentation. They reported significant lignin removal of about 84.2%, 59.7%, and 36.7% in *P. sinese*, oak, and camphor wood, respectively. The biohydrogen yield from the delignified biomass was found to be 152.3, 99.1, and 76.9 mL/g TS, respectively, in the cases of *P. sinese*, oak, and camphor wood.

11.18   LIQUID HOT-WATER (LHW) PRETREATMENT

As the name suggests, this hydrothermal process makes use of liquid water instead of steam for the cooking of biomass, involving essentially high temperatures (150°C–240°C), pressure (up to 5 MPa), and short exposure times ($\leq$ 50 min). Releasing pressure is not a necessity in this process, as it is in other explosion methods. Prevention of the evaporation of water is a critical factor, so pressure is applied (Baruah et al., 2018). The pressurized LHW employed in this method penetrates and disrupts the cell structure. Consequently, the solubilization of biomass forms two different streams, i.e., a liquid and a solid fraction. The liquid part is encompassed by hemicellulosic sugars, minerals, furfural from the degradation of pentoses, and acetic acid cleaved from hemicelluloses. Monomeric sugars are added at the lowest possible rate during the process. The solid part of the pretreated slurry is mostly rich in cellulose, lignin, and small amounts of remaining hemicellulose. The formation of inhibitors can be evaded by carrying out LHW treatment at a controlled pH of 4 and 7 (Li et al., 2014). Major changes, such as the enlargement of the exterior surface and pore volume made of individual segregated cellulose fibers, are seen in the feedstock undergoing this pretreatment. The lignin is re-polymerized over the surface of the cellulose matrix. The severity factor (Ro) determines the harshness of the process, which is decided by evaluating the combined effects of time and temperature on the recovery of sugar and other degradation products. Partial removal of hemicellulose and re-localization of lignin from the substrate in this pretreatment process is a significant dimension that dictates the digestibility of pretreated biomass rather than cell wall rupturing and alterations in crystallinity (Yu et al., 2016). Various feedstocks, including agricultural waste, woody materials, and industrial waste, are successfully pretreated by LHW. To overcome its inherent drawbacks and increase sugar recovery, LHW can be combined with other methods.

A major positive attribute of LHW is the enhanced enzymatic digestibility of the cellulose owing to the fractionation of hemicelluloses, which circumvents inhibitor formation. Moreover, it does not facilitate the use of any catalysts or chemicals, which is an added benefit of this technique. It uses a low-cost solvent for large-scale application, and the formation of any hazardous or toxic materials is nearly absent (Baruah et al., 2018). Large biomass flow rates can be handled efficiently in LHW pretreatment as the impact of particle size on hydrolysis is low. However, compared to the SE, LHW pretreatment contributes to higher pentose recovery and lowered levels of solubilized hemicellulose and lignin products. Despite various advantages, the method has several limitations or drawbacks. It causes the dissolution of a large part of the hemicellulose into oligomers. The lower xylose yields certainly influence the ethanol and sugar yields. The formation of acidic and furan-like by-products from sugar degradation at extreme treatment conditions is a significant risk during this method (Bensah and Mensah, 2018). Greater expenditure is associated with LHW to generate saturated liquid water. Also, the process is very energy-intensive.

Boakye-Boaten et al. (2015) revealed that LHW pretreatment of *Miscanthus* X *giganteus* at 200°C for 15 min under controlled pH conditions attained the highest ethanol production (71.8%). Imman et al. (2018) carried out LHW pretreatment of the corn cobs at 160°C using a residence time of 10 min and recovered higher levels of sugars (58.8%), along with an enzymatic saccharification yield of 73.1% and 60% lignin removal. Phuttaro et al. (2019) pretreated Napier grass with LHW and observed a nearly 35% rise in CH_4 yield relative to the untreated biomass. Ao et al. (2020) produced methane from rice husk pretreated by LHW at 170°C for 2 h and found an increased (225.7%) methane production in comparison to the untreated husk. Suriyachai et al. (2020) estimated the ethanol yield from acid-catalyzed LHW-pretreated corn stover. They observed that 162.4°C temperature, 29.5 min treatment time, and 0.45% v/v sulfuric acid were optimal conditions for the process that resulted in the production of 93.91% ethanol.

11.19   WET OXIDATION

The WO method has been found to offer immense benefits. It involves the use of an oxidizing agent in an aqueous solution. The treatment breaks the lignocellulose into its constituents (Ravindran and Jaiswal, 2016). The oxidation causes increased dissolution of hemicellulose and lignin removal from the biomass. This process has also been known as 'oxidative delignification'. Oxidizers, which were formerly used, are pressurized gases such as liquid-based peroxides and air or oxygen (5–30 bars). Working temperatures for this pretreatment in an oxygen-rich atmosphere, fall within a range between 120°C and 350°C, and residence times are typically in the range of 0.5–4 h (Taylor et al., 2019). But hydrogen peroxide-based pretreatments require a relatively much longer residence time of about 8 h at a far milder temperature of around 30°C. The percolation of hemicellulose during the WO of wheat straw has been found to be associated with temperature change. Water and carbon dioxide are formed upon complete oxidation. Furthermore, smaller carboxylic acids, aldehydes, and alcohols are created upon hydrolysis involving the decomposition of polymeric chains (Aftab et al., 2019). WO has been observed to be useful for high cellulose recovery from the lignocellulose when the process involves the use of a base, although it contributes to an increase in the process' cost (Schimdt, 1998). The outcome of this pretreatment can be significantly enhanced by optimizing factors like duration of treatment, oxygen levels, and temperature. The WO, however, is not suitable for industrial applications because oxygen is flammable and H_2O_2 is relatively costly (Bajpai, 2016).

Song et al. (2012) documented enhanced cellulose accessibility to the hydrolytic enzymes via the removal of lignin and hemicellulose from the rice straw pretreated by this method. They achieved a 50%–120% higher yield of methane relative to untreated substrate. Arvaniti et al. (2012) utilized wet oxidized pretreated rape straw for bioethanol generation through a simultaneous saccharification and fermentation (SSF) approach. The biomass was subjected to the pretreatment conditions of 205°C temperature, 3 min exposure time, and 12 bar oxygen pressure. The treatment recovered higher levels of cellulose, hemicelluloses (100%), lignin (86%), and 67% of the maximum ethanol yield from the pretreated whole slurry. Ayeni et al. (2016) carried out alkaline WO of shea tree sawdust, followed by enzymatic saccharification and fermentation. They recorded 22-fold rises in the reducing sugar yield from the treated biomass and achieved production of 12.73 g/L bioethanol. Ahaou et al. (2020) produced bioethanol from WO-pretreated cassava skin, Ulva (alga), and *Eichhornia crassipes* and reported a significant dissolution of holocellulose, which increased the sugar content to 1.5, 2.38, and 1.83 g/L in the case of water hyacinth, cassava skin, and Ulva, respectively. The fermentative bioconversion of pretreated biomass yielded 0.26 g/L ethanol from Ulva, 0.31 g/L from cassava peel, and 0.2 g/L from the water hyacinth.

11.20 CARBON DIOXIDE EXPLOSION

CO_2 explosion has been found to be a promising alternative to ammonia fiber explosion and SE. In AFEX, the use of highly corroding ammonia and its damaging effects on the environment limit the scope of this method, whereas high thermal energy required in SE also restricts its use (Baruah et al., 2018). In carbon dioxide explosion, supercritical CO_2 is used. Supercritical fluids (SCFs) are highly compressed fluids (compressed above their critical point) that can act as solvents and represent a state in which both liquid and gaseous forms can coexist. The CO_2 explosion process is primarily carried out by introducing CO_2 into a high-pressure barrel containing substrate, which is then agitated at an appropriate temperature for a specific time period (Shukla et al., 2023). Usually, a pressure between 30 bar and 275 bar (used for SEs and SC CO_2, respectively) and temperatures lower than those for SE (35°C–175°C) are employed (Taylor et al., 2019). This method appears to be a viable alternative compared to AFEX and SE because of the lesser energy requirements and use of greener chemicals (Bharathiraja et al., 2018). In this pretreatment, supercritical CO_2 exhibits its dual nature of gas and liquid, such that it can penetrate the biomass like gas and dissolve its components like a liquid, due to the existing features of a "gas-like" mass transfer and a "liquid-like" solvating power of SCFs (Rostagno et al., 2015). The formation of carbonic acid occurs upon the dissolution of CO_2 in water. It has been found to be less corrosive than its counterparts and is involved in the hydrolysis of hemicellulose (Arumugham et al., 2022). The smaller CO_2 molecules diffuse into the biomass interior easily (Aftab et al., 2019). Supercritical CO_2 can be used to induce both acid-based hydrolysis and an explosive shearing force. The LCB must be restrained from drying by maintaining an appropriate moisture content, as CO_2 exhibits its acidic effects when dissolved in water (Mood et al., 2013). The biomass structure is disrupted upon release of the pressurized gas, and cellulose fibers become more accessible (Capolupo and Faraco, 2016). It has been noted that the increase in pressure is linked to the increased penetration rate of the CO_2 molecules into the cellulosic pores, resulting in a high glucose yield. The lower operation temperature during the method accounts for sugar decomposition prevention, even in the presence of acid (Aftab et al., 2019).

Narayanaswamy et al. (2011) pretreated corn stover using supercritical CO_2, 24 MPa pressure, 150°C temperature, and a 60-min treatment time, which enhanced the glucose yield by 2.5 folds in comparison to the native biomass. Islam et al. (2017) pretreated soybean hulls with supercritical CO_2 under optimized conditions (8 MPa pressure, 130°C temperature, 30 min treatment time) and achieved a maximum of 97% glucose yield from the enzymatic saccharification of the pretreated substrate. The use of co-solvents (such as water and ethanol) during this method can cause lignin removal and improve enzymatic saccharification significantly (Serna et al., 2016). Rosera-Henao et al. (2019) reported that CO_2 explosion pretreatment of sugarcane bagasse under optimal conditions of 60°C per 200 kgf/cm^2 removed nearly 8.07% lignin. The treatment produced an accumulated methane of 0.6498 L_N Kg VS^{-1} (at 273.15 K, 1.013×10^5 Pa), which was 23.4% greater in comparison with the native material. Ge et al. (2020) subjected peanut shells to supercritical CO_2 pretreatment at a temperature and pressure of 190°C and 60 bar, respectively, during the process of bioethanol production. The treatment reduced the hemicellulose content from 12.4% to as low as 1.8% while producing 80.7% of the highest glucose yields in the subsequent step of enzymatic hydrolysis. A further 2.4% increase (compared to a glucose control) in bioethanol production was attained through the alcoholic fermentation of the hydrolyzed glucose. Utilization of supercritical CO_2 for lignocellulosic pretreatment involves a cheap cost of CO_2, nonflammability, low environmental damage, easy recovery, and no generation of toxins, making it a promising pretreatment method for bioenergy recovery processes (Baruah et al., 2018). Still, a prominent hindrance in its application at an industrial scale has been observed to be the capital inputs for the set-up capable of withstanding the conditions of high pressure used during this method. However, this is categorized among green methods of pretreatment from an eco-friendly stand point and also due to the ready recyclability of the CO_2 (Cha et al., 2014).

11.21   SPORL PRETREATMENT

SPORL stands for sulfite pretreatment to overcome recalcitrance of lignocellulose, and this method is essentially utilized to pretreat LCB. This method offers a feasible approach for the utilization of the forest thinning materials, which is crucial for lowering high costs involved in forest thinning operations (Zhu et al., 2011). Thus, SPORL makes healthy forest management operations viable. In this pretreatment process, the lignocellulosic materials are reacted with aqueous solutions of sulfite or bisulfite salts (of Na, Mg, or Ca) for around 30 min, at a pH of 2–4 and a temperature between 160°C and 180°C to remove lignin and hemicellulose fractions. Then, disc milling is done to produce a fibrous material for its subsequent processing (Galbe and Wallberg, 2019). The wood's recalcitrance is generally reduced as a result of various effects such as fractionation of hemicelluloses, enhanced surface area, partial delignification, cellulose de-polymerization, and the partial sulfonation of lignin (Aftab et al., 2019). Several studies have reported the use of the SPORL method in bioconversion processes for bioenergy recovery from LCB. Zhu et al. (2009) produced ethanol using lodgepole pine chips as feedstock. They subjected the biomass to the SPORL pretreatment at 180°C for 25 min, keeping the liquid to solid ratio at 3:1, in a lab-scale digester. The bioconversion process for ethanol production from the recovered cellulose-rich biomass resulted in an ethanol yield of 276 L/metric ton of biomass, which corresponds to 72% of the theoretical ethanol yield. Wang et al. (2012) reported higher ethanol and cellulose yields from the SPORL-pretreated native aspen (*Populus tremuloides*), exhibiting a higher enzymatic digestibility of over 80% after pretreatment compared to the dilute acid treatment of the biomass. Zhang et al. (2013) performed SPORL pretreatment on switchgrass and reported that it removed hemicellulose significantly, dissolved lignin partially, and reduced lignin hydrophobicity by its sulfonation. Together, these factors improved the digestibility of biomass. The pretreated biomass was saccharified with 15 FPU (filter paper unit) of cellulase and 30 CBU (cellobiose unit) of β-glucosidase/g cellulose, resulting in 83% hydrolysis in 48 h. Zhou et al. (2018) utilized three different substrates, i.e., beetle-killed lodgepole pine, poplar NE222, and Douglas-fir, subjected them to SPROL pretreatment, and obtained significantly higher yields of ethanol from the pretreated biomass.

11.22 BIOLOGICAL PRETREATMENT METHODS

Biological pretreatment utilizes microbial strains or microbial enzymes capable of disrupting LCB (Chaurasia, 2019). Biological pretreatment is an environmentally favorable method and is considered a better alternative to thermochemical pretreatment processes. The microorganisms and microbial enzyme pretreatments are very selective for their substrates. Therefore, biological pretreatment methods show relatively higher selectivity. No chemicals are used in this method, and unlike thermochemical methods, these methods involve the use of milder operating conditions. Thus, in contrast to other pretreatment processes, biological pretreatment is environmentally benign and economically feasible due to its nonchemical nature, use of moderate energy input, low severity, greater substrate reaction specificity, and elevated yields of sugar products.

The biological pretreatment process may involve the cultivation of selected microbes directly on the native biomass or the exposure of the biomass to specific enzyme concoctions (Ummalyma et al., 2019). Suitable microorganisms convert biomass into desirable products or biofuels and tend to have a greater yield in an economical manner. The selected microorganisms produce extracellular hydrolytic enzymes that attack the biomass components and alter lignocellulosic structure. Microbes capable of biomass pretreatment include bacteria, fungi, and actinomycetes. The selection of a suitable microbial strain depends on multiple factors, the most important being the biomass type. Fungi are known among the microbes capable of decomposing organic matter. They represent the most active group of hemicellulolytic microbes. The fungi such as *Trichoderma* and *Aspergillus* produce higher levels of diverse types of hemicellulases extracellularly that work in synergy while attacking their substrate (Chhetri et al., 2022). In addition, many white-, brown-, red-, or soft-rot fungi secrete cellulases, enzymes that attack the bonds joining cellulose to lignin and also reduce cellulose crystallinity. Thus, cellulose accessibility is enhanced along with biomass porosity, which facilitates subsequent hydrolysis and fermentation of cellulosic components (Kumar et al., 2017). The white-rot fungi have been more efficient in altering biomass structure as they secrete different lignolytic enzymes like laccases, lignin peroxidases, and manganese peroxidases, which carry out the oxidation of lignin (Nagai et al., 2007). These fungi have been found to degrade hemicelluloses and lignin at the same rate. This is because they produce hemi-cellulolytic enzymes as well as lignolytic enzymes. However, the relative proportion of enzymes varies among different fungal strains. It has been documented that brown- and soft-rot fungi primarily hydrolyze the cellulosic component of the biomass with minimum hydrolysis of lignin, while white-rot fungi primarily carry out lignin degradation with higher efficiency (Zhang et al., 2020).

The microorganisms commonly employed for biological pretreatment are *Clostridium* sp., *Cellulomonas* sp., *Bacillus* sp., *Thermomonospora* sp., *Streptomyces* sp., *Phanerochaete chrysosporium*, *Trichoderma reesei*, *T. viride*, *A. niger*, etc. (Sharma et al., 2019). The bacteria show rapid growth compared to fungi and high adaptability to environmental changes. Different bacterial strains, such as *Sphingobium* sp. SYK-6, *Rhodococcus* sp., *Ceriporiopsis* sp., *Pandoraea* sp., *Galactomyces* sp., and *Mycobacterium* sp., when utilized for biological pretreatment, showed significant dissolution of lignin prior to cellulose (Zhang et al., 2019). Several research studies have documented the use of microorganisms, including bacteria and fungi, to pretreat diverse biomass biologically. Some of them are listed in Table 11.3. However, there are several limitations to the biological method, the most important being the longer residence times and slower rate of the process. These factors make the process tedious and time-consuming. An alternative to this is to integrate biological pretreatment with other chemical or physicochemical pretreatment methods. Moreover, process optimization and other research efforts are in progress to overcome the challenges of making biological pretreatment feasible at the industrial level.

11.23   COMBINED PRETREATMENT METHODS

Integration of one or more methods of pretreatment into combined approaches can curtail the time limit and usage of chemicals. This classifies this method among the greener alternatives and can even corroborate the output of the final product. Competent digestibility and selective retrieval of biomass components are quite essential outputs that are obligatory for the effective conversion of biomass to

TABLE 11.3

Biological Pretreatment of Diverse Biomass for Bioenergy Production

Microbial Strain	Biomass	Process Conditions	Effect on Biomass/Process Outcome	References
		Fungal Pretreatment		
Ceriporiopsis subvermispora	Wheat straw	25°C, 7 weeks	Gas production @ 297.0 mL/g under *in vitro* conditions	Nayan et al., 2018
Pleurotus florida	Paddy straw	25°C–29°C temperature, treatment time of 28 days	75.3% of saccharification efficiency	Kumar et al., 2018
Fusarium oxysporum (zinc oxide nanoparticles)	Rice straw	Incubation at temperature of about 30°C for 3 days	Reduction of 48.8% cellulose, and 16.7% lignin relative to the native biomass, and ethanol recovery of 0.0359 g/g of biomass dry weight	Gupta and Chundawat, 2020
S. cerevisiae and *Actinomyces* co-culture	Apple pomace	Solid state fermentation at 30°C for 72 h	Bioethanol production @ 49.64 g/L	Kumar et al., 2020
Cephalotrichum stemonitis	Maize silage	Incubation at 25°C for 6 days	Degradation rate of lignin (55.2%), cellulose (25.0%), hemicellulose (24.5%) was higher, 70% rise in methane yield	Zanellati et al., 2021
Penicillium aurantiogriseum	Corn stover	Treatment at 37°C for 10 days	Methane yield of 281 mL$_N$/g TS or 16% increase	Kovacs et al., 2022
Microbial consortium of *S. cerevisiae* and *T. harzianum*	Empty fruit bunches of palm	Incubation at 30°C for 72 h	Production of 9.65 g/L of bioethanol (0.46 g/g ethanol yield), with conversion efficiency of 89.56%	Derman et al., 2022

(Continued)

TABLE 11.3 (*Continued*)
Biological Pretreatment of Diverse Biomass for Bioenergy Production

Microbial Strain	Biomass	Process Conditions	Effect on Biomass/Process Outcome	References
		Bacterial Pretreatment		
Bacterial consortium of *Methanosarcina* and *Methanobacterium*	Corn straw		VS removal efficiency @ 54.3% and methane yield of 216.8 mL/g VS$_{substrate}$	Fu et al., 2016
Bacillus subtilis	Cellulosic vegetable waste	Incubation for 24 h at 37°C, at 80 rpm	Ethanol recovery of 14.17%	Promon et al., 2018
Pandoraea sp. B-6	Corn stover	Pretreatment at 30°C temperature, under shaking conditions at 150 rpm, for 3 days	42.6% cellulose, 41.6% hemicellulose, and 12.9% lignin reduction compared to native biomass	Zhuo et al., 2018
Pseudomonas sp.	*Miscanthus sacchariflorus*	Pretreatment at 37°C, with agitation rate at 200 rpm for 96 h	50.1% lignin degradation, 2.2 folds increase in glucose yield	Guo et al., 2019
Cellulolytic microbial consortia	Sweet sorghum	Pretreatment at 37°C for 60 days	Improved enzymatic hydrolysis and enhanced bioethanol yield of 28.42 gL	Zhu et al., 2023

bioenergy. Hence, consolidations of two or more suitable pretreatment processes are being increasingly preferred (Galbe and Wallberg, 2019). Despite all the factors mentioned above that make integrated pretreatment processes a necessity, there are certain constraints. For example, if methods are different, there is an additional expense of implementing several vessels for different pretreatment methods. So, the use of widely different pretreatment methods is often discouraged. Generally, mechanical pretreatment, including size reduction, is commonly preceded by other steps. But there is also the possibility of a reverse-operative procedure. Several studies have documented the positive impact of combined pretreatment on biomass structural alterations and subsequent conversion to bioenergy. Pretreatment using LHW pretreatment technologies improves the digestibility of lignocelluloses while involving less expenditure on the treatment operation and more usage of the biomass. Asada et al. (2011) utilized fungal and SE-pretreated spent shiitake mushroom medium to achieve efficient ethanol production. They achieved a maximum 38.8 g/L ethanol yield using initial substrate levels of 300 g/L after 60 h of incubation. Kumar et al. (2020) reported reducing sugar yield to 370.23 mg/g by following a combination of biomass grinding (to a particle size of 0.4 mm) with 1% H_2SO_4 treatment and biological pretreatment with *Phanerochaete chrysosporium*. Here, the ground biomass, having a small size, exposed the maximum biomass surface area to the microbial cells and the acid, which together improved the pretreatment outcome. In a study, Yin et al. (2014) carried out supercritical CO_2 pretreatment combined with ultrasonic pretreatment for the corn cob and cornstalk biomass. They achieved an improvement in the enzymatic saccharification of the corn cob and corn stalk by 75% and 13.4%, respectively, at 20 MPa pressure and 170°C temperature. Alexandropoulou et al. (2017) used a mixed approach for the pretreatment of Willow sawdust for biogas production. They employed fungal pretreatment *(Leiotramete menziesii* and *Abortiporus biennis*) in the first step, followed by dilute (1% (w/v) NaOH) alkaline pretreatment. They reported that a methane yield of 142.2 L/kg of TS was obtained when *L. menziesii* was employed in the first step of fungal pretreatment and a 205.3 LN/kg TS recovery of methane using *A. biennis* in the same step. Here, alkali aids in lignin percolation and neutralization of the acidic products generated from the feedstock (Taherdanak et al., 2014). But the demerit of this method is the risk of Na^+ ion production, which can interfere in the methanogenic process and cause adverse impacts on the environment through disposal in the effluent (Zheng et al., 2014). Integration of fungal treatment with this method can improve the methane yield by dissipating the hazardous effects of NaOH effluent (Meenakshisundaram et al., 2022). In another study by Zhang et al. (2018), the integration of *Phanerochaete chrysosporium* treatment along with 1% H_2SO_4 acid treatment (100°C temperature, 1 h residence time) of water hyacinth followed by the SSF using *Saccharomyces cerevisiae* resulted in an 8.61% increase in the ethanol yield relative to acid treatment alone. Verardi et al. (2018) revealed that the average yield for glucose and xylose was increased by 12% and 34%, respectively, and cellobiose content was decreased by about 30% when sugarcane bagasse was subjected to an SE combined with hydrogen peroxide for producing bioethanol. Muthuvelu et al. (2019) employed alkaline pretreatment assisted with high frequency sound waves for four different lignocellulosic residues and obtained higher sugar release from all four types of biomass, resulting in a bioethanol yield of 72.6% from hardy sugar cane, 82.5% from giant reed, 77.4% from narrow-leaved cattail, and 85.0435% from the pink morning glory. This treatment is a promising option as it reduces the longer treatment time and the larger amounts of chemicals required in the case of alkali treatment alone. The use of ultrasound can speed up the process. John et al. (2019) reported the use of thermal pretreatment in combination with ultra-sonication for the hydrolysis of sweet lime peel and found 60% of the highest reducing sugar yield from the pretreated substrate. Yu et al. (2019) employed intense pulverization in combination with phosphoric acid pretreatment for the corn stover and achieved higher enzymatic digestibility of the biomass. In this case, pulverization aids in this pretreatment process as it reduces the size of the particle, which provides easily available reactive sites that are required for further processing. Besides this, more reducing sugars are released after the saccharification process. Mikulski et al. (2020) employed a combination of microwave treatment with dilute acid pretreatment and exposure of biomass to high pressure. The acid used in the method breaks the hemicellulosic structure, and the temperature raised by the microwave irradiation assists in the rupturing of cellulose. Delignification occurs due to conditions of raised temperature and pressure, which together lead to the breakdown of biomass structure. Lu et al. (2020) found that a combination of LHW and alkaline pretreatment ($NaOH/O_2$) improved the xylan recovery and lignin removal, thereby increasing the digestibility of the biomass. Ummalyma et al. (2020) utilized combined dilute acid, ultrasound, and alkali pretreatment for the ethanol production from *Phragmites karka* (Tall Red Grass). They achieved the highest, 79%, of the reducing sugar yield using 10% biomass loads, 1% w/v alkali, an ultrasonication frequency of 20 kHz, and 20 min of treatment time. The pretreatment led to 40% biomass deliginification and an ethanol recovery of 78%, compared to the untreated biomass. Nasir et al. (2020) applied an SE along with choline chloride, which removed 84.7% of the lignin, resulting in a 4.5 times higher glucose recovery compared to the individual pretreatment method. In this method, choline chloride is employed as a deep eutectic solvent. When dissolved in water, it acts as a Lewis acid. Moreover, it has the capacity to alter the crystallinity of cellulose as it becomes viscous at a higher temperature. David and Grzegorz (2021) made use of combined pretreatment with microwave radiation and sodium cumene sulfonate (microwave-assisted hydrotropic pretreatment) to generate cellulosic ethanol from maize distillery stillage. The pretreatment removed 44% of lignin, no fermentation inhibitors were obtained, and a high ethanol concentration of above 40 g/L (95% of ethanol) was obtained after the pretreatment. During this pretreatment, sodium cumene sulfonate acted as a hydrotrope, functioning like surfactants (due to their amphiphilic structure) that reduce the surface tension. Moreover, they reduce the lignin concentration in the biomass and are effective at elevated temperatures. They neither put a burden on the environment nor generate fermentation inhibitors. Hydrotropic treatment along with microwave irradiation ensures effective biomass fractionation, making the feedstock more responsive to enzymatic hydrolysis and subsequent fermentation. Cai et al. (2022) performed AFEX pretreatment combined with H_2O_2 pretreatment for rice straw, poplar, and pine biomass, which improved the efficiency of lignin degradation, saccharification (58.7%, 39.5%, and 20.6%), and hydrogen production through the dark fermentation process in the subsequent step (145.49, 80.75, and 57.52 mL/g) from rice straw, poplar, and pine biomass, respectively. Shukla et al. (2023) integrated SE with a H_2O_2/H_2O_2-citric acid mixture to pretreat rice straw and reported a 54.5% increase in the cellulosic content, a 46.5% reduction in the hemicellulose, and a 6.7% delignification at optimum conditions of 121°C temperature and 130 kPa pressure. Further, 220.05 and 273.21 g/L of reducing sugar yield were obtained from the sample pretreated with 0.05% (v/v) H_2O_2 and a 1:1 ratio of H_2O_2/citric acid in the same amount, respectively. H_2O_2 is utilized for lignin removal as it forms highly reactive hydroxyl (OH^-) and superoxide anion radicals (O_2^-), which actively take part in the oxidative process and accelerate the delignification of the LCB. This approach of SE with H_2O_2 is regarded as an eco-friendly and safe method to produce bioethanol, having the essential attributes of an economic and sustainable approach.

11.24 CONCLUSION AND FUTURE PROSPECTS

The process of pretreatment transforms the LCB at the fiber, fibril, and microfibril levels. Diverse methods, including physical, chemical, physicochemical, and biological, are known to be capable of biomass pretreatment. However, the varied composition and structure of different lignocellulosic materials makes it hard to find the most common, cost-effective, and favorable treatment for enhanced bioenergy recovery from all types of biomass. Also, several limitations associated with traditional pretreatment techniques limit their acceptance for various types of biomass. Therefore, no method can be claimed to be the best pretreatment method. The selection of an appropriate pretreatment method is mostly determined by its final use as well as desirable attributes such as lower power inputs, minimal generation of inhibitors and effluent streams, and minimum loss of useful components of the biomass. To achieve these characteristics, more elaborate and intriguing studies are needed aimed at optimizing the type and conditions of the pretreatment for different biomass types. The important points that need attention during these studies include little or no inhibitor formation, minimum chemical usage, energy requirements, and waste disposal costs, along with higher efficiency of saccharification, fermentation, and biomass fractionation. Various other approaches are also being explored for their potential effects on pretreatment outcomes. Novel integrated pretreatment processes can be developed aiming at time and energy savings, cost-effectiveness, and environmental viability. Furthermore, the use of nanobiocatalysts can offer advantages such as uninterrupted penetration into the biomass interior. Also, the nanoparticles could be utilized for biomass saccharification during biofuel production. In addition, active enzymes and microbes showing high stability can be developed through recombinant DNA technology, which can prove beneficial through reduced enzyme costs and the possibility of a one-step conversion process for bioenergy to biomass. Recently, plant genetic engineering technology has been tapped for the genetic modification of plants in order to modify their lignocellulosic structure and composition and make them more responsive to disruptive pretreatment processes. Further, any recommendation of a method for specific biomass must precede a comprehensive techno-economic assessment involving data collection from studies performed at a minimum pilot scale.

REFERENCES

Aftab, M.N., Iqbal, I., Riaz, F., Karadag, A. and Tabatabaei, M., 2019. Different pretreatment methods of lignocellulosic biomass for use in biofuel production. In: A. Abomohra (Ed.), *Biomass for Bioenergy-Recent Trends and Future Challenges*, pp. 1–24. Intechopen. DOI: 10.5772/intechopen.84995.

Ahou, Y.S., Christami, M.N.A., Awad, S., Priadi, C.R., Baba-Moussa, L., Moersidik, S.S. and Andres, Y., 2020. Wet oxidation pretreatment effect for enhancing bioethanol production from cassava peels, water hyacinth and green algae (Ulva). *AIP Conference of Proceedings*, 2255(1), p. 030039.

Alexandropoulou, M., Antonopoulou, G., Fragkou, E., Ntaikou, I. and Lyberatos, G., 2017. Fungal pretreatment of willow sawdust and its combination with alkaline treatment for enhancing biogas production. *Journal of Environmental Management*, 203, pp. 704–713.

Alizadeh, H., Teymouri, F., Gilbert, T.I. and Dale, B.E., 2005. Pretreatment of switchgrass by ammonia fiber explosion (AFEX). *Applied Biochemistry and Biotechnology*, 124, pp. 1133–1141.

Álvarez, C., González, A., Ballesteros, I. and Negro, M.J., 2021. Production of xylooligosaccharides, bioethanol and lignin from structural components of barley straw pretreated with a steam explosion. *Bioresource Technology*, 342, p. 125953.

Alvira, P., Tomás-Pejó, E., Ballesteros, M. and Negro, M.J., 2010. Pretreatment technologies for an efficient bioethanol production process based on enzymatic hydrolysis: A review. *Bioresource Technology*, 101(13), pp. 4851–4861.

Amin, F.R., Khalid, H., Zhang, H., Rahman, S.U., Zhang, R., Liu, G. and Chen, C., 2017. Pretreatment methods of lignocellulosic biomass for anaerobic digestion. *AMB Express*, 7, pp. 1–12.

Ao, T., Luo, Y., Chen, Y., Cao, Q., Liu, X. and Li, D., 2020. Towards zero waste: A valorization route of washing separation and liquid hot water consecutive pretreatment to achieve solid vinasse based biorefinery. *Journal of Cleaner Production*, 248, p. 119253.

Arce, C. and Kratky, L., 2022. Mechanical pretreatment of lignocellulosic biomass towards enzymatic/fermentative valorization. *Iscience*, 25, p. 104610.

Arumugham, T., AlYammahi, J., Rambabu, K., Hassan, S.W. and Banat, F., 2022. Supercritical CO_2 pretreatment of date fruit biomass for enhanced recovery of fruit sugars. *Sustainable Energy Technologies and Assessments*, 52, p. 102231.

Arvaniti, E., Bjerre, A.B. and Schmidt, J.E., 2012. Wet oxidation pretreatment of rape straw for ethanol production. *Biomass and Bioenergy*, 39, pp. 94–105.

Asada, C., Asakawa, A., Sasaki, C. and Nakamura, Y., 2011. Characterization of the steam-exploded spent Shiitake mushroom medium and its efficient conversion to ethanol. *Bioresource Technology*, 102(21), pp. 10052–10056.

Aski, A.L., Borghei, A., Zenouzi, A., Ashrafi, N. and Taherzadeh, M.J., 2019. Effect of steam explosion on the structural modification of rice straw for enhanced biodegradation and biogas production. *BioResources*, 14(1), pp. 464–485.

Ayeni, A.O., Omoleye, J.A., Hymore, F.K. and Pandey, R.A., 2016. Effective alkaline peroxide oxidation pretreatment of shea tree sawdust for the production of biofuels: Kinetics of delignification and enzymatic conversion to sugar and subsequent production of ethanol by fermentation using *Saccharomyces cerevisiae*. *Brazilian Journal of Chemical Engineering*, 33, pp. 33–45.

Bajpai, P., 2016. *Pretreatment of Lignocellulosic Biomass* (pp. 17–70). Springer, Singapore.

Bak, J.S., 2014. Electron beam irradiation enhances the digestibility and fermentation yield of water-soaked lignocellulosic biomass. *Biotechnology Reports*, 4, pp. 30–33.

Baruah, J., Nath, B.K., Sharma, R., Kumar, S., Deka, R.C., Baruah, D.C. and Kalita, E., 2018. Recent trends in the pretreatment of lignocellulosic biomass for value-added products. *Frontiers in Energy Research*, 6, p. 141.

Bensah, E.C. and Mensah, M.Y., 2018. Emerging physico-chemical methods for biomass pretreatment. In: A. Abomohra (Ed.), *Fuel Ethanol Production from Sugarcane* (pp. 1–22). Intechopen.

Bharathiraja, B., Jayamuthunagai, J., Chakravarthy, M. and Kumar, R.P., 2018. Bioprocessing of biofuels for green and clean environment. In: V. Sivasubramanian (Ed.), *Bioprocess Engineering for a Green Environment* (pp. 237–249). CRC Press, Boca Raton, FL.

Boakye-Boaten, N.A., Xiu, S., Shahbazi, A. and Fabish, J., 2015. Liquid hot water pretreatment of *Miscanthus* x *giganteus* for the sustainable production of bioethanol. *BioResources*, 10(3), pp. 5890–5905.

Bonfiglio, F., Cagno, M., Yamakawa, C.K. and Mussatto, S.I., 2021. Production of xylitol and carotenoids from switchgrass and *Eucalyptus globulus* hydrolysates obtained by intensified steam explosion pretreatment. *Industrial Crops and Products*, 170, p. 113800.

Brodeur, G., Yau, E., Badal, K., Collier, J., Ramachandran, K.B. and Ramakrishnan, S., 2011. Chemical and physicochemical pretreatment of lignocellulosic biomass: A review. *Enzyme Research*, 2011, p. 787532.

Cai, Z., Zhang, W., Zhang, J., Zhang, J., Ji, D. and Gao, W., 2022. Effect of ammoniated fiber explosion combined with H_2O_2 pretreatment on the hydrogen production capacity of herbaceous and woody waste. *ACS Omega*, 7(25), pp. 21433–21443.

Capolupo, L. and Faraco, V., 2016. Green methods of lignocellulose pretreatment for biorefinery development. *Applied Microbiology and Biotechnology*, 100, pp. 9451–9467.

Cha, Y.L., Yang, J., Ahn, J.W., Moon, Y.H., Yoon, Y.M., Yu, G.D., An, G.H. and Choi, I.H., 2014. The optimized CO 2-added ammonia explosion pretreatment for bioethanol production from rice straw. *Bioprocess and Biosystems Engineering*, 37, pp. 1907–1915.

Chaurasia, B., 2019. Biological pretreatment of lignocellulosic biomass (Water hyacinth) with different fungus for enzymatic hydrolysis and bio-ethanol production resource: Advantages, future work and prospects. *Acta Scientific Agriculture*, 5(3), pp. 89–96.

Chen, H. and Chen, H., 2014. Chemical composition and structure of natural lignocellulose. In: H. Chen (Ed.), *Biotechnology of lignocellulose: Theory and Practice* (pp. 25–71). Springer Dordrecht.

Chen, Y., Stevens, M.A., Zhu, Y., Holmes, J. and Xu, H., 2013. Understanding of alkaline pretreatment parameters for corn stover enzymatic saccharification. *Biotechnology for Biofuels*, 6, pp. 1–10.

Chhetri, B.R., Acharya, D., Gautam, A., Bajracharya, N., Shrestha, A. and Khadka, S., 2022. Microbial pre-treatment of lignocellulosic biomass for biofuel production: A review. *International Journal of Applied Sciences and Biotechnology*, 10(3), pp. 140–148.

Chundawat, S.P.S., Bals, B., Campbell, T., Sousa, L., Gao, D., Jin, M., Eranki, P., Garlock, R., Teymouri, F., Balan, V. and Dale, B.E., 2013. Primer on ammonia fiber expansion pretreatment. In: C. C. E. Wyman (Ed.), *Aqueous Pretreatment of Plant Biomass for Biological and Chemical Conversion to Fuels and Chemicals* (pp. 169–200). John Wiley & Sons, Ltd.

Corrales, R.C.N.R., Mendes, F.M.T., Perrone, C.C., Sant'Anna, C., de Souza, W., Abud, Y., Bon, E.P.P.D.S. and Ferreira-Leitão, V., 2012. Structural evaluation of sugar cane bagasse steam pretreated in the presence of CO_2 and SO_2. *Biotechnology for Biofuels*, 5(1), pp. 1–8.

Cotana, F., Cavalaglio, G., Gelosia, M., Nicolini, A., Coccia, V. and Petrozzi, A., 2014. Production of bioethanol in a second generation prototype from pine wood chips. *Energy Procedia*, 45, pp. 42–51.

Datar, R., Huang, J., Maness, P.C., Mohagheghi, A., Czernik, S. and Chornet, E., 2007. Hydrogen production from the fermentation of corn stover biomass pretreated with a steam-explosion process. *International Journal of Hydrogen Energy*, 32(8), pp. 932–939.

de Jong, E. and Gosselink, R.J., 2014. Lignocellulose-based chemical products. In: V. K. Gupta, C. P. Kubicek, J. Saddler, F. Xu, and M. G. Tuohy (Eds.), *Bioenergy Research: Advances and Applications* (pp. 277–313). Elsevier, Amsterdam.

Den, W., Sharma, V.K., Lee, M., Nadadur, G. and Varma, R.S., 2018. Lignocellulosic biomass transformations via greener oxidative pretreatment processes: Access to energy and value-added chemicals. *Frontiers in Chemistry*, 6, p. 141.

Derman, E., Abdulla, R., Marbawi, H., Sabullah, M.K., Gansau, J.A. and Ravindra, P., 2022. Simultaneous saccharification and fermentation of empty fruit bunches of palm for bioethanol production using a microbial consortium of *S. cerevisiae* and *T. harzianum*. *Fermentation*, 8(7), p. 295.

Deshavath, N.N., Mohan, M., Veeranki, V.D., Goud, V.V., Pinnamaneni, S.R. and Benarjee, T., 2017. Dilute acid pretreatment of sorghum biomass to maximize the hemicellulose hydrolysis with minimized levels of fermentative inhibitors for bioethanol production. *3-Biotech*, 7, pp. 1–12.

Domanski, J., Marchut-Mikołajczyk, O., Polewczyk, A. and Januszewicz, B., 2017. Ozonolysis of straw from *Secale cereale* L. for anaerobic digestion. *Bioresource Technology*, 245, pp. 394–400.

El-Naggar, N.E.A., Deraz, S. and Khalil, A., 2014. Bioethanol production from lignocellulosic feedstocks based on enzymatic hydrolysis: Current status and recent developments. *Biotechnology*, 13(1), pp. 1–21.

Emam and Seyed, F., 2013. Ammonia fiber expansion (Afex) treatment of wheat straw for production of bioethanol. Toronto Metropolitan University, Thesis.

Fu, S.F., Wang, F., Shi, X.S. and Guo, R.B., 2016. Impacts of microaeration on the anaerobic digestion of corn straw and the microbial community structure. *Chemical Engineering Journal*, 287, pp. 523–528.

Galbe, M. and Wallberg, O., 2019. Pretreatment for biorefineries: A review of common methods for efficient utilisation of lignocellulosic materials. *Biotechnology for Biofuels*, 12, pp. 1–26.

Ge, S., Wu, Y., Peng, W., Xia, C., Mei, C., Cai, L., Shi, S.Q., Sonne, C., Lam, S.S. and Tsang, Y.F., 2020. High-pressure CO_2 hydrothermal pretreatment of peanut shells for enzymatic hydrolysis conversion into glucose. *Chemical Engineering Journal*, 385, p. 123949.

Ghasemian, M., Zilouei, H. and Asadinezhad, A., 2016. Enhanced biogas and biohydrogen production from cotton plant wastes using alkaline pretreatment. *Energy & Fuels*, 30(12), pp. 10484–10493.

Golberg, A., Sack, M., Teissie, J., Pataro, G., Pliquett, U., Saulis, G., Stefan, T., Miklavcic, D., Vorobiev, E. and Frey, W., 2016. Energy-efficient biomass processing with pulsed electric fields for bioeconomy and sustainable development. *Biotechnology for Biofuels*, 9, pp. 1–22.

Guerrero, A.B., Ballesteros, I. and Ballesteros, M., 2017. Optimal conditions of acid-catalysed steam explosion pretreatment of banana lignocellulosic biomass for fermentable sugar production. *Journal of Chemical Technology & Biotechnology*, 92(9), pp. 2351–2359.

Guo, F., Fang, Z., Xu, C.C. and Smith Jr, R.L., 2012. Solid acid mediated hydrolysis of biomass for producing biofuels. *Progress in Energy and Combustion Science*, 38(5), pp. 672–690.

Guo, H., Zhao, Y., Chen, X., Shao, Q. and Qin, W., 2019. Pretreatment of Miscanthus with biomass-degrading bacteria for increasing delignification and enzymatic hydrolysability. *Microbial Biotechnology*, 12(4), pp. 787–798.

Gupta, K. and Chundawat, T.S., 2020. Zinc oxide nanoparticles synthesized using Fusarium oxysporum to enhance bioethanol production from rice-straw. *Biomass and Bioenergy*, 143, p. 105840.

Han, Y., Bai, Y., Zhang, J., Liu, D. and Zhao, X., 2020. A comparison of different oxidative pretreatments on polysaccharide hydrolyzability and cell wall structure for interpreting the greatly improved enzymatic digestibility of sugarcane bagasse by delignification. *Bioresources and Bioprocessing*, 7(1), pp. 1–16.

Heller, R., Roth, P., Hülsemann, B., Böttinger, S., Lemmer, A. and Oechsner, H., 2023. Effects of pretreatment with a ball mill on methane yield of horse manure. *Waste and Biomass Valorization*, 14, pp. 1–15.

Henniges, U., Hasani, M., Potthast, A., Westman, G. and Rosenau, T., 2013. Electron beam irradiation of cellulosic materials-opportunities and limitations. *Materials*, 6(5), pp. 1584–1598.

Ibbett, R., Gaddipati, S., Davies, S., Hill, S. and Tucker, G., 2011. The mechanisms of hydrothermal deconstruction of lignocellulose: New insights from thermal-analytical and complementary studies. *Bioresource Technology*, 102(19), pp. 9272–9278.

Imman, S., Laosiripojana, N. and Champreda, V., 2018. Effects of liquid hot water pretreatment on enzymatic hydrolysis and physicochemical changes of corncobs. *Applied Biochemistry and Biotechnology*, 184, pp. 432–443.

Isikgor, F.H. and Becer, C.R., 2015. Lignocellulosic biomass: A sustainable platform for the production of bio-based chemicals and polymers. *Polymer Chemistry*, 6(25), pp. 4497–4559.

Islam, S.M., Li, Q., Al Loman, A. and Ju, L.K., 2017. CO_2-H_2O based pretreatment and enzyme hydrolysis of soybean hulls. *Enzyme and Microbial Technology*, 106, pp. 18–27.

Jiang, D., Fang, Z., Chin, S.X., Tian, X.F. and Su, T.C., 2016. Biohydrogen production from hydrolysates of selected tropical biomass wastes with *Clostridium butyricum*. *Scientific Reports*, 6(1), p. 27205.

John, I., Pola, J. and Appusamy, A., 2019. Optimization of ultrasonic assisted saccharification of sweet lime peel for bioethanol production using Box-Behnken method. *Waste and Biomass Valorization*, 10, pp. 441–453.

Kabenge, I., Omulo, G., Banadda, N., Seay, J., Zziwa, A. and Kiggundu, N., 2018. Characterization of banana peels wastes as potential slow pyrolysis feedstock. doi:10.5539/jsd.v11n2p14

Kim, J.S., Lee, Y.Y. and Kim, T.H., 2016. A review on alkaline pretreatment technology for bioconversion of lignocellulosic biomass. *Bioresource Technology*, 199, pp. 42–48. doi:10.1016/j.biortech.2015.08.08.

Kim, S., 2018. Enhancing bioethanol productivity using alkali-pretreated empty palm fruit bunch fiber hydrolysate. *BioMed Research International*, 2018, p. 5273925.

Kim, S.B., Lee, J.H., Yang, X., Lee, J. and Kim, S.W., 2015. Furfural production from hydrolysate of barley straw after dilute sulfuric acid pretreatment. *Korean Journal of Chemical Engineering*, 32, pp. 2280–2284.

Kim, S.M., Dien, B.S. and Singh, V., 2016. Promise of combined hydrothermal/chemical and mechanical refining for pretreatment of woody and herbaceous biomass. *Biotechnology for Biofuels*, 9, pp. 1–15.

Konan, D., Koffi, E., Ndao, A., Peterson, E.C., Rodrigue, D. and Adjallé, K., 2022. An overview of extrusion as a pretreatment method of lignocellulosic biomass. *Energies*, 15(9), p. 3002.

Kovács, E., Szűcs, C., Farkas, A., Szuhaj, M., Maróti, G., Bagi, Z., Rákhely, G. and Kovács, K.L., 2022. Pretreatment of lignocellulosic biogas substrates by filamentous fungi. *Journal of Biotechnology*, 360, pp. 160–170.

Kulshreshtha, A., 2022. Sustainable energy generation from municipal solid waste. In: C. M. Hussain, S. Singh, and L. Goswami (Eds.), *Waste-to-Energy Approaches Towards Zero Waste* (pp. 315–342). Elsevier, Amsterdam.

Kumar, A.K. and Sharma, S., 2017. Recent updates on different methods of pretreatment of lignocellulosic feedstocks: A review. *Bioresources and Bioprocessing*, 4(1), pp. 1–19.

Kumar, D., Surya, K. and Verma, R., 2020. Bioethanol production from apple pomace using co-cultures with *Saccharomyces cerevisiae* in solid-state fermentation. *The Journal of Microbiology, Biotechnology and Food Sciences*, 9(4), p. 742.

Kumar, M.N., Ravikumar, R., Sankar, M.K. and Thenmozhi, S., 2018. New insight into the effect of fungal mycelia present in the bio-pretreated paddy straw on their enzymatic saccharification and optimization of process parameters. *Bioresource Technology*, 267, pp. 291–302.

Kumar, M.N., Ravikumar, R., Thenmozhi, S. and Sankar, M.K., 2017. Development of natural cellulase inhibitor mediated intensified biological pretreatment technology using *Pleurotus florida* for maximum recovery of cellulose from paddy straw under solid state condition. *Bioresource Technology*, 244, pp. 353–361.

Kumar, P., Kumar, V., Kumar, S., Singh, J. and Kumar, P., 2020. Bioethanol production from sesame (*Sesamum indicum* L.) plant residue by combined physical, microbial and chemical pretreatments. *Bioresource Technology*, 297, p. 122484.

Lamsal, B., Yoo, J., Brijwani, K. and Alavi, S., 2010. Extrusion as a thermo-mechanical pre-treatment for lignocellulosic ethanol. *Biomass and Bioenergy*, 34(12), pp. 1703–1710.

Lau, M.J., Lau, M.W., Gunawan, C. and Dale, B.E., 2010. Ammonia fiber expansion (AFEX) pretreatment, enzymatic hydrolysis and fermentation on empty palm fruit bunch fiber (EPFBF) for cellulosic ethanol production. *Applied Biochemistry and Biotechnology*, 162, pp. 1847–1857.

Li, B.Z., Balan, V., Yuan, Y.J. and Dale, B.E., 2010. Process optimization to convert forage and sweet sorghum bagasse to ethanol based on ammonia fiber expansion (AFEX) pretreatment. *Bioresource Technology*, 101(4), pp. 1285–1292.

Li, H.Q., Jiang, W., Jia, J.X. and Xu, J., 2014. pH pre-corrected liquid hot water pretreatment on corn stover with high hemicellulose recovery and low inhibitors formation. *Bioresource Technology*, 153, pp. 292–299.

Lizasoain, J., Trulea, A., Gittinger, J., Kral, I., Piringer, G., Schedl, A., Nilsen, P.J., Potthast, A., Gronauer, A. and Bauer, A., 2017. Corn stover for biogas production: Effect of steam explosion pretreatment on the gas yields and on the biodegradation kinetics of the primary structural compounds. *Bioresource Technology*, 244, pp. 949–956.

Lu, J., Liu, H., Xia, F., Zhang, Z., Huang, X., Cheng, Y. and Wang, H., 2020. The hydrothermal-alkaline/oxygen two-step pretreatment combined with the addition of surfactants reduced the amount of cellulase for enzymatic hydrolysis of reed. *Bioresource Technology*, 308, p. 123324.

Maurya, D.P., Singla, A. and Negi, S., 2015. An overview of key pretreatment processes for biological conversion of lignocellulosic biomass to bioethanol. *3-Biotech*, 5, pp. 597–609.

Meenakshisundaram, S., Fayeulle, A., Léonard, E., Ceballos, C., Liu, X. and Pauss, A., 2022. Combined biological and chemical/physicochemical pretreatment methods of lignocellulosic biomass for bioethanol and biomethane energy production: A review. *Applied Microbiology*, 2(4), pp. 716–734.

Meng, X., Bhagia, S., Wang, Y., Zhou, Y., Pu, Y., Dunlap, J.R., Shuai, L., Ragauskas, A.J. and Yoo, C.G., 2020. Effects of the advanced organosolv pretreatment strategies on structural properties of woody biomass. *Industrial Crops and Products*, 146, p. 112144.

Mikulski, D. and Kłosowski, G., 2020. Microwave-assisted dilute acid pretreatment in bioethanol production from wheat and rye stillages. *Biomass and Bioenergy*, 136, p. 105528.

Mkhize, T., Mthembu, L.D., Gupta, R., Kaur, A., Kuhad, R.C., Reddy, P. and Deenadayalu, N., 2016. Enzymatic saccharification of acid/alkali pre-treated, millrun and depithed sugarcane bagasse. *BioResources*, 11(3), pp. 6267–6285.

Monlau, F., Sambusiti, C. and Barakat, A., 2019. Comparison of dry versus wet milling to improve bioethanol or methane recovery from solid anaerobic digestate. *Bioengineering*, 6(3), p. 80.

Mood, S.H., Golfeshan, A.H., Tabatabaei, M., Jouzani, G.S., Najafi, G.H., Gholami, M. and Ardjmand, M., 2013. Lignocellulosic biomass to bioethanol, a comprehensive review with a focus on pretreatment. *Renewable and Sustainable Energy Reviews*, 27, pp. 77–93.

Moodley, P. and Trois, C., 2021. Lignocellulosic biorefineries: The path forward. In: R. C. Ray (Ed.), *Sustainable Biofuels* (pp. 21–42). Academic Press, London.

Muthuvelu, K.S., Rajarathinam, R., Kanagaraj, L.P., Ranganathan, R.V., Dhanasekaran, K. and Manickam, N.K., 2019. Evaluation and characterization of novel sources of sustainable lignocellulosic residues for bioethanol production using ultrasound-assisted alkaline pre-treatment. *Waste Management*, 87, pp. 368–374..

Nagai, M., Sakamoto, Y., Nakade, K. and Sato, T., 2007. Isolation and characterization of the gene encoding a manganese peroxidase from *Lentinula edodes*. *Mycoscience*, 48(2), pp. 125–130.

Nam, N.H., Linh, V.N., Dung, L.D. and Ha, V.T.T., 2020. Physico-chemical characterization of forest and agricultural residues for energy conversion processes. *Vietnam Journal of Chemistry*, 58(6), pp. 735–741.

Narayanaswamy, N., Faik, A., Goetz, D.J. and Gu, T., 2011. Supercritical carbon dioxide pretreatment of corn stover and switchgrass for lignocellulosic ethanol production. *Bioresource Technology*, 102(13), pp. 6995–7000.

Nasir, A., Chen, H.Z. and Wang, L., 2020. Novel single-step pretreatment of steam explosion and choline chloride to de-lignify corn stover for enhancing enzymatic edibility. *Process Biochemistry*, 94, pp. 273–281.

Nayan, N., Sonnenberg, A.S., Hendriks, W.H. and Cone, J.W., 2018. Screening of white-rot fungi for bioprocessing of wheat straw into ruminant feed. *Journal of Applied Microbiology*, 125(2), pp. 468–479.

Nazari, L., Sarathy, S., Santoro, D., Ho, D., Ray, M.B. and Xu, C.C., 2018. Recent advances in energy recovery from wastewater sludge. In: L. Rosendahl (Ed.), *Direct Thermochemical Liquefaction for Energy Applications* (pp. 67–100). Elsevier Science.

Ngernyen, Y., 2007. Effect of surface functional groups on water vapor adsorption of eucalyptus wood-based activated carbon. *Suranaree Journal of Science and Technology*, 14(1), pp. 9–23.

Nordin, N., Illias, R.M., Manas, N.H.A., Ramli, A.N.M. and Azelee, N.I.W., 2020, December. Efficient delignification of pineapple waste by low-pressure steam heating pre-treatment. In: *Third International Conference on Separation Technology 2020 (ICoST 2020)* (pp. 10–16). Atlantis Press, London.

Padrino, B., Lara-Serrano, M., Morales-delaRosa, S., Campos-Martín, J.M., Fierro, J.L.G., Martínez, F., Melero, J.A. and Puyol, D., 2018. Resource recovery potential from lignocellulosic feedstock upon lysis with ionic liquids. *Frontiers in Bioengineering and Biotechnology*, 6, p. 119.

Panagiotopoulos, I.A., Chandra, R.P. and Saddler, J.N., 2013. A two-stage pretreatment approach to maximise sugar yield and enhance reactive lignin recovery from poplar wood chips. *Bioresource Technology*, 130, pp. 570–577.

Panepinto, D. and Genon, G., 2016. Analysis of the extrusion as a pretreatment for the anaerobic digestion process. *Industrial Crops and Products*, 83, pp. 206–212.

Panneerselvam, A., Sharma-Shivappa, R.R., Kolar, P., Ranney, T. and Peretti, S., 2013. Potential of ozonolysis as a pretreatment for energy grasses. *Bioresource Technology*, 148, pp. 242–248.

Park, J.Y., Kang, M., Kim, J.S., Lee, J.P., Choi, W.I. and Lee, J.S., 2012. Enhancement of enzymatic digestibility of *Eucalyptus grandis* pretreated by NaOH catalyzed steam explosion. *Bioresource Technology*, 123, pp. 707–712.

Patil, R., Cimon, C., Eskicioglu, C. and Goud, V., 2021. Effect of ozonolysis and thermal pre-treatment on rice straw hydrolysis for the enhancement of biomethane production. *Renewable Energy*, 179, pp. 467–474.

Phuttaro, C., Sawatdeenarunat, C., Surendra, K.C., Boonsawang, P., Chaiprapat, S. and Khanal, S.K., 2019. Anaerobic digestion of hydrothermally-pretreated lignocellulosic biomass: Influence of pretreatment temperatures, inhibitors and soluble organics on methane yield. *Bioresource Technology*, 284, pp. 128–138.

Promon, S.K., Kamal, W., Rahman, S.S., Hossain, M.M. and Choudhury, N., 2018. Bioethanol production using vegetable peels medium and the effective role of cellulolytic bacterial (*Bacillus subtilis*) pre-treatment. *F1000Research*, 7, p. 271

Ravindran, R. and Jaiswal, A.K., 2016. A comprehensive review on pre-treatment strategy for lignocellulosic food industry waste: Challenges and opportunities. *Bioresource Technology*, 199, pp. 92–102.

Rego, F., Dias, A.P.S., Casquilho, M., Rosa, F.C. and Rodrigues, A., 2019. Fast determination of lignocellulosic composition of poplar biomass by thermogravimetry. *Biomass and Bioenergy*, 122, pp. 375–380.

Rizal, N.F.A.A., Ibrahim, M.F., Zakaria, M.R., Abd-Aziz, S., Yee, P.L. and Hassan, M.A., 2018. Pre-treatment of oil palm biomass for fermentable sugars production. *Molecules*, 23(6), p. 1381.

Rocha-Meneses, L., Otor, O.F., Bonturi, N., Orupõld, K. and Kikas, T., 2019. Bioenergy yields from sequential bioethanol and biomethane production: An optimized process flow. *Sustainability*, 12(1), p. 272.

Rosero-Henao, J.C., Bueno, B.E., de Souza, R., Ribeiro, R., Lopes de Oliveira, A., Gomide, C.A., Gomes, T.M. and Tommaso, G., 2019. Potential benefits of near critical and super-critical pre-treatment of lignocellulosic biomass towards anaerobic digestion. *Waste Management & Research*, 37(1), pp. 74–82.

Rostagno, M.A., Prado, J.M., Mudhoo, A., Santos, D.T., Forster-Carneiro, T. and Meireles, M.A.A., 2015. Subcritical and supercritical technology for the production of second generation bioethanol. *Critical Reviews in Biotechnology*, 35(3), pp. 302–312.

Roy, R., Rahman, M.S. and Raynie, D.E., 2020. Recent advances of greener pretreatment technologies of lignocellulose. *Current Research in Green and Sustainable Chemistry*, 3, p. 100035.

Schmidt, A.S. and Thomsen, A.B., 1998. Optimization of wet oxidation pretreatment of wheat straw. *Bioresource Technology*, 64(2), pp. 139–151.

Selvarajoo, A., Wong, Y.L., Khoo, K.S., Chen, W.H. and Show, P.L., 2022. Biochar production via pyrolysis of citrus peel fruit waste as a potential usage as solid biofuel. *Chemosphere*, 294, p. 133671.

Serna, L.D., Alzate, C.O. and Alzate, C.C., 2016. Supercritical fluids as a green technology for the pretreatment of lignocellulosic biomass. *Bioresource Technology*, 199, pp. 113–120.

Sharma, H.K., Xu, C. and Qin, W., 2019. Biological pretreatment of lignocellulosic biomass for biofuels and bioproducts: An overview. *Waste and Biomass Valorization*, 10, pp. 235–251.

Shukla, A., Kumar, D., Girdhar, M., Kumar, A., Goyal, A., Malik, T. and Mohan, A., 2023. Strategies of pretreatment of feedstocks for optimized bioethanol production: Distinct and integrated approaches. *Biotechnology for Biofuels and Bioproducts*, 16(1), p. 44.

Shukla, A., Kumar, D., Girdhar, M., Sharma, A. and Mohan, A., 2023. Steam explosion pre-treatment with different concentrations of hydrogen peroxide along with citric acid: A former step towards bioethanol production. *International Journal of Energy Research*, 2023, pp. 1–13

Sipos, B., Kreuger, E., Svensson, S.E., Reczey, K., Björnsson, L. and Zacchi, G., 2010. Steam pretreatment of dry and ensiled industrial hemp for ethanol production. *Biomass and Bioenergy*, 34(12), pp. 1721–1731.

Smith, M.D., 2019. An abbreviated historical and structural introduction to lignocellulose. In: *Understanding Lignocellulose: Synergistic Computational and Analytic Methods* (pp. 1–15). American Chemical Society, Washington, DC.

Song, Z., Yang, G., Guo, Y. and Zhang, T., 2012. Comparison of two chemical pretreatments of rice straw for biogas production by anaerobic digestion. *BioResources*, 7(3), 3223–3226.

Sui, W. and Chen, H., 2016. Effects of water states on steam explosion of lignocellulosic biomass. *Bioresource Technology*, 199, pp. 155–163.

Suriyachai, N., Weerasai, K., Upajak, S., Khongchamnan, P., Wanmolee, W., Laosiripojana, N., Champreda, V., Suwannahong, K. and Imman, S., 2020. Efficiency of catalytic liquid hot water pretreatment for conversion of corn stover to bioethanol. *ACS Omega*, 5(46), pp. 29872–29881.

Swiatek, K., Gaag, S., Klier, A., Kruse, A., Sauer, J. and Steinbach, D., 2020. Acid hydrolysis of lignocellulosic biomass: Sugars and furfurals formation. *Catalysts*, 10(4), p. 437.

Szwarc, D. and Szwarc, K., 2020. Use of a pulsed electric field to improve the biogas potential of maize silage. *Energies*, 14(1), p. 119.

Taherdanak, M. and Zilouei, H., 2014. Improving biogas production from wheat plant using alkaline pretreatment. *Fuel*, 115, pp. 714–719.

Taylor, M.J., Alabdrabalameer, H.A. and Skoulou, V., 2019. Choosing physical, physicochemical and chemical methods of pre-treating lignocellulosic wastes to repurpose into solid fuels. *Sustainability*, 11(13), p. 3604.

Travaini, R., Martín-Juárez, J., Lorenzo-Hernando, A. and Bolado-Rodríguez, S., 2016. Ozonolysis: An advantageous pretreatment for lignocellulosic biomass revisited. *Bioresource Technology*, 199, pp. 2–12.

Travaini, R., Otero, M.D.M., Coca, M., Da-Silva, R. and Bolado, S., 2013. Sugarcane bagasse ozonolysis pretreatment: Effect on enzymatic digestibility and inhibitory compound formation. *Bioresource Technology*, 133, pp. 332–339.

Ummalyma, S.B., Supriya, R.D., Sindhu, R., Binod, P., Nair, R.B., Pandey, A. and Gnansounou, E., 2019. Biological pretreatment of lignocellulosic biomass: Current trends and future perspectives. In: *Second and Third Generation of Feedstocks* (pp. 197–212). Elsevier, Amsterdam.

Ummalyma, S.B., Sahoo, D., Pudiyamadam, A., Adarsh, V.P., Sukumaran, R.K., Bhaskar, T. and Parida, A., 2021. Sono-assisted alkali and dilute acid pretreatment of *Phragmites karka* (tall reed grass) to enhance enzymatic digestibility for bioethanol conversion. *Frontiers in Energy Research*, 8, p. 594452.

Usmani, Z., Sharma, M., Gupta, P., Karpichev, Y., Gathergood, N., Bhat, R. and Gupta, V.K., 2020. Ionic liquid based pretreatment of lignocellulosic biomass for enhanced bioconversion. *Bioresource Technology*, 304, p. 123003.

Varma, A.K. and Mondal, P., 2017. Pyrolysis of sugarcane bagasse in semi batch reactor: Effects of process parameters on product yields and characterization of products. *Industrial Crops and Products*, 95, pp. 704–717.

Verardi, A., Blasi, A., Marino, T., Molino, A. and Calabrò, V., 2018. Effect of steam-pretreatment combined with hydrogen peroxide on lignocellulosic agricultural wastes for bioethanol production: Analysis of derived sugars and other by-products. *Journal of Energy Chemistry*, 27(2), pp. 535–543.

Viola, E., Zimbardi, F., Morgana, M., Cerone, N., Valerio, V. and Romanelli, A., 2021. Optimized organosolv pretreatment of biomass residues using 2-Methyltetrahydrofuran and n-butanol. *Processes*, 9(11), p. 2051.

Vivekanand, V., Olsen, E.F., Eijsink, V.G. and Horn, S.J., 2013. Effect of different steam explosion conditions on methane potential and enzymatic saccharification of birch. *Bioresource Technology*, 127, pp. 343–349.

Wang, Z.J., Zhu, J.Y., Zalesny Jr, R.S. and Chen, K.F., 2012. Ethanol production from poplar wood through enzymatic saccharification and fermentation by dilute acid and SPORL pretreatments. *Fuel*, 95, pp. 606–614.

Waqas, M., Aburiazaiza, A.S., Miandad, R., Rehan, M., Barakat, M.A. and Nizami, A.S., 2018. Development of biochar as fuel and catalyst in energy recovery technologies. *Journal of Cleaner Production*, 188, pp. 477–488.

Xia, M., Peng, M., Xue, D., Cheng, Y., Li, C., Wang, D., Lu, K., Zheng, Y., Xia, T., Song, J. and Wang, M., 2020. Development of optimal steam explosion pretreatment and highly effective cell factory for bioconversion of grain vinegar residue to butanol. *Biotechnology for Biofuels*, 13, pp. 1–17.

Yin, J., Hao, L., Yu, W., Wang, E., Zhao, M., Xu, Q. and Liu, Y., 2014. Enzymatic hydrolysis enhancement of corn lignocellulose by supercritical CO_2 combined with ultrasound pretreatment. *Chinese Journal of Catalysis*, 35(5), pp. 763–769.

Yu, H., Xiao, W., Han, L. and Huang, G., 2019. Characterization of mechanical pulverization/phosphoric acid pretreatment of corn stover for enzymatic hydrolysis. *Bioresource Technology*, 282, pp. 69–74.

Yu, Q., Liu, J., Zhuang, X., Yuan, Z., Wang, W., Qi, W., Wang, Q., Tan, X. and Kong, X., 2016. Liquid hot water pretreatment of energy grasses and its influence of physico-chemical changes on enzymatic digestibility. *Bioresource Technology*, 199, pp. 265–270.

Zadeh, Z.E., Abdulkhani, A., Aboelazayem, O. and Saha, B., 2020. Recent insights into lignocellulosic biomass pyrolysis: A critical review on pretreatment, characterization and products upgrading. *Processes*, 8(7), p. 799.

Zanellati, A., Spina, F., Poli, A., Rollé, L., Varese, G.C. and Dinuccio, E., 2021. Fungal pretreatment of non-sterile maize silage and solid digestate with a Cephalotrichum stemonitis strain selected from agricultural biogas plants to enhance anaerobic digestion. *Biomass and Bioenergy*, 144, p. 105934.

Zhang, D.S., Yang, Q., Zhu, J.Y. and Pan, X.J., 2013. Sulfite (SPORL) pretreatment of switchgrass for enzymatic saccharification. *Bioresource Technology*, 129, pp. 127–134.

Zhang, J., Zhang, W., Cai, Z., Zhang, J., Guan, D., Ji, D. and Gao, W., 2022. Effect of ammonia fiber expansion combined with NaOH pretreatment on the resource efficiency of herbaceous and woody lignocellulosic biomass. *ACS Omega*, 7(22), pp. 18761–18769.

Zhang, J., Zhou, H., Liu, D. and Zhao, X., 2020. Pretreatment of lignocellulosic biomass for efficient enzymatic saccharification of cellulose. In: *Lignocellulosic Biomass to Liquid Biofuels* (pp. 17–65). Academic Press, London.

Zhang, K., Si, M., Liu, D., Zhuo, S., Liu, M., Liu, H., Yan, X. and Shi, Y., 2018. A bionic system with Fenton reaction and bacteria as a model for bioprocessing lignocellulosic biomass. *Biotechnology for Biofuels*, 11(1), pp. 1–14.

Zhang, K., Xu, R., Abomohra, A.E.F., Xie, S., Yu, Z., Guo, Q., Liu, P., Peng, L. and Li, X., 2019. A sustainable approach for efficient conversion of lignin into biodiesel accompanied by biological pretreatment of corn straw. *Energy Conversion and Management*, 199, p. 111928.

Zhang, Q., Hu, J. and Lee, D.J., 2017. Pretreatment of biomass using ionic liquids: Research updates. *Renewable Energy* 111, 77–84. doi:10.1016/j.renene.2017.03.093.

Zhang, Y., Oates, L.G., Serate, J., Xie, D., Pohlmann, E., Bukhman, Y.V., Karlen, S.D., Young, M.K., Higbee, A., Eilert, D. and Sanford, G.R., 2018. Diverse lignocellulosic feedstocks can achieve high field-scale ethanol yields while providing flexibility for the biorefinery and landscape-level environmental benefits. *GCB Bioenergy*, 10(11), pp. 825–840.

Zhang, Z., Fan, X., Li, Y., Jin, P., Jiao, Y., Ai, F., Zhang, H. and Zhang, Q., 2021. Photo-fermentative biohydrogen production from corncob treated by microwave irradiation. *Bioresource Technology*, 340, p. 125460.

Zhao, W., Yang, R., Zhang, Y. and Wu, L., 2012. Sustainable and practical utilization of feather keratin by an innovative physicochemical pretreatment: High density steam flash-explosion. *Green Chemistry*, 14(12), pp. 3352–3360.

Zheng, B., Yu, S., Chen, Z. and Huo, Y., 2022. A consolidated review of commercial-scale high-value products from lignocellulosic biomass. *Frontiers in Microbiology*, p. 3139.

Zheng, J. and Rehmann, L., 2014. Extrusion pretreatment of lignocellulosic biomass: A review. *International Journal of Molecular Sciences*, 15(10), pp. 18967–18984.

Zheng, Y., Zhao, J., Xu, F. and Li, Y., 2014. Pretreatment of lignocellulosic biomass for enhanced biogas production. *Progress in Energy and Combustion Science*, 42, pp. 35–53.

Zhou, H., Gleisner, R., Zhu, J.Y., Tian, Y. and Qiao, Y., 2018. SPORL pretreatment spent liquors enhance the enzymatic hydrolysis of cellulose and ethanol production from glucose. *Energy & Fuels*, 32(7), pp. 7636–7642.

Zhu, J.Y., Luo, X., Tian, S., Gleisner, R., Negrone, J. and Horn, E., 2011. Efficient ethanol production from beetle-killed lodgepole pine using SPORL technology and Saccharomyces cerevisiae without detoxification. *TAPPI Journal*, 10(5), pp. 9–18.

Zhu, J.Y., Pan, X.J., Wang, G.S. and Gleisner, R., 2009. Sulfite pretreatment (SPORL) for robust enzymatic saccharification of spruce and red pine. *Bioresource Technology*, 100(8), pp. 2411–2418.

Zhu, Y.X., Zhang, X., Yang, W.C. and Li, J.F., 2023. Enhancement of biomass conservation and bioethanol production of sweet sorghum silage by constructing synergistic microbial consortia. Microbiology Spectrum, 11, pp. e03659–22.

Zhuo, S., Yan, X., Liu, D., Si, M., Zhang, K., Liu, M., Peng, B. and Shi, Y., 2018. Use of bacteria for improving the lignocellulose biorefinery process: Importance of pre-erosion. *Biotechnology for Biofuels*, 11, pp. 1–13.

Ziaei-Rad, Z., Fooladi, J., Pazouki, M. and Gummadi, S.N., 2021. Lignocellulosic biomass pre-treatment using low-cost ionic liquid for bioethanol production: An economically viable method for wheat straw fractionation. *Biomass and Bioenergy*, 151, p. 106140.

Zoghlami, A. and Paes, G., 2019. Lignocellulosic biomass: Understanding recalcitrance and predicting hydrolysis. *Frontiers in Chemistry*, 7, p. 874.

12 Moving Bed Biofilm Reactor

A Promising Approach for Wastewater Treatment and Bioenergy Generation

Sourav Debsarma Biswas, Sunaina Nag,
Roshni Raj, and P. Sankar Ganesh
Birla Institute of Technology and Science,
Pilani, Hyderabad Campus

12.1 INTRODUCTION

Water, food, and energy are significant resources for India and the world. As the population density drastically increases daily, the country is experiencing mild to severe water shortages resulting from rapid industrialization, agricultural growth, and urbanization. To overcome this crisis, it is essential to treat the wastewater so that it can be reused. The wastewater treatment plants (WWTP) can be classified based on the type of wastewater treated: sewage, industrial, and agricultural. In WWTPs, solid particles are removed by grit removal, suspended and settled organic materials are digested, and the microbes are disinfected.

Treating wastewater is necessary to remove the chances of waterborne infections and reuse the water. Several physicochemical and biological processes are available for this purpose. Some physicochemical processes include reverse osmosis, flocculation, activated carbon filtration (ACF), pressure sand filtration (PSF), and ozone treatment. However, these physicochemical processes are expensive and require proper maintenance and operation. Biological processes such as activated sludge (ASP), trickling filters, rotating biological contactors (RBC), and membrane bioreactors (MBRs) were developed to overcome these issues. The biological processes can be further classified into suspended and attached growth processes. In the suspended growth process, the microorganisms are free-floating in the system, and examples include the ASP process and aerated lagoons. However, in the attached growth process, the microorganisms attach to a specific membrane or a carrier to facilitate the treatment of the wastewater, and examples include trickling filters, RBC, and MBRs.

However, there are certain limitations in attached growth processes, which affect the system's efficiency. The problem of membrane clogging is one such limitation.

DOI: 10.1201/9781003406501-12

With time, as more and more wastewater is allowed to pass through the membrane, the membrane's pores get clogged with suspended solids or excessive biomass. Due to this, there is a marked loss of pressure head on the system, and the treatment efficiency drops significantly. Frequent backwashing of the membrane is required to avoid the problem of clogging. Most attached growth systems collapse at high organic loads due to increased membrane clogging and sludge accumulation. Hence, they have to be operated at low organic loads.

Due to the disadvantages of the attached growth processes, there is a need to develop wastewater treatment systems (i) that are compact and require lesser space, (ii) with lesser Solid retention time (SRT), (iii) that do not need sludge return, and (iv) perform with high efficiency. Such a system that incorporates attached microorganisms using carrier elements was named as moving bed biofilm reactor (MBBR) and was first developed in the late 1980s by Prof. Hallvard Odegaard at the Norwegian University of Science and Technology in collaboration with a company called Kaldnes Miljoteknoloji.

MBBR consists of an aeration tank with cylindrical polyethylene carrier elements on which the biofilm can attach and grow. The carrier elements containing the biomass are suspended throughout the reactor. They are constantly re-circulated throughout the tank by aeration spargers (for an aerobic system) or rotors (for an anoxic system). Therefore, the biofilm 'attaches' to the carrier elements, while the carrier elements are 'suspended' and re-circulated throughout the tank. Thus, it is a tandem of both suspended and attached growth processes. The carrier elements are the primary working units of an MBBR (Odegaard et al., 1994). The carrier elements are made up of polyethylene (density 0.95 g/cm^3) with a cylindrical shape (about 7 mm in height and 10 mm in diameter). The maximum volumetric filling of carrier elements in an empty reactor should be 70%. Using carrier elements has reduced bio-clogging with no need for frequent backwashing. The reactor's capacity for any given volume can be altered by quickly changing the filling rate.

12.2 MECHANISM OF BIOFILM FORMATION

Biofilm formation is a process that is a summation of attachment, growth, and detachment of microorganisms. The biofilm formation process includes the following steps:

1. Preliminary attachment of the free-floating planktonic phase forms a conditioning layer after several hours of adding carrier elements.
2. After the conditioning layer has formed, the microorganisms attach to it and accumulate because of the extracellular polymeric substance (EPS) bridge between the microbes and conditioning. Eventually, as the bond strengthens, the attachment becomes irreversible.
3. Eventually, the biofilm detaches due to changes in microbial community and fluctuating biomass concentration.
4. As the biofilm matures, the biomass concentration and the microbial community stabilizes.

Multiple factors such as the temperature maintained, microbial activity, pH, nutrient levels, iron content, and EPS concentration play a significant role in biofilm formation

(Gu et al., 2018). The inoculation and modification in the carrier elements influence the initial attachment of the free-floating planktonic phase, but it plays a significant role in bacterial enrichment. The EPS protects the biomass against recalcitrant and inhibitory compounds (di Biasse et al., 2019). EPS majorly consists of two components: proteins and polysaccharides. Higher protein content increases cell adhesion capability, thus enhancing cellular aggregation (Liu et al., 2015).

Liquid flow diffusion and substrate penetration are the two significant factors that affect the MBBR's performance (Stewart et al., 2003). A biofilm thickness of less than 100 µm has been reported to be optimal for effective substrate diffusion through the biofilm (Dezotti et al., 2011). Hence, a uniform biofilm of less thickness achieved by turbulence and shear forces due to homogeneous mixing is preferred. For this purpose, MBBRs have aeration spargers for aerobic systems and mechanical rotors for anoxic systems to provide mixing. A substrate diffusion schematic (Figure 12.1) can better explain the substrate diffusion mechanism inside the carrier elements (di Biase et al., 2019).

The nutrients are first adsorbed onto the biofilm surface and stored in the region where EPS is present in a loosely bound form. As the nutrients accumulate on the biofilm surface, hydrolysis and solute diffusion commence bringing these nutrients to the dense inner layers of the biofilm for growth and proliferation. As the microorganisms in the dense layers of the biofilm start metabolizing the nutrients, the end products are diffused back to the liquid phase via the loose layer. In such a scenario, with increasing nutrient accumulation, the dense layers can keep increasing in thickness until the phenomenon of sloughing can set in, resulting in biofilm loss. Therefore, providing optimum shear stress on the biofilm is necessary to scrape off the unnecessary biomass and keep it at its optimum thickness to facilitate the abovementioned process. The substrate diffusion process is the heart of the unit operation inside the carrier elements suspended in the MBBR. The optimum shear stress is thus provided by aeration for aerobic MBBR and rotor system in anoxic MBBR to promote:

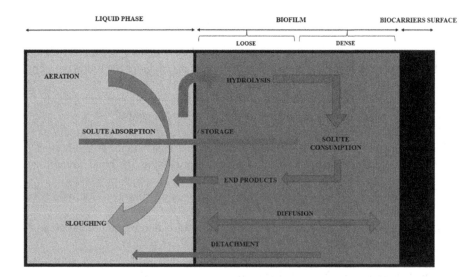

FIGURE 12.1   Substrate diffusion mechanism.

1. Strong and stable biofilm with increased EPS content (Liu and Tay, 2001)
2. High-density and compact biofilm (Li et al., 2016)
3. Biofilm with optimum thickness (Odegaard et al., 2006)

12.3 CARRIER ELEMENTS

Biofilm formation depends on five (biotic and abiotic) factors:

1. Microbial community diversity
2. Carrier surface physical properties (electrostatic interactions and surface energy properties)
3. Surface roughness of carriers (topological properties)
4. Chemical properties of the carrier surface
5. System environments such as pH of the substrate and ambient temperature (Zhao et al., 2019)

Traditionally, biofilm carriers are divided into inorganic and organic types based on the material used. Under the organic material category, there can be reactive and inert types. In general, inorganic materials such as zeolite, ceramics, and activated carbon were used as biofilm carriers (Dong et al., 2011; Lameira et al., 2008; El-Shafai et al., 2013). The use of inorganic materials provides excellent mechanical strength, an extensive range of specific surface area for growth (Müller-Renno et al., 2013), protection of biomass against shock loads, and a better environment for growth due to rough surface and wide available pores (Zhang et al., 2010). However, these materials also pose certain disadvantages like low permeation ability, prolonged rate of growth of biofilm, considerable resistance to the flow due to low permeation, and high clogging rate (Misaelides, 2011; Kvetková et al., 2012; Inam et al., 2011)

Reactive organic-based carrier elements, conversely, have good biocompatibility and hydrophilicity while being very cheap (Yang et al., 2015; Xiao and Chu, 2015). Some examples of such materials are alginate and bamboo plant fibre (Xiao and Chu, 2015; Liu et al., 2017; Behera et al., 2010). Agricultural wastes of corncob, peanut shell, wheat straw, and cotton stalk have been used as biocarriers in MBRs (Yang et al., 2015). They reported enhanced denitrification by 20%–40% and controlled COD levels in the effluent. These materials facilitate microorganism adherence due to their surface structure, are environmentally friendly, and are not toxic to the cells (Fan et al., 2012; Yang et al., 2015). Another point of advantage is that the solid carbon source of the carrier material promotes microorganism growth and helps in the biological removal of nitrogen by acting as an electron donor (Liu et al., 2017; Yang et al., 2015; Reyes-Alvarado et al., 2017). However, these materials have low mechanical strength and mass transfer properties. Microorganisms rapidly degrade them, so the carrier media cannot be reused.

Inert organic-based carrier elements provide a significant advantage over the previous two types and are commercially used in the wastewater treatment industry. These materials include polyester (Lim et al., 2011; Guo et al., 2010), polyethylene (Tang et al., 2017; Ooi et al., 2018; Chen et al., 2012; Shore et al., 2012), and polyolefins (Makarevich et al., 2000). High NH_4-N removal was reported in MBBR using

High-density polyethylene (HDPE) as carrier element material. Commercially HDPE-based carrier elements (AnoxKaldane K5 or AnoxKaldane DMS) are widely used in MBBR systems. The carrier elements belonging to this group show low density, high stability, and strong mechanical strength. They are also resistant to biodegradation, therefore increasing re-usability. They also have high biomass retention ability. Polyurethane-based sponges, when used as biocarriers, showed higher total organic content removal and NH_4-N removal due to high biomass concentration (Chu and Wang, 2011a). However, these show low bio-affinity and hydrophilicity. The water contact angle is said to be a decisive factor for hydrophilicity. Materials with contact angles less than 90° are hydrophilic, while those having greater than 90° are hydrophobic (Cerca et al., 2005; Liu and Zhao, 2005; Chu et al., 2014). Polyurethane, polypropylene, polyethylene, and high-density polyethylene have contact angles of 90°, 88.7°, 92°, and 94.33°, respectively (Chen et al., 2012; Chu et al., 2014; Liu et al., 2018b; Mao et al., 2017; Zhu, 2017). Hence, they have low hydrophilicity. Therefore, new biofilm carriers are fabricated with superior surface properties by altering the chemical properties of the base material (di Biasse, 2019).

12.3.1 CARRIER ELEMENTS DESIGN

The first carrier element used in the MBBR wastewater treatment process was the K-series AnoxKaldnes and high-density polyethylene (density 0.95–0.98 g/cm³) (di Biasse, 2019). The K-series biocarriers differ by dimensions (height and diameter), potential surface area for biofilm development, and the grooves within which the biofilm grows. Depending upon the process, whether aerobic or anoxic, the use of the carrier elements will differ. In the aerobic process, the air is sparged to keep the carrier elements in suspension. However, the carrier elements are mixed to keep it suspended in an anoxic process.

To have a better performance of a carrier element, the following criteria are required:

1. The carrier element's design should maintain the biofilm produced and supplement it with the required material for growth and development.
2. A proper mixing speed is required to maintain the biofilm's specific velocity and thickness.
3. The carriers should not have dead spaces with limited oxygen and mass transfer rates.
4. The carrier should allow the maturation of the biofilm and stabilize it (Kruszelnicka et al., 2018).
5. The grids are used instead of the openings to maintain a certain thickness.

To understand the use of the proper design of any carrier elements in the system, the relationship between the shape of the carrier and the velocity of the internal groove where the biofilm develops should be known. Simulation studies using varying heights and diameters of carrier elements helped observe the following conclusions (Kruszelnicka et al., 2018):

1. The increase in the length of the carrier element reduces the high velocity in the internal region, so the biofilm does not detach easily and develops a thicker surface.
2. The velocity field around the carriers is crucial as it influences mass transfer, oxygen transfer, and biofilm stability. It is a difficult situation as the carrier elements will have different confirmations and velocity distribution accordingly, leading to the formation of dead spaces. When the length of the carrier element is increased, it provides asymmetry in velocity distribution. This reduces the number of possible confirmations of the carrier elements, thereby reducing the ability to form dead spaces (Tables 12.1 and 12.2).

Commercial Carrier Elements:

12.3.2 Criteria for Proper Use of Carrier Elements in MBBR

12.3.2.1 Proper Mixing and Mass Transfer

Aeration or mechanical mixing is an essential requirement in MBBR. It was reported that the density of the carrier elements should be close to that of water for proper mixing (McQuarrie and Boltz, 2011). The carrier fill ratio has to be maintained in the range of 30%–45% to give enough space for the carrier elements to move throughout the reactor and thereby allow proper mixing (Barwal and Chaudhary, 2014). With the appropriate mixing intensity, good turbulence will be achieved inside the reactor, which can improve the mass transfer of nutrients to the core of the biofilm by removing the stagnant boundary layer surrounding the outer surface of the biofilm. However, this requires the carrier elements to be large enough to allow good water passage. Turbulence also helps maintain a thin biofilm layer by shearing off the extra biofilm layers (Nogueira et al., 2015).

TABLE 12.1
Physical Dimensions of AnoxKaldnes Carriers

Veolia AnoxKaldnes	Surface Area (m^2/m^3)	Nominal Diameter	Nominal Height
K1	500	7.1	7.2
K2	350	15	15
K3	500	25	10
K5	800	25	3.5
C2	220	36	30
F3	200	46	37
BiofilmChip M	1,200	48	2.2
BiofilmChip P	900	45	3
Z-200	-	30	-
Z-400	-	30	-

Source: di Biase et al. (2019).

TABLE 12.2

Commercially Available Carrier Elements

Manufacturer Biocarrier's Name	Surface Area (m²/m³)
Headworks AC450	402
Headworks AC515	485
Headworks AC920	680
Biowater Technology BWT 15	828
Biowater Technology BWT X	650
Biowater Technology BWT S	650
Warden Biomedia Biofil	135
Warden Biomedia Bioball	220
Warden Biomedia Biomarble	310
Warden Biomedia Biopipe	600
Warden Biomedia Biopipe⁺	500
Warden Biomedia Bioflo⁺	800
Warden Biomedia Biotube	1,000
Createch Aqua Curler Advance X-1	800
Createch Aqua Cylinder Plus	350–550
Createch Aqua Cylinder X-0	600–900
BCP 750-0.93	750
BCP 750-1.00	750
BCP 750-1.2	750
BCP 175	175
BCP 115	115
BCP 100	100
BCN 020	610
BCN 040	340
BCN 030	320
BCN 060	229
EVOQUA CM-10D™	750
P.E.W.E ASO™	650
Mutag BioChip™	5,500
BNC 011	790

Source: de Biasse et al. (2019).

12.3.2.2 Carrier Clogging

A clogged carrier is one where all the pores and grooves are entirely packed with biofilm. In such a scenario, partial penetration of substrate occurs inside the biofilm, which results in reduced biofilm surface area of growth, anaerobic layers leading to biofilm sloughing or odour, and increased carrier weight (Boltz and Daigger, 2010), which would hamper the proper mixing requirement as discussed in the previous section.

The phenomenon of biofilm scaling also has to be avoided. With time, the deposition of inorganic precipitates begins inside the biofilm, which can cause excessive carrier weight and clogging. Therefore, carriers with large, open grooves and pores should facilitate better detachment (Forrest et al., 2016).

12.3.2.3 Rapid Biofilm Attachment and Growth

Carriers with deep pores and cavities are often better for rapid initial biofilm attachment and growth (Chen et al., 2015; Li et al., 2016b). The addition of pre-seeded carriers is a practice that can accelerate the startup time. Enhancing the carrier surface roughness is another proven method for better attachment rates.

12.3.2.4 Proper Oxygen Transfer Efficiency

Oxygen transfer is an essential criterion for proper MBBR operation. The addition of carrier elements has been proven to improve oxygen transfer efficiency (OTE). Generally, for all aerobic reactors, the OTE is dependent on various factors like the type of bubble diffusion, size, shape, and surface area of bubble distribution nozzles, depth of the diffusers inside the reactor, gas–liquid interfacial area, size of bubbles, and bubble retention time (Jing et al., 2009). Regarding the MBBR, the carrier fills percentage is essential in determining the OTE apart from those factors. It was reported that OTE increased up to 40% of the carrier fill ratio, after which the OTE fell. That increase in OTE was attributed to (i) increased bubble retention time by the prolonged path of the rising bubble and by the capture of bubbles inside the carrier voids, (ii) breakage of bubbles, thereby increasing the gas–liquid interface, and (iii) continuously increased renewal of gas–liquid interface resulting from increased turbulence, due to bubble breakage (Piculell, 2016).

For the OTE to be positively affected by the carrier fill percentage, the aeration has to be maintained by coarse bubble diffusers. Using fine bubble diffusers will lead to bubble coalescence instead of bubble breakage.

Carrier element designs also have a significant effect on the OTE. Flat carriers were reported to follow a parallel orientation for the airflow, thus resulting in less bubble retention time due to the fast passing of the bubbles, finally leading to less OTE. On the other hand, three-dimensional carrier elements were reported to continuously change orientation on hitting the bubbles, resulting in increased reactor turbulence and bubble collisions (Piculell, 2016). Hence, the design of Z-shaped carriers was suggested, which were reported to spin on being hit by bubbles, thereby leading to improved OTE.

12.3.3 Novel Carrier Elements

As described in Section 12.3, certain surface modifications are done on existing carrier element materials to enhance their surface properties. Some processes (di Biase et al., 2019) were reported to be carried out for carrier material surface characteristics modification: Wet chemical oxidation on HDPE AnoxKaldane K5 carriers using potassium permanganate, Fenton reagent, and ozone oxidation were reported to show accelerated de-ammonification (Klaus et al., 2016).

Polymer grafting is done by attaching monomers on polymer chains by covalent binding, thereby attaining surface modifications. Introduction of an amino functional group on polyethylene and polypropylene plastic carriers enhanced thickness of biofilm, homogenous nature, and mechanical strength against shear forces of the nitrifying community (Bhattacharya et al., 2009).

Polymer blending involves blending original material with toluene diisocyanate, polyether glycol, foam stabilizer, etc. This results in an increased amount of net positive charges and hydrophilicity (Chu et al., 2014). This is specifically advantageous because this can remove the electrostatic repulsion between the biocarrier surface and microorganism cell wall due to negative charges on both. Modifying HDPE with polyquaternium 10 (PQAS-100) and cationic polysaccharides helped reduce biofilm growth time in startup (Mao et al., 2017).

12.4   PROCESS PARAMETERS OF MBBR

12.4.1   BOD/COD REMOVAL

Biochemical oxygen demand (BOD) and Chemical oxygen demand (COD) removal analyses are done to determine the removal of the organics. The degradation of organic matter can be carried out with or without a biomass separation step. Odegaard et al. (2006) reported degradation of organic matter without biomass separation step is given by soluble COD (SCOD). As a result, the maximum removal rate of COD is $30\,g$ SCOD/m^2d. This does not provide the solution because the soluble organic matter is also produced during hydrolysis. Another alternative method is used, i.e., "obtainable removal rate," where the removal rate of COD is done by using a 100% biomass separation step. It also states that high loading rates can have high removal rates if a proper biomass separation process is carried out.

Extrapolating from the information given above for optimization of BOD/COD removal, a report suggested the use of 7% (m/v) aluminium sulphate and 3% (m/v) ferric chloride and 1 g/m^3 of an anionic polymer as the coagulant. The coagulant addition was done after the treatment of influent having COD of 360 mg/L in three MBBRs placed in series, each having Hydraulic retention time (HRT) of 40 min (Dezotti et al., 2018). As a result of the coagulant addition, they reported a 78% COD reduction as opposed to a 45% COD reduction when reactors were operated without adding coagulants.

12.4.2   NITRIFICATION OR AMMONIA REMOVAL

The process of conversion of ammonia to nitrate or nitrite is nitrification. *Nitrosomonas* sp. carries out the oxidation of ammonia to nitrite, while the oxidation to nitrate is done by *Nitrobacter* sp. In a review article (di Biasse et al., 2019), it was reported that the initial process of conversion of ammonia to nitrite consumes about 75% of the total oxygen demand, while the rest 25% is used during the conversion of nitrite to nitrate.

Initially, the nitrification rate depends on organic load, influent ammonium concentration, and oxygen concentration.

12.4.3 Denitrification

Denitrification is the process that occurs during the post-nitrification step (removal of ammonia). The nitrates or nitrites are reduced to nitrogen gas. This process is anoxic and requires a readily biodegradable carbon source. Forty per cent of the carbon source is used to convert nitrite to nitrate, while the rest 60% is used in the final conversion to N_2 gas. Intermediate products are NO and N_2O gases (Pan et al., 2015).

Nitrate concentration, biodegradable organic matter concentration, or oxygen concentration are the major limiting factors to denitrification rates. When more than $3\,mg\ NO_3$-N/L is present, the denitrification process will be controlled by the kind and amount of carbon sources available (Odegaard et al., 2006). Based on this, various configurations have been developed.

12.4.4 Phosphorous Removal

The addition of chemical coagulants generally does the removal of phosphorus. In the biological mode of removal, this is done by implementing an Integrated Fixed film Activated System (IFAS). Four configurations have been described below for efficient biological phosphorous removal.

12.5 MBBR DESIGN PROCEDURE

For the fabrication of MBBR, several design components have to be calculated. These design components consider the system's aerobic nature (for aerobic MBBR) and the ultimate goal of COD/BOD reduction. Those components are shown in Figure 12.2.

All design calculation formulae and principles have been taken from Metcalf and Eddy (2005).

The design procedure is enlisted below.

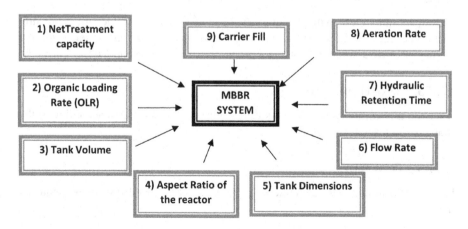

FIGURE 12.2 Moving bed biofilm reactor components.

12.5.1   Determination of the Treatment Capacity

The treatment capacity determination considers various factors like the area's population, the average wastewater generation per day, the net flow rate of sewage at peak hours, etc.

12.5.2   The Organic Loading Rate (OLR)

This is the rate at which the organic pollutants are loaded onto the system. This parameter is the crucial point that will be used to calculate the tank volume. According to Chapter 5 of CPHEEO (Central Public Health and Environmental Engineering Organisation) Manual on Sewerage and Sewage treatment systems (2013), for systems using synthetic media for biofilm attachment, the OLR should be between 0.8 and 6 kg BOD/m³/d¹. So after the determination of the net treatment capacity of the plant, it is essential to select an OLR value that fits the given range.

12.5.3   Tank Volume

The input parameters required for the tank volume calculation are the average wastewater flow at peak times, influent BOD/COD, and the OLR.

The total BOD/COD to be applied to the system is first calculated using the formula:

$$\text{Total BOD/COD applied} = \text{Avg. influent flow at peak} \times \text{Influent BOD/COD}$$

$$(12.1)$$

Using the OLR value set in the previous step, the tank volume will be calculated by:

$$\text{Tank volume} = \text{Total BOD/COD applied/organic loading rate} \qquad (12.2)$$

12.5.4   Aspect Ratio

The aspect ratio refers to the height-to-diameter ratio or the tank's length to breadth. The aspect ratio of a reactor should be optimum so that the wastewater inside the system can have a proper height and a proper base for the mixing to occur correctly.

12.5.5   Tank Dimensions

Once the aspect ratio is set, the tank dimensions can be calculated according to the shape of the reactor, as required.

12.5.6   Flow Rate

For the influent flow rate calculation, the input parameter needed is the net required treatment capacity.

$$\text{Flow rate: total capacity/24 (h)} \qquad (12.3)$$

12.5.7 Hydraulic Retention Time

Calculation of the HRT requires only the flow rate and the tank volume.

$$\text{HRT: Tank volume / Flow rate} \tag{12.4}$$

12.5.8 Aeration Rate

The calculation of the oxygen supply needed involves the determination of four critical factors. The formulae for determining the four factors are given below.

12.5.8.1 Mass Transfer Coefficient for Oxygen, K_La

Apparent mass transfer coefficient equation: $(C_z - C_x)/(C_x - C_o) = e(-K_La)t$

$$\tag{12.5}$$

K_La = the overall coefficient for liquid film; C_x = the liquid bulk phase concentration at time t, mg/L; C_z = the equilibrium concentration with gas as given by Henry's law; and C_o = concentration at the initial point.

In the case of biological systems for wastewater treatment, the K_La value calculation requires the consideration of the uptake of oxygen by microorganisms. This is given by:

$$dC/dT = K_La(C_z - C) - r_m \tag{12.6}$$

C = oxygen concentration in solution and r_m = oxygen uptake rate by microorganisms. The respirometer can be used to determine the value of r_m:

If the oxygen level is consistent, $K_La = r_m/(C_s - C)$ (12.7)

12.5.8.2 Oxygen Transfer Correction Factor (α Factor)

This takes into account the effect of mixing intensity and tank geometry. Its value is given by:

$$\alpha = K_La(\text{wastewater})/K_La\ (\text{tap water with low TDS}) \tag{12.8}$$

The value of α varies with basin geometry, type of aerator, wastewater characteristics, and degree of mixing. The value of α factor ranges from 0.3 to 1.2.

12.5.8.3 Salinity-Surface Tension Correction Factor (β-Factor)

This is used to correct the oxygen requirement of the system by taking into account the oxygen solubility due to the presence of salts, particulates, etc.

$$\beta = C_s(\text{waste water})/C_s(\text{clean water}) \tag{12.9}$$

The value of β factor ranges from 0.7 to 0.98. But for wastewater, this value is generally 0.95.

12.5.8.4 The Average Saturation Concentration for Dissolved Oxygen in an Aeration Tank with Clean Water at Altitude H and Temperature T ($C_{a1,b1,H}$)

$$C_{a1,b1,H} = C_{a,b,H} X \left(P_t / P_{atm.h} + O_t / 21 \right) / 2 \qquad (12.10)$$

For a surface aerator $C_{a1,b1,H} = C_{a,b,H}$

The factor $(P_t/P_{atm.h} + O_t/21)$ is the pressure at the mid-point position of the tank. This accounts for oxygen loss to biological sources. If this factor is not taken into consideration, then

$$C_{a1,b1,H} = C_{a,b,H} \left(P_{atm.H} + P_{w,d/2} \right) / P_{atm.h} \qquad (12.11)$$

$C_{a,b,H}$ = the saturation concentration of oxygen in clean water at temperature T and altitude H, mg/L; P_t = pressure at a depth of air release, kPa; $P_{atm.h}$ = atmospheric pressure at altitude H; $P_{w,d/2}$ = pressure at mid-dept, above the point of air release, due to the water column; O_t = percentage of oxygen concentration leaving tank (18%–20%); C_{OOC} = Operational Oxygen concentration (mg/L); $C_{DO.2\ 0}$ = dissolved oxygen concentration in clean water at 20°C and 1 atm. pressure, mg/L; T = temperature, °C; α = oxygen transfer correction factor; and F = Fouling factor.

These four factors, combined, give the Actual Oxygen Required (AOR), which is the required variable.

The oxygen amount required is obtained by applying all the factors (mentioned below) to the standard oxygen amount requirement:

1. Effects of salinity-surface tension (β factor)
2. Effect of mixing intensity and tank geometry (α factor)
3. Temperature, T
4. Elevation, H
5. Diffusion aerator depth
6. Desired operational oxygen level.

Combining all the four factors given above, the AOTR is given by:

$$AOR = SOR \left(\beta . C_{a1,b1,H} - C_{OOC} \right) / C_{DO,20} \times 1.024^{(T-20)} \alpha\ F \qquad (12.12)$$

SOR = Standard rate of Oxygen Transfer in tap water at 20°C and zero DO, kg O_2/h

β = Salinity-surface tension correction factor $(0.95 - 0.98$ for wastewater$)$

AOR = Actual rate of oxygen transfer in field conditions, kg O_2/h

Once the AOR is determined, the aeration system has to be chosen concerning the requirements. Generally, there are two types of aeration systems: One with air diffused from the bottom and one with aeration done mechanically.

The first type uses diffusers that are submerged in water and diffuse the air from the bottom of the reactor. They can be of two types: fine bubble and coarse bubble diffusers.

The second type involves mechanically agitating the wastewater to promote air dissolution from the atmosphere. These can be of three types: Vertical Axis Surface Mechanical Aerators, Vertical Axis Submerged Mechanical Aerators, and Horizontal Axis Mechanical Aerators.

The air input required can be calculated from the oxygen required calculation given above, using the relation that oxygen occupies 21% of air.

12.5.9 CARRIER ELEMENT FILL RATIO

As discussed in Section 12.3, the optimum carrier fill ratio is essential for maintaining proper OTE. Typically, the fill ratio should range from 30% to 45% for OTE to be maximum. Post 45%, and the OTE seems to drop or flatten out. The optimum carrier fill is also used for proper mixing and mass transfer. If the fill fraction is high, the reactor will be clogged as there will not be any free space for the carrier elements to move, leading to improper mixing.

A table enlisting various MBBR design configurations in 100 research articles on MBBR has been given in the supplementary section.

12.6 BIOFUEL GENERATION FROM THE MBBR SYSTEM

Most of the biomass in an MBBR system comprises microorganisms, such as bacteria and fungi, which develop on the biofilm carriers or media. These microorganisms utilize biological processes to decompose organic matter in effluent. While the primary objective is to remove contaminants from effluent, the process's by-product, sludge, can be harvested and utilized for bioenergy production.

Using anaerobic digestion to convert the sludge into biogas is a viable option. Microorganisms decompose organic matter without oxygen, producing methane as a by-product during anaerobic digestion. Biogas is a mixture of methane and carbon dioxide that can be utilized as a renewable energy source. Biomethane, comparable to natural gas and used as a fuel for transportation or heating, can be produced through the processing and upgrading of methane.

Another option is to utilize the sludge as a feedstock for biofuel production through processes such as fermentation or pyrolysis. Fermentation converts organic matter into biofuels such as ethanol, whereas pyrolysis is a thermal decomposition process that can produce bio-oil, which can then be refined into transportation fuels.

Notably, the suitability and effectiveness of these conversion processes are contingent on the composition and characteristics of the biomass produced by the MBBR system. Depending on the effluent being treated, the sludge composition can vary, and additional preprocessing processes may be necessary to maximize the biofuel production potential (Figure 12.3).

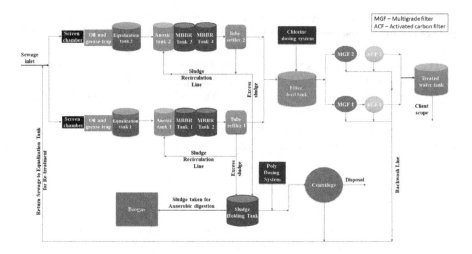

FIGURE 12.3   Biogas production from sludge in moving bed biofilm reactor system.

12.7   CONCLUSION

The MBBR is an efficient wastewater treatment strategy that has been used successfully worldwide since the 1990s. Using carrier elements to replace conventional membranes proves to be an efficient innovation over its predecessors. The most important
part of MBBR is its carrier element onto which the biofilm gets attached. A critical
process parameter is the aeration requirement, which dictates the biofilm's stability and
community structure. One of the significant drawbacks of this technology is its slow
startup time. Hence, most of the research has targeted the carrier elements (the primary
working unit of MBBR) for developing efficient strategies to reduce startup time. Other
areas of focus in research on MBBR lie in developing strategies to combine MBBR
with an existing treatment plant to increase the overall efficiency of the systems. Apart
from domestic sewage, MBBR has also been used to treat and clarify pharmaceuticals,
industrial effluents, chemical dyes, etc. Even though MBBR alone is a highly efficient
secondary treatment method, future directives always exist to optimize the process
parameters to gain higher efficiency and better effluent qualities.

REFERENCES

Barwal, A., & Chaudhary, R. (2014). To study the performance of biocarriers in moving bed
        biofilm reactor (MBBR) technology and kinetics of biofilm for retrofitting the exist-
        ing aerobic treatment systems: A review. *Reviews in Environmental Science and Bio/
        Technology*, 13(3), 285–299.
Barwal, A., & Chaudhary, R. (2016). Application of response surface methodology to opti-
        mize the operational parameters for enhanced removal efficiency of organic matter and
        nitrogen: Moving bed biofilm reactor. *Environmental Science and Pollution Research*,
        23(10), 9944–9955.

Behera, S., Kar, S., Mohanty, R. C., & Ray, R. C. (2010). Comparative study of bio-ethanol production from mahula (*Madhuca latifolia* L.) flowers by Saccharomyces cerevisiae cells immobilized in agar agar and Ca-alginate matrices. *Applied Energy*, 87(1), 96–100.

Boltz, J. P., & Daigger, G. T. (2010). Uncertainty in bulk-liquid hydrodynamics and biofilm dynamics creates uncertainties in biofilm reactor design. *Water Science and Technology*, 61(2), 307–316.

Cerca, N., Pier, G. B., Vilanova, M., Oliveira, R., & Azeredo, J. (2005). Quantitative analysis of adhesion and biofilm formation on hydrophilic and hydrophobic surfaces of clinical isolates of Staphylococcus epidermidis. *Research in Microbiology*, 156(4), 506–514.

Chen, S., Cheng, X., Zhang, X., & Sun, D. (2012). Influence of surface modification of polyethylene biocarriers on biofilm properties and wastewater treatment efficiency in moving-bed biofilm reactors. *Water Science and Technology*, 65(6), 1021–1026.

Chen, X., Kong, L., Wang, X., Tian, S., & Xiong, Y. (2015). Accelerated startup of moving bed biofilm reactor by using a novel suspended carrier with porous surface. *Bioprocess and Biosystems Engineering*, 38(2), 273–285.

Chu, L., & Wang, J. (2011a). Comparison of polyurethane foam and biodegradable polymer as carriers in moving bed biofilm reactor for treating wastewater with a low C/N ratio. *Chemosphere*, 83(1), 63–68.

Chu, L., & Wang, J. (2011b). Nitrogen removal using biodegradable polymers as carbon source and biofilm carriers in a moving bed biofilm reactor. *Chemical Engineering Journal*, 170(1), 220–225.

Chu, L., Wang, J., Quan, F., Xing, X. H., Tang, L., & Zhang, C. (2014). Modification of polyurethane foam carriers and application in a moving bed biofilm reactor. *Process Biochemistry*, 49(11), 1979–1982.

Dezotti, M., Lippel, G., & Bassin, J. P. (2018). Advanced biological processes for wastewater treatment. *İsviçre*, 10, 978. doi:10.1007/978-3-319-58835-3

Di Biase, A., Kowalski, M. S., Devlin, T. R., & Oleszkiewicz, J. A. (2019). Moving bed biofilm reactor technology in municipal wastewater treatment: A review. *Journal of Environmental Management*, 247, 849–866. doi:10.1016/j.jenvman.2019.06.053.

Di Trapani, D., Christensso, M., & Ødegaard, H. (2011). Hybrid activated sludge/biofilm process for the treatment of municipal wastewater in a cold climate region: A case study. *Water Science and Technology*, 63(6), 1121–1129.

Dong, Z., Lu, M., Huang, W., & Xu, X. (2011). Treatment of oilfield wastewater in moving bed biofilm reactors using a novel suspended ceramic biocarrier. *Journal of Hazardous Materials*, 196, 123–130.

El-Shafai, S. A., & Zahid, W. M. (2013). Performance of aerated submerged biofilm reactor packed with local scoria for carbon and nitrogen removal from municipal wastewater. *Bioresource Technology*, 143, 476–482.

Fan, Z., Hu, J., & Wang, J. (2012). Biological nitrate removal using wheat straw and PLA as substrate. *Environmental Technology*, 33(21), 2369–2374.

Forrest, D., Delatolla, R., & Kennedy, K. (2016). Carrier effects on tertiary nitrifying moving bed biofilm reactor: An examination of performance, biofilm and biologically produced solids. *Environmental Technology*, 37(6), 662–671.

Gu, Y. Q., Li, T. T., & Li, H. Q. (2018). Biofilm formation monitored by confocal laser scanning microscopy during startup of MBBR operated under different intermittent aeration modes. *Process Biochemistry*, 74, 132–140.

Guo, W., Ngo, H. H., Dharmawan, F., & Palmer, C. G. (2010). Roles of polyurethane foam in aerobic moving and fixed bed bioreactors. *Bioresource Technology*, 101(5), 1435–1439.

Helness, H. (2007). Biological phosphorus removal in a moving bed biofilm reactor. Doctoral theses, Norwegian University of Science and Technology, ISBN 978-82-471-3893-9 (electronic version).

Inam, F., Peijs, T., & Reece, M. J. (2011). The production of advanced fine-grained alumina by carbon nanotube addition. *Journal of the European Ceramic Society*, 31(15), 2853–2859.

Jing, J. Y., Feng, J., & Li, W. Y. (2009). Carrier effects on oxygen mass transfer behavior in a moving-bed biofilm reactor. *Asia-Pacific Journal of Chemical Engineering*, 4(5), 618–623.

Klaus, S., McLee, P., Schuler, A. J., & Bott, C. (2016). Methods for increasing the rate of anammox attachment in a sidestream deammonification MBBR. *Water Science and Technology*, 74(1), 110–117.

Kruszelnicka, I., Kramarczyk, D. G., Poszwa, P., & Stręk, T. (2018). Influence of MBBR carriers' geometry on its flow characteristics. *Chemical Engineering and Processing-Process Intensification*, 130, 134–139.

Kvetková, L., Duszová, A., Hvizdoš, P., Dusza, J., Kun, P., & Balázsi, C. (2012). Fracture toughness and toughening mechanisms in graphene platelet reinforced Si3N4 composites. *Scripta Materialia*, 66(10), 793–796.

Lameiras, S., Quintelas, C., & Tavares, T. (2008). Biosorption of Cr (VI) using a bacterial biofilm supported on granular activated carbon and on zeolite. *Bioresource Technology*, 99(4), 801–806.

Li, C., Wagner, M., Lackner, S., & Horn, H. (2016). Assessing the influence of biofilm surface roughness on mass transfer by combining optical coherence tomography and two-dimensional modeling. *Biotechnology and Bioengineering*, 113(5), 989–1000.

Lim, J. W., Seng, C. E., Lim, P. E., Ng, S. L., & Sujari, A. N. A. (2011). Nitrogen removal in moving bed sequencing batch reactor using polyurethane foam cubes of various sizes as carrier materials. *Bioresource Technology*, 102(21), 9876–9883.

Liu, X., Wang, L., & Pang, L. (2018). Application of a novel strain Corynebacterium pollutisoli SPH6 to improve nitrogen removal in an anaerobic/aerobic-moving bed biofilm reactor (A/O-MBBR). *Bioresource Technology*, 269, 113–120.

Liu, Y., Kang, X., Li, X., & Yuan, Y. (2015). Performance of aerobic granular sludge in a sequencing batch bioreactor for slaughterhouse wastewater treatment. *Bioresource Technology*, 190, 487–491.

Liu, Y., & Tay, J. H. (2001). Metabolic response of biofilm to shear stress in fixed-film culture. *Journal of Applied Microbiology*, 90(3), 337–342.

Liu, Y., & Zhao, Q. (2005). Influence of surface energy of modified surfaces on bacterial adhesion. *Biophysical Chemistry*, 117(1), 39–45.

Makarevich, A. V., Dunaitsev, I. A., & Pinchuk, L. S. (2000). Aerobic treatment of industrial wastewaters by biofilters with fibrous polymeric biomass carrier. *Bioprocess Engineering*, 22(2), 121–126.

Mao, Y., Quan, X., Zhao, H., Zhang, Y., Chen, S., Liu, T., & Quan, W. (2017). Accelerated startup of moving bed biofilm process with novel electrophilic suspended biofilm carriers. *Chemical Engineering Journal*, 315, 364–372.

McQuarrie, J. P., & Boltz, J. P. (2011). Moving bed biofilm reactor technology: Process applications, design, and performance. *Water Environment Research*, 83(6), 560–575.

Metcalf, L., Eddy, H. P., & Tchobanoglous, G. (1979). *Wastewater Engineering: Treatment, Disposal, and Reuse* (Vol. 4). New York: McGraw-Hill.

Misaelides, P. (2011). Application of natural zeolites in environmental remediation: A short review. *Microporous and Mesoporous Materials*, 144(1–3), 15–18.

Müller-Renno, C., Buhl, S., Davoudi, N., Aurich, J. C., Ripperger, S., Ulber, R., ... & Ziegler, C. (2013). Novel materials for biofilm reactors and their characterization. In: *Productive Biofilms* (pp. 207–233). Cham: Springer.

Nogueira, B. L., Pérez, J., van Loosdrecht, M. C., Secchi, A. R., Dezotti, M., & Biscaia Jr, E. C. (2015). Determination of the external mass transfer coefficient and influence of mixing intensity in moving bed biofilm reactors for wastewater treatment. *Water Research*, 80, 90–98.

Ødegaard, H. (2006). Innovations in wastewater treatment:-the moving bed biofilm process. *Water Science and Technology*, 53(9), 17–33.

Ødegaard, H., Rusten, B., & Westrum, T. (1994). A new moving bed biofilm reactor: Applications and results. *Water Science and Technology*, 29(10–11), 157–165. doi:10.2166/wst.1994.0757.

Ooi, G. T., Tang, K., Chhetri, R. K., Kaarsholm, K. M., Sundmark, K., Kragelund, C., ... & Christensson, M. (2018). Biological removal of pharmaceuticals from hospital wastewater in a pilot-scale staged moving bed biofilm reactor (MBBR) utilising nitrifying and denitrifying processes. *Bioresource Technology*, 267, 677–687.

Pan, Y., Ni, B. J., Lu, H., Chandran, K., Richardson, D., & Yuan, Z. (2015). Evaluating two concepts for the modelling of intermediates accumulation during biological denitrification in wastewater treatment. *Water Research*, 71, 21–31.

Piculell, M. (2016). *New Dimensions of Moving Bed Biofilm Carriers: Influence of Biofilm Thickness and Control Possibilities*. Lund: Lund University.

Reyes-Alvarado, L. C., Camarillo-Gamboa, Á., Rustrian, E., Rene, E. R., Esposito, G., Lens, P. N., & Houbron, E. (2017). Lignocellulosic biowastes as carrier material and slow release electron donor for sulphidogenesis of wastewater in an inverse fluidized bed bioreactor. *Environmental Science and Pollution Research*, 25(6), 5115–5128.

Shore, J. L., M'Coy, W. S., Gunsch, C. K., & Deshusses, M. A. (2012). Application of a moving bed biofilm reactor for tertiary ammonia treatment in high temperature industrial wastewater. *Bioresource Technology*, 112, 51–60.

Stewart, P. S. (2003). Diffusion in biofilms. *Journal of Bacteriology*, 185(5), 1485–1491.

Tang, K., Ooi, G. T., Litty, K., Sundmark, K., Kaarsholm, K. M., Sund, C., ... & Andersen, H. R. (2017). Removal of pharmaceuticals in conventionally treated wastewater by a polishing moving bed biofilm reactor (MBBR) with intermittent feeding. *Bioresource Technology*, 236, 77–86.

Xiao, J., & Chu, S. (2015). A novel bamboo fiber biofilm carrier and its utilization in the upgrade of wastewater treatment plant. *Desalination and Water Treatment*, 56(3), 574–582.

Yang, X. L., Jiang, Q., Song, H. L., Gu, T. T., & Xia, M. Q. (2015). Selection and application of agricultural wastes as solid carbon sources and biofilm carriers in MBR. *Journal of Hazardous Materials*, 283, 186–192.

Zhang, Y., Liu, H., Shi, W., Pu, X., Zhang, H., & Rittmann, B. E. (2010). Photobiodegradation of phenol with ultraviolet irradiation of new ceramic biofilm carriers. *Biodegradation*, 21(6), 881–887.

Zhao, Y., Liu, D., Huang, W., Yang, Y., Ji, M., Nghiem, L. D., ... & Tran, N. H. (2019). Insights into biofilm carriers for biological wastewater treatment processes: Current state-of-the-art, challenges, and opportunities. *Bioresource Technology*, 121619.

13 Process Design of Various Biomass Gasification Processes Using Aspen Plus and Its Effects on Syngas and Hydrogen Production

Deepanshu Awasthi, Bhautik Gajera,
Rakesh Godara, and Arghya Datta
Sardar Swaran Singh National Institute of Bio-Energy

Nikhil Gakkhar
Ministry of New & Renewable Energy,
Atal Akshay Urja Bhavan

Tapas Kumar Patra
Sardar Swaran Singh National Institute of Bio-Energy

13.1 INTRODUCTION

Biomass accounted for approximately 10% of the global total primary energy supply during 2021 (Popp et al., 2021). In this, India has a substantial share and around 32% of India's total primary energy supply comes primarily from conventional biomass. India possesses around 750 million metric tons (MMT)/year of biomass, of which 230 MMT/year are surplus biomass after domestic uses, and use as animal feed, manure, etc. (Ministry of New and Renewable Energy, 2023). In addition to the abundance and versatility of biomass, it is also an extremely viable clean energy source. Carbon emitted as carbon dioxide or any other form of combustion product is stabilized by the amount of carbon that is utilized during biomass plantation and rearing. There are broadly three methods for transforming biomass into an energy source which are biochemical (fermentation, anaerobic digestion, etc.), thermochemical (gasification, pyrolysis, etc.), and chemical conversions (hydrothermal liquefaction, catalytic reactions, etc.). However, due to certain advantages such as the versatility of feedstocks,

DOI: 10.1201/9781003406501-13

low processing cost, ease of scaling-up of operations, etc., gasification is an excellent process for biomass to produce high grade of energy compared to other processes.

Biomass gasification is the breakdown of the organic physiology of biomass in a high temperature reducing environment to produce a gas mixture known as syngas. Syngas comprises carbon monoxide, hydrogen, carbon dioxide, and traces of other gases, with many different uses in different industrial segments including electricity generation, heating, and transportation fuel. Syngas can also be further converted and cleaned to make high-calorific-value products such as hydrogen, ammonia, etc.

A potential non-carbon energy source that could replace non-renewable energy sources is hydrogen. Given that it can be produced from sustainable and environmentally friendly sources such as biomass gasification, hydrogen is regarded as a viable replacement for fossil fuels (Qureshi et al., 2021; Qureshi et al., 2022a; Yusuf et al., 2020). Hydrogen is a clean and efficient transportation fuel as it converts into water and energy during its reaction with oxygen. Also, no carbon dioxide or other pollutants are produced during this process. Also, hydrogen is considered to be an excellent fuel because of its high calorific value of 142 MJ/kg (Qureshi et al., 2022b) as compared to other fossil fuels. Presently, world hydrogen production is about 7.7 EJ/year with a 5–10% annual increase (Hydrogen Council, 2017). In this, natural gas reforming and petroleum-based oil reforming accounts for 48% and 30% production, respectively. The rest is produced by gasification of coal (18%) and water electrolysis (4%) (Qureshi et al., 2022b). Hydrogen production by gasification is still an unexplored avenue that will be beneficial in the coming future due to the need to adopt low-carbon processes to produce hydrogen. Hydrogen is produced in the gasification process after methane conversion and its separation from the syngas.

The syngas produced during the gasification process is the result of four major steps occurring in different types of gasifiers, which are drying, pyrolysis, oxidation, and reduction. The syngas composition produced as a result of these four processes in any gasifier is affected mainly by three main factors that are temperature (slagging or non-slagging temperatures) (Bridgwater, 2003), pressure (atmospheric pressure or high pressure) (Hosseini et al., 2012; Lacey, 1967), and gasification agent (water, air, or oxygen) (Lee, 1996; Motta et al., 2018; Shayan et al., 2018). Gasifiers are classified based on these three factors as well as other factors such as fluid flow behaviour (updraft, downdraft, fluidized bed, and entrained flow) (Bridgwater, 2003; Reed, 1985), solid/gas contact (fixed or moving bed, fluidized bed, entrained bed, and molten bath bed) (Davis, 2002; Lee, 1996), and heat supplied to the gasifier (all thermal and autothermal) (Motta et al., 2018).

In any typical gasifier, various reactions take place during gasification operation and are discussed in Table 13.1. Due to the occurrence of such a large number of simultaneous reactions at different zones inside a typical gasifier, it becomes extremely complex to model the gasifier to predict the syngas composition. Due to such complexity, it becomes imperative to employ simulation software to design the gasification process and model the gasifier accurately. Also, in order to make an accurate gasification process model, experimental data sourced from industry and pilot plants can prove to be highly beneficial. This can be achieved by using sophisticated modelling tools which can also pre-determine experimental bounds and operating limitations. Using modelling tools also saves time as well as effort which can be utilized in designing a process which is robust and is immune to experimental vulnerability.

TABLE 13.1

List of Reactions in the Gasification Process

Chemical Reaction	Reaction Heat (kJ/kmol)	Reaction Name	Zone
$C + CO_2 \leftrightarrow 2CO$	+172	Boudouard	Reduction
$C + H_2O \leftrightarrow CO + H_2$	+131	Water gas	Reduction
$C + 2H_2 \leftrightarrow CH_4$	−75	Hydrogasification	Reduction
$C + 0.5O_2 \leftrightarrow CO$	−111	Char partial oxidation	Oxidation
$CO + H_2O \leftrightarrow CO_2 + H_2$	−41	Water gas shift	Reduction
$H_2 + 0.5O_2 \leftrightarrow H_2O$	−242	H_2 oxidation	Oxidation
$CO + 0.5O_2 \leftrightarrow CO_2$	−283	CO oxidation	Oxidation
$CH_4 + H_2O \leftrightarrow CO + 3H_2$	+206	Steam-methane reforming	Reduction
$H_2 + S \leftrightarrow H_2S$	NA	H_2S formation	NA
$0.5N_2 + 1.5H_2 \leftrightarrow NH_3$	NA	NH_3 formation	NA

Source: Han et al. (2017) and Timsina et al. (2019).

Aspen Plus functions as one of the most popularly used process design programmes for simulating and optimizing biomass gasification systems. The tool provides a powerful platform for modelling the complex chemical reactions and transport phenomena involved in the gasification process. It offers a range of modelling capabilities, including thermodynamic modelling, reaction kinetics, heat transfer, and fluid flow. Using Aspen Plus, engineers and researchers can calculate the impacts of various process variables on the biomass gasification process's efficiency and effectiveness. For example, the software can be used to optimize the temperature and pressure of the gasifier, determine the optimum composition of the gasifying agent, and evaluate the impact of multiple feedstocks on the composition and quality of the syngas.

Aspen Plus has a significant advantage due to its ability to provide flexibility with different process configurations, allowing for the optimization of operational conditions and the identification of limitations in such conditions. Furthermore, this makes it possible to create different process designs, such as by incorporating heat exchangers and additional process components, on which techno-economic evaluation can be performed to check the real-life production suitability.

There are broadly two ways in which Aspen Plus addresses the modelling of a chemical process like gasification. To begin, there is the equilibrium method, which presupposes that the reaction has reached a state of equilibrium and disregards the rate at which the reaction is occurring. Consequently, it is assumed that there is no char and tar produced during gasification. This way of modelling is important when a quick estimate of the gas composition is needed, or when the gasification process is assumed to have reached thermodynamic equilibrium. The second way to represent the modelling of a gasification process is following the kinetic route. The kinetic approach in Aspen Plus involves modelling the gasification process by considering the rates and kinetics of reactions. This method requires experimental data on reaction rate constants and activation energies. This approach facilitates a comprehensive and precise depiction of the gasification mechanism and can be useful for optimizing it and predicting the gasifier's performance under various operating conditions.

13.2 BIOMASS GASIFICATION IN ASPEN PLUS

The mathematical modelling of the biomass gasification process is feasible using Aspen Plus software by defining biomass and its properties. In addition, all other gaseous products and their properties are defined in the tool. Based on inputs regarding biomass and all the other participating chemical species, the process flowsheet is drawn including all the necessary reactors and separators.

13.2.1 BIOMASS DESCRIPTION AND PROCESS FLOW SHEETING

It can be observed that the properties of biomass are highly heterogeneous, resulting in variations in the characteristics among various biomass samples, and even within an identical species of biomass. Without proper elemental signature of the biomass, Aspen Plus will not be able to include biomass in phase or equilibrium calculations. Thus, any particular biomass is first and foremost described in the Aspen Plus by using the properties that pertain to its physico-chemical makeup such as proximate, ultimate, sulphate, etc., analysis data. Afterwards, an appropriate property method is chosen based on the process needs. Common property methods used for gasification modelling include PENG-ROB, PR-BM, IDEAL, RKS, and RKS-BM. Biomass density and enthalpy are typically calculated with the help of pre-built model in the software named "DCOALIGT" and "HCOALGEN," respectively.

After properly describing the biomass and the property methods, a flowsheet of the process is drawn. In case of biomass gasification, the flowsheet is very closely and accurately based on the different processes (drying, pyrolysis, reduction, and combustion) of the gasifier. It is done by choosing appropriate blocks which are summarized in Table 13.2 and putting them together in certain order based on the already available information. Figure 13.1 illustrates a usual flowchart of a gasification process. Biomass gasification typically involves several steps, namely, drying, decomposition, gasification, and partial combustion. In Aspen Plus, the RStoic block is employed to represent the moisture drying step, which postulates that the elimination of water from biomass is a chemical reaction. In case the biomass has a high moisture content, other dryer blocks such as convective dryer or contact dryer could be essential (Han et al., 2017; Nikoo and Mahinpey, 2008). The dried biomass is then allowed to undergo thermal breakdown into its basic elemental (C, H, N etc.) form. The fuel's proximate and ultimate analyses can be used to determine the output and makeup of the elements formed from decomposed biomass in the RYield block. Next, biomass gasification is modelled in Aspen Plus using either the RGibbs or RCSTR block (Ramzan et al., 2011). In the RGibbs model, the gasification process forward and reverse reactions are hypothesized to proceed at the same rate, resulting in a stable composition, while the RCSTR model accounts for the kinetics of the gasification reactions as well as heat and mass transfer. Subsequently, the RStoic block included in Aspen Plus is used to model the process of partial combustion. More specifically, this modelling exercise concentrates on the burning of char when oxygen or air is present. After that step, the stream that was produced is sent to the

TABLE 13.2
Various Blocks in Aspen Plus for Biomass Gasification

Units	Description	Role in Gasification
RStoic	• Drying of fuel	• Removes moisture from feed
RYield	• Breakdown the biomass according to its proximate and ultimate analysis	• Transforms biomass into elemental form
RGibbs	• Gibbs free energy reactor • Gasification and combustion of fuel	• Simulates the reforming of higher hydrocarbons and CH_4 and shifting of CO to H_2 • Determines the chemical makeup of syngas
RCSTR	• Addresses solid reactions along with kinetic and equilibrium reactions using existing kinetic data by integrating a user-defined FORTRAN code • Assumes ideal mixing and negligible concentration gradient inside the reactor	• As a gasifier by providing the ideal mixing conditioner • As distinct hydrodynamic elements, supporting the gasification process
RPLUG	• Handles solid reactions along with kinetic and equilibrium reactions using existing kinetic data by including a user-defined FORTRAN code • Hypothesizes perfect radial mixing and in the absence of any axial mixing	• As a gasifier to analyse axial concentration profiles of the product gases • Allows the examination of the ideal gasification design parameters, including length and diameter
Flash2	• Models flashes, evaporators, and other single-stage separators • Calculates the equilibrium between the vapour–liquid or vapour–liquid–liquid phases	• To eliminate the moisture and segregate the dried biomass and water
SSplit	• Splits a stream into two or more different flow lines	• Separation of inert ash from the product hot gas stream
Mixer	• Adds two or more streams	• Mixes elements formed during decomposition with air, steam, or oxygen in gasifier • Mixes biomass containing moisture with dry air to remove water • Mixes other unreacted streams with other streams to reduce
Heater	• Represents heaters, coolers, valves as well as pumps and compressors	• A heat exchanger can be used to warm up unreacted carbon before it enters the gasifier or cool down hot syngas in gasification processes for better efficiency

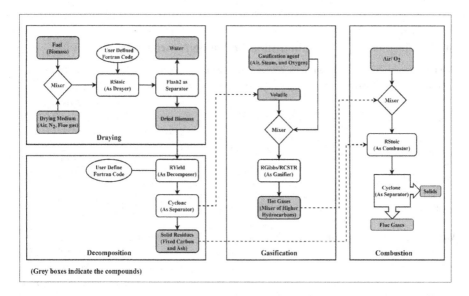

FIGURE 13.1   A typical flowsheet of a gasification process in Aspen Plus.

cyclone block so that the finished gaseous components and particulates may be more easily segregated.

It is important to note that these steps may contain extra blocks or customized calculation blocks depending on the complexity of the overall process and the specific design or aim of the simulation (Ahmed et al., 2015). The dryer models, for example, may require additional information such as length, residence time, flow direction, particle size, or drying kinetics. Ultimately, the specific parameters and models used will depend on the biomass feedstock and gasification conditions being studied.

13.2.2   Biomass Gasification Models

The efficient operation of a biomass gasifier is dependent on several complex chemical reactions involving fast pyrolysis, partial oxidation, gasification of char, and conversion of tar and lower hydrocarbons, as well as the water gas shift reaction. The composition of the products of gasification is dependent on heating rate and the time spent by the reactants inside the gasifier. Thus, it becomes imperative to develop mathematical models to simulate the gasification process. In order to analyse the impact of the different parameters such as moisture, heating values of the output gas, etc., these models are of utmost importance. Some models take into account the final chemical equilibrium composition, whereas others assume zonal (two or more zones) processes through the geometry of the gasifier. These models can be categorized into kinetic rate models and thermodynamic equilibrium models. Some models combine thermodynamic and kinetic rate models using process simulators like Aspen Plus.

Two primary methods for modelling biomass gasification are equilibrium and kinetic approaches. The equilibrium approach is divided into stoichiometric and non-stoichiometric models, with the assumption that the reactants are perfectly

TABLE 13.3

Studies Performed with Equilibrium Modelling Approach in Aspen Plus

Feedstock's	Model	Property Method	Parametric Study	References
Municipal solid waste	RYield, RGibbs	RKS-BM	Air-to-fuel ratio, temperature	Begum et al. (2014)
Pine sawdust	RYield, RGibbs	PR-BM	Temperature, steam amount	Zhai et al. (2016)
Pellets and rubber wood	RYield, RGibbs	PENG-ROB	ER and moisture content	Gagliano et al. (2017)
Hardwood chips	RYield, RGibbs	IDEAL	Temperature, ER, moisture content	Han et al. (2017)
Microalgae	RYield, RGibbs	PENG-ROB	ER, steam addition	Adnan and Hossain (2018)
Wood Chips	RYield, RGibbs	PR-BM	Temperature, ER	Safarian et al. (2019)
Hazelnut shells	RYield, RGibbs	PR-BM	SBR, temperature	Marcantonio et al. (2019)
Coconut coir pith	RYield, RGibbs	PR-BM	Temperature, steam-to-feed ratio, feedstock type	Al-Nouss et al. (2020)
Rubber wood	RYield, RGibbs	RKS-BM	Temperature, SBR, H_2 production, LHV	Tavares et al. (2020)

mixed and react indefinitely over time in order to forecast the composition of syngas. Stoichiometric models require the values of equilibrium constants and the choice of appropriate reactions to estimate the final product composition (Baruah and Baruah, 2014). However, they can lead to errors due to limited information, especially for complex processes. Non-stoichiometric models, which centre around the concept of minimizing the Gibbs free energy, are better suited for such cases as they require minimal system details. Aspen Plus uses the RGibbs block for equilibrium modelling, which minimizes the Gibbs free energy of the different compounds in the system present in all phases and further information on this approach can be found elsewhere (Basu, 2018; Kuo et al., 2014). Some of the previous reported studies are listed in Table 13.3.

Kinetic modelling is a more realistic method of estimating the yield and composition of syngas in a gasifier, as it considers specific operating conditions and gasifier configurations. It is particularly useful when the temperature is low in the gasifier and gasification process is relatively faster. Common gasification reactions used in this approach are described in various sources and are discussed in Table 13.4.

Aspen Plus software provides the capability to integrate reaction kinetics, bed hydrodynamics, and gasifier configuration through a user-defined FORTRAN subroutine (Mutlu and Zeng, 2020). A survey conducted by Safarian et al. (2019) showed that most of the gasification models with biomass use equilibrium model, whereas only about one-third of all the simulations in the literature use kinetic models. The pros and cons of these methods are enlisted in Table 13.5.

TABLE 13.4

Studies Performed with Kinetics Modelling Approach in Aspen Plus

Feedstock's	Model	Property Method	Parametric Study	References
Olive kernel	RYield, RGibbs, RCSTR	PENG–ROB[a]	Temperatures, air equivalence ratios, and energetic assessment	Damartzis et al. (2012)
Birch wood	RGibbs, RCSTR	RKS–BM[b]	Steam flow, temperature, residence time	Eikeland et al. (2015)
Coal and waste wood	RYield, RGibbs, RPLUG	RKS[c]	Gasifier diameter, height, fuel type	Adeyemi and Janajreh (2015)
Pine sawdust, rice husk, corn core, legume straw and woodchip	RYield, RGibbs, RCSTR		Temperature, equivalence ratio, oxygen enrichment	Cao et al. (2019)
Woody biomass	RYield, RGibbs, RPLUG	PR-BM[d]	Air-to-fuel ratio, temperature, heating rate	Rabea et al. (2022)
Biomass	RYield, RPLUG	IDEAL	Temperature, gasifying agent and Equivalence ratio	Catalanotti et al. (2022)

[a] PENG-ROB: Peng-Robinson cubic equation-of-state.
[b] RKS-BM: Redlich-Kwong-Soave cubic equation-of-state with Boston-Mathias alpha function.
[c] RKS: Redlich-Kwong-Soave cubic equation-of-state.
[d] PR-BM: Peng-Robinson cubic equation-of-state with the Boston-Mathias alpha function.

13.3 PARAMETERS FOR BIOMASS GASIFICATION

After the designing of model for biomass gasification, a number of parameters are used to evaluate the effect of these factors that have a direct impact on the syn gas yield and quality, which leads to hydrogen yield. These factors are temperature, equivalence ratio, gasification agent, biomass particle size, and bed material/catalyst/ bed additives. The process to get a value where the best results for biomass gasification are obtained is called optimization. When working within a set of restrictions, optimization can be used to determine what those parameters should be in order to achieve the maximum value of a given objective function. Highest yield, economic gain, and process efficiency along with the lowest production time and minimal greenhouse gas emission are all examples of possible objectives in an optimization problem (Ahmad et al., 2016; Han et al., 2017).

13.3.1 Effect of Temperature

The gasifier temperature is essential for syngas and hydrogen production. In order to produce less char and tars, more H_2 and more yield, it's important to control the temperature. According to Le Chatelier's principle, endothermic processes produce more forward products with an increase in temperature and vice-verse. It follows in reactions like Boudouard, water gas, and steam-methane reforming reactions that are

TABLE 13.5

Characteristics of Gasification Modelling Approaches

Features	Weakness
Equilibrium Modelling	
• Simple method	• Very general that species involved are completely mixed over infinitely long time period
• Not needed to take chemical reactions into account	• Hypothesizes that reactor volume and geometry doesn't affect the final composition at all
• Estimates the highest quantity of products possible	• Tars are not considered
• Beneficial in designing downdraft	• Not completely reliable for low range of temperatures
• Efficient for a quick assessment of different fuels and process variables	• Fast but less accurate results
• Unrelated to gasifier design	• Heat loss is neglected
• Adaptable to different feedstocks and process variables	• Not suitable for fluidized bed gasifiers
• Suitable for temperatures higher than 750°C and 1,000°C	
• Gives better results when taken into consideration	
Kinetics Modelling	
• Considers the real volume and time	• Dependent on the gas–solid contacting process
• Works well at low temperatures	• Computationally intensive
• Consists of kinetics and as well as hydrodynamics inside the gasifier	• Provides hindrance to being used in industrial plants
• Takes into consideration the effect of particle size	
• Works excellent for fluidized bed systems	
• Can estimate the composition at different geometrical points of the reactor	

endothermic reactions where high temperature produced more H_2 and CO concentration and decreased CH_4 concentration. On the other hand, CO production was not favourable due to exothermic char partial combustion. Generally, hydrogen production as well as gas yield is simultaneous with higher temperature but not so much for gas heating value (Li et al., 2009; Lv et al., 2004). Table 13.6 shows the comparison of literature where the effect of temperature was studied on the composition of syngas.

13.3.2 Effect of Gasification Agent

Many researchers are presently investigating the effects of gasifying the biomass with air, oxygen, steam, or a combination of these agents. Steam is generally favoured to produce H_2 in biomass gasification. The equivalent ratio (ER) value is typically used

TABLE 13.6

Effect of Temp and Gasifying Agent on Molar Fraction of Syngas

Model Type	Biomass	Gasifying Temperature	Gasifying Agent		Molar Fraction % Age					LHV of Syn Gas	References
			Agent Name	Ratio	CO	H$_2$	CH$_4$	CO$_2$			
Equilibrium (Downdraft)	Rubber wood	700°C–800°C	Steam	SBR 1.2	25%–28%	20%–23%	1–2%	10–13%		5.5–5.7 MJ/m^3	Tavares et al. (2020)
Equilibrium (Downdraft)	Hardwood chips	650°C–800°C	Air	ER 0.2–0.3	18%–23%	18%–16%	1–2%	13%–11%		3 MJ/m^3	Han et al. (2017)
Semi-kinetic	Biomass	800°C–950°C	Air-steam	ER 0.10,0.14,0.50 & SBR 0, 0.2, 0.7 respectively	33%–27%	22%–35%	13%–08%	31%–29%			Doranehgard et al. (2017)
Equilibrium	Palm kernel shell	625°C	Steam	SBR 2.5	3.90%	81.76%	8.96%	5.94%			Hussain et al. (2021)
Equilibrium	Different biomass	900°C	Steam	SBR 0.2	26%–29%	59%–63%	0.1%–0.2%	7%–13%			Pala et al. (2017)
Semi-kinetic(Fluidized bed)	MSW	600°C–700°C	Steam	SBR 1.9	15%–10%	70%–75%	17%–10%	12%–5%		13 MJ/m^3–11 MJ/m^3	Shafiq et al. (2021)
Equilibrium	Coconut coir pith	850°C	Steam	SBR 0.7	14%	35%	46%	0.7%		5.17 MJ/m^3	Al-Nouss et al. (2020)
Kinetic (Fluidized bed)	Different biomass	800°C–820°C	Air	ER 0.26	17%–18%	8%–12%	3%–5%	15%–16%			Puig-Gamero et al. (2021)
Equilibrium (Bubbling fluidized bed)	Almond shells	850°C–900°C	Steam	SBR 1.5	28%–30%	58%–60%	12%–10%	2%–1%		12 MJ/Nm^3–13 MJ/Nm^3	Acar and Böke (2019)
Equilibrium (Downdraft)	Sawdust	800°C	Steam	SBR 0.6	25%	20%	0.1–1%	10%		5.03 MJ/m^3	Safarian et al. (2022).

in air gasification processes as a key measure for determining the condition of the syngas. There are two distinct ways in which the ER can impact gasification behaviour.

Lower char and tar yields, along with decreased LHV, lower CO concentration, and greater CO_2 concentration were all effects of increased ER. With a higher ER, more oxidation processes take place, decreasing the concentrations of H_2 and CO while increasing the concentration of CO_2 in the exhaust gas as a result of the partial combustion reactions for H_2 and CO. However, when the magnitude of ER is very small, the gasifier temperature is small, which acts as a hindrance for the reactions involving gas produced from gasifying the biomass. Yet, the elimination of tar during the gasification process, enabled by the increased heat provided by the higher equivalency ratios, can improve product quality to some amount (Mohammed et al., 2011; Skoulou et al., 2008, 2009). Table 13.6 displays the studies in which the impact of the gasifying agent on the constituents of syngas was investigated.

13.3.3 Effect of Steam to Biomass Ratio

The gasification yield is also affected by the steam to biomass (SB) ratio. Since increasing the SB ratio increases the steam concentration and improves the steam reforming reactions, it leads to greater total gas yields, H_2 yield, and H_2 composition, while decreasing the CO and CH_4 composition. This pattern is consistent with Le Chatelier's principle, which states that resistance to change will occur on the opposite side of the system. The reaction is swayed in the direction of decreasing steam concentration as the SB ratio increases. Consequently, the high steam rate drives the water/gas shift reaction forward, resulting in an increasing trend of CO_2, and accelerates gasification and methane reforming, resulting in more H_2. It is inefficient for the system to expend a lot of energy on heating the steam if there is a large quantity of steam in the gasifier. The low quality of the gas could have been caused by the presence of too much steam in the gasifier, which would have caused the reaction temperature to drop (Inayat et al., 2010; Li et al., 2009; Lv et al., 2007).

13.3.4 Effect of Biomass Particle Size

In addition to increasing overall gas yields and H_2 concentrations, smaller particle sizes also reduced char and tar yields. Since the centre of the feedstock particle is cooler than the particle surface at any given time, the yields of char and liquids are increased, while the yields of gases are decreased as the particle size increases. The greater surface area of smaller particles allows for a quicker heating rate. More light gases and less char and condensation are created at higher heating rates. Also, the increased chemical reaction rates and sufficient gasification reactions can be attributed to the smaller particle size because of the increased contact area of biomass and steam. Particles with a larger diameter take longer to react, but gasification may proceed more quickly at higher temperatures (Li et al., 2009; Mohammed et al., 2011).

13.3.5 Effect of Bed Material/Additive/Catalyst

Secondary cracking of tar in vapour and of hydrocarbons like CH_4 and C_nH_m in gaseous products can boost gas yields, and a catalyst can help. In addition to an increase in tar cracking, the presence of a catalyst also stimulated tar reforming processes,

TABLE 13.7
Effect of Catalyst or Bed Material

Catalyst/Bed Material	Effect	References
CaO	Acts as CO_2 sorbent; reduced CO_2 emission and increased H_2 yield	Doranehgard et al. (2017)
CaO	As an adsorbent for CO_2 adsorption boosts the water gas shift reaction as well as the steam-methane reforming reaction	Hussain et al. (2021)
Dolomite	As a sorbent contributes to an increase in the production of synthetic gas and hydrogen (H_2) by reducing the amount of carbon dioxide (CO_2) produced	Shafiq et al. (2021)

which led to a rise in H_2 content. During the gasification process, the inclusion of bed material or an additive such as calcium oxide (CaO) can reduce the amount of tar that is produced while simultaneously boosting the overall gas production, H_2 gas composition, and carbon conversion efficiency. Also, the water gas shift reaction produces more hydrogen in the presence of additives. As the tars and the hydrocarbons are reformed, the additive or bed material performs the simultaneous job of a sorbent and a catalyst, thereby producing additional hydrogen (Hanaoka et al., 2005; Li et al., 2009; Mahishi and Goswami, 2007). Various studies on the effect of bed material/catalyst with respect to syngas composition.

13.4 HYDROGEN PRODUCTION FROM SYNGAS

After a series of steps to remove the impurities in the gasifier output, the stream is cleaned and water gas shift reactions are performed to increase the content of hydrogen to the maximum. After the hydrogen-rich syngas is produced, hydrogen is separated from the syngas by several separation techniques such as pressure swing adsorption (PSA), membrane separation, cryogenic distillation, and chemical absorption. After separation, hydrogen is generally stored at high pressure for transportation. In Aspen Plus, researchers generally design a conceptual process of the gasification process using different gasifiers and hydrogen separation techniques. Babatabar and Saidi (2021) studied the production of maximum hydrogen from three generations of biomass using steam gasification. In the first generation, corn cob was used, whereas in the second generation, wood residue and rice husk were used. In the third generation, spirulina algae were used as the biomass source. They developed a conceptual process design in Aspen Plus using a fluidized bed gasifier. Ash and char produced are separated and tar was converted to syngas to in a tar reformer. Afterwards inorganic materials such as HCl, NH_3, etc., are removed in a gas cleaning unit. After the separation of impurities, they used high- and low-temperature water gas shift reactor to increase the amount of hydrogen produced. Lastly, a hydrogen separation unit was used to obtain pure hydrogen. They also determined the consequence of gasification temperature, steam-to-biomass ratio on the production of carbon monoxide and hydrogen, water gas shift reaction temperature, and the steam to carbon ratio in water gas shift reactor. Safarian et al. (2021) also studied the hydrogen production similar to Babatabar and Saidi (2021) but instead of fluidized bed they used a downdraft gasifier. They also produced syngas and cleaned as well as separated hydrogen in a similar way. Similarly, in many studies, as shown in Table 13.8, a conceptual process design is developed and pure hydrogen is produced by separation from gasifier syngas.

13.5   GASIFICATION AND FUTURE OF HYDROGEN IN THE INDIAN SCENARIO

Hydrogen production through gasification of biomass is an important and promising technology for renewable energy production. With its various advantages such as being a clean, abundant, and versatile energy carrier, hydrogen has the potential to significantly reduce our dependence on fossil fuels. Hydrogen is produced by separating the unwanted species (CO_2, H_2S, N_2, NH_3, ash, tar, etc.) from the syngas produced from the gasifier. In order to design, simulate, and optimize an accurate gasification process to produce syngas and thus, hydrogen, Aspen Plus is an excellent software. Its flexibility in modelling different process configurations and analysing process limitations and hazards makes it an indispensable tool for the development of efficient and effective gasification processes. The combination of gasification technology and Aspen Plus software can contribute to the development of sustainable and economically viable energy systems. In this age where fossil fuel sources are depleting at an alarming rate, it is extremely important to switch to bio-hydrogen as an energy source.

Hydrogen has all the characteristics to be the fuel of the future and it is also evident from many countries in the world making necessary policies to push hydrogen as the fuel for the future. Globally, the European Union has set a target to install at least 40 GW of electrolysers in the EU by 2030 (European Commission, 2023), while Japan targets to produce 3 million tons of hydrogen per year by 2030. Japan aims to become a "hydrogen society" by 2050, with a goal of producing 300,000 fuel cell vehicles and installing 900 hydrogen refuelling stations by that year (Ohno, 2022). The country also plans to use hydrogen to power homes and offices, as well as for industrial applications such as steel production (Sieler, 2021). Similarly, the Hydrogen Shot initiative, recently launched by the US Department of Energy, seeks to achieve a significant reduction in the cost of clean hydrogen by 80% to $1 per kilogram within a decade (U.S. Department of Energy, 2021). South Korea aims to have 6.2 million fuel cell electric vehicles and 1,200 refuelling stations by 2040, with a goal of producing 5.26 million tons of hydrogen per year by that year (Kan, 2020). The country is also financing heavily in the growth of fuel cell technology for use in ships, trains, and power generation. Germany aims to become a global leader in hydrogen technology, with a goal of having 5 GW of installed electrolyser cap (The National Hydrogen Strategy, 2020). The country also plans to have 400 hydrogen refuelling stations and 5,000 hydrogen-powered buses on the road by 2030. Australia aims to become a major exporter of hydrogen, with a goal of producing hydrogen at a cost of less than AUD 2 per kilogram by 2030 (COAG Energy Council Hydrogen Working Group, 2019). The country is also

TABLE 13.8

Studies Based on the Production of Hydrogen from Syngas in Aspen Plus

Feedstocks	Model	Gasifying Agent	Property Method	Hydrogen Production from Syngas	Hydrogen Content	References
Corn cob, wood residue, rice husk, and spirulina algae	RYield, RGibbs, REquil	Steam	Peng-Robinson – Boston-Mathias	1. Conversion of tar into syngas 2. Ash and char removal 3. Inorganic compound removal 4. High- and low-temperature water gas shift reaction 5. Hydrogen separation in pressure swing adsorption (PSA)	50% mole fraction of hydrogen in the gasifier output	Babatabar and Saidi (2021)
Fifty different woody and agricultural biomass	RStoic, RYield, RGibbs	Steam and air	NA	1. High- and low-temperature water gas shift reaction 2. Hydrogen separation in pressure swing adsorption (PSA)	Maximum hydrogen content is 100 g/kg of biomass	Safarian et al. (2021)
Rubber wood	RStoic, RYield, RGibbs	Air	Redlich-Kwong-Soave equation-of-state with Boston-Mathias alpha	1. Hydrogen-rich syngas was produced	19% mole fraction of hydrogen in the gasifier output	Tavares et al. (2020)
Rubber wood	RStoic, RYield, RGibbs	Steam and air	NA	1. High- and low-temperature water gas shift reaction 2. Hydrogen separation in pressure swing adsorption (PSA)	15% mole fraction of hydrogen in the gasifier output	Safarian et al. (2022)
Saw dust	RYield, RGibbs	Steam	NA	1. Hydrogen separation	61% mole fraction of hydrogen in the gasifier output	Tan and Zhong (2010)
Tar from biomass gasification	RGibbs	Steam	NA	1. Water gas shift reactor and gas compression 2. Hydrogen purification in PSA unit 3. Carbon dioxide absorption unit	29% mole fraction of hydrogen in the tar reformer output	Shamsi et al. (2022)

financing in the advancement of a national hydrogen infrastructure, including the construction of refuelling stations and pipelines.

Several industries have also started to embrace hydrogen as a clean energy source. In the transportation sector, automakers such as Toyota, Hyundai, and General Motors have announced plans to launch hydrogen fuel cell vehicles, which emit only water vapour and have a longer driving range than battery-electric vehicles. In the shipping sector, several companies such as Maersk and MSC have announced plans to use hydrogen fuel cells to power their vessels. In the industrial sector, hydrogen is being explored as a potential feedstock for steelmaking and other heavy industries.

India has also recognized the potential of hydrogen as a clean energy source and has launched several initiatives to promote its production and use. To make India a leader in the production, application, and export of Green Hydrogen, as well as its by-products, the Indian government has launched the National Hydrogen Energy Mission. The mission aims to produce at least 5 MMT of Green Hydrogen per year by 2030, with the possibility to reach 10 MMT per year by expanding export markets (MNRE, 2023). The mission also seeks to displace fossil fuels and feedstocks that rely on fossil fuels with green hydrogen based renewable fuels and feedstocks used in t the synthesis of ammonia, refining of petroleum, manufacturing of steel and for transportation purposes. The mission will be implemented in two phases. Phase I, from 2022–23 to 2025–26, will focus on creating demand for Green Hydrogen in sectors already using hydrogen, such as refineries, fertilizers, and city gas, while enabling adequate supply by increasing domestic electrolyser manufacturing capacity. Initiating the green transition in steel manufacturing, long-haul heavy-duty mobility, and shipping will also begin with pilot projects.

Phase II, from 2026–27 to 2029–30, will explore and investigate the possibility of large-scale Green Hydrogen initiatives in sectors like steel, mobility, shipping, railways, and aviation, as well as enhancing R&D activities for continuous development of products. The mission aims to drive deep decarbonization of the economy while establishing India as a global hub for Green Hydrogen production and usage. In the transportation sector, India has set a target to achieve 30% electric vehicle penetration by 2030, which comprises both hydrogen-fuelled fuel cell vehicles and battery-electric vehicles. The government has declared intentions to build a network of hydrogen refuelling stations in significant cities and along major highways in order to meet this goal. Several automakers, including Tata Motors and Mahindra, have also announced plans to launch fuel cell vehicles in the Indian market.

To hasten India's adoption of hydrogen, a number of obstacles need to be overcome, such as lowering the cost of hydrogen production, creating infrastructure for hydrogen use, and ensuring the supply of low-carbon hydrogen. Nonetheless, with growing political support and increasing investments, the scenario of hydrogen in India looks promising.

13.6   CONCLUSION

The current study discussed the gasification technologies for hydrogen production. Using a process simulation software like Aspen Plus provides an effective platform for modelling and analysing biomass gasification processes. This study explained the different steps involved in designing a biomass gasification process. It involves initializing the biomass feed to be used and drawing an appropriate process flow sheet to represent the gasification process. Afterwards, considering an appropriate kinetic or equilibrium model, the gasification process is simulated. Subsequently, considering the various parameters affecting the gasification process, such as temperature, gasification agent, steam-to-biomass ratio, biomass particle size, and bed material/additive/catalyst, the optimal conditions for hydrogen production can be determined. The potential of hydrogen production from syngas and its future in the Indian scenario was also explored after the description of the process of gasification and production of syngas/hydrogen. The future of hydrogen is promising, and several countries, including India, have launched ambitious plans to scale up the production and use of hydrogen in their energy systems. The National Hydrogen Energy Mission in India aspires to make the nation a major centre for the export, use, and manufacturing of Green Hydrogen and its derivatives, contributing to India's self-reliance through clean energy and inspiring the global clean energy transition. In order to realize the ambition to transition to a hydrogen economy, it is imperative to design better processes for hydrogen production from gasification. In this process, using the services of a process design and simulation software like Aspen Plus can prove to be extremely beneficial.

REFERENCES

Acar, M. C., & Böke, Y. E. (2019). Simulation of biomass gasification in a BFBG using chemical equilibrium model and restricted chemical equilibrium method. *Biomass and Bioenergy, 125*, 131–138.

Adeyemi, I., & Janajreh, I. (2015). Modeling of the entrained flow gasification: Kinetics-based ASPEN Plus model. *Renewable Energy, 82*, 77–84.

Adnan, M. A., & Hossain, M. M. (2018). Gasification performance of various microalgae biomass-A thermodynamic study by considering tar formation using Aspen plus. *Energy Conversion and Management, 165*, 783–793.

Ahmad, A. A., Zawawi, N. A., Kasim, F. H., Inayat, A., & Khasri, A. (2016). Assessing the gasification performance of biomass: A review on biomass gasification process conditions, optimization and economic evaluation. *Renewable and Sustainable Energy Reviews, 53*, 1333–1347.

Ahmed, A. M. A., Salmiaton, A., Choong, T. S. Y., & Azlina, W. W. (2015). Review of kinetic and equilibrium concepts for biomass tar modeling by using Aspen Plus. *Renewable and Sustainable Energy Reviews, 52*, 1623–1644.

Alnouss, A., Parthasarathy, P., Shahbaz, M., Al-Ansari, T., Mackey, H., & McKay, G. (2020). Techno-economic and sensitivity analysis of coconut coir pith-biomass gasification using ASPEN PLUS. *Applied Energy, 261*, 114350.

Babatabar, M. A., & Saidi, M. (2021). Hydrogen production via integrated configuration of steam gasification process of biomass and water-gas shift reaction: Process simulation and optimization. *International Journal of Energy Research, 45*(13), 19378–19394.

Baruah, D., & Baruah, D. C. (2014). Modeling of biomass gasification: A review. *Renewable and Sustainable Energy Reviews, 39*, 806–815.

Basu, P. (2018). *Biomass Gasification, Pyrolysis and Torrefaction: Practical Design and Theory*, 3rd edn. Academic Press, London.

Begum, S., Rasul, M. G., & Akbar, D. (2014). A numerical investigation of municipal solid waste gasification using aspen plus. *Procedia Engineering, 90*, 710–717.

Bridgwater, A. V. (2003). Renewable fuels and chemicals by thermal processing of biomass. *Chemical Engineering Journal, 91*(2–3), 87–102.

Cao, Y., Wang, Q., Du, J., & Chen, J. (2019). Oxygen-enriched air gasification of biomass materials for high-quality syngas production. *Energy Conversion and Management, 199*, 111628.

Catalanotti, E., Porter, R. T. J., & Mahgerefteh, H. (2022). An Aspen plus kinetic model for the gasification of biomass in a downdraft gasifier. *Chemical Engineering Transactions, 92*, 679–684.

COAG Energy Council Hydrogen Working Group. (2019). *Australia's National Hydrogen Strategy.* https://www.dcceew.gov.au/sites/default/files/documents/australias-national-hydrogen-strategy.pdf

Damartzis, T., Michailos, S., &Zabaniotou, A. (2012). Energetic assessment of a combined heat and power integrated biomass gasification-internal combustion engine system by using Aspen Plus(r). *Fuel Processing Technology, 95*, 37–44.

Davis, B. H. (2002). Overview of reactors for liquid phase Fischer-Tropsch synthesis. *Catalysis Today, 71*(3-4), 249–300.

Doranehgard, M. H., Samadyar, H., Mesbah, M., Haratipour, P., & Samiezade, S. (2017). High-purity hydrogen production with in situ CO2 capture based on biomass gasification. *Fuel, 202*, 29–35.

Eikeland, M. S., Thapa, R. K., & Halvorsen, B. (2015). *Aspen plus simulation of biomass gasification with known reaction kinetic.* In: *Proceedings of 56th SIMS*, Sweden. doi:10.3384/ecp15119149

European Commission. (2023). *Communication from the Commission to the European Parliament, the Council, the European Economic and Social Committee and the Committee of the Regions on the European Hydrogen Bank.* https://eur-lex.europa.eu/legal-content/EN/TXT/PDF/?uri=CELEX:52023DC0156.

Gagliano, A., Nocera, F., Bruno, M., & Cardillo, G. (2017). Development of an equilibrium-based model of gasification of biomass by Aspen Plus. *Energy Procedia, 111*, 1010–1019.

Han, J., Liang, Y., Hu, J., Qin, L., Street, J., Lu, Y., & Yu, F. (2017). Modeling downdraft biomass gasification process by restricting chemical reaction equilibrium with Aspen Plus. *Energy Conversion and Management, 153*, 641–648.

Hanaoka, T., Yoshida, T., Fujimoto, S., Kamei, K., Harada, M., Suzuki, Y., Hatano, H., Yokoyama, S., & Minowa, T. (2005). Hydrogen production from woody biomass by steam gasification using a CO_2 sorbent. *Biomass and Bioenergy, 28*(1), 63–68.

Hosseini, M., Dincer, I., & Rosen, M. A. (2012). Steam and air fed biomass gasification: Comparisons based on energy and exergy. *International Journal of Hydrogen Energy, 37*(21), 16446–16452.

Hussain, M., Zabiri, H., Uddin, F., Yusup, S., & Tufa, L. D. (2021). Pilot-scale biomass gasification system for hydrogen production from palm kernel shell (part A): Steady-state simulation. *Biomass Conversion and Biorefinery, 5*, 1–14.

Hydrogen Council. (2017). *Hydrogen Scaling up: A Sustainable Pathway for the Global Energy Transition.* https://hydrogencouncil.com/wp-content/uploads/2017/11/Hydrogen-Scaling-up_Hydrogen-Council_2017.compressed.pdf

Inayat, A., Ahmad, M. M., Yusup, S., & Mutalib, M. I. A. (2010). Biomass steam gasification with in-situ CO_2 capture for enriched hydrogen gas production: A reaction kinetics modelling approach. *Energies, 3*(8), 1472–1484.

Kan S. (2020). *South Korea's Hydrogen Strategy and Industrial Perspectives.* https://www.ifri.org/sites/default/files/atoms/files/sichao_kan_hydrogen_korea_2020_1.pdf

Kuo, P.-C., Wu, W., & Chen, W.-H. (2014). Gasification performances of raw and torrefied biomass in a downdraft fixed bed gasifier using thermodynamic analysis. *Fuel, 117*, 1231–1241.

Lacey, J. A. (1967). Gasification of coal in a slagging pressure gasifier. In: *Advances in Chemistry,* Vol. 69. ACS Publications, London. doi:10.1021/ba-1967-0069.ch004

Lee, S. (1996). *Alternative Fuels: Applied Energy Technology Series.* CRC Press, Washington.

Li, J., Yin, Y., Zhang, X., Liu, J., & Yan, R. (2009). Hydrogen-rich gas production by steam gasification of palm oil wastes over supported tri-metallic catalyst. *International Journal of Hydrogen Energy, 34*(22), 9108–9115.

Lv, P., Yuan, Z., Ma, L., Wu, C., Chen, Y., & Zhu, J. (2007). Hydrogen-rich gas production from biomass air and oxygen/steam gasification in a downdraft gasifier. *Renewable Energy, 32*(13), 2173–2185.

Lv, P. M., Xiong, Z. H., Chang, J., Wu, C. Z., Chen, Y., & Zhu, J. X. (2004). An experimental study on biomass air-steam gasification in a fluidized bed. *Bioresource Technology, 95*(1), 95–101.

Mahishi, M. R., & Goswami, D. Y. (2007). An experimental study of hydrogen production by gasification of biomass in the presence of a CO_2 sorbent. *International Journal of Hydrogen Energy, 32*(14), 2803–2808.

Marcantonio, V., Bocci, E., & Monarca, D. (2019). Development of a chemical quasi-equilibrium model of biomass waste gasification in a fluidized-bed reactor by using Aspen Plus. *Energies, 13*(1), 53.

Ministry of New and Renewable Energy, GOI. (2023). *Bioenergy Overview.* https://mnre.gov.in/bio-energy/current-status

MNRE. (2023). *National Green Hydrogen Mission.* https://mnre.gov.in/img/documents/uploads/file_f-1673581748609.pdf

Mohammed, M. A. A., Salmiaton, A., Azlina, W. W., Amran, M. S. M., &Fakhru'l-Razi, A. (2011). Air gasification of empty fruit bunch for hydrogen-rich gas production in a fluidized-bed reactor. *Energy Conversion and Management, 52*(2), 1555–1561.

Motta, I. L., Miranda, N. T., Maciel Filho, R., & Maciel, M. R. W. (2018). Biomass gasification in fluidized beds: A review of biomass moisture content and operating pressure effects. *Renewable and Sustainable Energy Reviews, 94*, 998–1023.

Mutlu, Ö. Ç., & Zeng, T. (2020). Challenges and opportunities of modeling biomass gasification in Aspen Plus: A review. *Chemical Engineering & Technology, 43*(9), 1674–1689.

Nikoo, M. B., & Mahinpey, N. (2008). Simulation of biomass gasification in fluidized bed reactor using ASPEN PLUS. *Biomass and Bioenergy, 32*(12), 1245–1254.

Ohno, T. (2022). *Re-examining Japan's Hydrogen Strategy Moving Beyond the "Hydrogen Society" Fantasy.* Tokyo, Japan: REI.

Pala, L. P. R., Wang, Q., Kolb, G., & Hessel, V. (2017). Steam gasification of biomass with subsequent syngas adjustment using shift reaction for syngas production: An Aspen Plus model. *Renewable Energy, 101*, 484–492.

Popp, J., Kovács, S., Oláh, J., Divéki, Z., & Balázs, E. (2021). Bioeconomy: Biomass and biomass-based energy supply and demand. *New Biotechnology, 60*, 76–84.

Puig-Gamero, M., Pio, D. T., Tarelho, L. A. C., Sánchez, P., & Sanchez-Silva, L. (2021). Simulation of biomass gasification in bubbling fluidized bed reactor using aspen plus(r). *Energy Conversion and Management, 235*, 113981.

Qureshi, F., Yusuf, M., & Abdullah, B. (2021). A brief review on hydrogen production to utilization techniques. In: *2021 Third International Sustainability and Resilience Conference: Climate Change.* IEEE, New York, pp. 66–71.

Qureshi, F., Yusuf, M., Kamyab, H., Zaidi, S., Khalil, M. J., Khan, M. A., Alam, M. A., Masood, F., Bazli, L., & Chelliapan, S. (2022). Current trends in hydrogen production, storage and applications in India: A review. *Sustainable Energy Technologies and Assessments, 53*, 102677.

Qureshi, F., Yusuf, M., Pasha, A. A., Khan, H. W., Imteyaz, B., & Irshad, K. (2022). Sustainable and energy efficient hydrogen production via glycerol reforming techniques: A review. *International Journal of Hydrogen Energy, 47*, 41397–41420.

Rabea, K., Michailos, S., Akram, M., Hughes, K. J., Ingham, D., & Pourkashanian, M. (2022). An improved kinetic modelling of woody biomass gasification in a downdraft reactor based on the pyrolysis gas evolution. *Energy Conversion and Management, 258*, 115495.

Ramzan, N., Ashraf, A., Naveed, S., & Malik, A. (2011). Simulation of hybrid biomass gasification using Aspen plus: A comparative performance analysis for food, municipal solid and poultry waste. *Biomass and Bioenergy*, *35*(9), 3962–3969.

Reed, T. B. (1985). Principles and technology of biomass gasification. *Advances in Solar Energy: An Annual Review of Research and Development*, *2*, 125–174.

Safarian, S., Ebrahimi Saryazdi, S. M., Unnthorsson, R., & Richter, C. (2021). Modeling of hydrogen production by applying biomass gasification: Artificial neural network modeling approach. *Fermentation*, *7*(2), 71.

Safarian, S., Unnþórsson, R., & Richter, C. (2019). A review of biomass gasification modelling. *Renewable and Sustainable Energy Reviews*, *110*, 378–391.

Safarian, S., Unnthorsson, R., & Richter, C. (2022). Performance investigation of biomass gasification for syngas and hydrogen production using Aspen Plus. *Open Journal of Modelling and Simulation*, *10*(2), 71–87.

Shafiq, H., Azam, S. U., & Hussain, A. (2021). Steam gasification of municipal solid waste for hydrogen production using Aspen Plus(r) simulation. *Discover Chemical Engineering*, *1*, 1–16.

Shamsi, M., Obaid, A. A., Farokhi, S., & Bayat, A. (2022). A novel process simulation model for hydrogen production via reforming of biomass gasification tar. *International Journal of Hydrogen Energy*, *47*(2), 772–781.

Shayan, E., Zare, V., & Mirzaee, I. (2018). Hydrogen production from biomass gasification; a theoretical comparison of using different gasification agents. *Energy Conversion and Management*, *159*, 30–41.

Sieler RE., C. L., S. B., B. S., T. BL. (2021). *Hydrogen Factsheet*. https://adelphi.de/en/system/files/mediathek/bilder/H2%20Factsheet%20Japan.pdf

Skoulou, V., Swiderski, A., Yang, W., & Zabaniotou, A. (2009). Process characteristics and products of olive kernel high temperature steam gasification (HTSG). *Bioresource Technology*, *100*(8), 2444–2451.

Skoulou, V., Zabaniotou, A., Stavropoulos, G., & Sakelaropoulos, G. (2008). Syngas production from olive tree cuttings and olive kernels in a downdraft fixed-bed gasifier. *International Journal of Hydrogen Energy*, *33*(4), 1185–1194.

Tan, W., & Zhong, Q. (2010). Simulation of hydrogen production in biomass gasifier by ASPEN PLUS. In: *2010 Asia-Pacific Power and Energy Engineering Conference*. IEEE, New York, pp. 1–4.

Tavares, R., Monteiro, E., Tabet, F., & Rouboa, A. (2020). Numerical investigation of optimum operating conditions for syngas and hydrogen production from biomass gasification using Aspen Plus. *Renewable Energy*, *146*, 1309–1314.

The National Hydrogen Strategy. (2020). https://www.bmwk.de/Redaktion/EN/Publikationen/Energie/the-national-hydrogen-strategy.pdf?__blob=publicationFile&v=6

Timsina, R., Thapa, R. K., & Eikeland, M. S. (2019). *Aspen Plus simulation of biomass gasification for different types of biomass*. doi:10.3384/ecp20170151.

U.S. Department of Energy. (2021). *Hydrogen Shot: An Introduction. Energy earth Shots*. https://www.energy.gov/eere/fuelcells/articles/hydrogen-shot-introduction

Yusuf, M., Farooqi, A. S., Keong, L. K., Hellgardt, K., & Abdullah, B. (2020). Latest trends in Syngas production employing compound catalysts for methane dry reforming. *IOP Conference Series: Materials Science and Engineering*, *991*(1), 012071.

Zhai, M., Guo, L., Wang, Y., Zhang, Y., Dong, P., & Jin, H. (2016). Process simulation of staging pyrolysis and steam gasification for pine sawdust. *International Journal of Hydrogen Energy*, *41*(47), 21926–21935.

14 Implementing Targeted Total Soluble Product Recovery during Food Waste Biomethanation for Enhanced Recovery of Energy and Value-Added Products

S. Hemapriya and P. Sankar Ganesh
Birla Institute of Technology and Science,
Pilani, Hyderabad Campus

14.1 INTRODUCTION

Biomethanation is an effective method for treating organic wastes, which additionally produce biogas for renewable energy in a controlled environment. Examples of organic-rich wastes include food wastes, organic fractions of municipal solid wastes (MSWs), pruning wastes, poultry and slaughterhouse wastes (SHWs), brewery wastes, etc. Food waste contributes majorly to MSW. Nowadays, with an increase in population and urbanization, food wastage has increased sharply. According to The Food and Agricultural Organisational (FAO, 2013) report, around 1.3 billion tonnes of food is wasted globally. Management of MSW and SHW, including collection, treatment, and disposal, is a significant challenge in India and generates about 208 million metric tons of MSW every year. All the aforementioned wastes are considered as MSWs and end up in landfill sites or undergo incineration treatment. Especially, food wastes are unavoidable and irrepressible. The food produced but never consumed represents an inappropriate use of valuable natural resources.

These conventional techniques of handling solid wastes are inefficient as they contribute to increasing greenhouse gas emissions, high chances of polluting the groundwater, air pollution, leachate generation, and space constraints. Problems associated with improper handling of waste have resulted in massive resource waste and prompted hazards. Therefore, organic waste management using biomethanation is crucial because it protects the environment and generates bioenergy.

DOI: 10.1201/9781003406501-14

Biomethanation is a significant replacement for conventional and less significant treatment options. It is considered a cost-effective technology that enables the production of methane-rich biogas via microbial digestion. It results in a sustainable and promising solution for the problem of untreated food waste disposal in the environment. In addition, it provides a useful remaining residue in the form of sludge from anaerobic digestion called digestate, which can be bioprocessed to produce biofertilizers.

During biomethanation, a number of intermediates are produced, and some of them will enhance the process whereas some will inhibit the process. Such intermediates are collectively called total soluble products (TSPs); among them, the most dominant is the volatile fatty acid (VFA). The VFA accumulates in excess and hinders anaerobic digestion due to acidification. It is important to implement a precise technique like adsorption using ion exchange resin (IER) to counteract the product inhibition imposed by the undissociated acids that predominate at low pH. Excess VFA can be adsorbed during adsorption to maintain the optimal concentration for enhanced bioenergy production. The extracted VFAs can also be desorbed from the resins and they can be downstream processed and purified. They support the circular bioeconomy and have a significant economic value. The regenerated resin can be again used in the adsorption–desorption cycle.

14.2 BIOMETHANATION TREATMENT

Biomethanation of organic substrates has numerous advantageous over traditional waste treatment techniques, including a reduction in the formation of biomass sludge (Lalak et al., 2015), minimal odor emission, and generation of bioenergy in the form of methane-rich biogas (Alastair et al., 2008). Furthermore, a high level of compliance with various waste regulations was established to lower the quantity of biodegradable waste going to landfills. Also, anaerobic fermentation is the only commercially proven process capable of producing energy from organics. Organic substances including polysaccharides, proteins, and lipids are hydrolyzed or broken down into various intermediate products during anaerobic digestion. These compounds are then oxido-reduced to generate methane-rich biogas. It is indeed suitable owing to its less sludge production and energy requirement. Anaerobic digestion involves four biochemical processes, as shown in Figure 14.1. Different types of microbial populations govern these four reaction stages in anaerobic digestion. In the first step, the high molecular weight and granular complex organic polymers such as proteins, lipids, and carbohydrates are hydrolyzed by hydrolytic bacteria to simple soluble monomers such as glucose and amino acids. This initial step is regarded as the rate-determining step since synergistic relationship persists between different microorganisms. In acidogenesis, the soluble monomers are degraded into acetic acid, propionic acid, butyric acid, and valeric acid, which are collectively called volatile fatty acids. In the further step, long-chain VFAs are oxidized further to acetate along with carbon dioxide and hydrogen for further reaction, and this process of conversion to acetate is called acetogenesis. In methanogenesis, methane is generated by two groups of bacteria; acetoclastic methanogens utilize acetate to generate methane, while the other group, called hydrogenotrophic methanogens, utilizes hydrogen and carbon dioxide to form methane.

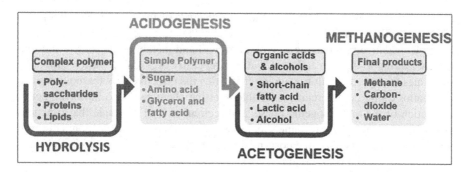

FIGURE 14.1   Biochemical conversion pathway in anaerobic digestion.

14.3   INTERMEDIATE PRODUCTS

Anaerobic digestion is essential for converting organic-rich wastes to renewable fuels and electricity. Various products are produced during the process and some are already present in the reactor that includes water, biomass, organic waste, TSPs, particulates, salt, and dissolved gases. Alcohols (ethanol, propanol, butanol), solvents (acetone), lactate, and fatty acids (acetate, propionate, iso-butyrate, valerate, and caproate) all are formed as intermediate products and are collectively called total soluble products (TSPs), which have an impact on methane production (Chakraborty et al., 2022).

Among various intermediates produced during the process of anaerobic digestion, some will function as enhancers and some will function as inhibitors. The stability of the process necessitates having control over the generated intermediates. Due to the syntropic correlation existing between the microbes of anaerobic digestion, the product of one microbe will act as the substrate for another. If the inhibitory intermediate has high economic value, then it can be recovered as it can act as a source for generating value-added products, which will contribute to the circular bioeconomy and concomitantly maintain the optimal concentration for normal functioning of anaerobic digesters.

The aforementioned intermediates are produced as by-products of various metabolic pathways during biomass conversion (Zhou et al., 2018). Each kind of VFA generated during this process follows a distinct metabolic pathway. With varied environmental factors and microbial community dominance, the subsequent metabolic pathway changes. During acidogenesis, the formation and distribution of VFAs in anaerobic digestion depend on the breakdown of metabolic pathways through multiple routes. Therefore, for enhancing the performance and regulation, it is important to comprehend the difference along with the syntrophic relationship of the microorganisms to maintain the co-existence of different pathways in the reactor (Zhou et al., 2018). Pyruvate is the most essential component since it can result in the generation of different types of VFAs and its alcoholic counterparts.

VFAs will be formed as one of the major intermediates during the anaerobic process of converting the food waste to methane-rich biogas, as easily hydrolyzable waste, such as food waste, results in increased production of VFAs by acidogens and acetogens. However, the ability of the methanogens to consume the produced VFAs occurs at a lower rate when compared to the production rate, leading to the accumulation of VFA and concomitant acidification of the reactor. The lipophilic VFA can easily cross the cell barrier and dissociate within cells, decreasing intracellular pH, negatively impacting microbial growth and metabolic pathways, and ultimately leading to digester failure. Co-digesting easily degradable wastes with other substrates that are difficult to degrade maintains the effective C/N (carbon to nitrogen) ratio. Using commercially available alkaline chemicals as a buffering agent, elutriation of the methanogenic phase, and recirculation of digestate that functions as a self-buffering agent are some specific ways to overcome acidification. However, all these methods increase the cost involved in biogas production and just delay the acidification without preventing it.

14.3.1 ALCOHOLS

Ethanol is produced through the process of fermentation of organic compounds such as glucose. Acetaldehyde is produced as the result of pyruvate decarboxylation and is then reduced to generate ethanol. However, there is an alternate pathway to generate ethanol from pyruvate containing three steps producing acetyl-CoA as an intermediate and acetaldehyde as a by-product by a certain population of microorganism (Chaganti et al., 2011). Some microbial species additionally synthesize acetone and butanol from glucose in addition to ethanol (Yen et al., 2011). However, the metabolic route involved in the conversion process of glucose to acetyl-CoA has equivalent reaction steps. After that, as shown in Figure 14.2, it branches into numerous paths; acetyl-CoA undergoes transformation into acetocetyl-CoA in order to make acetone, whereas butanol is produced from acetoacetyl-CoA via the generation of 3-hydroxy butyl-CoA, crotonyl-CoA, and butyryl-CoA as intermediates.

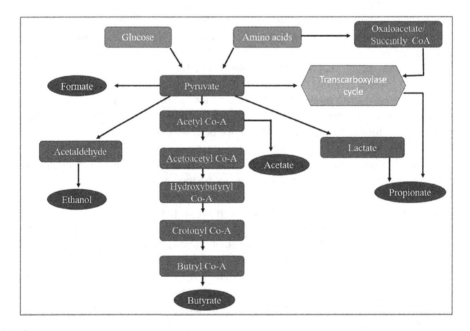

FIGURE 14.2   Metabolic pathways involved in volatile fatty acid production.

14.3.2 Lactate

Lactic acid is generated as one of the intermediates during the acidogenesis stage, and also acts as the precursor of the VFAs beforehand of methane generation, the final end product. During the process of acidogenesis, two molecules of lactic acid are generated from one molecule of glucose. The microorganisms involved in the generation of lactic acid are potent degraders of starch because they are capable of producing crucial enzymes like amylase and maltase. Lactate dehydrogenase results in the production of lactate. Similarly, in the presence of propionate dehydrogenase, the lactate is reduced to generate propionate (Lee et al., 2008).

It is reported that when a high concentration of easily degradable substrate is available, lactic acid will be dominantly produced along with the undefined mixed culture (Vishnu et al., 2006). The production of lactic acid will lead to the inhibition of various other microorganisms involved in the production of certain alcohols such as ethanol and also VFA producers since the lactic acid producers will reduce the pH to extremely low levels and suppress the growth of other anaerobic microorganisms (James et al., 1985).

14.3.3 Volatile Fatty Acid: An Overview

Organic acids like formic, acetic, propionic, butyric, valeric, caproic, lactic, succinic, fumaric, citric, gluconic, ascorbic, etc. are among the group of aliphatic mono- and dicarboxylic acids known as "carboxylic acids". Carboxylic acids have significant applications in the many industries, namely, pharmaceutical, food, and fuel industries. Interestingly, some carboxylic acids have already met biochemical production at an industrial scale. When compared to other carboxylic acids, VFSs are mono-carboxylic acids with 1–6 carbons and have comparatively high volatility (Bhatia and Yang, 2017).

Each kind of VFA is produced by different pathways from either single or different substrate such as pyruvate. The function of different enzymes involved in the acetyl-CoA pathway influences the production of acetate and also oxidation of other VFAs (Müller et al., 2010). Propionate can be produced via two different pathways, one of which involves pyruvate reduction directly with lactate as intermediate and the other which involves the transcarboxylase cycle. In the second pathway, propionate is generated through a series of intermediates including oxaloacetate, succinate, fumarate, propionyl CoA, etc. The production of propionate is highly influenced by various environmental factors such as pH and inoculum.

The process involves generating acetyl-CoA from pyruvate, which is then subsequently transformed to butyryl-CoA and acetoacetyl-CoA that further results in production of intermediates such as of 3-hydroxy butyryl-CoA and crotonyl-CoA (Chaganti et al., 2011). Ultimately, butyrate kinase and phosphotransbutyrylase catalyze the transformation of butyryl-CoA into butyrate (Vital et al., 2014). When compared with the production of acetate, extra molecules of $NADH_2$ are needed for the reduction reaction. Thus, it lowers the efficiency of butyrate production.

Production of commercial VFAs is highly dependent on chemical synthesis utilizing non-renewable petroleum as raw material; however, these petroleum-based methods are not sustainable. A better alternative for VFA production would be microbial fermentation. The simplest methods rely on using sugars like glucose or xylose, which allow for higher production and fewer by-products, but there are also drawbacks such as a higher cost for the raw materials. In contrast, waste biomass such as food waste can be used as a substrate, as they contain highly available and abundant carbon sources that can be fermented to VFAs (Figures 14.3 and 14.4).

VFA is having larger market value and it is the fastest-growing biochemical. VFAs are demanded by chemical industries for the production of alcohol, ketones, esters, olefins, or aldehydes. Furthermore, VFA has the potential to replace petrochemicals addressing the concern of plastic accumulation in the ecosystem. This would also offset the requirement for petrochemicals to produce biodegradable PHA (polyhydroxyalkanoates) and PHB (polyhydroxy butyrate). The cost of production of PHA and PHB is strongly correlated with substrate cost. If VFAs could be produced from waste, then production costs could be more than halved. Thus, VFA production by microbial fermentation is considered an effective alternative to chemical routes. Figure 14.5 shows the diverse application of VFA in industries.

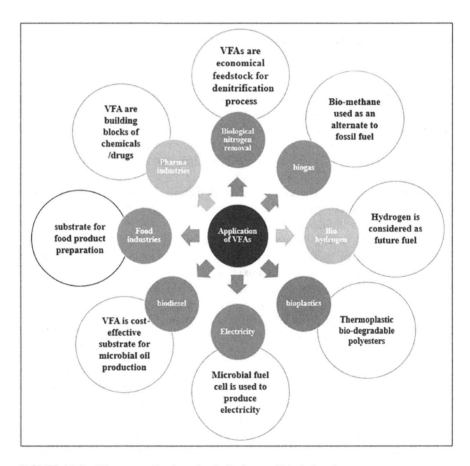

FIGURE 14.3 Diverse application of volatile fatty acid in industries.

Repeating Unit
NaA Solution
NaA Solution
AEM
CEM
AEM
A⁻
Na⁺
Na⁺
A⁻
Anode
Cathode
diluate
Concentrated NaA
Δv

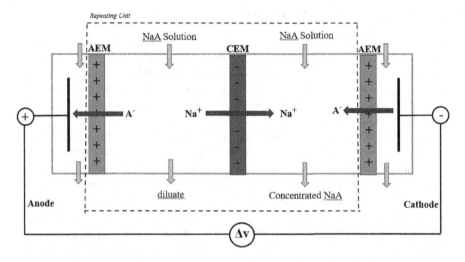

FIGURE 14.4 Principle of common electrodialysis in a two-compartment configuration, concentration of a sodium carboxylate (NaA) solution. CEM, cation exchange membrane; AEM, anion exchange membrane.

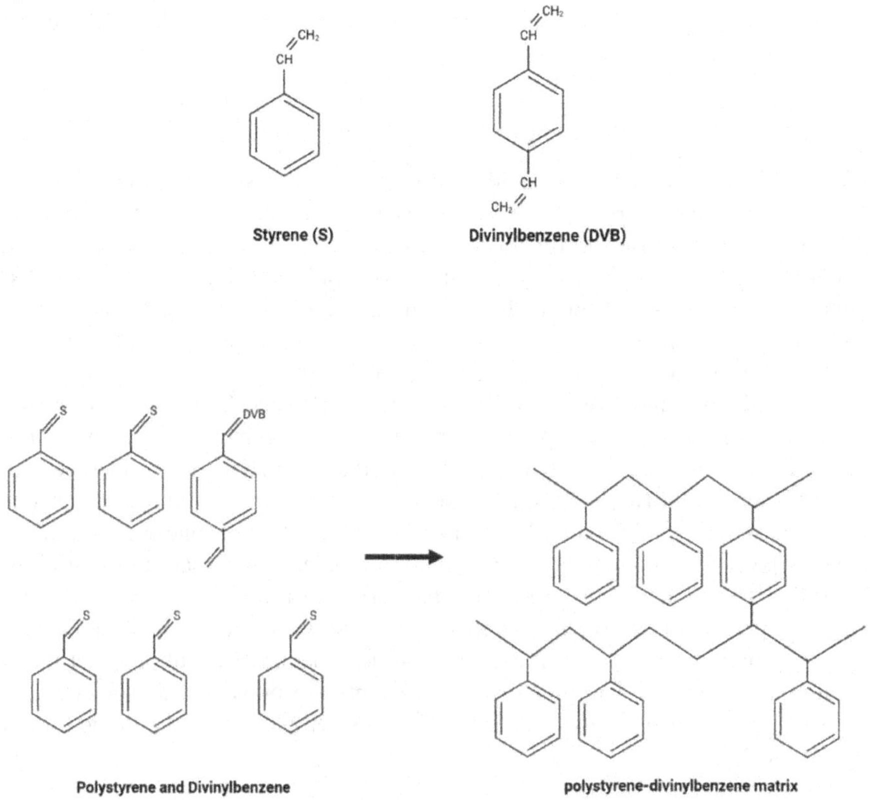

Styrene (S)
Divinylbenzene (DVB)
Polystyrene and Divinylbenzene
polystyrene-divinylbenzene matrix

FIGURE 14.5 Direct interspecies electron transfer.

14.4 VFA RECOVERY METHODS

The intermediate metabolites such as VFSs can be valuable and compete economically with the methane and carbon dioxide end products (Silva et al., 2013). The wide range of organic acids and other molecules that can be obtained during anaerobic digestion will depend on the raw material composition and operational conditions, which will control the equilibrium between different organisms and determine the final microbial community.

Regardless of the financial potential of VFA production from the anaerobic process, their extraction and recovery from the heterogeneous reactor mixture are still significant challenges. Existing conventional processes for the recovery of organic acids include stripping, electrodialysis, direct distillation, solvent extraction, evaporation, chromatographic methods, precipitation, liquid–liquid extraction ultrafiltration, and drying. Although some of the conventional techniques are simple to set up and have a high extraction yield but the usage of chemical solvents and the generation of solid wastes make them unsuitable and non-environment friendly. Thus, a viable solution would be the removal of the excess VFAs, and maintaining the ideal concentration inside the reactor, which leads to continuous reactor operation. Techniques such as membrane-based separation and IER adsorption will assist with the selective recovery of the desired VFA. In this study, we compare the aforementioned methods to determine the most effective one for enriching VFA recovery. The different techniques used to recover carboxylic acids are discussed in detail below.

14.4.1 LIQUID–LIQUID EXTRACTION

Solvent selection is crucial in liquid–liquid extraction. Also, it operates as one of the affinity separation procedures, and the characteristics of the extractant have a substantial impact on the function. Standard extractants such as alcohols and ketones have a weak ability to extract carboxylic acids because of their low distribution coefficients (Eda et al., 2017). Thus, extractants with higher distribution coefficients and greater efficiency such as amines or organo-phosphoric compounds are used (Mungma et al., 2019). The physical extraction technique uses a variety of diluents and offers efficient extraction outcomes. To reach equilibrium, the sample is kept at the appropriate temperatures in a mechanical stirring incubator for a period of time. Following incubation, the samples were provided certain time to reach equilibrium. Then, using separating funnel the organic and aqueous layers are separated (Sprakel et al., 2019). The residual VFA concentration in the aqueous layer was then analyzed, and the mass balance calculation was used to determine how much VFA had been transported to the organic layer. Another type of liquid–liquid extraction is reactive extraction, where the VFA can sometimes be changed to a different compound with a better distribution co-efficient for simple extraction. The most used extractants are phosphorous solvents and aliphatic amines. Nowadays, ionic liquids are also employed for the liquid–liquid extraction process.

14.4.2 Membrane-Based Separation

Recent investigations have demonstrated that membrane-based separation methods show better yield and selectivity of VFAs with less energy input. Membranes enable continuous VFA extraction, which controls the pH-independent solid retention time. Continuous VFA extraction is accomplished by membranes, which concomitantly adjusts the pH independent of solid retention period, which reduces the fermentation inhibition.

Many membrane-based technologies using various transport mechanisms are available to separate VFAs from the fermented waste stream, including reverse osmosis (RO) and nanofiltration (NF). Both NF and RO are well-established pressure-driven membrane technologies for downstream biotechnology as well as the chemical and water treatment industries (Bona et al., 2020). They are considered a feasible technique because of the less molecular weight and chemical characteristics of VFAs. The RO membrane consists of relatively minimal pore size for the process and benefits from high ion selectivity, which enables it to reject other ionic species. NF has larger pore size and surface charge, which makes it more advantageous than RO and is anticipated to enhance the selectivity (Zhu et al., 2021). In addition, NF has a low molecular weight cut-off ranging from 100 to 1,000 Da. Despite all of its merits, the membrane fouling that occurs in industrial-scale reactors is a drawback of the membrane-based bioreactors.

14.4.3 Electrodialysis

Electrodialysis is a crucial technology that enables carboxylic acid product recovery by application of electrical field–assisted ion transport toward electrodes for the optimum acid removal. It can result in high acid purity and a study refers to the possibility of removing up to 99% of acetic acid and butyric acid (Tartakovsky et al., 2011). However, one of the most limiting factors for using electrodialysis at an industrial scale is electricity consumption as it can result in high processing costs (Figure 14.4).

A new approach for VFA and nitrogen recovery is by combining the bio-electrochemical system with anaerobic digestion in order to improve its performance (Cerrillo et al., 2016). In microbial electrolysis cells, the microbes cling to the electrodes present inside the system and catalyze the oxidation/reduction reactions is one of the bio-electrochemical systems.

14.4.4 Adsorption

Adsorption is a surface phenomenon that leads to the transfer of molecule/ions (adsorbate) from gas or liquid phase to solid surface (adsorbent). The adsorption process can occur through physical forces or chemical bond formation between the adsorption medium and the adsorbate. This procedure is generally reversible, and desorption is the term used to describe the release of adsorbate from an adsorbent. Chemical compounds can be easily separated from either diluted or complicated solutions using the technique of adsorption. One of the most efficient and significant ways of recovering VFS produced during fermentation with the adsorption and desorption cycle is by using IERs.

Ion exchange for VFAs also relies on chemisorption, where the process is driven by the creation of ionic interactions between molecules of ionized carboxylic acid and the cationic functional group of the anion exchange resin.

The adsorption occurs between the oppositely charged carboxylic acid and the functional group present over the solid matrixes. It is a simple method to carry out and grants high selectivity (Reyhanitash et al., 2017). The aforementioned benefits of ion exchange adsorption make it an efficient technique to implement in industrial-level VFA extraction. The electronic properties of the adsorbents in ionic and non-ionic materials, as well as their functionality and support structure, can be used to categorize them in relation to the recovery of carboxylates. The common functional groups that operate as reactive sites in these adsorbents are primary, secondary, and tertiary amines and quaternary ammonium (López-Garzón et al., 2014). IERs are mostly generated from the polymerization reaction and consist of a uniform matrix.

14.5   ION EXCHANGE RESIN

The IERs are generated by the process of covalent bonding between the cross-linked polystyrene (PS) matrix and charged functional groups. Porous structure and a highly stable resin is achieved by cross-linking divinylbenzene (DVB) along with the PS. However, the higher cross-linkage will reduce the adsorption capacity and increase the contact time. Figure 14.6 shows the polymer matrix formation using polymerization of styrene (S) with DVB produces a cross-linked PS-DVB matrix.

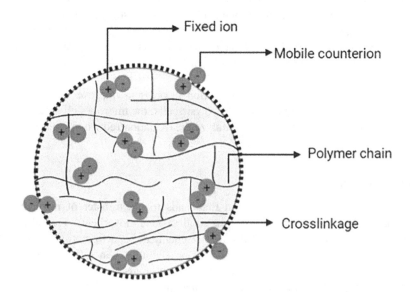

FIGURE 14.6   Copolymerization of styrene (S) with divinylbenzene (DVB) produces a cross-linked polystyrene-divinylbenzene matrix.

14.5.1 STRUCTURE OF RESIN

The structure of the host polymer must facilitate hydrated ion diffusion to confer ion exchange capabilities. Modern organic IERs are synthesized by polymerization mechanisms. Monomers with vinyl double bonds are used to create polymers; one of the monomers has two or more vinyl double bonds to enable cross-linking. Without cross-linking, the polymerization would result in linear polymers, and the ion exchange material would be soluble (Gerin et al., 2018).

There are two distinct structure types for resins: the first kind is a continuous polymeric phase–based gel with a uniform structure and nonporous. The second kind consists of a bead-like structure and is referred to as microporous beads. These beads are heterogeneous; they are made up of interconnected macropores surrounded by gel-like microbeads (Figure 14.7).

14.5.2 TYPES OF RESIN

Ion-active sites are dispersed evenly throughout the cross-linked polymer matrix that makes up an ion exchange structure. The IER has a spherical morphology with specific size and uniformity.

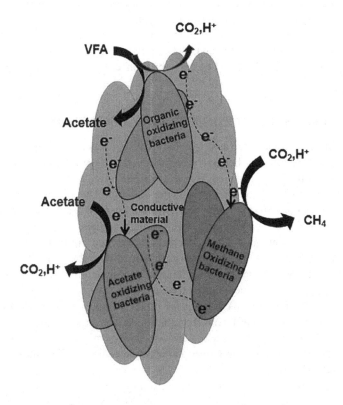

FIGURE 14.7 Schematic of Anion exchange resin.

14.5.2.1 Weak Base Resins

Weak base resins serve as acid adsorbents because they lack exchangeable ionic sites. These resins are easily regenerable with caustic soda and possess a high capacity for adsorbing strong acids (López-Garzón et al., 2014). The pH has a significant impact on this reaction cycle's adsorption capacity because a lower pH makes more protons accessible to permit ion pairing between the soluble acid and the amine.

Recovery of a more concentrated carboxylate solution results from the use of a smaller volume of a more concentrated desorbant. However, recovering carboxylic acid from them requires extra processes. As a result, mineral acids may be employed to desorb carboxylic acid rather than carboxylate. It is anticipated that mineral acids with lower pKa values than carboxylic acids will interact with resin more strongly.

14.5.2.2 Strong Base Resins

Strong base resins are functionalized with a quaternary amine, a permanently charged functional group, and contain negative counterion to maintain the electroneutrality (López-Garzónet al., 2014). When carboxylates are absorbed by ionic paring, the counterion (OH-) is released. When compared to other amine groups, the interaction between the quaternary amine and the carboxylate is stronger.

The quaternary ammonium is the most fundamental strong base resin commonly available and has a stronger affinity for weak acids. It is more stable and can tolerate applications involving high temperatures. Another type of resin is obtained by the reaction of the S-DVB copolymer with dimethylethanolamine (Wheaton and Lefevre, 2003). The regeneration efficiency is more in this type of resin.

14.6   ION EXCHANGE RESIN FOR VFA ADSORPTION

Effective carboxylic acid recovery is achieved using anion exchange resins. Since the weak anion exchangers become charged over a narrow window of pH range, the capacity to function is reduced. However, strong anion exchangers perform their function over a wide range of pH. Out of different types of adsorbents present according to their functional groups, quaternary ammonium-based adsorbents come under the strong basic adsorbents category. Even though a quaternary ammonium-based adsorbent seems advantageous for recovering carboxylate from fermented waste, the replacement anion of the ammonium was exchanged with the mineral acid (Cerrillo et al., 2016). The VFA obtained after adsorbent regeneration must go through an additional chemical and processing stage, and it is not pure and is present in an aqueous solution with a sizable amount of mineral impurities. PS-DVB resins are used in recent days for an efficient process, which was either nonfunctionalized or functionalized with a primary, secondary, or tertiary amine. Characteristics of certain PS-DVBs like Lewatit, Amberlyst, Dowex Amberlite, and Purolite are shown in Table 14.1 (Rizzioli et al., 2021).

To maintain the neutral charge anion exchange, resins are preferred as their amine group will adsorb the organic acids. A list of various commercially available anion exchange resins for VFA adsorption are listed in Table 14.2. Dowex

TABLE 14.1
Characterization of Ion Exchange Resins and Carbon-Based Adsorbents

Description	Lewatit	Amberlyst	Amberlite	Purolite	Dowex	GAC	PAC
Chemical composition	Styrene-divinyl benzene primary amine (benzyl amine)	Styrene-divinyl benzene tertiary amine (benzyl amine)	Styrene-divinyl benzene (weak base anion exchange resin)	Macroporous polystyrene cross-linked with divinyl benzene (tertiary amine functionalized resin)	Micro porous copolymer of styrene and divinylbenzene	carbon	carbon
Particle size (μm)	0.47–0.57	0.49–0.69	0.3–1.18	0.42–1.2	0.2–0.3	0.4–1.2	0.001–0.15
Approx. pore volume (cm^3/g)	0.27	0.10	0.10	0.6	0.72–0.79	0.97	0.65
Approx. surface area (m^2/g)	0.005	0.0035	35	450	1,048	950–2,000	0.000012

TABLE 14.2
Commercially Available Anion Exchange Resin for VFA Adsorption

Resin Name	Matrix	Type	Particle Size (µm)	Capacity (meq/mL)	Density (g/mL)	Active Group
Dowex 1×8–100	Styrene-divinyl benzene	Strongly basic	50–100	1.7	3.325	Trimethyl benzyl ammonium
Dowex 21K XLT	Styrene-divinyl benzene	Strongly basic	525–625	1.4	1.08	Trimethyl benzyl ammonium
Reillex 425	Poly(4-inylpyridine), 25% cross-linked with divinylbenzene	-	400–800	-	0.65	Pyridine
Dowex marathon A2	Styrene-divinyl benzene	Weakly basic	500–600	1.2	1.09	Dimethyl-2-hydroxyethyl benzyl ammonium,
Amberlite IRA-67 free base	Styrene-divinylbenzene	Weakly basic	651	1.70	1.60	Polyamine
Amberlite IRA-400	Styrene-divinyl benzene	Strongly basic	330–900	2.6–3	--	Trialkyl benzyl ammonium
Amberlite IRA-96 free base	Styrene-divinylbenzene	Weakly basic	550–750	1.25	--	Polyamine
Amberlite IRA-900	Styrene-divinylbenzene copolymer	Strongly basic	650–820	-	1.06	Trialkyl benzyl ammonium
Dowex Optipore L-493	Polystyrene-divinylbenzene (PS-DVB)	-	550–900	-	0.68	Polymer
Diaion HP-20	Styrene-divinylbenzene	-	250	1.01	-	None

21K XLT, Reillex 425, Dowex IER, Amberlite IRA-400, and Amberlite IRA-900 showed about <30% efficiency, whereas the resins Amberlite IRA-67 free base, Amberlite IRA-96 free base, and Dowex Optipore L-493 showed adsorption efficiencies >50%. The length of the VFA's aliphatic chain affects the percentage adsorption. Based on the studies, Dowex's VFA adsorption was dependent on the length of the aliphatic chain, while Amberlite demonstrated the highest adsorption for all VFA (Eregowda et al., 2019).

To encourage VFA consumption during anaerobic digestion, carbon-based conductive materials are currently exploited. These materials enhance the syntropic interactions between microorganisms by encouraging the growth of biofilms on their surface (Capson-Tojo et al., 2018). Activated carbon (granular and powdered) has been extensively developed in favor of reactor performance by utilizing VFA at a faster rate which leads to quicker methane production. The potent benefit of activated carbon is the high-surface-area and porous structure. In addition, activated carbon has high electrical conductivity and enhances direct interspecies electron transfer (DIET) as shown in Figure 14.6. In DIET, there will be a free flow of electrons without being carried by reduced molecules unlike MIET (mediated interspecies electron transfer). The electrons are transferred with the aid of conductive materials between electron-donating and -accepting partners (Capson-Tojo et al., 2018). DIET is anticipated to be a more rapid and efficient electron transfer mechanism than MIET. Thus, the application of activated carbon materials favors DIET and avoids VFA accumulation during the anaerobic digestion process (Table 14.3).

TABLE 14.3
Different Aspects of VFA Adsorption Reported in Different Studies

Adsorption Resin	Adsorption Medium	Adsorbed Compound	Objectives	Results
Amberlite IRA-67	Fermentation broth/wastewater streams	Acetic and glycolic acids	Efficiency of the adsorbent at different initial concentrations and contact time to remove monocarboxylic acids from waste stream	At maximum resin dosage of 2 g, the acetic acid and glycolic acid adsorption capacity of 86.29% and 61.36% is obtained
Tuision A-8×MP, Indion 810 and Indion 850	Organic solvents (mixture of ethyl acetate and ethanol and dark fermentation effluent)	Acetic acid	To remove acetic acid impurities from organic solvents by mechanism of sorption reinforced specific interaction	Equilibrium studies revealed selective adsorption of acetic acid with high loading capacity

(Continued)

TABLE 14.3 (*Continued*)
Different Aspects of VFA Adsorption Reported in Different Studies

Adsorption Resin	Adsorption Medium	Adsorbed Compound	Objectives	Results
Purolite A133S and granular activated carbon	Organic solvents	Acetic, propionic, and butyric acids	Efficiency of activated carbon and an ion exchange resin as adsorbents for organic acids To evaluate the thermodynamic parameters to comprehend the mechanism involved	Resin showed about 35% higher adsorption capacity when compared with the carbon adsorbent and 64% of propionic acid recovery is achieved by optimized combination of resin and solvent
Amberlite IRA-67 and activated carbon	Dark fermentation broth of food waste	Acetic, lactic, and butyric acids	Adsorbent dosage and the pH on the recovery of carboxylic acid in batch mode from a dark fermentation broth of food waste	It is reported that when pH is below their pKa of carboxylic acids, it is advantageous for both the adsorbents The carboxylic acid removal was about 74% and 63% from the resin and activated carbon, respectively.
Sepra NH2, Amberlyst A21, Sepra SAX, Sepra ZT-SAX	Liquid effluent of grape pomace (GP) acidogenic anaerobic digestion	Acetic, propionic, butyric, isovaleric, valeric, and caproic acids	Adsorption mechanism using four different resins and subsequent desorption techniques. Also, predicting the ion exchange and adsorption capacity of the resin using an ion exchange model	Ion exchange performance increased with increasing chain length of VFA. Amberlyst has better efficiency. Better recovery was obtained at pH 3–4.5 in the presence of acetic acid, and at pH 6.5 for GP digestate

Sources: Hasan Uslu et al. (2010), Anasthas et al. (2001), Ahasa Yousuf et al. (2016), Stefano Rebecchi et al. (2016), and Alan Henrique da Silva (2013).

14.7   ADSORPTION ISOTHERMS AND KINETICS

The distribution of VFAs on the surface of the adsorbent can be described and characterized through isotherm models. The VFA adsorption process can be elucidated using Langmuir and Freundlich models. The Langmuir adsorption isotherm is based on the assumption that each discrete adsorption site on an adsorbent surface can retain only one adsorbate. The adsorption sites available on the adsorbent surface have equal energy and adsorbate forms a monolayer on the adsorbent surface. The Freundlich adsorption isotherm model describes the empirical relationship between pressure at a given temperature and the amount of adsorbate that is absorbed by a unit mass of adsorbent. In addition, it provides an exponential distribution of adsorption sites and energy and explains the surface heterogeneity of adsorbents. This isotherm model assumes that multiple molecular layers of adsorbate will form on the adsorbent surface (Eregowda et al., 2019).

An important factor in any adsorption process is the impact of contact time and concentration of resins. The total number of sites that are available for exchange makes up the total capacity of the resin. Kinetics also determines the speed at which the ion exchange takes place. This speed is related to the diffusion through the film of solution in close contact with the beads called the Nernst film, and diffusion in the porosity of the resin. The intra-particle mass transfer diffusion model is commonly used to describe diffusion-controlled processes.

14.8   DESORPTION AND PURIFICATION TECHNIQUES

Regeneration of the loaded adsorbents is necessary for product recovery (Reyhanitash et al., 2017). Certain techniques involved in desorption techniques are solvent wash. It is required to recover the organic acids that have been adsorbed in the IER. The desorbants can be water or alkali solution. Using such desorbants has the disadvantage of increasing the water content of the resultant stream. Thermal operation is a significant alternative method. In order to condense and recover the VFAs in the vapor phase in a concentrated form, thermal desorption employs the application of heat. The condensate samples were analyzed by using high-performance liquid chromatography (HPLC) or gas chromatography (GC), which represents evaporated species that were condensed in the condenser. Desorption can also be done using organic solvents like methanol, and ethanol directly, as the hydroxide ion replaces the carboxylic acid in the ion-active site of the resin.

Either distillation or crystallization, which often includes the evaporation of water, is required for final VFA purification. Distillation is a method of separating a mixture based on differences in volatility of the components (Gerbaud et al., 2019). Even if there are other methods for separating organic acids from fermentation broths, distillation is still a significant alternative technology, particularly in the last stages of refinement. Water will evaporate first during distillation of a VFA–water mixture because all VFAs have boiling points greater than water (Yalkowsky et al., 2010). Water has a high latent heat of vaporization, so distillation of VFAs is only economically attractive as a final step of purification, after the VFAs in aqueous phase have already been concentrated (Gangadwala et al., 2003). Figure 14.7 depicts the adsorption–desorption process to regulate the VFA concentration using IER.

14.9 CONCLUSION

According to the Environmental Performance Index, 2022, India gets the lowest ranking among 180 countries. Greater attention is needed to achieve the spectrum of sustainable living by maintaining air and water quality, preserving natural resources, and reducing greenhouse gas emissions. Management of solid wastes is vital since it continues to increase as population and industrialization increase. The majority of solid wastes are rich sources of organic material, making them significant resources that can be exploited to produce energy. In recent decades, the generation of biogas using anaerobic digestion with organic-rich substrate is becoming a popular practice worldwide.

Anaerobic digestion is a complex biological process thus, when dealing with highly concentrated and rapidly hydrolyzable organic wastes, it is susceptible to failure if not properly regulated. The reactor can readily get overloaded due to the easily digestible carbohydrates that make up most of food waste, leading to excessive build-up of the inhibitory intermediate product VFA. Methanogens, being the most delicate organism, are unable to tolerate the sudden drop in pH and will ultimately fail to produce biogas. Recovering the VFA enables precise pH regulation and maintains the stability of the process, subsequently enhancing the methanogenesis which leads to high biomethane production. The IER and carbon-based adsorbents are used for the recovery of carboxylic acids. Since the resins made from PS and DVB exhibit very high selectivity for VFAs, regeneration of these resins showed that they are remarkably robust during adsorption and desorption cycles. Ion exchange resins can produce high-purity biobased VFA in general. The adsorbent can be used in the continuous cycle of the adsorption–desorption process without losing its capacity for a longer period. Thus, biorefinery can be established by practicing the management of organic wastes, an excellent method to mitigate the adverse effects of disposing the wastes and concomitant resource recovery along with generation of VFAs, a value-added product which benefits both society and the environment positively.

REFERENCES

Anasthas, H.M., V.G. Gaikar (2001). Adsorption of acetic acid on ion-exchange resins in non-aqueous conditions. *Reactive & Functional Polymers*. 47, 23

Begum, Sameena, Vijayalakshmi Arelli, Gangagni Rao Anupoju, S. Sridhar, Suresh K. Bhargava, Nicky Eshtiaghi (2020). Optimization of feed and extractant concentration for the liquid-liquid extraction of volatile fatty acids from synthetic solution and landfill leachate. *Journal of Industrial and Engineering Chemistry*. doi:1016/j.jiec.2020.07.011

Bhatia, Shashi Kant and Yung-Hun Yang (2017). Microbial production of volatile fatty acids: Current status and future perspectives. *Reviews in Environmental Science and Bio/Technology*. doi:1007/s11157-017-9431-4

Bona, Aron, Peter Baknoyi, Ildiko Galambos (2020). Separation of volatile fatty acids from model anaerobic effluents using various membrane technologies. *Membranes*. 10, 252.

Capson-Tojo, Gabriel, Roman Moscoviz, Diane Ruiz (2018). Addition of granular activated carbon and trace elements to favour volatile fatty acid consumption during anaerobic digestion of food waste. Bioresource Technology. doi:10.1016/j.biortech.2018.03.097

Cerrillo, Miriam, Marc Vinas and Agust Bonmati (2016). Removal of volatile fatty acids and ammonia recovery from unstable anaerobic digesters with a microbial electrolysis cell. *Bioresource Technology*. doi:10.1016/j.biortech.2016.07.103

Chaganti, S.R., D.-H. Kim, J.A. Lalman, 2011. Flux balance analysis of mixed anaerobic microbial communities: Effects of linoleic acid (LA) and pH on biohydrogen production. *International Journal of Hydrogen Energy*. 36 (21), 14141–14152. doi:10.1016/j.ijhydene.2011.04.161

Chakraborty, Debkumar, Sankar Ganesh Palani, M.M. Ghangrekar, N. Anand, Pankaj Pathak (2022). Dual role of grass clippings as buffering agent and biomass during anaerobic co-digestion with food waste. *Clean Technologies and Environmental Policy*. doi:1016/j.biortech.2021.126396

da Silva, Alan Henrique and Everson Alves Miranda (2013). Adsorption/desorption of organic acids onto different adsorbents for their recovery from fermentation broths. *Journal of Chemical Engineering and Data*. doi:10.1021/je3008759

Dennehy, Conor, Peadar G. Lawlor, Zhenhu Hu, Matthew McCabe, Paul Cormicand, Xinmin Zhan (2018). Inhibition of volatile fatty acids on methane production kinetics during dry co-digestion of food waste and pig manure. *Waste Management*. doi:1016/j.wasman.2018.07.049

Eda, Sumalatha, Alka Kumari, Prathap Kumar Thella, Bankupalli Satyavathi, Parthasarathy Rajarathinam (2017). Recovery of volatile fatty acids by reactive extraction using tri-n-octylamine and tri-butyl phosphate in different solvents: Equilibrium studies, pH and temperature effect, and optimization using multivariate taguchi approach. *The Canadian Journal of Chemical Engineering*. doi:1002/cjce.22803

Eregowda, Tejaswini, Eldon R. Rene, Jukka Rintala, Piet N.L. Lens (2019). Volatile fatty acid adsorption on anion exchange resins: Kinetics and selective recovery of acetic acid. *Separation Science and Technology*. doi:1080/01496395.2019.1600553

Eregowda, Tejaswini, Eldon R. Rene, Jukka Rintala, Piet N.L. Lens (2019). Volatile fatty acid adsorption on anion exchange resins: Kinetics and selective recovery of acetic acid. doi:1080/01496395.2019.1600553

FAO (2013). Food Wastage Footprint Impact on Natural Resources: Summary Report. Food and Agricultural Organization.

Food Material-Specific Data (2018). *EPA United States Environmental Protection Agencies*. Food Material-Specific Data.

Gangadwala, Jignesh, Surendra Mankar, Sanjay Madhusudan Mahajani (2003). Esterification of acetic acid with butanol in the presence of ion-exchange resins as catalysts. *Industrial & Engineering Chemistry Research*. doi:10.1021/ie0204989

Gerbaud, Vincent, Ivonne Rodriguez-Donis, Laszlo Hegely, Peter Lang, Ferenc Denesd, Xin Qiang You (2019). Review of extractive distillation: Process design, operation, optimization and control. *Chemical Engineering Research and Design*. doi:1016/j.cherd.2018.09.020

Gerin, Patrick, Guillaume Castel (2018). *Extraction and Purification of Volatile Fatty Acids*. UCL.

Keshav, A., K.L. Wasewar, S. Chand (2008). *Extraction of Propionic Acid with Tri-N-Octyl Amine in Different Diluents*: Separation and Purification Technology. doi:10.1016/j.seppur.2008.04.012

Keshav, A., S. Chand, K.L. Wasewar (2009). Recovery of propionic acid from aqueous phase by reactive extraction using quarternary amine (Aliquat 336) in various diluents. *Chemical Engineering Journal*. doi:1016/j.cej.2009.03.037

Lalak, Justyna, Agnieszka Kasprzycka, Ewelina M. Paprota, Jerzy Tys, Aleksandra Murat (2015). Development of optimum substrate compositions in the methane fermentation process. *International Agrophysics*. 29, 313.

López-Garzón, S. Camilo, Adrie J.J. Straathof (2014). Recovery of carboxylic acids produced by fermentation. *Biotechnology Advances*. doi:10.1016/j.biotechadv.2014.04.002

Mungma, Nuttakul, Marlene Kienberger, Matthäus Siebenhofer (2019). Reactive extraction of lactic acid, formic acid and acetic acid from aqueous solutions with tri-n-octylamine/1-octanol/n-undecane. *ChemEngineering*. doi:10.3390/chemengineering3020043

Opatokun, Suraj Adebayo, Vladimir Strezov, Tao Kan (2015). Product-based evaluation of pyrolysis of food waste and its digestate. *Energy.* doi:10.1016/j.energy.2015.02.098

Rebecchi, Stefano, Davide Pinelli, Lorenzo Bertin, Fabiana Zama, Fabio Fava, Dario Frascari (2016). Volatile fatty acids recovery from the effluent of an acidogenic digestion process fed with grape pomace by adsorption on ion exchange resins. *Chemical Engineering Journal.* doi:10.1016/j.cej.2016.07.101

Reyhanitash, Esan, Sascha R.A. Kersten and Boelo Schuur (2017). Recovery of volatile fatty acids from fermented wastewater by adsorption. *ACS Sustainable Chemistry and Engineering.* doi:10.1021/acssuschemeng.7b02095

Rizzioli, Fabia, Federico Battista, David Bolzonella and Nicola Frison (2021). Volatile fatty acid recovery from anaerobic fermentate: Focussing on adsorption and desorption performances article. *Industrial and Engineering Chemistry Research.* doi:1021/acs. iecr.1c03280

Russell, James B., Tsuneo Hino (1985). Regulation of lactate production in *Streptococcus bovis*: A spiraling effect that contributes to rumen acidosis. *Journal of Dairy Science.* doi:10.3168/jds.s0022-0302(85)81017-1

Sprakel, L.M.J., B. Schuur (2019). Solvent developments for liquid-liquid extraction of carboxylic acids in perspective. *Separation and Purification Technology.* doi:1016/j. seppur.2018.10.023

Tartakovsky, B., P. Mehta, J.-S. Bourque, S.R. Guiot (2011). Electrolysis-enhanced anaerobic digestion of wastewater. Bioresource Technology. doi:10.1016/j.biortech.2011.02.097

Uslu, Hasan, Ismail Inci, Sahika Sena Bayazit (2010). Adsorption equilibrium data for acetic acid and glycolic acid onto amberlite IRA-67. *Journal of Chemical Engineering Journal.* 55, 1295.

Vishnu, C., B.J. Naveena, Md. Altaf M. Venkateshwar, Gopal Reddy (2006). Amylopullulanase-A novel enzyme of *L. amylophilus* GV6 in direct fermentation of starch to L(+) lactic acid. *Enzyme and Microbial Technology.* doi:1016/j.enzmictec.2005.07.010

Ward, Alastair J., Phil J. Hobbs, Peter J. Holliman, David L. Jones (2008). Optimisation of the anaerobic digestion of agricultural resources. *Bioresource Technology.* 97, 2166

Wheaton, R.M., L.J. Lefevre (2003). *Fundamental of Ion Exchange DOWEX Ion Exchange Resins.* Academia.

Yalkowsky, Samuel H., Yan He, Parijat Jain (2010). *Handbook of Aqueous Solubility Data,* 2nd edition. CRC Press

Yousuf, Ahasa, Fabian Bonk, Juan-Rodrigo Bastidas-Oyanedel, Jens Ejbye Schmidt (2016). Recovery of carboxylic acids produced during dark fermentation of food waste by adsorption on Amberlite IRA-67 and activated carbon. *Bioresource Technology.* doi:10.1016/j.biortech.2016.02.035

Zhang, Lei, Yong-Woo Lee, Deokjin Jahng (2011). Anaerobic co-digestion of food waste and piggery wastewater: Focusing on the role of trace elements. *Bioresource Technology.* doi:10.1016/j.biortech.2011.01.082

Zhang, Wanqin, Shubiao Wu, Jianbin Guo, Jie Zhou, Renjie Dong (2015). Performance and kinetic evaluation of semi-continuously fed anaerobic digesters treating food waste: Role of trace elements. *Bioresource Technology.* doi:10.1016/j.biortech.2014.08.046

Zhou, M., B. Yan, J.W. Wong, Y. Zhang, 2018. Enhanced volatile fatty acids production from anaerobic fermentation of food waste: A mini-review focusing on acidogenic metabolic pathways. Bioresource Technology. doi:1016/j.biortech.2017.06.121

Zhu, Xiaobo, Aaron Leininger, David Jassby, Nicolas Tsesmetzis, Zhiyong Jason Ren (2021). Will membranes break barriers on volatile fatty acid recovery from anaerobic digestion? *ACS ES&T Engineering.* doi:10.1021/acsestengg.0c00081

15 Torrefaction of Agriculture Residues and Municipal Solid Waste for Char Production

Sugali Chandra Sekhar, Bukke Vani,
and Sridhar Sundergopal
CSIR-Indian Institute of Chemical Technology
Academy of Scientific and Innovative Research (AcSIR)

15.1 INTRODUCTION

Torrefaction, a process that uses heat for transforming biomass as well as other combustible substances within a range of temperatures of 200-300 C in the absence of air [1], can be utilized to enhance the characteristics of MSW to be suitable for the combustion process. MSW is produced predominantly as a consequence of population increase, growth in the economy, and industrialization. This rising volume of MSW has given rise to significant issues with human health, the environment, exhausted natural resources, and social problems [2,3]. Biomass is a renewable energy source in addition to fossil fuels, which may be further transformed into fuels and chemicals. Such assets are constrained, though, and a 50-year shortfall has been projected [4–9]. In addition to the possibility of scarcity, fossil fuels have a significant detrimental influence on the surroundings.

Consequently, lowering the emission of carbon dioxide (CO_2), which is among the primary substances that contribute to causing global warming (greenhouse gases), by means of the implementation of energy from renewable sources constitutes the primary target, with goals of lowering greenhouse gases (GHG) emissions from 1990 to 2030 by 40% and to decrease GHG emission levels by 80%–95% by 2050 as the target date [9–13].

A mild thermochemical technique called torrefaction is employed to transform biomass into high-quality solid fuel. It is carried out in air circumstances without oxygen at a temperature ranging from 200°C to 300°C. Biomass characteristics are altered while torrefaction provides fuel with significantly higher quality for combustion and gasification purposes.

In the context of torrefaction, a hemicellulose portion, the largest and most active component that constitutes the biomass, undergoes breakdown followed by deterioration of the cellulose and lignin components to a lesser extent [6,13]. Torrefaction is a

DOI: 10.1201/9781003406501-15

slow, mild pyrolysis that takes place around 200°C and 350°C. To prevent feedstock oxidation and combustion, the process is typically run at ambient pressure with an atmosphere of inertia [14]. The total duration of the time can range from a few minutes and several hours.

The process of torrefaction is one of many methods for enhancing biomass and altering its chemical as well as physical constitution. This procedure involves gradually heating biomass at temperatures from 200°C to 300°C in a controlled environment without air (O) [15]. Biomass productivity is improved by torrefaction during co-combustion and gasification [16–18].

The outcome of the process is an evenly distributed solid substance of a dark-colored (torrefied biomass). Water, organics, and lipids make up the majority of the condensable gases, while carbon dioxide and carbon monoxide make up the majority of the non-condensable portions.

At the moment, there currently have been multiple investigations concerning the impact of various processes of tor factors on the end-product chemical as well as the physical substance of biomass to the generation of energy [19–23].

The generated power is capable of being utilized straight away; however, it may also be transformed into a supplementary resource of energy by means of a series of thermal and biochemical reactions, which include the torrefaction, gasification process, pyrolysis, anaerobic digestion, liquefaction, fermentation, and trans-esterification [22–24].

The aforementioned torrefaction procedure begins with the evaporation of moisture, which is followed by a partial de-volatilization. The char, which is known as the primary product, possesses a significantly greater amount of energy compared with the feedstock being used. Based on a number of factors, including the processing temperatures, the process of tor can be categorized into two categories: light (below 240°C) and severe (above 270°C) torrefaction [25]. Among the benefits of torrefaction are a rise in energy volume, enhanced grind ability, decreased level of moisture, and a lower vulnerability to microbial breakdown. The char is suitable for use to produce premium fuel in many different kinds of uses, such as small- to medium-combustion facilities, entrained flow gasification, and combustion in power stations [26]. It may be additionally utilized for in situ remediation of soil as well as an adsorbent that is used for the purification of water [27].

In comparison to the available research on gasification and pyrolysis, there are comparatively few published investigations on torrefaction techniques for MSW management. Studies have largely been performed to examine both the chemical and physical characteristics of torrefied MSW [28–31]. The various kinds of resources utilized as feedstock include Food waste (FW), Polyvinyl chloride (PVC) plastic, discarded tires, and wood residues. It is well-accepted that one of the key elements in torrefaction is temperature and it is known as the temperature rises, char production decreases [32,33].

When the consumption of electricity is rising and there are demands for waste utilization and reuse and recycling, waste conversation for power generation offers an efficient method of reusing. Making fuels using waste and sending it for later thermal recovery or repurposing is a possible technique to maximize the amount of heat and power that may be generated through waste.

Refuse-derived fuel is a recyclable material that has undergone mechanical processing to become combustible municipal and industrial waste, primarily composed of plastic, paper, and wood. The most common uses for derived fuel from waste materials are to generate heat and electricity.

As a consequence during the torrefaction procedure, a pair of items are found: Biochar and torrefaction gas via a weight-to-volume ratio of 70%–80% and 23%–30%, respectively. The substance is known as bio. When biomass from forestry or agriculture is utilized as a base for another substrate, the carbon is known as carbonate or biochar. The gas substance is known as tor gas.

The mechanical characteristics, chemical makeup, skeletal structure, and reactivity alterations of torrefied biomass have all been extensively studied in recent years. A significant amount of the moisture and low-weight volatiles in the biomass are also removed during torrefaction, which additionally enhances the feedstock's grind ability, material handling capabilities, bulk density, and energy density.

In comparison with alternative dumping methods, incinerating waste from municipalities has been demonstrated to offer numerous benefits, including the possibility of energy production and a substantial decrease in waste quantity and mass over the years.

Biomass serves as a flexible resource for energy which could potentially be utilized to generate environmentally friendly energy sources in solid, liquid, and gaseous forms in addition to other energy resources. The capacity to lessen the main drawbacks of biomass, such as variability, lower bulk density, reduced energy density, hygroscopic behavior, and fibrous character, is provided by the thermal pretreatment technique known as pyrolysis. Torrefaction is a process that uses high temperatures, an inert atmosphere, and air pressure to create high-quality solid biomass products. The fundamental idea behind the torrefaction process is the elimination of volatile s through various decomposition events. Torrefaction improves biomass quality and modifies combustion behavior, allowing for effective usage in co-firing power plants.

This article provides a thorough analysis of biomass torrefaction and its properties. Torrefaction is mostly used for thermochemical conversion processes since it can make biomass more hydrophobic, grindable, and energy-dense despite its many benefits. Torrefied biomass might also be pushed as a charcoal substitute and utilized in place of coal in the metallurgical process. Materials other than lignocellulose can also be considered biomass. The plant's starch and fiber portion, which is mostly made up of cellulose as well as hemicelluloses and lignin, is followed by the lignocellulose elements. To prevent potential harm from global warming and climate change, new and environmentally friendly sources of energy are being investigated in light of the rising energy demand and expanding environmental challenges.

Biomass is extensively utilized throughout the globe and can be affordable to purchase. Thus, it might develop into a significant source of renewable energy for the entire planet. Although it can produce biofuels such as biodiesel, methanol, and hydrogen through the Fischer–Tropsch Synthesis process, biomass could potentially substitute fossil fuels in both the transportation and power generation sectors. The use of biomass for the production of energy has so far been constrained due to several of its intrinsic constraints, such as low energy density, fibrous character, and hygroscopic nature. Currently, co-firing technology has acquired widespread recognition for lowering the utilization of fossil fuels and accompanying emissions in thermal power plants by substituting biomass for a portion of the fossil fuel.

15.2 EXPERIMENTS

15.2.1 SAMPLES AND MATERIALS

The MSW and the agriculture residue waste were collected from the local area in Hyderabad, India. The agriculture waste contains sugarcane biogases and sweet corn and the municipal waste contains plastic covers, cloths, some organic materials, etc. After collection samples were dried up to 6–7 days before the experiments. The sample was pulverized or ground or cut into a small particle size of 3–4 mm, which determined the gross calorific value (GCV) for raw feed samples and recorded it.

15.2.2 PROCEDURE

Feed samples of municipal waste and agriculture residue biomass were experimented with within the reactor, the reactor was made up of stainless steel material with a diameter of 38 mm and height of 600 mm the reactor was in close contact with an electrical preheater which covered the reactor temperature was controlled by varying the heating rate using a proportional-integral-derivative (PID) controller to the electrical preheater. The preheater consists of thermocouples T_1, $T2$, and $T3$ outside to reactor, and T_a and T_b are inside the reactor. Thermocouples are used for recording the temperature which is displayed on the control panel.

The reactor was filled with a known mass of feed samples (municipal waste or agriculture residue biomass) up to the heating zone. The remaining space is filled with porcelain beads to ensure complete packing (Figure 15.1).

A wire mesh was placed at the bottom to support the feed sample and beads. A nitrogen cylinder was placed beside the electrical preheater, and it connects to the reactor nozzle at the top side, and rotameter was also required to control or regulate the flow rate of nitrogen (0–0.2 bar) which is placed in between the reactor nozzle and valve of the cylinder. The nitrogen gas was spurge through the reactor to ensure an inert atmosphere inside the reactor by displacing air and nonvolatile gases from the reactor. The outlet of the reactor was connected to a bubbler or condenser, which is placed in a water bath or in cooling water to condense the vapors which consist of condensable and non-condensable gases liberated when the torrefaction process goes on. The liquid product is collected in the bubbler from which water and very little bio-oil phases were observed. The outlet was connected to a water scrubber. The main aim of the water scrubber was to remove SO_x, NO_x, and particulate matter before vapor is vented off into the air.

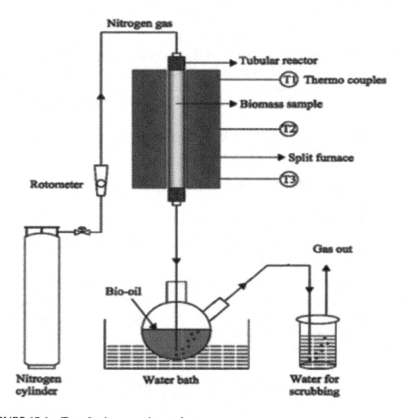

FIGURE 15.1 Torrefaction experimental setup.

During the process, heating of the feed sample of municipal waste and biomass was started at ambient temperature and continued at a slow heating rate of 20°C/min, until the required temperature of 200°C, 250°C, 300°C, and 330°C as per the experiment. The temperature was maintained for a constant retention time of 60 min. After completion of the experiment, the heating was stopped and the temperature of the reactor was allowed to decrease. While maintaining, the nitrogen flow didn't stop until the reactor temperature reached below 50°C. The product was collected in the sample covers or box and was sent for proximate and ultimate analysis.

15.3 METHODOLOGY

15.3.1 DETERMINATION OF GROSS CALORIFIC VALUE

The techniques of testing for coal and coke relevant to the calculation of calorific value are prescribed by IS: 1350 (Part II)-1970. The calorific value (CV) of coal can be determined using one of two methods. Using a calorimetric bomb submerged in an isothermal or static water jacket is one, and using a calorimetric bomb submerged in an adiabatic jacket is another.

The GCV is defined as the number of heat units released when a unit mass of fuel is burned at a constant volume in oxygen that has been saturated with water vapor, with both the initial substance and the finished result being at a temperature of 250°C. The residual products are taken as CO_2, SO_2, N_2, and H_2O. Combustion processes take place in a calorimeter under defined conditions. The operating mode is based on the isoperibolic principle at 250°C and a cooling water temperature of 12°C–230°C.

Equilibrated GCV is calculated from the ratio of equilibrated moisture and air-dried moisture, CV.

$$GCV = CV \left(100 - \text{equilibrated moisture } 60\% \text{ RH at } 400°C / 100 - \text{Air dried moisture}\right)$$

CV = Calorific value of the sample by bomb calorimeter

15.3.2 Determination of Ash

Ash content, one of the important environmental concerns, can be determined by weighing the residue remaining after burning a sample of coal under controlled conditions. A difficulty may be experienced in securing satisfactory determinations of ash for coals unusually high in calcite and pyrite.

Calculated the percent of ash in the analysis sample as follows:

$$\% \text{ ash} = \left(\frac{M3 - M4}{M2 - M1}\right) \times 100$$

where $M1$ = mass of the empty crucible in grams; $M2$ = mass of crucible and sample in grams; $M3$ = mass of crucible and ash in grams; and $M4$ = mass of the crucible after brushing out the ash and re-weighing in grams.

15.3.3 Determination of Moisture

This is an indirect method based on IS: 1350 (Part I)-1984 (second revision), in which, a known mass of the material is dried and the loss of mass is calculated as moisture. The moisture may be determined either by drying in one stage at 1,080°C±20°C or by a two-stage process in which the coal is first air-dried under atmospheric conditions and the remaining moisture removed by drying in an oven at 1,080°C±20°C. In the latter case, the total moisture is calculated from the loss during air-drying and that during oven-drying. Here we adopted the former case, in which, about a gram of sample is weighed and dried in an oven at 1,080°C for 1 h. The weight lost is equal to the amount of moisture present in the sample.

Calculated the percent of moisture in the analysis sample as follows:

$$\% \text{ Moisture} = \left(\frac{M2 - M3}{M2 - M1}\right) \times 100$$

where $M1$, $M2$, and $M3$ represent the mass of the empty crucible with lid, and the crucible with sample before and after heating, respectively.

15.3.4 DETERMINATION OF VOLATILE MATTER

Fresh coal sample is weighed, transferred to an enclosed crucible, and heated in a furnace at 900°C + 15°C. Refer to IS 1350 part I: 1984, parts III and IV for the techniques, which include those for carbon and ash. After cooling, the sample is measured. Moisture and volatile substances are represented by a weight loss of zero. The rest is coke, which is fixed carbon and ash. Methane, hydrocarbons, hydrogen, carbon monoxide, and non-combustible gases such as carbon dioxide and nitrogen found in coal are examples of volatile substances. The presence of gaseous fuels is, therefore, indicated by the volatile substance. Volatile matter usually varies from 20% to 35%. A volatile substance proportionally extends the flame and makes it simpler for coal to ignite. The minimum reduction in the furnace height and volume influences secondary air distribution and need factors has an impact on secondary oil support

$$\% \text{ of } VM = (100 \times (M2 - M3) / (M2 - M1)) - \text{Moisture in } \%$$

where $M1$, $M2$, and $M3$ represent the weight of the empty crucible without the sample, with the sample, and after heating, respectively.

15.4 RESULT AND DISCUSSION

The elemental analyses and proximate analyses of raw samples and torrefied samples of MSW and agriculture waste biomass were done. Calorific value, moisture, ash content, and volatile matter were determined by using IS: 1350 (Part II)-1970 methods (Figures 15.2–15.7).

Figure 15.8 shows that the GCV of torrified samples enhanced with torrefaction temperature. The calorific value is high at 300°C temperature for sugarcane biomass and sweet corn cob biomass. At 330°C, samples were showing a reduction in the GCV with increasing temperature due to some non-condensable gases such as CO, $CO_2 CH_4$. The elemental analysis of the raw feed samples and torrefied samples, heated with the temperatures 200°C, 250°C, 300°C, and 330°C at a retention time of 60 min was illustrated in Tables 15.1–15.3. It was found that the torrefied samples had a higher amount of carbon content than the non-torrefied samples (raw feed samples). The total amount of carbon amplified with the temperature of torrefaction. The carbon content had reduced with the temperature increases above 300°C and 330°C.

15.5 CONCLUSIONS

In conclusion, as per the data, the 300°C samples had better results in ultimate and proximate analysis, and the torrefied samples had higher calorific values as well as better combustion characteristics. All the results recommend that the torrefied

FIGURE 15.2   Raw sugar cane sample

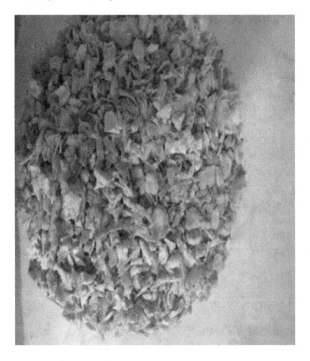

FIGURE 15.3   Raw sweet corn.

FIGURE 15.4 Raw MSW.

FIGURE 15.5 Torrified sweet corn.

FIGURE 15.6   Torrified sugar cane.

FIGURE 15.7   Torrified MSW.

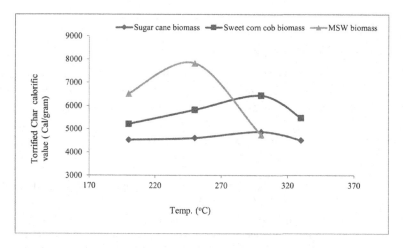

FIGURE 15.8   Calorific values of torrified char at different temperatures.

TABLE 15.1
Sugar Cane Samples Data

Sample	Proximate Analysis				Ultimate Analysis					Gross Calorific Value (cal/g)
	Moisture	Ash	VM	FC	C	H	O	N	S	
Raw sample	5.61	2.45	80.46	11.48	44.9	6.24	45.96	0.1	0.05	4,261
Torrified at 200°C	4.8	2.8	60.8	31.6	45.8	5.88	47.67	0.6	0.05	4,532
Torrified at 250°C	1.74	8.4	51.7	38.2	47.99	5.21	32.59	0.63	0.05	4,608
Torrified at 300°C	0.87	16.61	48.07	34.45	60.3	4.52	35.14	0.29	0.04	4,869
Torrified at 330°C	5.52	3.59	70.39	20.76	48	5.44	47.67	0.29	0.04	4,513

TABLE 15.2
Sweet Corn Cob Samples Data

Sample	Proximate Analysis (%)				Ultimate Analysis (%)					Gross Calorific Value (cal/g)
	Moisture	Ash	VM	FC	C	H	O	N	S	
Raw sample	6.87	78.17	1.56	13.4	0.19	19.14	46.57	0.0	0.0	4,111
Torrified at 200°C	6.83	45.68	15.05	32.44	35.2	6.43	57.72	0.65	0.0	5,216
Torrified at 250°C	3.11	38.07	22.58	36.25	51.53	5.78	37.99	0.16	0.0	5,825
Torrified at 300°C	4.09	39.13	9.93	46.86	68.95	4.66	30.16	0.84	0.0	6,439
Torrified at 330°C	3.16	66.73	9.75	20.36	50.55	3.62	20.45	1.07	0.06	5,474

TABLE 15.3
Municipal Solid Waste Samples Data

Sample	Proximate Analysis (%)				Ultimate Analysis (%)					Gross Calorific Value (cal/g)
	Moisture	Ash	VM	FC	C	H	O	N	S	
Torrified at 200°C	0.79	67.02	27.07	5.12	34.4	3.7	23.32	0.94	0.11	6,516
Torrified at 250°C	0.78	66.43	31.98	0.81	22.22	2.13	10.53	0.4	0.1	7,825
Torrified at 300°C	0.62	72.65	16.53	10.21	14.71	2.06	10.83	0.39	0.28	4,735

samples of MSW and biomass were better altered to be utilized as fuel and torrefaction is an attractive process of MSW pretreatment.

Torrefied samples of agriculture residues biomass led to the improvement of the properties of the torrified biomass samples. The amount of carbon was found to be higher in 300°C torrefied samples such as MSW 7,825 cal/g, sweet corn 6,439 cal/g, and sugar cane 4,869 cal/g.

REFERENCES

1. van der Stelt, M.J.C., Gerhauser, H., Kiel, J.H.A., Ptasinski, K.J. Biomass upgrading by torrefaction for the production of biofuels: A review. *Biomass and Bioenergy* 2011,35:3748–3762.
2. Siritheerasasa, P., Waiyanatea, P. Torrefaction of municipal solid waste (MSW) pellets using microwave irradiation with the assistance of the char of agricultural residues. In: *2017 International Conference on Alternative Energy in Developing Countries and Emerging Economies*, AEDCEE, 25–26 May 2017, Bangkok, Thailand.
3. Sukholthaman, P., Shirahada, K. Technological challenges for effective development towards sustainable waste management in developing countries: Case study of Bangkok, Thailand. *Technology in Society* 2015,43:231–239.
4. Chen, W.H., Kuo, P.C. Torrefaction and co-torrefaction characterization of hemicellulose, cellulose and lignin as well as torrefaction of some basic constituents in biomass. *Energy* 2011,36:803–811.
5. Van der Stelt, M.J.C., Gerhauser, H., Kiel, J.H.A., Ptasinski, K.J. Biomass upgrading by torrefaction for the production of biofuels: A review. *Biomass Bioenergy* 2011,35:3748–3762.
6. Vassilev, S.V., Vassileva, C.G., Vassilev, V.S. Advantages and disadvantages of composition and properties of biomass in comparison with coal: An overview. *Fuel* 2015,158:330–350.
7. Iwabuchi K., A new approach to stabilize waste biomass for valorization using an oxidative process at 90°C. *PLoS One* 2018,13(4):e0196249.
8. Elizondo, A., Pérez-Cirera, V., Strapasson, A., Fernández, J.C., Cruz-Cano, D. Mexico's low carbon futures: An integrated assessment for energy planning and climate change mitigation by 2050. *Futures* 2017,93:14–26.
9. Meeus, L., Azevedo, I., Marcantonini, C., Glachant, J.-M., Hafner, M. EU 2050 low-carbon energy future: Visions and strategies. *The Electricity Journal* 2012,25:57–63.
10. Saidur, R., Abdelaziz, E.A., Demirbas, A., Hossain, M.S., Mekhilef, S. A review on biomass as a fuel for boilers. *Renewable and Sustainable Energy Reviews* 2011,15:2262–2289.

11. Fragkos, P., Tasios, N., Paroussos, L., Capros, P., Tsani, S. Energy system impacts and policy implications of the European Intended Nationally Determined Contribution and low-carbon pathway to 2050. *Energy Policy* 2017,100:216–226.

12. Corradini, M., Costantini, V., Markandya, A., Paglialunga, E., Sforna, G. A dynamic assessment of instrument interaction and timing alternatives in the EU low-carbon policy mix design. *Energy Policy* 2018,120:73–84.

13. Schanes, K., Jäger, J., Drummond, P. Three scenario narratives for a resource-efficient and low-CarbonEurope in 2050. *Ecological Economics* 2018,155:70–79

14. Van der Stelt, M.J.C., Gerhauser, H., Kiel, J.H.A., Ptasinski, K.J. Biomass upgrading by torrefaction for the production of biofuels: A review. *Biomass Bioenergy* 2011,35:3748–3762. doi:10.1016/j.biombioe.2011.06.023

15. Nunes, L.J.R., Matias, J.C.O., Catalão, J.P.S. A review on torrefied biomass pellets as a sustainable alternative to coal in power generation. *Renewable and Sustainable Energy Reviews* 2014,40:153–160.

16. Bergman, P.C.A., Kiel, J.H.A. Torrefaction for biomass upgrading. In: *Proceedings of the 14th European Biomass Conference*, October 2005, Paris, France, 17–21.

17. Chen, Q., Zhou, J.S., Liu, B.J., Mei, Q.F., Luo, Z.Y. Influence of torrefaction pretreatment on biomass gasification technology. *Chinese Science Bulletin* 2011,56:1449–1456.

18. Xin, S., Mi, T., Liu, X., Huang, F. Effect of torrefaction on the pyrolysis characteristics of high moisture herbaceous residues. *Energy* 2018,152:586–593.

19. Van der Stelt, M.J.C., Gerhauser, H., Kiel, J.H.A., Ptasinski, K.J. Biomass upgrading by torrefaction for the production of biofuels: A review. *Biomass Bioenergy* 2011,35:3748–3762.

20. Shankar Tumuluru, J., Sokhansanj, S., Hess, J.R., Wright, C.T., Boardman, R.D. REVIEW: A review on biomass torrefaction process and product properties for energy applications. *Industrial Biotechnology* 2011,7:384–401.

21. Nunes, L.J.R., Matias, J.C.O., Catalão, J.P.S. A review on torrefied biomass pellets as a sustainable alternative to coal in power generation. *Renewable and Sustainable Energy Reviews* 2014,40:153–160.

22. Acharya, B., Sule, I., Dutta, A. A review on advances of torrefaction technologies for biomass processing. *Biomass Conversion and Biorefinery* 2012,2:349–369.

23. Chen, W., Peng, J., Bi, X.T. A state-of-the-art review of biomass torrefaction, densification and applications. *Renewable and Sustainable Energy* 2015,44:847–866.

24. Resultat Kontrakt (RK) *ReportTorrefaction of Unutilized Biomass Resources and Characterization of Torrefaction Gasses*. Resultat Kontrakt

25. Joshi, Y., De Vries, H., Woudstra, T., De Jong, W. Torrefaction: Unit operation modelling and process simulation. *Applied Thermal Engineering* 2015,74:83–88.

26. Bilgic,E,YamanS,Haykiri-AcmaH,KucukbayrakS.Istorrifactionofpolysaccharides-rich biomass equivalent to carbonization of lignin-rich biomass? *Bioresource Technology* 2016,200:201–207. https://dx. doi.org/10.1016/j.biortech.2015.10.032.

27. Uslu, A, Faaij APC, Bergman PCA. Pre-treatment technologies and their effect on international bioenergy supply chain logistics: Techno-economic evaluation of torrefaction, fast pyrolysis and pelletisation. *Energy* 2008,33:1206–1223. doi:10.1016/j.energy.2008.03.007.

28. Basu, P. *Biomass Gasification, Pyrolysis and Torrefaction: Practical Design and Theory*, 2nd ed. Elsevier Inc., 2013.

29. Poudel, J., Ohm, T.I., Oh, S.C. A study on torrefaction of food waste. *Fuel* 2015,140:275–281. doi:10.1016/j.fuel.2014.09.120.

30. Yuan, H., Wang, Y., Kobayashi, N., Zhao, D., Xing, S. Study of fuel properties of torrefied municipal solid waste. *Energy Fuel* 2015,29:4976–4980. doi:10.1021/ef502277u.

31. Yu, J., Sun, L., Wang, B., Qiao, Y., Xiang, J., Hu, S. et al. Study on the behavior of heavy metals during thermal treatment of municipal solid waste (MSW) components. *Environmental Science and Pollution Research* 2016,23:253–265. doi:10.1007/s11356-015-5644-7.

32. An, D., Wang, Z., Zhang, S., Yang, H. Low-temperature pyrolysis of municipal solid waste: Influence of pyrolysis temperature on the characteristics of solid fuel. *International Journal of Energy Research* 2006,30:349–357. doi:10.1002/er.1152.

33. Matsakas, L., Gao, Q., Jansson, S., Rova, U., Christakopoulos, P. A Green conversion of municipal solid wastes into fuels and chemicals. https://creativecommons.org/licenses/by-nc-nd/4.0/.

16 The Circular Bioeconomy Concept

Larissa Echeverria, Carina Contini Triques,
Keiti Lopes Maestre,
Jacqueline Ferandin Honório,
Veronice Slusarski-Santana, and
Gabriela Eduarda Zeni
State University of West Parana

Fabio Augusto Gubiani
Federal University of Santa Catarina

Leila Denise Fiorentin-Ferrari
and Mônica Lady Fiorese
State University of West Parana

16.1 INTRODUCTION

The first warnings that the economic linear system of extracting, transforming, consuming, and discarding would have impacts on all aspects of society and the environment were issued by Boulding (1966). Boulding (1966) described the economic system based on the extraction of raw materials, which subsequently were used as inputs in technological transformations, generating marketable products with a pre-determined life cycle, and later discarded in landfills or passed through disintegration processes without the aim of posterior use. Even so, a great volume of these materials is incorrectly discarded in rivers, forests, oceans, and other inappropriate landscapes, thus harming the environment (Andersen, 2007).

This pattern, still very much in use, has as a negative consequence, the constant need to extract raw materials to produce new materials goods, causing the depletion of primary resources, and intensifying economic and social problems. For instance, the system leads to a constant loss of biodiversity, promotes the increase of hydric stress, and generates emissions of greenhouse gases, for example, the emission of pollutants when productive activities occur and when final material goods are used, besides making climate change progress more quickly. It is necessary to adopt measures that change this economic system (Corona et al., 2019).

Thenceforth, concepts such as circular economy, bioeconomy, and its intercept with circular bioeconomy concept (CBC) emerged as fundamental strategic areas

DOI: 10.1201/9781003406501-16

to reach an economy with neutral impacts in both climate and society spheres, by promoting a transition from the actual economic system, essentially linear, to a more sustainable one (European Commission, 2015; Juan et al., 2019).

The concept of circular economy is approached in different ways by different people (Tan and Lamers, 2021). However, it is a consensus that the circular economy searches to overcome the limitations of the current linear economic model (extract-produce-discard). It aims to reduce the dependence on a new extraction of natural resources, creating conditions for more efficient use of materials and flow optimization, through the use of alternative cycles that will increase the lifespan of extracted raw materials. Those principles are coupled to reuse, recycle, redraw, remanufacture, and reform, among others, offering strategies related to sustainability (Stefanakis and Nikolaou, 2021).

The concept of circular economy and bioeconomy can complement each other, creating an intersection between approaches. Bioeconomy, according to FAO (Juan et al., 2022), is based on the knowledge and the use of resources, processes, and biological methods to provide products and services in a sustainable manner in all economic sectors, that is, economic activities related to invention, development, production, and the use of products and biological processes. Thus, the intersection of the concepts of circular economy and bioeconomy results in what is called "circular bioeconomy" (Stegmann et al., 2020; Tan and Lamers, 2021; Girard, 2022).

The advantages of circular bioeconomy are varied and include environmental aspects, such as reducing food wastage, protecting the environment, and mitigating climate changes; and social aspects, presenting business advantages, improving the revenue of a company and its growth rate (WBCSD, 2020), increasing the economic value of biomasses and residues generated in the industry sector, and providing renewable products to several segments from society. Hence, when using renewable resources in a conscious, sustainable, and integrated way, with zero waste, a rigorous realist comprehension of sustainable circular bioeconomy is adopted (Banu et al., 2021; Holden et al., 2022).

In this context of circular bioeconomy, biomass is an important subject. Since it is a renewable carbon source that enables attaining bioproducts and bioenergy, it is considered the most important resource from an economical aspect (Ubando et al., 2020; Kardung et al., 2021). The conversion of biomass into bioproducts happens in what is called a biorefinery, also regarded as the place where a circular bioeconomy can be practiced (Awasthi et al., 2020). Data from 2018 shows that 23.1 billion tons of biomass (19 billion tons from agriculture, 4 billion tons from forestry, and 0.2 billion tons from aquaculture) were destined for the processing and production of basic biological products (WBCSD, 2020).

Despite cases of industries and processes that apply the concept of circular bioeconomy being found, the methodology and data are still too scarce and standardized (Kardung et al., 2021) to measure the real influence of bioeconomy on regional growth and circular economy. This occurs because, in most countries, national statistic agencies rarely distinguish between products with biological and non-biological basis (Jander et al., 2020). Such a fact demonstrates that the whole potential of bioeconomy has not been reached and challenges and barriers still need to be overcome (WBCSD, 2020). One example, considered the biggest challenge, is waste degradation (Awasthi et al., 2020). But one cannot suggest that this could hamper the circular bioeconomy, since, if correctly established, the system would be able to grant all of the above-mentioned benefits, while also enabling a lot of opportunities.

The topics previously mentioned will be better discussed over the course of this chapter, which aims to contribute to the current knowledge on circular bioeconomy by presenting its definitions, contributions, data, examples, use of biomass, its diversity, the current challenges, and barriers, as well as its opportunities.

16.2 DEFINITIONS AND CONCEPTS

The concept of circular economy was first proposed by Kenneth Boulding in 1996 in his work "The economics of the coming spaceship Earth". In his text, the author compares contrasting views on economic systems, referred to by him as "cowboy" and "spaceman" economy, based on their attitude toward consumption. In that sense, a successful "cowboy" economy is the result of large consumption and production, whereas a "spaceman" one is focused on sustainable stock maintenance.

Through time, the concept has evolved into what is now known as circular economy; however, the principle has remained the same: to eliminate waste and ensure the continued use of resources, thus creating a loop system that minimizes the use of resource inputs and reduces pollution (Geissdoefer et al., 2017). As such, this economy strives toward the renewability of its resources and converting alternative matter into energy that can be looped back into the system, perpetuating the cycle (Venkatesh, 2022).

Alternatively, the definition of bioeconomy is closely related to a country's technological capacity, natural resource base, and economic and trade policies (Gallo, 2022). For this reason, while the general concept of bioeconomy is linked to the use of renewable resources, different countries may adapt the definition to better portray their reality. As an example, Brazil refers to bioeconomy as "the generation of innovative products and services based on the country's natural resources and ecosystem services" (German Bioeconomy Council, 2018). On the other hand, in 2019 the Office of Science and Technology, United States, stated that "the bioeconomy represents the infrastructure, innovation, products, technology, and data derived from biologically-related processes and science that drive economic growth, improve public health, agricultural, and security benefits" (Office of Science and Technology Policy, 2019).

Despite the lack of a standard definition for bioeconomy, it is possible to distinguish the respect for the environment and the need for an ever-decreasing reliance on non-renewable resources as a common objective. In addition, regardless of the definition, some advantages are common, such as the introduction of healthy food, new materials being created, the use of bioenergy and biofuels in substitution to fossil fuels, and the contribution to the environment and the economy themselves (Carus and Dammer, 2018). In addition, consumer habits have changed, and their acceptance is fundamental for the progress of bioeconomy, which has been highlighted as a market opportunity since 73% of global consumers stated that they would change their purchasing habits for environmental reasons (WBCSD, 2020).

However, bioeconomy also sometimes presents the conventional linear economic model of producing, consuming, and discarding (Holden et al., 2022). The linear model represents about 90% of the economy and urgent changes are needed (WBCSD, 2020). Statistics highlight that only 8.65 billion tons of raw materials (8.6% of the total extracted material) are circularly returned to the economy, leaving 92 billion tons of raw materials that are processed and, after use, discarded without any opportunity to recover the materials (WBCSD, 2020). Thus, there is a need to include the bioeconomy in the circular economy, which, by itself, is not necessarily sustainable (Lewandovski, 2018).

In this context, both circular and bioeconomy are complementary to each other. Indeed, it is important to be selective regarding the resources being used as raw materials for various processes nowadays (Geissdoefer et al., 2017). However, making sure that these same resources are explored wholly, and that process by-products are recycled back into the loop is what makes this economic system up to date (Salvador et al., 2021). As such, circular bioeconomy is the all-encompassing concept that strives for the replacement of bulk (non-renewable) materials by scalable, bio-based, low-carbon products in both technical and biological cycles (Palahì et al., 2020).

The idea of bioeconomy relies on the recycling of products, previously treated as waste, as a new resource for second-generation processes. In this scenario, agro-industrial, aquaculture, and fishery by-products take the lead and prove advantageous for lowering greenhouse gas footprints and industries' reliance on non-renewable energy sources (Carus and Dammer, 2018). In particular, these transformations are facilitated by a process known as biorefining. According to Leong et al. (2021), the technique consists of a newer and better approach to waste management and can be defined as a sustainable bioprocess that turns biomass into marketable products with high added value.

Overall, circular economy, bioeconomy, bio-based economy, circular bioeconomy, and biorefining make up the green economy. The UN Environment Program (UNEP) states that a green economy is structured into three main pillars – social, economic, and environmental development – that intertwine with the reduction of environmental risks and ecological scarcities. This concept became especially significant in 2012 at the Rio+20 Conference, where solutions to poverty eradication and sustainable development were built on the implementation of an international green economy (IISD and UNEP, 2014).

In that sense, UNEP affirms that "the Green Economy is one in which the vital linkages among the economy, society, and environment are taken into account and in which the transformation of production processes, and consumption patterns while contributing to a reduced waste, pollution, and the efficient use of resources, materials, and energy, will revitalize and diversify economies, create decent employment opportunities, promote sustainable trade, reduce poverty, and improve equity and income distribution". In other words, the economy is viewed as a tool that shapes social, environmental, and economic systems. As it stands, the green economy provides the link between an improved economic system and the impact it has on the two other pillars of development, approaching the cycle to perpetuate it.

16.2.1 Contribution of the Bioeconomy to the Circular Economy (and Vice Versa)

As previously mentioned, the application of circular economy with bioeconomy, where both are green economies, promotes the sustainable use of natural resources and reduces environmental impacts. Therefore, it can be said that one kind of economy contributes to the other – that is, because their purposes are similar, and the bioeconomy contributes to the circular economy and vice versa. Besides, establishing a common ground between both economies is necessary to achieve full sustainability purposes, because although bioeconomy is renewable, it can generate waste, thus the necessity of including the circular economy in this context, for example (Carus and Dammer, 2018). More examples of these contributions include biodegradable products that return to their production cycles, recyclability that is enhanced using innovative additives, and different sectors of industries that benefit through their union (Carus and Dammer, 2018).

16.2.2 Data and Examples of Bioeconomy, Circular Economy, and Circular Bioeconomy

Ten countries included bioeconomy practices in their NDC (Nationally Determined Contribution): Brazil, El Salvador, India, Mauritania, Namibia, Nigeria, Pakistan, Rwanda, Tunisia, and Venezuela (Juan et al., 2022). In addition, actions are presented as opportunities in the transition process for a more sustainable economy such as the Agenda 2030, where the sustainable development objectives (SDO) from United Nations Organization are shown, released in 2015, the commitments from the European Union regarding the climate with goals for 2050 and the Paris Agreement in 2016 under the United Nations Framework Convention on Climate Change to regulate the global temperature increase (Abad-Segura et al., 2021).

The SDO indicators are highlighted because most of them aid in the measurement of bioeconomy contribution by focusing on industrial and human behavior change. The SDO that promote sustainable technologies and practices are clean energy (ODS no. 7), industrial innovation and infrastructure (ODS no 9), sustainable cities and communities (ODS no 11), responsible consumption and production (ODS no. 12), and climate action (ODS no. 13).

Based on the actions cited, business opportunities are highlighted, as it is expected that the total market of food, animal feed, products, and energy with biological bases grow from US$ 10.3 trillion in 2018 to US$ 12.8 trillion in 2030, representing an annual increase of 1.8% in the period mentioned (WBCSD, 2020). It is important to note, however, that statistics and methods that measure the contribution of bioeconomy to reach global social goals are relatively well equipped and developed for traditional products such as food, animal feed, paper and cellulose, and bioenergy chains, but gaps exist for innovative sectors with biological basis (Holden et al., 2022).

In 2020, 70% of the raw material with biological basis was used in food, drinks, and animal feed, while the remaining 30% was used to produce energy and products

(WBCSD, 2020). However, alternative sources of raw material should be considered, among them, lignocellulosic materials.

The modern technology of biorefinery for lignocellulosic materials makes it possible to fraction lignin, cellulose, and hemicellulose using sustainable and low-cost technologies (Lask et al., 2019) and obtain valuable products, such as bioethanol, biodiesel, biopolymers, bioplastics, biochip, biogas, bio-oil, and biochar (Awasthi et al., 2020). Globally, the lignin market, for example, is valued at 954.5 million dollars with the expectation of expanding about 2% in revenue between the years of 2020 and 2027, making up a total of 34% of the total market participation in Europe (Banu et al., 2021).

Among the uses of lignin, it is possible to cite (Banu et al., 2021; Bajwa et al., 2019):

1. Production of biopolymers, which totaled US$ 10.5 billion in 2021 and is estimated to increase to US$ 27.9 billion by 2025. About 80% of non-combustible chemical products produced by the petrochemical industry are sold for plastic production (Holden et al., 2022), thus, biopolymers are of great importance for the bioeconomy.
2. Adhesives production that possessed a growth projection from 58.9 billion dollars to 73.8 billion dollars between the years 2019 and 2024.
3. Phenol production, representing 8 million tons with a growth rate of 3.9% in the next decades.
4. Fiber production from lignin to substitute carbon fiber, which possesses a demand of 98,000 metric tons, and is expected to be replaced by 15%–20% in the next years.
5. Valine production (flavoring, medicine, and perfume agent), which currently originates (approximately 85%) from the petrochemical industry.
6. Lignin conversion for hydrocarbon derivatives, adding more value to circular bioeconomy.

When cellulose and hemicellulose are used as raw materials in the production of combustible products and bioenergy, these products are labeled second generation. Cellulose fiber had a global market evaluated at 21.5 billion dollars in 2017 with projections to reach 41.5 billion dollars by 2025 (Banu et al., 2021).

The implementation of biorefineries in different regions has the potential to add value to culturally specific residues regionally produced (Andrade et al., 2022), and one example that deserves to be highlighted is bioethanol. Several policies were established in different countries to implement biorefinery processes from different sectors but the one that is highlighted is the biofuel sector, according to a report from the United Nations Organization (Banu et al., 2021).

Among these biofuels, there is corn ethanol, representing 15.25 billion gallons produced in the United States, and generating 22.65 billion corn fiber pounds. These fibers are used as an ingredient in the formulation of animal feed and have recently been destined to produce ethanol by the lignocellulosic conversion process, which has the potential to use up an additional 1.06 billion gallons (Awasthi et al., 2020). Brazil is the second largest global producer of ethanol, representing 29.6% according to the

Renewable Fuels Association, and has its production based mainly on sugarcane, coming second only to the United States, which represents 54.4% of the total (Vidal, 2021).

The Food and Agriculture Organization from the United States estimated that about 1.4 billion tons of food waste are produced every year, which represents approximately one-third of the total food originating for human consumption (Awasthi et al., 2020). Still, approximately 181.5 billion tons of lignocellulosic biomass are annually produced, and only 3% were efficiently used and incorporated into circular bioeconomy (Dahmen et al., 2019).

Through this topic, the relevance of bioproducts and natural raw materials was evidenced, showing the great potential and importance of a circular bioeconomy. The next topic will deepen the knowledge of circular bioeconomy regarding renewable biomass.

16.3 CIRCULAR BIOECONOMY BASED ON THE VALORIZATION OF RENEWABLE BIOMASS FEEDSTOCKS

Worldwide, fossil fuels are the main source of energy (Martins et al., 2019). However, they are also the source of many problems. A few of these are air pollution and global warming, but the main one is its depletion as a non-renewable resource, and what makes fossil fuels contrast Circular Economy (Martins et al., 2019).

The substitution of fossil fuels for renewable sources, therefore, agrees with sustainable goals and has piqued the interest of the government and researchers. Government efforts have been made to substitute petroleum derivatives to guarantee access to energy in the future, mitigate climate changes, maintain agricultural activities, and ensure food safety (Bušić et al., 2018). For that, technologies have been developed by research, and the best approach found has been the processing of biomass, responsible for supplying 15% of the global energy demand (Kohli et al., 2019; Saratale et al., 2019).

Therefore, biomass, which has the potential to generate energy and products, is the primary emerging alternative to fossil fuels (Awasthi et al., 2020). By definition, renewable biomass is a renewable source of organic carbon from plants and animals, which can offer advantages such as carbon sequestration, production of bioenergy and bioproducts, being, thus, considered rich, renewable, and green (Ubando et al., 2020; Salvador et al., 2021; Luo et al., 2021).

The sources of biological material from several residues (agriculture, forest, confined animal waste, domestic and industrial sewage, and discarded food) are the main resources with the potential to be used as raw material to produce bioenergy and several bioproducts that move circular bioeconomy (Banu et al., 2021).

Currently, the main sources of renewable biomass come from agriculture, forestry, marine, and environmental waste (Kardung et al., 2021), and 28% of this biomass is composed of agricultural materials such as wheat, corn, and barley, while 28% are plants harvested green, such as green corn (WBCSD, 2020). As examples of forestry biomass, it is possible to cite lumber and wood pulp, and for aquiculture, fish capture, and bycatch (WBCSD, 2020). A few other examples worth mentioning are municipal solid waste, algal feedstocks, waste-activated sludge, dairy waste, and glycerol (Saratale et al., 2019).

Ideally, biomass would provide high yield, have low cost, and require low energy and nutrients (Mckendry, 2002). In this context, what is even more environmentally friendly in a circular bioeconomy and renewable biomass context is the use of a certain waste as biomass, especially agricultural residues, and food waste (Kardung et al., 2021). This alternative enables the reduction of waste while value-added bio-based products are achieved (Lange et al., 2020). Food waste, for example, can be used to produce protein, animal feed, enzymes, organic acids, flavors, colorants, bio-fertilizers, bioplastics, and biofuels, thus avoiding the emission of greenhouse gas, the loss of potential nutrients, and the necessity to treat the wastes, all in consonance with the circular bioeconomy (Tsegaye et al., 2021). Agricultural residues, also of great interest, include stover, straws, stem, stalk, bagasse, husk, and cobs (Awasthi et al., 2020).

Among the composition of renewable biomass carbohydrates (cellulose, hemicellulose, and lignin), lipids, and proteins are found, but their composition is highly varied (Agbor et al., 2011; Carpenter et al., 2014; Awasthi et al., 2020). These constituents are then digested into glucose, fatty acids, and amino acids, for example (Awasthi et al., 2020). It is worth mentioning that its variable composition allows for different materials to be obtained (Luo et al., 2021). Thus, each kind of biomass can be destined for a different goal, such as the production of fuels, chemical products, and energy (Casau et al., 2022).

Several processes that use renewable biomass are discussed in Section 3.1.

16.3.1 Biorefineries and Biomass Conversion Processes

Circular bioeconomy can become a reality through biorefineries, the approach to converting biomass into bioproducts and bioenergy in a sustainable way (Awasthi et al., 2020). Biorefineries are integrated and multifunctional processes that aim the formulation of several products instead of only one product to ensure sustainable processes, collaborating with the pillars of the economy, society, and the environment (Lappa et al., 2019; Ubando et al., 2019). Thus, from different raw materials, several products can be generated, such as chemicals and biofuels, allowing for the maximum use of each raw material, the destination of the co-products, for example, food or feed (Wagemann, 2012). As such, biorefineries emerged to substitute petroleum refineries and to value a circular economy (Andrade et al., 2022; Aworunse et al., 2023).

In a biorefinery, the biomass will be pre-treated and prepared, its components will be separated and then converted into the desired products, which will finally be purified (Bušić et al., 2018). Biorefineries can be classified as bottom-up biorefineries or top-down refineries, being that the bottom-up one will broaden its facilities that are used to produce one or a few products to produce more, and the top-down approach uses various biomass fractions to obtain different products aiming to zero-waste generation (Bušić et al., 2018).

Regarding the main process of conversion of biomass properly said, the routes that compose a biorefinery are thermochemical, biological, chemical, and mechanical processes (Ubando et al., 2019). Their main processes and products are described in sequence (Ubando et al., 2019; Tsegaye et al., 2021):

- The thermochemical conversion uses high temperatures and chemicals and includes torrefaction (biofuel), pyrolysis (biofuel and biochemicals), gasification (syngas, olefins, methanol), and combustion (electricity and heat);
  - Gasification uses temperatures higher than 700°C and slow pyrolysis uses temperatures between 350°C and 700°C, while fast pyrolysis uses between 450°C and 700°C, combustion uses temperatures higher than 1,200°C, and torrefaction uses temperatures between 200°C and 300°C (Awasthi et al., 2020).
- The biological conversion uses enzymes or microorganisms and includes anaerobic digestion (biomethane, methanol, and olefins), enzymatic hydrolysis (bioethanol, ethylene, and ethylamine), and fermentation (bioethanol, ethylene, and ethylamine).
- The chemical conversion uses chemicals as solvents and/or as catalysts and includes acid hydrolysis (fermentable sugars, biofuels, and ethanol), supercritical conversion of biomass (fermentable sugars, biofuels, and ethanol), and solvent extraction (bioethanol, and extractive waxes).
- The mechanical conversion includes separation (biofuels, electricity, heat, and biomaterials), extraction (biofuels and oleochemicals), drying, and pelletizing (electricity and heat).
- Other technologies involved may include fermentation, anaerobic digestion, and esterification (Awasthi et al., 2020).

Among these, combustion used to be the main process of the 1970s. At the time, other industrial processes started to be studied, and in the 1980s, great interest was given to the conversion of biomass to ethanol. Nowadays, there is a huge expansion of academic studies regarding the use of biomass and new conversion processes that focus on sustainability instead of profitability (Casau et al., 2022). It is important to mention, however, that every route has its advantages and disadvantages, and the choice will be based on the feedstock, the technology available, and the requirements regarding energy, or special inputs such as enzymes or microbes (Guragain and Vadlani, 2021).

Despite the differences in methods, it is important to mention that all the integrated biorefinery decreases waste generation and brings economic and environmental benefits, being coherent with the circular bioeconomy (Awasthi et al., 2020). Examples of different research concerning biorefinery with different renewable biomass can be found in Table 16.1. Moreover, Figure 16.1 summarizes some of the main aspects involved in the use of renewable biomass.

16.4 DIVERSITY OF BIOECONOMIES AROUND THE WORLD WITH A FOCUS ON THE STATE OF THE ART

Aiming to reduce the dependence on fossil fuels, reaching food sovereignty, and increasing the use of renewable biomass, the development of biorefineries is becoming a focal point. Table 16.2 introduces some examples of the diversity of bioeconomies around the world.

TABLE 16.1

Different Bioproducts that can be Obtained in Biorefineries with Different Raw Materials and Bioprocesses

Raw Material	Bioprocess	Bioproduct	References
Cheese whey permeate	Sequential fermentations	Ethanol, acetic acid, cellular biomass, galacto-oligosaccharides, and organic acids	Maestre et al. (2021)
Grape waste	Solid-state fermentation	Bioactive compounds (antioxidant)	Martínez-Ávila et al. (2012)
Wheat straw	Saccharification and co-fermentation, distillation, and dehydration	Ethanol	Hasanly et al. (2018)
Ulva prolifera macroalgae	Hydrothermal liquefaction	Bio-oil	Yan et al. (2019)
Potato peels waste	Thermal and chemical pretreatment and enzymatic hydrolysis or combinations followed by saccharification	Ethanol	Atitallah et al. (2019)
Poplar and masson pine (woody biomass)	2-naphthol assisted organosolv pretreatment and enzymatic hydrolysis	High-strength self-healing adhesives and sugar	Wang et al. (2022)
Soybean molasses and glycerin (effluent from biodiesel production)	Anaerobic digestion (Anaerobic Sequencing Batch Biofilm Reactor)	Methane	Paulinetti et al. (2022)
Sweet potato waste	Simultaneous hydrolysis and fermentation	Ethanol and distilled beverage	Weber et al. (2020)
Rice straw	Extrusion and fermentation	Methane	Chen et al. (2014)

As presented in Table 16.2, Nigeria, since 2016, presents a sustainable bioeconomy and is oriented to the knowledge on the premise of substituting petroleum for alternative raw materials; besides, it emphasizes creating jobs and contributing globally to society. The generation of new jobs was also detected in Finland (Aworunse et al., 2023).

In general (Table 16.2), a biorefinery of lignocellulosic biomass (renewable biological source) can generate chemical products (biochemical), biomaterials (biosorbents and bioplastics), bioenergy (biofuels and bioelectricity), enzymes, and nutraceutical compounds, organic acids, and other bioproducts, which are valuable to the market (Andrade et al., 2022).

The seaweed biomass provides benefits such as a lower need for land and can be classified into microalgae (Sharma et al., 2021) and macroalgae (Kostas et al., 2021), both serve as raw material in the development of sustainable biorefineries processes; however, studies are scarce in comparison to lignocellulosic biomass (Kostas et al., 2021).

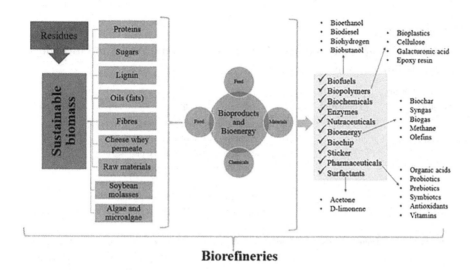

Biorefineries

FIGURE 16.1   Representation of a biorefinery.

(Source: the authors.)

TABLE 16.2

Biorefineries of Varied Biomass Examples around the World and Involving Circular Bioeconomy

Installation or Locality	Feedstock	Description	References
Stora Enso, Sunila Mill, Finland	Softwood	It produces several levels of cellulose to attend to the demand for paper, fabric, and cardboard producers. For example, cellulose, lignin, liquid resin, turpentine, heat, and electricity. They are the leader in the production of renewable products, such as packages, biomaterials, and construction in wood and paper.	Sunila Mill (2023)
Finland	Several biomass	The Finland government impulsions the bioeconomy since 2014, and since then has valued several biomasses to produce sustainable and safe products.	Aworunse et al. (2023)
Finnish Lapland	Metal waste and separation of remaining materials from stony materials.	Evaluated the development of and piloting of sustainability assessment of slag processing in Finland, the collaboration occurred between university and industry. The biorefinery aimed at using residues from the process of stainless-steel production and separating the metallic remaining elements from stony materials to be again incorporated into the processes. The evaluation indicated as viable the implementation of the bioeconomy because it applied residues, and because it separated metallic materials.	Husgafvel et al. (2017)

(Continued)

TABLE 16.2 (*Continued*)

Biorefineries of Varied Biomass Examples around the World and Involving Circular Bioeconomy

Installation or Locality	Feedstock	Description	References
Lenzing, Austria	Beech wood	The industrial plant processes wood (renewable source material) as a universal substitute to petroleum, from dissolving sulfite pulping of wood pulp for textiles, nonwovens, and industrial applications such as dissolving pulp, acetic acid, furfural, xylose, lignosulfonate, sodium sulfate, sodium carbonate, heat, and electricity.	Lenzing (2023)
Austrian	Lignocellulosic biomass from wood, cereal grains and straw or green grasses.	Pilot plants are located especially in the United States, Europe, and other countries, and the study aims to implement Austrian Green Biorefinery to produce products with a biological basis, such as proteins, lactic acid, fibers, and biogas from different biomass.	Bušić et al. (2018)
Domtar, EUA	Pine wood	Kraft paper is produced in the cellulose industry with lignin from separation. Among the bioproducts, there are cellulose and paper, especially pulp, nanocrystalline cellulose, liquid resin, turpentine, lignin, heat, and electricity.	Domtar (2023)
EUA	Several biomasses	Bioeconomy has contributed to the direct growth in information and computer science, energy and biofuels, agriculture, biomedical science, and other sectors. The evolution, besides environmental, is also social, due to lots of new opportunities for jobs generated.	Nasem (2020)
Nigeria	Biomass from agriculture, forestry and woody resources, residues from harvesting, and municipal residues, such as food wasted and lignocellulosic biomass.	Nigeria is years into the post-petroleum era as it values its sustainable potential by harnessing approximately 200 billion kilograms of biomass to produce biofuels, plastics, and/or medicines through integrated biorefineries and microbial conversion.	Aworunse et al. (2023)
Borregaard, Norway	Spruce wood, woodchips	In the pulp mill industry using sulfite, different wood components are used to produce lignin products, especially cellulose, vanillin, bioethanol, and microfibrillar cellulose.	Borregaard (2023)

(Continued)

TABLE 16.2 (Continued)

Biorefineries of Varied Biomass Examples around the World and Involving Circular Bioeconomy

Installation or Locality	Feedstock	Description	References
Muswellbrook, New South Wales	Lignocellulosic material	The pilot plant from Ethtec is developing and commercializing new technologies to produce ethanol and xylitol sustainably from structural sugars from lignocellulosic materials (non-food, woody, or fibrous), such as sugarcane bagasse, crop stubble, and forest material. In addition, it aims to recover and recycle the acids formed.	Muswellbrook (2018)

The macroalgae do not have a definite function in the coastal ecosystems; however, they do pose the potential for sustainable cultivation. For this reason, they can serve as raw material to produce food, and food additives, and to promote a circular bioeconomy when valued by the biorefineries, because this material has a high yield of biomass, quick growth rate, and does not need any land for its cultivation (Kostas et al., 2021).

Biotechnology is essential to reach technological advances and innovations in the context of the bioeconomy (Aworunse et al., 2023). Among the advances reached by the biotechnological processes, there are improvements promoted by computer and information science that, coupled with engineering segments, changed lives in the last years. In this scenario, it was possible to read genetic codes, edit genomes, or create new synthetic genomes, impacting DNA sequencing, molecular operations of high yield, the capacity of reading, coding, and modifying genomes, and metabolic pathways boosting bioeconomy (Frisvold et al., 2021).

The diversity of bioeconomies is also present in the variety of chemical products, biofuels, and materials that can be produced from food residues, composting the biorefinery of food waste (Tsegaye et al., 2021).

As such, in the food industry, the circular economy allows the introduction of food components produced in the food chain, or even, residues and subproducts generated, targeting the development of a biorefinery, and so, in the end, the products generated have a higher economic value. This industrial concept aims to add economic value to the raw materials with low importance with the goal of meeting the global demand for food and promoting health with the formed bioproducts (Lappa et al., 2019; Dragone et al., 2020).

Aiming for an auto-sustainable process to dairy residues, Lappa et al. (2019) evaluated an integrated project of a biorefinery and transition from the circular economy, implementing a bioprocess with "zero wastage" and increasing sustainability to convert cheese whey into products with economic value, focusing on food applications. It stands out that the traditional use of cheese whey protein is like a health promoter, for both humans and animals, and when employed in bioprocesses, it produces several organic compounds, such as ethanol, microbial biomass, cellular protein, lactic acid, microbial lipids, and enzymes, among others. However, the study highlighted new bioprocesses, which look into the fabrication of functional food by synthesizing lactose to obtain prebiotic oligosaccharides and fatty acids esters.

Biofuels can also be produced from food waste, as is the case for cheese whey protein and probiotic yeast. Pendón et al. (2021) introduce the production of bioethanol from cheese whey or cheese whey permeate as a promising alternative. The interest in valuing these subproducts is given by their high pollutant rate due to the high lactose content, which represents a risk of environmental degradation if not adequately treated. The authors evaluated thirty microbial strains from the *Kluyveromyces marxianus* yeast, on a laboratory and a miniature-industrial scale. The research indicated that there is a sustainable economy and an integrated process, since, besides the bioethanol obtention, it was possible to produce yeast biomass with probiotic properties in the same bioprocess.

Considering what was exposed, several technological possibilities are noted to apply renewable resources in a conscious, sustainable, and integrated manner, as an alternative to substitute fossil fuels and/or non-renewable raw materials.

16.5 BARRIERS AND CHALLENGES TO THE CIRCULAR BIOECONOMY

Circular bioeconomy offers potential benefits for an organization. However, its adoption needs adequate structure and approach, because the transition to this economic model faces several barriers and challenges that discourage organizational actions (De Mattos and De Albuquerque, 2018; Gedam et al., 2021).

As proposed in this chapter, the approach of circular bioeconomy results in a more sustainable structure (Stegmann et al., 2020; Tan and Lamers, 2021; Girard, 2022; Salvador et al., 2022). From the sustainability point of view, a non-sustainable bioeconomy can cause an increase in the biofuels demand, and consequently an increase in the biomass demand. This increase promotes competition through the use of arable land to produce biomass raw materials and increases greenhouse gas emission due to changes in the use of land, resulting in social agitation and worries about social sustainability (Tan and Lamers, 2021).

According to the review of Salvador et al. (2022), nine barriers and nineteen challenges were identified in the circular bioeconomy. The costs of transportation/logistics and management, limitations of infrastructure resources and storage, lack of financial resources/capital excessive regulation or inadequate regulation, and lack of government support are the most cited barriers. For the challenges, the most cited were scale increase, the need for investments to integrate biorefineries, ensuring sustainability and safety of biomass supply in long-term, lack of conscientization from the consumer, collaboration, lack of adequate technology, and lack of inputs standardization.

Patermann and Aguilar (2021) also identified some challenges that bioeconomy faces, the main ones being an increase of social impact to reach not only the academia but also elite circles; ensure the union between bioeconomy, SDO, circular economy initiatives, and the Green Deal from European Union; the formation of bonds with society in general in a similar way of the already established with the academia, the industry, and financial market; ensure that bioeconomies promote a responsible and sustainable production to avoid wastage and extend the lifespan of the bioproducts and consolidate bioproducts in niches to expand its portfolio.

For this reason, it stands that the implementation of a circular bioeconomy, despite bringing several advantages, is not something trivial; several barriers and challenges are found, being necessary studies and adequate approaches to reach it.

16.6 CURRENT OPPORTUNITIES

The joint application of circular economy and bioeconomy can favor both the transformation of the economic model as well as change in current consumption habits. For this reason, the joint practice of those two concepts encourages the sustainable use of natural resources and, consequently, the efficiency of these resources is improved and the load on the environment is reduced (Tan and Lamers, 2021).

Bioeconomy possesses advantages such as minimized generation of residues and costs when compared with linear processes (Aguilar and Twardowski, 2022). Its products range from low-cost products and high demand for biomass (biofuels) up to high-added-value products that require a reduced quantity of biomass (chemical products or composts with biological basis) (D'amato and Korhonen, 2021). In addition, according to Aguilar and Twardowski (2022), bioeconomy can prevent or reduce global crisis, ensuring food safety, reducing climate changes effects, preserving natural ecosystems, and maintaining biodiversity (Aguilar and Twardowski, 2022).

The opportunities for the circular bioeconomy to grow can be highlighted by sectors. According to the European Commission, Oxford Economics and World Business Council for Sustainable Development reviews (WBCSD), and the Boston Consulting Group (BCG), the products with biological basis and energy, excluding the consumption of food and animal feed, are highlighted with a market growth even higher, of about 2.4% per year until 2030, reaching US$ 7.7 trillion in 2030 (WBCSD, 2020).

However, the business opportunity of US$ 7.7 trillion of products with a biological basis, energy, and food waste is based on the potential of some industries that are highlighted:

Food and animal feed waste are the main categories, estimated at US$ 2.6 trillion in 2030. Among the products with biological basis generated, there are pharmaceuticals, and it is hoped that they have the highest potential, with US$ 750 billion in 2030, followed by textile and clothing products, as well as building materials and building with US$ 700 billion each. Packages, motor vehicles, and components are in the sequence with an opportunity of US$ 550 billion, respectively. Forestry products, electronic and electric products, machines, and equipment, as well as biomass energy and biofuels, are seen as additional opportunities, varying from US$ 100 billion to US$ 200 billion each (WBCSD, 2020).

Since chemical products with biological basis are essential for many industries, the chemical industry develops a crucial role in the circular bioeconomy. Two chains of custody models are applied in the chemical industry: in the segregated chains, one raw material with a biological basis is used in a specific plant and the product of this production plant contains the carbon atoms of the renewable raw material. In the mass balance approach, renewable raw material is introduced in chemical production and allocated to several products. The substitution of fossil-basis raw materials for biological-basis raw materials aids in closing the cycle in the chemical industry because the recovery of needed basic molecules from the recycling of complex finished products is usually hard and consumes a lot of energy (WBCSD, 2020).

It is possible to find case examples of companies that have implemented circular bioeconomy strategies in the literature, which are highlighted below:

Clariant and Neste have developed bio-based plastic additives (the bio-based plastic additives developed by Clariant and Neste), which have assisted the electronics and electrical industry by making their products more sustainable through the use of such products developed by the companies (WBCSD, 2020).

DSM has used bio-based raw materials to produce high-tech fibers, which are known as Dyneema®, the strongest fiber in the world. Clothing made with Dyneema® lasts longer than traditional products, and the use of fossil-based resources can be reduced by using bio-based raw materials. According to DSM, for every metric ton of bio-based Dyneema® produced, approximately 5 metric tons of CO_2 is saved compared with fossil-based Dyneema®. In addition, DSM aims for at least 60 percent of the raw material for Dyneema® fiber to be bio-based by 2030 (WBCSD, 2020).

BASF uses the biomass balancing approach to produce various products for different types of industries and also develops bio-based products such as surfactants with advanced properties. In addition, in 1990 BASF was able to produce vitamin B2 by fermentation, which for nearly five decades had been produced by chemical synthesis. An eco-efficiency analysis conducted by BASF in 2003 determined that the fermentation process is more sustainable than the chemical route, helping to reduce CO_2 emissions and hazardous substances by 30%, and has an economic advantage, reducing production costs by 40%. Currently, 100% of vitamin B2 on the global market is produced by fermentation using biotechnology (WBCSD, 2020).

Mondi has partnered with Austrian fruit canning company Darbo to develop more sustainable packaging. Packaging that eliminates the use of plastic was developed, packaging produced only with corrugated cardboard. The new packaging has some advantages, among them, because it is mono-material, the probability of incorrect disposal is reduced; it can be recycled, in which the packaging itself is made of 65% recycled content and 35% fresh fiber. In addition, the robust design is more resilient in transportation, reducing food waste due to damage during transportation (WBCSD, 2020).

Goodyear has replaced petroleum-based oil with soybean oil for tire production. This innovation proves that research and development of sustainable materials can benefit not only the environment but also the performance of tires, which has been approved by the company, showing satisfactory benefits. Therefore, Goodyear aims to replace all petroleum-based oils by 2040 for tire production (WBCSD, 2020).

The Navigator Company recycles carbonate sludges, a waste product of their pulp processing and production stage. The waste is used as raw material in the production of Precipitated Calcium Carbonate (PCC), one of the main components of Navigator's wood-free paper. According to the company, such practice has several benefits: minimization of sludge disposal in industrial landfills and cost reduction; reduction of limestone mining, that is, fossil raw material, in addition to avoiding transportation costs and emissions related to traditional sludge disposal (WBCSD, 2020).

IFF has partnered with a Dutch startup, PeelPioneers, to use fresh orange peels that would otherwise be discarded for extraction of highly valued essential oils. IFF uses essential oil to create unique flavors, including natural orange flavors and extracts. In this way, the use of circular design principles saves resources and creates a marketing advantage for IFF's Re-Imagine Citrus Upcycled Orange product line (WBCSD, 2020).

Thus, it is observed that the circular bioeconomy offers business benefits and opportunities in addition to sustainable development.

16.7 CONCLUSIONS

This chapter sought to elucidate the concept of "circular bioeconomy" since this is a relatively new term with many approaches and intersections with the concepts of circular economy, bioeconomy, and biorefinery. Despite the differences found regarding its concept, all of them agree with the need for an immediate restructuring in the linear economic model practiced up to now. This is due to the current model being unsustainable and not providing economic, environmental, and social benefits for the current generation and the future ones since climate changes have been constantly occurring and they directly impact the population and economy, and, for example, food safety of growing populations. In addition, "biomass" was presented and discussed, which is an important resource for circular bioeconomy, for being renewable and for its potential to use residues as raw material to generate bioproducts, renewable biofuels, and bioenergy, which will aid in reducing, for example, the demand for fossil fuels, working inside the natural biogenic cycle of carbon. However, it is a fact that, besides the use of renewable raw materials, for a system of circular bioeconomy to work, it is necessary to develop disruptive technologies in all supply chains, where conventional concepts of production and consumption need to be reorganized and projected not only for the current user but also for the future use. They should encompass sustainable processes of biomass conversion into sustainable products, environmental-friendly packages, logistics, and transportation equalized to reduce pollutants emissions and sustainable or cyclic discards. Thus, sustainable systems in the format of biorefineries were presented which can include seaweed and lignocellulosic biorefineries, for example. Lastly, it was approached that the implementation of a new kind of sustainable economy is not a trivial task, because there are big barriers and challenges that need to be overcome. But, despite the barriers that circular bioeconomy may find, still it is evidenced by its need and how much it can contribute to a sustainable and economic industrial environment. The last subject presented in this chapter showed the opportunities that a circular bioeconomy system can create, in the development of new products with biological basis or even to substitute already existing conventional products. Its market is estimated for the next decades as trillion, with a high potential to grow in the biopharmaceutical, biofuels, energy, and food industries.

REFERENCES

Abad-Segura, E., Batlles-Delafuente, A., González- Zamar, M., Belmonte-Ureña, L. J. Implications for sustainability of the joint application of bioeconomy and circular economy: A worldwide trend study. *Sustainability*, v. 13, 2021. doi:10.3390/su13137182.

Agbor, V. B., Cicek, N., Sparling, R., Berlin, A., Levin, D. B. Biomass pretreatment: Fundamentals toward application. *Biotechnology Advances*, v. 29, 675–685, 2011. doi:10.1016/j.biotechadv.2011.05.005.

Aguilar, A., Twardowski, T. Bioeconomy in a changing word. *EFB Bioeconomy Journal*, v. 2, 100041, 2022. doi:10.1016/j.bioeco.2022.100041.

Andersen, M. S. An introductory note on the environmental economics of the circular economy. *Sustainability Science*, v. 2, 133–140, 2007. doi:10.1007/s11625-006-0013-6.

Andrade, M. C., Silva, C. O. G., Moreira, L. R. S., Ferreira Filho, E. X. Crop residues: applications of lignocellulosic biomass in the context of a biorefinery. *Frontiers in Energy*, v. 16, 224–245, 2022. doi:10.1007/s11708-021-0730-7.

Atitallah, I. B., Antonopoulou, G., Ntaikou, I., Alexandropoulou, M., Nasri, M., Mechichi, T., Lyberatos, G. On the evaluation of different saccharification schemes for enhanced bioethanol production from potato peels waste via a newly isolated yeast strain of *Wickerhamomyces anomalus*. *Bioresource Technology*, v. 289, 121614, 2019. doi:10.1016/j.biortech.2019.121614.

Awasthi, M. K., Sarsaiya, S., Patel, A., Juneja, A., Singh, R. P., Yan, B., Awasthi, S. K., Jain, A., Liu, T., Duan, Y., Pandey, A., Zhang, Z., Taherzadeh, M. J. Refining biomass residues for sustainable energy and bio-products: An assessment of technology, its importance, and strategic applications in circular bio-economy. *Renewable and Sustainable Energy Reviews*, v. 127, 109876, 2020. doi:10.1016/j.rser.2020.109876.

Aworunse, O. S., Olorunsola, H. A., Ahuekwe, E. F., Obembe, O. O. Towards a sustainable bioeconomy in a post-oil era Nigeria. *Resources, Environment and Sustainability*, v. 11, 100094, 2023. doi:10.1016/j.resenv.2022.100094.

Bajwa, D. S., Pourhashem, G., Ullah, A. H., Bajwa, S. G. A concise review of current lignin production, applications, products and their environment impact. *Industrial Crops and Products*, v. 139, 111526, 2019. doi:10.1016/j.indcrop.2019.111526.

Banu, J. R., Preethi, Kavitha, S., Tyagi, V. K., Gunasekaran, M., Karthikeyan, O. P., Kumar, G. Lignocellulosic biomass based biorefinery: A successful platform towards circular bioeconomy. *Fuel*, v. 302, 121086, 2021. doi:10.1016/j.fuel.2021.121086.

Borregaard. The Sustainable Biorefinery. https://www.borregaard.com/.

Boulding, K. The economy of the coming spaceship earth. In: Jarret H (ed) *Environmental Quality in a Growing Economy*. Johns Hopkins Press, Baltimore, pp. 3–14, 1966. https://arachnid.biosci.utexas.edu/courses/thoc/readings/boulding_spaceshipearth.pdf.

Bušić, A., Marđetko, N., Kundas, S., Morzak, G., Belskaya, H., Šantek, M. I., Komes, D., Novak, S., Šantek, B. Bioethanol production from renewable raw materials and its separation and purification: A review. *Food Technology & Biotechnology*, v. 56, 289–311, 2018. doi:10.17113/ftb.56.03.18.5546.

Carpenter, D., Westover, T. L., Czernik, S., Jablonski, W. Biomass feedstocks for renewable fuel production: a review of the impacts of feedstock and pretreatment on the yield and product distribution of fast pyrolysis bio-oils and vapors. *Green Chemistry*, v. 16, 384–406, 2014. doi:10.1039/C3GC41631C.

Carus, M., Dammer, L. The "circular bioeconomy"-concepts, opportunities and limitations. In: *Bio-Based Economy 2018-01*. Nova-Institut, Hürth, 2018. doi:10.1089/ind.2018.29121.mca.

Casau, M., Dias, M. F., Matias, J. C. O., Nunes, L. J. R. Residual biomass: A comprehensive review on the importance, uses and potential in a circular bioeconomy approach. *Resources*, v. 11, 2022. doi:10.3390/resources11040035.

Chen, X., Zhang, Y., Gu, Y., Liu, Z., Shen, Z., Chu, H., Zhou, X. Enhancing methane production from rice straw by extrusion pretreatment. *Applied Energy*, v. 122, 34–41, 2014. doi:10.1016/j.apenergy.2014.01.076.

Corona, B., Shen, L., Reike, D., Carreón, J. R., Worrell, E. Towards sustainable development through the circular economy: A review and critical assessment on current circularity metrics. *Resources, Conservation and Recycling*, v. 151, 104498, 2019. doi:10.1016/j.resconrec.2019.104498.

D'amato, D., Korhonen, J. Integrating the green economy, circular economy and bioeconomy in a strategic sustainability framework. *Ecological Economics*, v. 188, 107143, 2021. doi:10.1016/j.ecolecon.2021.107143.

Dahmen, N., Lewandowski, I., Zibek, S., Weidtmann, A. Integrated lignocellulosic value chains in a growing bioeconomy: Status quo and perspectives. *GCB Bioenergy*, v. 11, 107–117, 2019. doi:10.1111/gcbb.12586.

De Mattos, C. A., De Albuquerque, T. L. M. Enabling factors and strategies for the transition toward a circular economy (CE). *Sustainability*, v. 10, 2018. doi:10.3390/su10124628.

Domtar, P. M. https://www.domtar.com/en/who-we-are/all-locations/plymouth-mill, 2023.

Dragone, G., Kerssemakers, A. A. J., Driessen, J. L. S. P., Yamakawa, C. K., Brumano, L. P., Mussatto, S. I. Innovation and strategic orientations for the development of advanced biorefineries. *Bioresource Technology*, v. 302, 122847, 2020. doi:10.1016/j.biortech.2020.122847.

European Commission. *Closing the Loop: An EU Action Plan for the Circular Economy*, European Commission, Brussels, Belgium, 2015.

Frisvold, G. B., Moss, S. M., Hodgson, A., Maxon, M. E. Understanding the US bioeconomy: A new definition and landscape. *Sustainability*, v. 13, 1627, 2021. doi:10.3390/su13041627.

Gallo, M. E. The Bioeconomy: A Prime. *Congressional Research Service*, 2022. https://crsreports.congress.gov/product/pdf/R/R46881

Gedam, V. V., Raut, R. D., Jabbour, A. B. L. S., Tanksale, A. N., Narkhede, B. E. Circular economy practices in a developing economy: Barriers to be defeated. *Journal of Cleaner Production*, v. 311, 127670, 2021. doi:10.1016/j.jclepro.2021.127670.

Geissdoerfer, M., Savaget, P., Bocken, N. M. P., Hultink, E. J. The circular economy - a new sustainability paradigm? *Journal of Cleaner Production*, v. 143, 757–768. 2017 doi:10.1016/j.jclepro.2016.12.048.

German Bioeconomy Council. Bioeconomy Policy (Part III): Update Report of National Strategies around the World. German Bioeconomy Council, Berlin, 2018.

Girard, G. Does circular bioeconomy contain singular social science research questions, especially regarding agriculture: Industry nexus? *Cleaner and Circular Bioeconomy*, v. 3, 100030, 2022. doi:10.1016/j.clcb.2022.100030.

Guragain, Y. N., Vadlani, P. V. Renewable biomass utilization: A way forward to establish sustainable chemical and processing industries. *Clean Technologies*, v. 3, 243–259, 2021. doi:10.3390/cleantechnol3010014.

Hasanly, A., Talkhoncheh, M. K., Alavijeh, M. K. Techno-economic assessment of bioethanol production from wheat straw: a case study of Iran. *Clean Technologies and Environmental Policy*, v. 20, 357–377, 2018. doi:10.1007/s10098-017-1476-0.

Holden, N. M., Neill, A. M., Stout, J. C., O'Brien, D., Morris, M. A. Biocircularity: A framework to define sustainable, circular bioeconomy. *Circular Economy and Sustainability*, 2022. doi:10.1007/s43615-022-00180-y.

Husgafvel, R., Poikela, K., Honkatukia, J., Dahl, O. Development and piloting of sustainability assessment metrics for arctic process industry in Finland-the biorefinery investment and slag processing service cases. *Sustainability*, v. 9, 1693, 2017. doi:10.3390/su9101693.

IISD & UNEP. *International Institute for Sustainable Development & United Nations Environment Programme. Trade and Green Economy: A Handbook.* International Institute for Sustainable Development, Geneva, 2014

Jander W., Wydra, S., Wackerbauer, J., Grundmann, P., Piotrowski, S. Monitoring bioeconomy transitions with economic-environmental and innovation indicators: Addressing data gaps in the short term. *Sustainability*, v. 12, 4683, 2020. doi:10.3390/su12114683.

Juan, M. G. S., Bogdanski, A., Dubois, O. Towards sustainable bioeconomy-lessons learned from case studies. *Environment and Natural Resources Management*, 73, 132, Licence: CC BY-NC-SA 3.0 IGO, 2019

Juan, M. G. S., Harnett, S., Albinelli, I. *Sustainable and Circular Bioeconomy in the Climate Agenda: Opportunities to Transform Agrifood Systems.* Rome, FAO, 2022. doi:10.4060/cc2668en.

Kardung, M., Cingiz, K., Costenoble, O., Delahaye, R., Heijman, W., Lovri C, M., Leeuwen, M. V., M'barek, R., Meijl, H. V., Piotrowski, S., Ronzon, T., Sauer, J., Verhoog, D., Verkerk, P. J., Vrachioli, M., Wesseler, J. H. H., Zhu, B. X. Development of the circular bioeconomy: drivers and indicators. *Sustainability*, v. 13, 2021. doi:10.3390/su13010413.

Kohli, K., Prajapati, R., Sharma, B. K. Bio-based chemicals from renewable biomass for integrated biorefineries. *Energies*, v. 12, 2019. doi:10.3390/en12020233.

Kostas, E. T., Adams, J. M. M., Ruiz, H. A., Durán-Jiménez, G., Lye, G. J. Macroalgal biorefinery concepts for the circular bioeconomy: A review on biotechnological developments and future perspectives. *Renewable and Sustainable Energy Reviews*, v. 151, 111553, 2021. doi:10.1016/j.rser.2021.111553.

Lange, L., Connor, K. O., Arason, S., Bundgård-Jørgensen, U., Canalis, A., Carrez, D., Gallagher, J., Gøtke, N., Huyghe, C., Jarry, B., Llorente, P., Marinova, M., Martins, L. O., Menga, P., Paiano, P., Panoutsou, C., Rodrigues, L., Stenge, D. B., Van Der Meer, Y., Vieira, H. Developing a sustainable and circular bio-based economy in EU: By partnering across sectors, upscaling and using new knowledge faster, and for the benefit of climate, environment & biodiversity, and people & business. *Frontiers in Bioengineering and Biotechnology*, v. 8, 2020. doi:10.3389/fbioe.2020.619066.

Lappa, I. K., Papadaki, A., Kachrimanidou, V., Terpou, A., Koulougliotis, D., Eriotou, E., Kopsahelis, N. Cheese whey processing: integrated biorefinery concepts and emerging food applications. *Foods*, v. 8, 2019. doi:10.3390/foods8080347.

Lask, J., Wagner, M., Trindade, L. M., Lewandowski, I. Life cycle assessment of ethanol production from miscanthus: A comparison of production pathways at two European sites. *GCB-Bioenergy*, v. 11, 269–288, 2019. doi:10.1111/gcbb.12551.

Lenzing. Biorefinery: Innovative by Nature. https://www.lenzing.com/sustainability/production/biorefinery..

Leong, H. Y., Chang, C., Khoo, K. S., Chew, K. W., Chia, S. R., Lim, J. W., Chang, J., Show, P. L. Waste biorefnery towards a sustainable circular bioeconomy: a solution to global issues. *Biotechnology for Biofuels and Bioproducts*, v. 14, 2021. doi:10.1186/s13068-021-01939-5.

Lewandowski, I. *Bioeconomy: Shaping the Transition to A Sustainable, Biobased Economy.* Springer, Cham, Switzerland, 2018. doi:10.1007/978-3-319-68152-8.

Luo, X., Chen, S., Hu, T., Chen, Y., Li, F. Renewable biomass-derived carbons for electrochemical capacitor applications. *SusMat*, v. 1, 211–240, 2021. doi:10.1002/sus2.8.

Maestre, K. L., Passos, F. R., Triques, C. C., Fiorentin-Ferrari, L. D., Slusarski-Santana, V., Garcia, H. A. Silva, E. A., Fiorese, M. L. Cheese whey permeate valorization using sequential fermentations: case study performed in the Western Region of Paraná. *Research, Society and Development*, v. 10, e212101321082, 2021. doi:10.33448/rsd-v10i13.21082.

Martínez- Ávila, G. C., Aguilera- Carbó, A. F., rodríguez-Herrera, R., Aguilar, C. N. Fungal enhancement of the antioxidant properties of grape waste. *Annals of Microbiology*, v. 62, 923-930, 2012. doi:10.1007/s13213-011-0329-z.

Martins, F., Felgueiras, C., Smitkova, M., Caetano, N. Analysis of fossil fuel energy consumption and environmental impacts in European countries. *Energies*, v. 12, 2019. doi:10.3390/en12060964.

Mckendry, P. Energy production from biomass (part 1): Overview of biomass. *Bioresource Technology*, v. 83, 37–46, 2002. doi:10.1016/S0960-8524(01)00118-3.

Muswellbrook Shire Council & APACE. Hunter Pilot Biorefinery. https://www.ethtec.com.au/updates

Nasem-National Academies of Sciences, Engineering, and Medicine. *Safeguarding the Bioeconomy*. *National Academies*. The National Academies Press, Washington, DC, 2020. doi:10.17226/25525.

Office of Science and Technology Policy. *Summary of the 2019 White House Summit on America's Bioeconomy*. Office of Science and Technology Policy, Washington, DC, 2019.

Palahí, M., Pantsar, M., Costanza, R., kubiszewski, I., potočnik, J., Stuchtey, M., Nasi, R., Lovins, H., Giovannini, E., fioramonti, L., Dixson-Declève, S., Mcglade, J., Pickett, K., Wilkinson, R., Holmgren, J., Trebeck, K., Wallis, S., Ramage, M., Berndes, G., Akinnifesi, F., Ragnarsdóttir, K. V., Muys, B., Safonov, G., Nobre, A., Nobre, C., Ibañez, D., Wijkman, A., Snape, J., Bas, L. *Investing in Nature as the True Engine of Our Economy: A 10-Point Action Plan for a Circular Bioeconomy of Wellbeing* (Knowledge to Action 02). European Forest Institute, Finland, 2020. doi:10.36333/k2a0.

Patermann, C., Aguilar, A. A bioeconomy for the next decade. *EFB Bioeconomy Journal*, v. 1, 100005, 2021. doi:10.1016/j.bioeco.2021.100005.

Paulinetti, A. P., Augusto, I. M. G., Batista, L. P. P., Tavares, A. G. B., Albanez, R., Ratusznei, S. M., Lovato, G., Rodrigues, J. A. D. Anaerobic digestion as a core process for sustainable energy production in the soybean biorefinery: A techno-economic assessment. *Sustainable Horizons*, v. 3, 100024, 2022. doi:10.1016/j.horiz.2022.100024.

Pendón, M. D. Madeira, J. R., J. V., Romanin, D. E., Rumbo, M., Gombert, A. K., Garrote, G. L. A biorefinery concept for the production of fuel ethanol, probiotic yeast, and whey protein from a by-product of the cheese industry. *Applied Microbiology and Biotechnology*, v. 105, 3859–3871, 2021. doi:10.1007/s00253-021-11278-y.

Salvador, R., Barros, M. V., Donner, M., Brito, P., Halog, A., DE Francisco, A. C. How to advance regional circular bioeconomy systems? Identifying barriers, challenges, drivers, and opportunities. *Sustainable Production and Consumption*, v. 32, 248–269, 2022. doi:10.1016/j.spc.2022.04.025.

Salvador, R., Puglieri, F. N., Halog, A., De Andrade, F. G., Piekarski, C. M., de Francisco, A. C. Key aspects for designing business models for a circular bioeconomy. *Journal of Cleaner Production*, v. 278, 124341, 2021. doi:10.1016/j.jclepro.2020.124341.

Saratale, G. D., Saratale, R. G., Banu, J. R., Chang, J. Biohydrogen production from renewable biomass resources. In: *Biohydrogen*, 2nd edn. pp. 247–277, 2019. doi:10.1016/B978-0-444-64203-5.00010-1.

Sharma, P. Gaur, V. K., Sirohi, R., Varjani, S., Kim, S. H., Wong, J. W. C. Sustainable processing of food waste for production of bio-based products for circular bioeconomy. *Bioresource Technology*, v. 325, 124684, 2021. doi:10.1016/j.biortech.2021.124684.

Stefanakis, A. I., Nikolaou, I. Circular economy and sustainability. *Environmental Engineering*, v. 2, 2021. doi:10.1016/C2019-0-04146-5.

Stegmann, P., Londo, M., Junginger, M. The circular bioeconomy: Its elements and role in European bioeconomy clusters. *Resources, Conservation & Recycling*, 6, 100029, 2020. doi:10.1016/j.rcrx.2019.100029.

Sunila Mill. About Us: Stora Enso. https://www.storaenso.com/en/about-stora-enso/stora-enso-locations/sunila-mill

Tan, E. C. D., Lamers, P. Circular bioeconomy concepts: A perspective. *Frontiers in Sustainability*, v. 2, 701509, 2021. doi:10.3389/frsus.2021.701509.

Tsegaye, B., Jaiswal, S., Jaiswal, A. K. Food waste biorefinery: Pathway towards circular bioeconomy. *Foods*, v. 10, 2021. doi:10.3390/foods10061174.

Ubando, A. T., Felix, C. B., Chen, W. Biorefineries in circular bioeconomy: A comprehensive review. *Bioresource Technology*, v. 299, 122585, 2020. doi:10.1016/j.biortech.2019.122585.

Venkatesh, G. Circular bio-economy-paradigm for the future: Systematic review of scientific journal publications from 2015 to 2021. *Circular Economy and Sustainability*, v. 2, 231–279, 2022. doi:10.1007/s43615-021-00084-3.

Vidal, M. Ethanol Production and Market. ETENE Sectorial Section, Technical Office for Economic Studies of the Northeast. https://www.bnb.gov.br/s482-dspace/bitstream/123456789/906/1/2021_CDS_159.pdf

Wagemann, K., et al. Biorefineries Roadmap: As Part of the German Federal Government Action Plans for the Material and Energetic Utilisation of Renewable Raw Materials. Federal Government of Germany. https://www.etipbioenergy.eu/databases/reports/157-biorefineries-roadmap-as-part-of-the-german-federal-government-action-plans-for-the-material-and-energetic-utilisation-of-renewable-raw-materials.

Wang, K., Gao, S., Lai, C., Xie, Y., Sun, Y., Wang, J., Wang, C., Yong, Q., Chu, F., Zhang, D. Upgrading wood biorefinery: An integration strategy for sugar production and reactive lignin preparation. *Industrial Crops and Products*, v. 187, 115366, 2022. doi:10.1016/j.indcrop.2022.115366.

WBCSD—World Business Council for Sustainable Development. Circular Bioeconomy: The Business Opportunity Contributing to a Sustainable World. https://www.wbcsd.org/Archive/Factor-10/Resources/The-circular-bioeconomy-A-business-opportunity-contributing-to-a-sustainable-world.

Weber, C. T., Trierweiler, L. F., Trierweiler, J. O. Food waste biorefinery advocating circular economy: Bioethanol and distilled beverage from sweet potato. *Journal of Cleaner Production*, v. 268, 121788, 2020. doi:10.1016/j.jclepro.2020.121788.

Yan, L., Wang, Y., Li, J., Zhang, Y., Ma, L., Fu, F., Chen, B., Liu, H. Hydrothermal liquefaction of Ulva prolifera macroalgae and the influence of base catalysts on products. *Bioresource Technology*, v. 292, 121286, 2019. doi:10.1016/j.biortech.2019.03.125.

17 A Circular Economy Approach to Valorisation of Lignocellulosic Biomass-Biochar and Bioethanol Production

Sai Shankar Sahu and Subodh Kumar Maiti
Indian Institute of Technology (Indian School of Mines), Dhanbad

17.1 INTRODUCTION

Renewable bioresources–based circular economy has attracted a lot of interest as civilization has developed (Brodin et al., 2017). Lignocellulosic biomass (LCB) resources, such as wood residues, grasses, agricultural wastes, and municipal solid wastes (MSW), bear a great potential for valorisation and commercialization, as they are available in huge quantities and at low costs (Couto & Herrera, 2006; Bilal & Asgher, 2016). A variety of value-added products such as enzymes, biomaterials, efficient biopolymers, bioenergy, and other useful compounds can all be made using different LCBs as the raw material (Albashabsheh & Stamm, 2021). However, a large amount of LCB has not been properly utilised in India. Utilising these lignocellulosic materials can efficiently cut manufacturing costs and resource waste, as well as the severe environmental issues brought on by the vast disposal of biomass materials. It can also lower CO_2 emissions. However, the biomass that has been used for biofuel production is categorised into four generations, which are discussed in Table 17.1.

The production of biochar and bioethanol from LCB is relevant, as it is advantageous from both economic and environmental points of view (Donato et al., 2019; Godlewska et al., 2017). Production of bioethanol can be carried out by the hydrolysis of cellulose and hemicellulose (Bobleter, 1994; Marzo et al., 2019). The cellulose fibres in the native biomass are covered by a complicated structure, called lignin. Lignin is a recalcitrant natural polymer that supports structural integrity and safeguards cells. It contributes significantly to the lignocellulosic component, which makes up 15%–30% of it (Khan & Ahring, 2019; Szalaty et al., 2020).

The lignin can be degraded by pre-treating LCB which results in the yielding of more fermentable sugars. The most expensive and challenging phase in the production of bioethanol is pretreatment, though (Bhutto et al., 2017). Biological, chemical,

TABLE 17.1

Different Generations of Feedstocks and Their Distinct Features

First Generation: Food Based Feedstocks	Second Generation: Lignocellulosic Biomass (Non-food Sources)	Third Generation: Microalgae	Fourth Generation: Genetically Modified Microalgae
1. Crops such as corn, sugar cane, and soybean	1. Wood, agricultural and organic waste	1. Microalgae, seaweed, macroalgae.	1. Genetically modified or engineered biomass.
2. Food security is compromised.	2. Food security is not hampered.	2. No effect on food security.	2. Food security is not hampered.
3. It needs cultivable land for feedstock generation.	3. It requires cultivable lands or forests.	3. Arable land is not needed as it can be grown in waste lagoon, sea etc.	3. No loss of biodiversity.
4. Loss of biodiversity due to monoculture of feedstocks.	4. No loss of biodiversity.	4. No loss of biodiversity.	4. It can be grown in rough conditions.
5. Due to the use of fertilizers and pesticides, it may have impact on environment negatively.	5. Its growth may be affected by the environmental factors.	5. Eutrophication can be a concern.	5. Genetically modified strains are released, which can be an environmental concern.

and physical pretreatment techniques are among the ones, which are commonly being used to break down the walls of plant cells within LCB. In accordance with the structural and physicochemical properties of LCB, the method of pretreatment is being chosen. Pretreatment can be used to modify the microscopic structure of the cellulose matrix and reduce the degree of crystallinity. In addition, it expands the biomass' surface area and makes the cellulose and hemicellulose components more accessible to hydrolytic enzymes (Mood et al., 2013). The simplified operational procedure and lower energy inputs would both benefit from the optimised pretreatment approach (Sørensen et al., 2008; Wyman et al., 2005). However, a lot of fermentation inhibitors are unavoidably produced during the pretreatment stage, and these inhibitors may prevent microbial development and have an impact on the succeeding bioethanol production process (Guo et al., 2018; Fernandes et al., 2017). As a result, many detoxification techniques have been developed to get rid of these fermentation inhibitors. The methodology used depends on the makeup of the various LCB sources and primarily uses commercial adsorbents and resins. The simplified operational procedure and lower energy inputs would both benefit from the optimised pretreatment approach (Martín et al., 2017; Sun et al., 2018; Yu & Christopher, 2017).

Due to its numerous sources and advantageous economics, biochar has been utilised widely in the detoxification process as an excellent adsorbent. Biochar is effective for the removal of both organic and inorganic pollutants as it has a tendency to attract nonpolar compounds, and increase the porosity, surface area, and surface functional groups (Kołodyńska et al., 2012; Son et al., 2018; Zhang et al., 2019). In addition, the use of biochar in the fermentation of bioethanol is advantageous for enhancing the environment for fermentation and immobilising the fermentative microorganisms, which enhances bioethanol production (Kirdponpattara & Phisalaphong, 2013; Sun et al., 2018). However, there have been few reports of using biochar to improve the production of bioethanol by detoxifying a hydrolysate that has already been treated.

In this chapter, the production of biochar and bioethanol from LCB, as well as the potential of their combined production in a single system is discussed in a circular economy model. In particular, the techniques and technologies used to ferment various LCB materials and produce biochar and bioethanol are given. In addition to that, the insights of whether biochar can be effectively combined with the bioethanol production process to increase the bioethanol output because of its unique adsorption capability and physicochemical characteristics are discussed. A study that is particularly noteworthy attempts to produce bioethanol from LCB while also attempting to detoxify fermentation inhibitors in situ by creating a self-sufficient system using biochar made from the same LCB at low temperatures.

17.2 BIOETHANOL PRODUCTION FROM LIGNOCELLULOSIC BIOMASS

Bioethanol is a liquid fuel that has a significant capacity to replace conventional fuel to a certain extent and can be incorporated with the present framework for the dispensation of fuel. In the process of production of bioethanol, there are several challenges that need to be encountered. First, to ensure the availability of cellulosic feedstocks on a large scale. Second, to extract cellulose from the LCB and alter it into ethanol with an optimized cost. A substitute for the presently used fossil fuel is the need of the hour as the supply of oil is decreasing, the price being inconsistent, and the adverse effect on the environment increasing day by day (Isikgor & Becer, 2015). However, bioethanol production from LCB has been gaining more attention in the field of research and development over the recent years because of many environmental and economic well-being. The main issue for the utilization of LCB lies in the breakdown of lignin using different pretreatment techniques, as it is recalcitrant in nature. One of the most recommended approaches for acquiring fermentable sugars out of cellulose is enzymatic hydrolysis, as it is more cost-viable than the other approaches such as chemical hydrolysis (Paz et al., 2019).

The term "lignocellulosic" refers to a variety of cellulose-containing plant molecules and biomass that differ in lignin content, polymerization levels and chain length. LCB can be used for the production of advanced diesel and cellulosic ethanol through biochemical conversion techniques (Cheng & Timilsina, 2011). Lignocellulosic biopolymers – typically cellulose, hemicelluloses, and lignin – extracted from wood and plants have long been acknowledged as the most plentiful and promising alternative sources for making bioethanol and other value-added products. The agricultural wastes have 45%–56% of a homopolymer of glucose, that is, cellulose, 10%–25% of a heteropolymer (i.e., hemicellulose), which consist of xylose and other important components such as mannose, galactose, arabinose, and glucose and remaining part of 18%–30% is lignin with components such as minerals, phenolic and acetyl groups (Zhou et al., 2011). Depending on the kind of plant, these elements may have different compositions.

The feedstock's accessibility and transportation expenses continue to be crucial factors in the production of ethanol from sources that contain sucrose. On the other hand, because of the competition for the limited amount of arable land caused by the widespread farming of crop-based feedstocks for ethanol production, it is not considered a sustainable approach (Donato et al., 2019). As a possible alternative feedstock for the production of bioethanol, lower value lignocellulosic materials from forestry, agricultural leftovers, as well as urban and industrial wastes, are proposed. Lignocellulosic materials have undergone extensive research and development over the past few decades, and they will likely play a significant role as a feedstock for ethanol generation in the future.

LCBs resist enzymes' intervention as it is not so permeable in nature, which makes it difficult for them to be broken down into their component sugar molecules. The conversion of LCB to ethanol requires three steps: pretreatment, hydrolysis, and fermentation (Zabed et al., 2016). Among all, pretreatment is an expensive affair and it can be made efficient enough with respect to its cost viability through extensive research and development.

17.2.1 Pretreatment of Lignocellulosic Biomass

Pretreatment is an important step in enabling enzymatic or acidic hydrolysis of cellulose by disrupting the crystalline structure of lignin and making it more porous. The conversion of sugars should increase after a successful pretreatment. It should prevent the loss or degradation of carbohydrates and the production of by-products, which impede both fermentation and hydrolysis (Mood et al., 2013). It should be economical and energy-efficient. The development of an efficient pretreatment technique under ideal operating conditions is desired to make the process cost-effective. The main concerns for a cost-effective pretreatment are limited inhibitor development, preservation of cellulose and hemicellulose, and reduction in chemical and energy use (Kumar et al., 2020). A physical, physicochemical, chemical, or biological procedure can be used as a pretreatment approach for LCB. However, a schematic diagram for the effect of pretreatment on LCB is presented in Figure 17.1.

17.2.1.1 Physical Pretreatment

Several techniques, including grinding, shredding, chipping, milling, extrusion, irradiation, and others, can be used to comminute biomass in this process. Chipping's primary objectives are to minimise heat and restrict mass transfer. The biomass's size and crystallinity are efficiently reduced by milling and grinding. Studies have revealed that grinding and milling efficiently decrease the size of biomass to 0.2–2 mm, whereas chipping reduces biomass size to 10–30 mm. However, it was discovered that reducing the size of the biomass beyond 0.4 mm had little to no impact on the pace and yield of hydrolysis (Sarkar et al., 2012). During the treatment procedure, physical pretreatment does not result in harmful by-products. However, this approach has a drawback in terms of energy usage. The energy requirement for mechanical

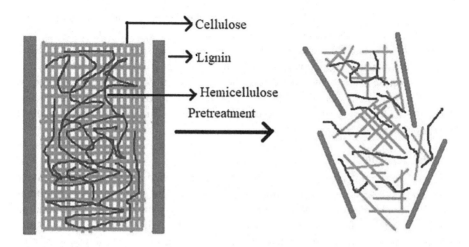

FIGURE 17.1   A schematic representation of the effect of pretreatment on lignocellulosic biomass.

comminution depends on the size and properties of LCB. This technology is not economically viable because of the enormous demand for energy. Another drawback is the inability to remove lignin and restrict the enzymes' accessibility to cellulose. The overall yield of the reducing sugars was, therefore, positively impacted by optimising the pretreatment process conditions and enzyme concentrations (Bhutto et al., 2017).

17.2.1.2   Chemical Pretreatment

17.2.1.2.1 Acid Pretreatment

To increase the accessibility of cellulose to enzymes, acid pretreatment is used to solubilize the lignin and hemicellulose part of the biomass. Both concentrated and diluted acids can be used for this purpose, although concentrated acid should not be utilised because it tends to create chemicals that hinder the synthesis of ethanol (Kumar et al., 2020). The primary benefit of utilising a diluted acid is the quick reaction time, which makes it possible to treat the biomass continuously. The time it takes to hydrolyse the sugars might range from a few minutes to many hours, depending on the pretreatment conditions. In addition, it has the benefit of solubilizing hemicelluloses and turning those hemicelluloses into fermentable sugars. Higher sugar recovery, higher porosity, and improved enzymatic digestibility are the outcomes (Balat, 2011). For diluted acid pretreatment, both inorganic and organic acids (such as fumaric acid and maleic acid) are utilised (Mood et al., 2013). On an industrial basis, two different types of acid pretreatment techniques are frequently used. Pretreatment at high temperature (above 180°C) for a brief period of time (1–5 min) and at low temperature (below 120°C) for a lengthy period of time (30–90 min). Enzyme hydrolysis of biomass could be omitted after acid pretreatment because the biomass gets hydrolysed by the acid itself and turn into fermentable sugars. Prior to the fermentation of sugars, however, the acid must be thoroughly washed away. The pretreatment of reed canarygrass (*Phalaris arundinacea* L.), switch grass (*Panicum virgatum* L.), and maize stover is frequently done using diluted sulfuric acid (Yoshioka et al., 2018). The use of acids causes excessive corrosion in the pretreatment reactors, which necessitates the need for specialised reactors having resistance to corrosion. Such reactors are typically very costly and thus make it impractical to procure. Hence, it is a key disadvantage of utilising diluted acid. Due to its corrosiveness and toxicity, acid pretreatment also has significant disadvantages. Due to the production of phenolic acids, aldehydes, and fermentation inhibitors such as furfural and hydroxymethyl furfural, it has received less attention.

17.2.1.2.2 Alkaline Pretreatment

Due to its many appealing qualities, alkaline pretreatment has drawn increased attention. The biomass is soaked in an alkaline solution and put at the necessary temperature for a predetermined period. The procedure is to soak the biomass in the alkaline solution and mix it for a predetermined amount of time at the necessary temperature. The amount of lignin in the biomass determines how the alkaline treatment performs (Guo et al., 2018). Alkaline pretreatment is carried out at lower temperatures and pressures than acid pretreatment, and it also results in less sugar degradation. Most of the chemicals used in it are non-corrosive and non-polluting, such as sodium, potassium, ammonia, and calcium hydroxides (Balat, 2011). The cheapest alkali is lime (CaO or Ca(OH)$_2$). They offer a less expensive alternative for removing lignin at higher pH levels. In addition, it suppresses lignin condensation, resulting in high lignin solubility for biomass having relatively less lignin. Moreover, the acetyl groups present in hemicellulose are eliminated and the resistance of the enzyme is lowered by this pretreatment (Zabed et al., 2016). The pretreatment of switchgrass, wheat grass, and poplar wood has been successfully accomplished with lime (Zhang et al., 2019).

The use of sodium hydroxide (NaOH) in the pretreatment of biomass is also extensively researched. By rupturing ether and ester bonds and inflating cellulose, it breaks down lignin. The removal of acetyl and different uronic acid substitutions on hemicelluloses further improves the accessibility of the enzymes to the cellulose with little loss of polysaccharides. Low-concentration and high-concentration NaOH pretreatments are the two categories. On agricultural waste, herbaceous plants, hardwood, and the treatment of spruce at low temperatures, NaOH pretreatment is more successful (Wyman et al., 2005). The disadvantage of alkaline pretreatment is that alkali is transformed into irrecoverable salts in the biomass, which may need additional treatment. Alkali pretreatment of highly resistant biomass is difficult and necessitates harsh pretreatment conditions. Hemicellulose is lost as a result of this situation, and inhibitors are produced. Lower pressure and temperature are used for NaOH pretreatment. But when solid NaOH dissolves in water, a lot of heat is generated, which might be harmful to the environment (Gismatulina & Budaeva, 2017). Neutralisation is necessary before enzymatic saccharification to get rid of lignin and inhibitors.

17.2.1.3 Physicochemical Pretreatment

17.2.1.3.1 Steam Explosion Method

The structure of the lignocellulosic material has been broken down using both chemical and physical methods in this procedure, which has been utilised to pre-treat various biomass feedstocks. By using this technique, the enzyme activity's surface area is increased, which accelerates the rate of enzyme hydrolysis. For a brief period of time (10 min), the biomass undergoes a temperature of 160°C–260°C and high pressure before being abruptly depressurized, which causes the biomass's structure to be quickly disrupted (Sørensen et al., 2008). This process has a number of appealing qualities, including the ability to recover all of the sugar produced, the use of little capital investment, few negative effects on the environment, and the use of less dangerous chemicals and working conditions. In any case, catalysts such as H_2SO_4, CO_2 and SO_2 can effectively increase the efficiency of steam pretreatment (Behera et al., 2014). The process is catalysed by autohydrolysis without the use of a catalyst. It has been discovered that a steam explosion with an acid catalyst is highly effective in minimizing the generation of chemicals that act as inhibitors and thereby enhance enzymatic hydrolysis (Balat, 2011). For the pretreatment of a variety of raw materials, including olive tree wood, wheat straw, and poplar wood, the steam explosion method is frequently utilised (Kapoor et al., 2015).

High energy demand and creation of inhibitors during the process are certain drawbacks of the steam explosion method. Another drawback is the partial hemicellulose breakdown that results in the production of hazardous chemicals, which may interfere with the hydrolysis and fermentation processes. In addition to contributing to acidification, formic and levulinic acids produced as a result of sugar degradation as well as acetic and uronic acid released from hemicellulose can also hinder further metabolic reactions (Mood et al., 2013).

17.2.1.4 Biological Pretreatment

When compared with traditional procedures such as chemical and physical pretreatments, biological treatment is regarded as an effective, low-energy process that is also environmentally benign. Microorganisms that can produce biomass-degrading enzymes, such as soft-rot fungi and white-brown fungi, are typically used in this approach (Khan & Ahring, 2019). On LCB, fungi have unique capabilities for degrading it (Zhang et al., 2019). Despite significant lignin degradation, it is a safe and environmentally benign approach that does not necessitate expensive machinery or high energy consumption. When compared with other procedures, this process produces the desired goods with a better yield. White and soft rots break down both cellulose and lignin, whereas brown rots mostly break down cellulose (Khan & Ahring, 2019). Peroxidases and laccase, two lignin-degrading enzymes, are responsible for the degradation. Enzymes that can degrade lignin such as laccases and peroxidases are present in white-rot fungi (Couto & Herrera, 2006). Phanerochaete chrysosporium, a white-rot fungus, is renowned for providing the most effective pretreatment due to its rapid development and capacity to break down lignin. P. chrysosporium's simultaneous pretreatment and saccharification of rice husk were investigated (Pérez et al., 2002). *Irpex lacteus*, a white-rot fungi, has been prominent for the biological pretreatment of maize stalks. It can produce a variety of extracellular hydrolytic and oxidative enzymes during the course of the treatment.

A brief description of different pretreatment methods and their advantages and disadvantages has been discussed in detail in Table 17.2.

TABLE 17.2

A Brief Discussion of Different Pretreatment Methods, and Its Advantages and Disadvantages

Pretreatment Methods	Brief Description	Advantages	Disadvantages
Physical Pretreatments			
Mechanical communition	A process that combines milling, grinding, and chipping to reduce the material's final particle size to 10–30 mm (10–30 mm after chipping and 0.2–2 mm after milling or grinding).	1. Reduction in the crystallinity of cellulose. 2. A larger surface area. 3. Reduction in polymerization intensity. 4. Simple to handle.	1. High intake of energy. 2. Removal of lignin is low.
Irradiation	Treatment of biomass using high-energy radiation, such as microwave heating, gamma rays, ultrasound, electron beams, and pulsed electrical fields.	1. Surface area increases 2. Polymerization decreases. 3. The crystallinity of cellulose gets reduced. 4. Degradation of lignin occurs partially.	1. Very costly. 2. Reaction rate is slow. 3. Energy exhaustive. 4. Not an eco-friendly approach.
Chemical Pretreatments			
Dilute acid treatment	1. Less than 4% acid concentration. 2. Works at low temperature (for a longer retention time like 30–90 min) or high temperature (for example, 180°C) for a short time. 3. Use of fumaric and maleic acids as well as organic acids (H_2SO_4, HCl, H_3PO_4, and HNO_3).	1. Surface area increases. 2. Rate of reaction is high. 3. Hemicellulose removal. 4. Lignin structure gets altered. 5. Digestibility improves.	1. Less lignin removal. 2. Inhibitors form in the process. 3. Requires neutralization.

(Continued)

TABLE 17.2 (*Continued*)

A Brief Discussion of Different Pretreatment Methods, and Its Advantages and Disadvantages

Pretreatment Methods	Brief Description	Advantages	Disadvantages
Concentrated acid pretreatment	1. Acid concentrations range from 70% to 77%, and its temperature is 40°C–100°C. 2. The main acids utilised are inorganic acids (H_2SO_4 and H_3PO_4).	1. Cellulose crystalline structure gets completely removed. 2. Amorphous cellulose is achieved. 3. Surface area increases. 4. Rate of reaction is high.	1. Inhibitors formation. 2. Equipment corrosion. 3. Recovery of acid is needed. 4. Cost is high.
Alkali pretreatment	1. Base such as NaOH, NH_4OH, KOH, and $Ca(OH)_2$ are used. 2. Perform at room temperature. 3. Reaction time is seconds to days.	1. Temperature and pressure is low. 2. Increase in surface area. 3. Hemicellulose gets removed. 4. Lignin is removed. 5. Lignin structure is altered. 6. Degradation of sugar is lower as compared with acid pretreatment.	1. Alkali is converted to salts which is not recoverable. 2. pH adjustment is needed.
Ozonolysis	1. At room temperature and pressure, treated with ozone gas. 2. Time of reaction is a few hours.	1. Cellulose and hemicellulose are not significantly affected by the selective lignin breakdown. 2. Inhibitor forms in a less quantity. 3. Works at room temperature and pressure. 4. Production and direct use of ozone on-site.	1. Risky, because ozone is highly reactive, combustible, corrosive, and toxic in nature. 2. Large energy demand results in high generation costs. 3. Processes with exothermic features could need cooling systems.

(*Continued*)

TABLE 17.2 (*Continued*)

A Brief Discussion of Different Pretreatment Methods, and Its Advantages and Disadvantages

Pretreatment Methods	Brief Description	Advantages	Disadvantages
Ionic liquid (IL) pretreatment	1. ILs are salts that are often made up of small inorganic anions and big organic cations. 2. ILs frequently exist as liquids at room temperature and have a propensity to stay liquid throughout a wide temperature range.	1. There is no formation of explosive or toxic gases. 2. Lignin and carbohydrates can both dissolve at the same time. 3. Less degradation of products which are desired. 4. Less temperature. 5. Biomass loading capacity is high. 6. Solubility of lignin is more. 7. Significantly improve enzymatic delignification.	1. Costly. 2. Washing is necessary before reuse. 3. Inadequately developed commercial IL recovery techniques. 4. During IL pretreatment, temperature and biomass input had a significant impact on the hydrolysis rate.
Organosolv	1. During IL pretreatment, temperature and biomass input had a significant impact on the hydrolysis rate. 2. To break hemicellulose bonds, solvent can be coupled with acid catalysts such as HCl, H_2SO_4, oxalic, or salicylic.	1. Recovery of lignin that is comparatively pure as a byproduct. 2. Cellulose loss of no more than 2%. 3. High yield of pretreated material. 4. Little decomposition of sugar.	1. High expenses of chemicals needed to remove the solvent from the system. 2. Development of inhibitors.
Physicochemical Pretreatments			
Steam explosion (Uncatalyzed)	1. Steam treatment is performed at high pressures and temperatures of 0.69–4.83 MPa and 160°C–260°C, respectively. 2. Several seconds to a few minutes for a reaction. 3. Chemical free treatment.	1. Hemicellulose is removed. 2. Reduced particle size – improved enzyme accessibility. 3. Volume of pores increases.	1. Removal of lignin is low. 2. Sugars get decomposed. 3. Requires high energy.

(*Continued*)

TABLE 17.2 (Continued)

A Brief Discussion of Different Pretreatment Methods, and Its Advantages and Disadvantages

Pretreatment Methods	Brief Description	Advantages	Disadvantages
Steam explosion (Acid catalysed)	1. H_2SO_4 or SO_2 addition catalyses steam explosion. 2. Performance under the temperature range of 160°C–220°C.	1. Hemicellulose is removed. 2. Surface area increases. 3. Accessibility of enzymes increases. 4. Impact on the environment is low.	1. Inhibitors formation. 2. Hydrolysis of hemicellulose is partial.
Liquid hot water	1. Biomass treated hydrothermally with quick decompression. 2. In order to keep water in a liquid form at high temperatures (160°C–240°C), pressure is applied. 3. 4–7 pH range and up to 15 min reaction time.	1. Lignin has undergone structural and chemical change. 2. No requirement of catalyst. 3. Accessible surface area increases. 4. Hemicellulose is removed. 5. Formation of inhibitor is less. 6. Reactor cost is low.	1. Requires high water. 2. Energy demand is high.
Ammonia fibre explosion (AFEX)	Rapid decompression is followed by treatment with anhydrous liquid ammonia at 60°C–120°C and above 3 MPa for 30–60 min.	1. Surface area increases. 2. Crystallinity of cellulose reduces. 3. Hemicellulose is removed. 4. Removes lignin. 5. Inhibitors don't form.	1. Less effective for biomass based on softwoods. 2. Environment-related issues.
Wet oxidation	1. A pretreatment that involves oxidation using air or oxygen as the catalyst. 2. At 170°C–200°C and 10–12 bars, the oxidation lasts for around 10–15 min.	1. Lignin and hemicellulose solubilisation. 2. Cellulose digestibility increases. 3. Inhibitors formation is less. 4. Lignin removal.	1. Costly. 2. Needs high pressure and temperature.
CO_2 explosion	As a supercritical fluid, CO_2 is utilised to pre-treat biomass.	1. Removes lignin efficiently. 2. Digestibility Increases. 3. Accessible surface area increases. 4. Degradation of sugar is low. 5. Works at low temperature.	Costly.

(Continued)

TABLE 17.2 (*Continued*)

A Brief Discussion of Different Pretreatment Methods, and Its Advantages and Disadvantages

Pretreatment Methods	Brief Description	Advantages	Disadvantages
		Biological Pretreatment	
Pretreatment using microorganisms	Treatments utilising microbes (bacteria, fungus, and actinomycetes).	1. Degradation of lignin. 2. Reduction in the level of cellulose and hemicellulose polymerization. 3. No requirement of chemical. 4. Cost is low. 5. Energy requirement is low. 6. Less inhibitor formation.	1. Rate of degradation is very slow. 2. Loss of carbohydrates due to microbial consumption. 3. Time of residence is very long.

17.2.2 HYDROLYSIS OF LIGNOCELLULOSIC BIOMASS

In this process, the hydrogen bonds in LCB are split off into monosaccharides such as pentose, hexose, and other sugar compounds. These monosaccharides are further fermented and bioethanol is produced (Mohapatra et al., 2017). Chemical hydrolysis (dilute and concentrated acid hydrolysis) and enzymatic hydrolysis are two commonly utilised hydrolysis techniques (Marzo et al., 2019). Pretreatment and hydrolysis can both be done in one step during chemical hydrolysis. This process makes it inevitable that glucose will break down into HMF and other undesirable chemicals (Zhou et al., 2011). Cellulolytic enzymes are currently used to carry out enzymatic hydrolysis of cellulose, and the results are superior for further fermentation because no breakdown products of glucose are produced. Enzymatic hydrolysis is preferred over acid or alkaline hydrolysis as the condition required for it is mild, that is, a pH of 4.8 and temperature of 40°C–50°C (Martín et al., 2017).

17.2.3 FERMENTATION TECHNOLOGY FOR BIOETHANOL PRODUCTION

The four primary categories of fermentation methods for bioethanol production are as follows: (i) consolidated bioprocessing (CBP), (ii) simultaneous saccharification and cofermentation (SSCF), (iii) solid state fermentation (SSF), and (iv) separate hydrolysis and fermentation (SHF). In CBP, microorganisms transform the raw materials directly, entailing all the processes such as production of enzymes, hydrolysis and fermentation in the same reactor and bioethanol is produced. In SSCF, a single reactor is used for the hydrolysis and fermentation processes to be carried out concurrently. SSF refers to an organised process where a solid matrix growth (i.e., nearly absence of water but in moist condition) of microorganisms is performed to carry out the fermentation of reducing sugars. Enzymatic hydrolysis and fermentation are carried out one after the other using the SHF technique (Balat, 2011).

17.2.4 Difficulties in the Process of Production of Bioethanol from LCB

Pretreatment of LCB plays a major role in getting more monosaccharides by breaking down lignin and exposing cellulose and hemicellulose, which further enhances the efficiency of subsequent fermentative processes (Hou et al., 2019). The majority of pretreatment techniques, however, have serious drawbacks, including their impact on the environment, their energy usage, and particularly the formation of inhibitors that prevent the growth of fermentative microorganisms (Hou et al., 2017). Several weak acids and compounds such as furans and phenols are formed by the degradation of hemicellulose and cellulose, which inhibit the fermentation process (Palmqvist & Hahn-Hägerdal, 2000; Hou et al., 2017). These inhibitors have the power to compromise the health of microbial cells' intracellular pH environment, metabolic enzymes, and cell membrane integrity (Bhattarai et al., 2015; Kapoor et al., 2015). The antioxidant defence mechanism can be destroyed by phenolic inhibitors, which can also cause mutations and harm cellular integrity (Maalouly, 2013). An excessive amount of intracellular adenosine triphosphate can be consumed by acidic inhibitors, which can also upset the pH (Palmqvist et al., 1999). The metabolic activity of enzymes in the cell gets affected significantly by furan inhibitors as it reduces permeability.

The removal of such inhibitors is necessary to lessen the deleterious effects on the synthesis of bioethanol. Various techniques have been used to accomplish this. The use of adsorbents, chemical agents, and others is the main detoxification technique. Other techniques use enzymatic treatment, extraction technology, and membrane separation, but they are ineffective, expensive, and energy-intensive. Most detoxification techniques have the potential to lower the hydrolysate's sugar concentration, thus lowering the bioethanol output (Pan et al., 2019).

17.3 BIOCHAR PRODUCTION FROM LIGNOCELLULOSIC BIOMASS

Biochar made from LCB can be employed to treat unanticipated environmental contaminants, such as heavy metals, antibiotics, and organic pollutants (Wang & Wang, 2019). The efficiency of producing bioethanol can be greatly increased by using biochar to remove inhibitors of fermentation. In addition, it might be utilised to immobilise microorganisms, enhance the fermentation environment, and remove the pretreatment bottleneck (Kyriakou et al., 2019).

Torrefaction, gasification, and pyrolysis are the processes used to produce biochar (Cha et al., 2016). An outline of biochar production processes is presented in Figure 17.2. Pyrolysis is carried out in an atmosphere free of oxygen and produces biochar by heating to temperatures between 300°C and 900°C in a matter of seconds to days (Heidari et al., 2014). The pyrolysis process can be categorised into three types, that is, slow pyrolysis, fast pyrolysis, and flash pyrolysis. Pyrolysis temperature and time period play a significant role in these processes (Ghosh & Maiti, 2021). However, a slow rate of heating with a low temperature for an extended duration can be the most favourable pyrolysis condition (Yang et al., 2019). Biochar can also be produced by gasification at very high temperatures (600°C–1,200°C). The majority of the LCB will be converted into CO and H_2 in this process within a few

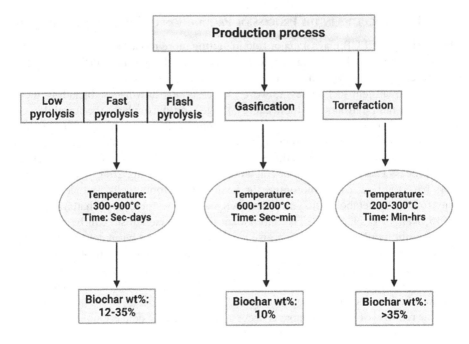

FIGURE 17.2   An outline of biochar production processes.

seconds to minutes due to incomplete combustion. In addition, the biomass ratio, gasification pressure and temperature all have an impact on the gasification process (Gómez-Barea et al., 2013). Gasification produces less biochar (about 10%) than pyrolysis (10%–35%) and torrefaction (>35%) process do. As torrefaction process needs a low temperature and a time period of some minutes to hours (2–3 h), it is regarded as a more cost-effective and environmentally friendly approach than gasification (Cha et al., 2016).

17.4   APPLICATION OF BIOCHAR PRODUCED FROM LIGNOCELLULOSIC BIOMASS IN THE PROCESS OF BIOETHANOL PRODUCTION: CIRCULAR ECONOMY APPROACHES

17.4.1   APPLICATION OF BIOCHAR IN THE REMOVAL OF FERMENTATION INHIBITORS

A high bioethanol yield can only be attained by removing fermentation inhibitors from the pretreated hydrolysate since they harm the fermentative microorganisms (Pan et al., 2019; Zhang et al., 2017; Hou et al., 2017). This is achievable by using the right detoxification procedure. Biochar made from LCB has already been the subject of substantial investigation. Heavy metals, organic, inorganic, and gas pollutants have all been successfully removed from the environment using biochar. Biochar adsorbs particular contaminants using ion exchange, coprecipitation, and functional group treatments (Qiu et al., 2018). Studies on biochar's ability to remove such compounds, which act as inhibitors and affect the fermentative environment in the process of bioethanol production, have been reported rarely.

Fermentation inhibitors have already been successfully eliminated using biochar made from LCB. One study found that the fermentation-inhibitory action of furan compounds might be lessened by the pyrolysis (>400°C)-derived LCB-derived biochar (Klasson et al., 2011). Some researchers have suggested making biochar via torrefaction (200°C–300°C) to minimise the overall economic input of bioethanol detoxification (Doddapaneni et al., 2018).

17.4.2 APPLICATION OF BIOCHAR IN THE IMMOBILIZATION OF MICROBES HELPFUL FOR FERMENTATION

This is another perspective to increase the capacity of fermentation even though the inhibitors are present. This approach includes the immobilisation of fermentative microorganisms by stabilising the cells using different techniques such as the use of alginate gel beads, and so on; however, these have performed poorly. To lessen the harmful effects that heavy metals have on soil-dwelling organisms and plants; the use of biochar is proven to be a practical and affordable method. In addition, it has been applied to enhance the anaerobic digestion process's ability to produce methane. Meanwhile, few studies have used biochar to transport and immobilise fermentative microorganisms. Another study reveals that *Kluyveromyces marxianus*, *S. cerevisiae*, and *Pichia kudriavzevii* fermentative microbes were successfully immobilised using LCB-derived biochar, which further improves the fermentative environment for bioethanol production (Kyriakou et al., 2019).

17.4.3 APPLICATION OF BIOCHAR FOR A SELF-SUFFICIENT SYSTEM FOR BIOETHANOL PRODUCTION

Reducing sugars, that is, glucose, xylose, and others are to be obtained from LCB by an adequate enzymatic hydrolysis to produce bioethanol (Paz et al., 2019). However, the LCB left over after the hydrolysis is typically not used efficiently, and its treatment also incurs significant costs. Therefore, the wise use of such enzymatic wastes would considerably aid in the low-energy and pollution-free production of bioethanol. A study reported that the biochar could most effectively lessen the inhibitory effects in the hydrolysate when compared with a commercial activated carbon. Moreover, in the process of making bioethanol from *Eucalyptus globulus* wood, the lignin leftover after hydrolysis was successfully used to produce biochar. Then the biochar produced was used as an adsorbent for the removal of inhibitors, which results in a self-sufficient system for bioethanol production, as no additional adsorbent was needed. As a result, the overall cost of producing bioethanol was decreased (Yoshioka et al., 2018). According to the studies stated above, LCB can be used to produce and consume both biochar and bioethanol.

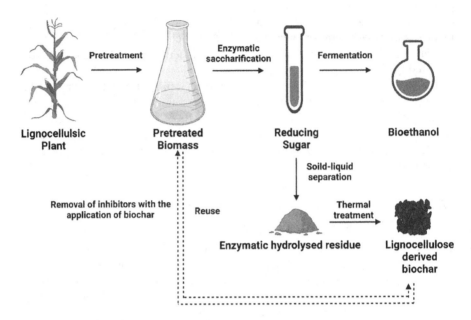

FIGURE 17.3 A circular economy approach for biochar and bioethanol production from LCB in a self-sufficient system.

Such a system based on a circular economy will encourage the efficient use of resources, lower the expense associated with the use of extra adsorbents, increase the widespread usage of bioethanol and ultimately result in subsequent environmental well-being. A system that is self-sufficient for biochar and bioethanol production from LCB is presented in Figure 17.3.

17.5 CONCLUSIONS

LCB can be a great source for making biofuels as a substitute for fossil fuels and bioproducts as adsorbents to fulfil the energy demand and minimize environmental pollution. Due to their numerous benefits, biochar and bioethanol have been emphasised specifically. The main elements of LCBs that could be used to produce biochar and ethanol were covered in this chapter. The potential of bioethanol and biochar production from LCB was also discussed, as well as the basic production techniques, technologies, and challenges. Most importantly, the chapter focused on the use of LCB-derived biochar to improve bioethanol production. Biochar can be used as an effective adsorbent for the detoxification of the fermentation inhibitors in the bioethanol production process. Combining the production of bioethanol and biochar into a single self-sufficient system has the potential to reduce resource waste and environmental pollution, supporting the circular economy model.

REFERENCES

Albashabsheh, N. T., & Heier Stamm, J. L. (2021). Optimization of lignocellulosic bio-mass-to-biofuel supply chains with densification: Literature review. *Biomass and Bioenergy, 144*, 105888. doi:10.1016/j.biombioe.2020.105888

Balat, M. (2011). Production of bioethanol from lignocellulosic materials via the biochemical pathway: A review. *Energy Conversion and Management, 52*(2), 858–875. doi:10.1016/j.enconman.2010.08.013

Behera, S., Arora, R., Nandhagopal, N., & Kumar, S. (2014). Importance of chemical pretreat-ment for bioconversion of lignocellulosic biomass. *Renewable and Sustainable Energy Reviews, 36*, 91–106. doi:10.1016/j.rser.2014.04.047

Bhattarai, S., Bottenus, D., Ivory, C. F., Gao, A. H., Bule, M., Garcia-Perez, M., & Chen, S. (2015). Simulation of the ozone pretreatment of wheat straw. *Bioresource Technology, 196*, 78–87. doi:10.1016/j.biortech.2015.07.022

Bhutto, A. W., Qureshi, K., Harijan, K., Abro, R., Abbas, T., Bazmi, A. A., Karim, S., & Yu, G. (2017). Insight into progress in pre-treatment of lignocellulosic biomass. *Energy, 122*, 724–745. doi:10.1016/j.energy.2017.01.005

Bilal, M., & Asgher, M. (2016). Biodegradation of agrowastes by lignocellulolytic activity of an oyster mushroom, Pleurotus sapidus. *Journal of the National Science Foundation of Sri Lanka, 44*(4), 399–407. doi:10.4038/jnsfsr.v44i4.8022

Bobleter, O. (1994). Hydrothermal degradation of polymers derived from plants. *Progress in Polymer Science, 19*(5), 797–841. doi:10.1016/0079-6700(94)90033-7

Brodin, M., Vallejos, M., Opedal, M. T., Area, M. C., & Chinga-Carrasco, G. (2017). Lignocellulosics as sustainable resources for production of bioplastics - A review. *Journal of Cleaner Production, 162*, 646–664. doi:10.1016/j.jclepro.2017.05.209

Cha, J. S., Park, S. H., Jung, S. C., Ryu, C., Jeon, J. K., Shin, M. C., & Park, Y. K. (2016). Production and utilization of biochar: A review. *Journal of Industrial and Engineering Chemistry, 40*, 1–15. doi:10.1016/j.jiec.2016.06.002

Cheng, B. H., Zeng, R. J., & Jiang, H. (2017). Recent developments of post-modification of biochar for electrochemical energy storage. *Bioresource Technology, 246*, 224–233. doi:10.1016/j.biortech.2017.07.060

Cheng, J. J., & Timilsina, G. R. (2011). Status and barriers of advanced biofuel technologies: A review. *Renewable Energy, 36*(12), 3541–3549. doi:10.1016/j.renene.2011.04.031

Di Donato, P., Finore, I., Poli, A., Nicolaus, B., & Lama, L. (2019). The production of second generation bioethanol: The biotechnology potential of thermophilic bacteria. *Journal of Cleaner Production, 233*, 1410–1417. doi:10.1016/j.jclepro.2019.06.152

Doddapaneni, T. R. K. C., Praveenkumar, R., Tolvanen, H., Rintala, J., & Konttinen, J. (2018). Techno-economic evaluation of integrating torrefaction with anaerobic digestion. *Applied Energy, 213*, 272–284. doi:10.1016/j.apenergy.2018.01.045

El-Naggar, A., Lee, S. S., Rinklebe, J., Farooq, M., Song, H., Sarmah, A. K., Zimmerman, A. R., Ahmad, M., Shaheen, S. M., & Ok, Y. S. (2019). Biochar application to low fertil-ity soils: A review of current status, and future prospects. *Geoderma, 337*, 536–554. doi:10.1016/j.geoderma.2018.09.034

Ghosh, D., & Maiti, S. K. (2021). Biochar assisted phytoremediation and biomass dis-posal in heavy metal contaminated mine soils: A review. *International Journal of Phytoremediation, 23*(6), 559–576. doi:10.1080/15226514.2020.1840510

Gismatulina, Y. A., & Budaeva, V. V. (2017). Chemical composition of five Miscanthus sinen-sis harvests and nitric-acid cellulose therefrom. *Industrial Crops and Products, 109*, 227–232. doi:10.1016/j.indcrop.2017.08.026

Godlewska, P., Schmidt, H. P., Ok, Y. S., & Oleszczuk, P. (2017). Biochar for composting improvement and contaminants reduction. A review. *Bioresource Technology, 246*, 193–202. doi:10.1016/j.biortech.2017.07.095

Gómez-Barea, A., Ollero, P., & Leckner, B. (2013). Optimization of char and tar conversion in fluidized bed biomass gasifiers. *Fuel, 103*, 42–52. doi:10.1016/j.fuel.2011.04.042

Guo, H., Chang, Y., & Lee, D. J. (2018). Enzymatic saccharification of lignocellulosic biorefinery: Research focuses. *Bioresource Technology, 252*, 198–215. doi:10.1016/j.biortech.2017.12.062

Haghighi Mood, S., Hossein Golfeshan, A., Tabatabaei, M., Salehi Jouzani, G., Najafi, G. H., Gholami, M., & Ardjmand, M. (2013). Lignocellulosic biomass to bioethanol, a comprehensive review with a focus on pretreatment. *Renewable and Sustainable Energy Reviews, 27*, 77–93. doi:10.1016/j.rser.2013.06.033

Heidari, A., Stahl, R., Younesi, H., Rashidi, A., Troeger, N., & Ghoreyshi, A. A. (2014). Effect of process conditions on product yield and composition of fast pyrolysis of Eucalyptus grandis in fluidized bed reactor. *Journal of Industrial and Engineering Chemistry, 20*(4), 2594–2602. doi:10.1016/j.jiec.2013.10.046

Hou, J., Tang, J., Chen, J., Deng, J., Wang, J., & Zhang, Q. (2019). Evaluation of inhibition of lignocellulose-derived by-products on bioethanol production by using the QSAR method and mechanism study. *Biochemical Engineering Journal, 147*, 153–162. doi:10.1016/j.bej.2019.04.013

Hou, J., Zhang, S., Qiu, Z., Han, H., & Zhang, Q. (2017). Stimulatory effect and adsorption behavior of rhamnolipids on lignocelluloses degradation system. *Bioresource Technology, 224*, 465–472. doi:10.1016/j.biortech.2016.11.108

Isikgor, F. H., & Becer, C. R. (2015). Lignocellulosic biomass: A sustainable platform for the production of bio-based chemicals and polymers. *Polymer Chemistry, 6*(25), 4497–4559. doi:10.1039/c5py00263j

Jaqueline Maalouly, J. A. (2013). Phenolic compounds from diluted acid hydrolysates of olive stones: Effect of overliming. *Advances in Crop Science and Technology, 01*(01). doi:10.4172/2329-8863.1000103

Kapoor, R. K., Rajan, K., & Carrier, D. J. (2015). Applications of Trametes versicolor crude culture filtrates in detoxification of biomass pretreatment hydrolyzates. *Bioresource Technology, 189*, 99–106. doi:10.1016/j.biortech.2015.03.100

Khan, M. U., & Ahring, B. K. (2019). Lignin degradation under anaerobic digestion: Influence of lignin modifications: A review. *Biomass and Bioenergy, 128*. doi:10.1016/j.biombioe.2019.105325

Kirdponpattara, S., & Phisalaphong, M. (2013). Bacterial cellulose-alginate composite sponge as a yeast cell carrier for ethanol production. *Biochemical Engineering Journal, 77*, 103–109. doi:10.1016/j.bej.2013.05.005

Klasson, K. T., Uchimiya, M., Lima, I. M., & Boihem, L. L. (2011). Feasibility of removing furfurals from sugar solutions using activated biochars made from agricultural residues. *BioResources, 6*(3), 3242–3251.

Kołodyńska, D., Wnetrzak, R., Leahy, J. J., Hayes, M. H. B., Kwapiński, W., & Hubicki, Z. (2012). Kinetic and adsorptive characterization of biochar in metal ions removal. *Chemical Engineering Journal, 197*, 295–305. doi:10.1016/j.cej.2012.05.025

Kumar, B., Bhardwaj, N., Agrawal, K., Chaturvedi, V., & Verma, P. (2020). Current perspective on pretreatment technologies using lignocellulosic biomass: An emerging biorefinery concept. *Fuel Processing Technology, 199*. doi:10.1016/j.fuproc.2019.106244

Kyriakou, M., Chatziiona, V. K., Costa, C. N., Kallis, M., Koutsokeras, L., Constantinides, G., & Koutinas, M. (2019). Biowaste-based biochar: A new strategy for fermentative bioethanol overproduction via whole-cell immobilization. *Applied Energy, 242*, 480–491. doi:10.1016/j.apenergy.2019.03.024

Marzo, C., Díaz, A. B., Caro, I., & Blandino, A. (2019). Valorization of agro-industrial wastes to produce hydrolytic enzymes by fungal solid-state fermentation. *Waste Management and Research, 37*(2), 149–156. doi:10.1177/0734242X18798699

Meyer, S., Glaser, B., & Quicker, P. (2011). Technical, economical, and climate-related aspects of biochar production technologies: A literature review. *Environmental Science and Technology*, *45*(22), 9473–9483. doi:10.1021/es201792c

Mohapatra, S., Mishra, C., Behera, S. S., & Thatoi, H. (2017). Application of pretreatment, fermentation and molecular techniques for enhancing bioethanol production from grass biomass - A review. *Renewable and Sustainable Energy Reviews*, *78*, 1007–1032. doi:10.1016/j.rser.2017.05.026

Palmqvist, E., Grage, H., Meinander, N. Q., & Hahn-Hägerdal, B. (1999). Main and interaction effects of acetic acid, furfural, and p- hydroxybenzoic acid on growth and ethanol productivity of yeasts. *Biotechnology and Bioengineering*, *63*(1), 46–55. doi:10.1002/(SICI)1097-0290(19990405)63:1<46::AID-BIT5>3.0.CO;2-J

Palmqvist, E., & Hahn-Hägerdal, B. (2000). Fermentation of lignocellulosic hydrolysates. II: Inhibitors and mechanisms of inhibition. *Bioresource Technology*, *74*(1), 25–33. doi:10.1016/S0960-8524(99)00161-3

Pan, L., He, M., Wu, B., Wang, Y., Hu, G., & Ma, K. (2019). Simultaneous concentration and detoxification of lignocellulosic hydrolysates by novel membrane filtration system for bioethanol production. *Journal of Cleaner Production*, *227*, 1185–1194. doi:10.1016/j.jclepro.2019.04.239

Paz, A., Outeiriño, D., Pérez Guerra, N., & Domínguez, J. M. (2019). Enzymatic hydrolysis of brewer's spent grain to obtain fermentable sugars. *Bioresource Technology*, *275*, 402–409. doi:10.1016/j.biortech.2018.12.082

Pérez, J., Muñoz-Dorado, J., De La Rubia, T., & Martínez, J. (2002). Biodegradation and biological treatments of cellulose, hemicellulose and lignin: An overview. *International Microbiology*, *5*(2), 53–63. doi:10.1007/s10123-002-0062-3

Qiu, Z., Chen, J., Tang, J., & Zhang, Q. (2018). A study of cadmium remediation and mechanisms: Improvements in the stability of walnut shell-derived biochar. *Science of the Total Environment*, *636*, 80–84. doi:10.1016/j.scitotenv.2018.04.215

Rocha-Martín, J., Martinez-Bernal, C., Pérez-Cobas, Y., Reyes-Sosa, F. M., & García, B. D. (2017). Additives enhancing enzymatic hydrolysis of lignocellulosic biomass. *Bioresource Technology*, *244*, 48–56. doi:10.1016/j.biortech.2017.06.132

Rodríguez Couto, S., & Toca Herrera, J. L. (2006). Industrial and biotechnological applications of laccases: A review. *Biotechnology Advances*, *24*(5), 500–513. doi:10.1016/j.biotechadv.2006.04.003

Sarkar, N., Ghosh, S. K., Bannerjee, S., & Aikat, K. (2012). Bioethanol production from agricultural wastes: An overview. *Renewable Energy*, *37*(1), 19–27. doi:10.1016/j.renene.2011.06.045

Silva-Fernandes, T., Santos, J. C., Hasmann, F., Rodrigues, R. C. L. B., Izario Filho, H. J., & Felipe, M. G. A. (2017). Biodegradable alternative for removing toxic compounds from sugarcane bagasse hemicellulosic hydrolysates for valorization in biorefineries. *Bioresource Technology*, *243*, 384–392. doi:10.1016/j.biortech.2017.06.064

Son, E. B., Poo, K. M., Chang, J. S., & Chae, K. J. (2018). Heavy metal removal from aqueous solutions using engineered magnetic biochars derived from waste marine macro-algal biomass. *Science of the Total Environment*, *615*, 161–168. doi:10.1016/j.scitotenv.2017.09.171

Sørensen, A., Teller, P. J., Hilstrøm, T., & Ahring, B. K. (2008). Hydrolysis of Miscanthus for bioethanol production using dilute acid presoaking combined with wet explosion pre-treatment and enzymatic treatment. *Bioresource Technology*, *99*(14), 6602–6607. doi:10.1016/j.biortech.2007.09.091

Sun, X., Atiyeh, H. K., Kumar, A., & Zhang, H. (2018). Enhanced ethanol production by Clostridium ragsdalei from syngas by incorporating biochar in the fermentation medium. *Bioresource Technology*, *247*, 291–301. doi:10.1016/j.biortech.2017.09.060

Sun, X., Atiyeh, H. K., Kumar, A., Zhang, H., & Tanner, R. S. (2018). Biochar enhanced ethanol and butanol production by Clostridium carboxidivorans from syngas. *Bioresource
Technology, 265*, 128–138. doi:10.1016/j.biortech.2018.05.106

Szalaty, T. J., Klapiszewski, Ł., & Jesionowski, T. (2020). Recent developments in modification of lignin using ionic liquids for the fabrication of advanced materials-A review.
Journal of Molecular Liquids, 301. doi:10.1016/j.molliq.2019.112417

Wang, J., & Wang, S. (2019). Preparation, modification and environmental application of
biochar: A review. *Journal of Cleaner Production, 227*, 1002–1022. doi:10.1016/j.
jclepro.2019.04.282

Wyman, C. E., Dale, B. E., Elander, R. T., Holtzapple, M., Ladisch, M. R., & Lee, Y. Y. (2005).
Coordinated development of leading biomass pretreatment technologies. *Bioresource
Technology, 96*, 1959–1966. doi:10.1016/j.biortech.2005.01.010

Yang, C., Li, R., Zhang, B., Qiu, Q., Wang, B., Yang, H., Ding, Y., & Wang, C. (2019). Pyrolysis
of microalgae: A critical review. *Fuel Processing Technology, 186*, 53–72. doi:10.1016/j.
fuproc.2018.12.012

Yoshioka, K., Daidai, M., Matsumoto, Y., Mizuno, R., Katsura, Y., Hakogi, T., Yanase, H., &
Watanabe, T. (2018). Self-sufficient bioethanol production system using a lignin-derived
adsorbent of fermentation inhibitors. *ACS Sustainable Chemistry and Engineering, 6*(3),
3070–3078. doi:10.1021/acssuschemeng.7b02915

Yu, Y., & Christopher, L. P. (2017). Detoxification of hemicellulose-rich poplar hydrolysate by
polymeric resins for improved ethanol fermentability. *Fuel, 203*, 187–196. doi:10.1016/j.
fuel.2017.04.118

Zabed, H., Sahu, J. N., Boyce, A. N., & Faruq, G. (2016). Fuel ethanol production from lignocellulosic biomass: An overview on feedstocks and technological approaches. *Renewable
and Sustainable Energy Reviews, 66*, 751–774. doi:10.1016/j.rser.2016.08.038

Zhang, N., Wang, L., Zhang, K., Walker, T., Thy, P., Jenkins, B., & Zheng, Y. (2019).
Pretreatment of lignocellulosic biomass using bioleaching to reduce inorganic elements.
Fuel, 246, 386–393. doi:10.1016/j.fuel.2019.02.138

Zhang, Q., Huang, H., Han, H., Qiu, Z., & Achal, V. (2017). Stimulatory effect of in-situ detoxification on bioethanol production by rice straw. *Energy, 135*, 32–39. doi:10.1016/j.
energy.2017.06.099

Zhang, Z., Zhu, Z., Shen, B., & Liu, L. (2019). Insights into biochar and hydrochar production
and applications: A review. *Energy, 171*, 581–598. doi:10.1016/j.energy.2019.01.035

Zhou, C. H., Xia, X., Lin, C. X., Tong, D. S., & Beltramini, J. (2011). Catalytic conversion of
lignocellulosic biomass to fine chemicals and fuels. *Chemical Society Reviews, 40*(11),
5588–5617. doi:10.1039/c1cs15124j

Zhou, N., Wang, Y., Huang, L., Yu, J., Chen, H., Tang, J., Xu, F., Lu, X., Zhong, M. e., & Zhou, Z.
(2019). In situ modification provided by a novel wet pyrolysis system to enhance surface
properties of biochar for lead immobilization. *Colloids and Surfaces A: Physicochemical
and Engineering Aspects, 570*, 39–47. doi:10.1016/j.colsurfa.2019.03.012

18 Global Research Trends in Biomass as Renewable Energy

N. Premalatha
Rajalakshmi Engineering College

S. R. Saranya
Anna University

18.1 INTRODUCTION

In recent years, the need for sustainable and renewable energy sources has become increasingly important due to the environmental challenges posed by fossil fuel consumption and climate change. Biomass, derived from organic materials such as agricultural residues, forestry wastes, dedicated energy crops, and organic municipal waste, has emerged as a promising alternative to traditional energy sources. Biomass has considerable promise as a renewable energy resource due to its ability to reduce greenhouse gas emissions, promote rural development, and enhance energy security. As a result, worldwide research efforts have been concentrated on discovering and enhancing biomass utilization technologies, conversion processes, and the development of efficient and sustainable biomass supply chains. Renewable energy sources are gaining popularity as a way to combat climate change, reduce greenhouse gas emissions, and achieve energy security. Because of its promise as a sustainable and diverse energy resource, biomass has emerged as a prominent contender among these sources. Biomass, which is obtained from organic materials such as agricultural residues, forestry wastes, and specific energy crops, is a possible alternative to fossil fuels.

In recent years, global research efforts have been directed toward exploring and advancing biomass utilization technologies, conversion processes, and sustainable supply chains. This has led to the identification of key trends and advancements that are shaping the biomass energy sector. This chapter provides an overview of the current global trends in biomass as renewable energy, shedding light on the most significant developments and highlighting the transformative potential of biomass in the renewable energy landscape.

Researchers and policymakers are exploring a wide range of biomass feedstocks to diversify the sources of renewable energy. This includes agricultural residues, forestry by-products, energy crops, and organic waste materials. By leveraging a diverse range of biomass feedstocks, the global biomass sector aims

DOI: 10.1201/9781003406501-18

to optimize resource availability and improve sustainability. Significant progress has been made in developing efficient and cost-effective biomass conversion technologies. Thermochemical processes such as combustion, gasification, and pyrolysis, as well as biochemical processes such as anaerobic digestion and fermentation, are being refined and optimized. These advancements enhance energy recovery, improve conversion efficiencies, and enable the production of a range of bioenergy products.

Biomass is being integrated with other renewable energy technologies to create hybrid systems that enhance overall efficiency and flexibility. Combining biomass with solar, wind, or geothermal energy can provide continuous power generation, as biomass can act as a dispatchable energy source. Researchers and industry stakeholders are focusing on developing sustainable biomass supply chains to ensure responsible sourcing, minimize environmental impacts, and promote social and economic benefits. Efforts include optimizing logistics, promoting local sourcing, and implementing certification schemes to ensure the sustainable production and utilization of biomass feedstocks.

Governments worldwide are enacting policies and providing incentives to promote biomass utilization for renewable energy. This includes feed-in tariffs, renewable energy targets, tax incentives, and supportive regulatory frameworks. These policy measures encourage investment in biomass energy projects and foster the growth of the biomass sector. Biomass presents an opportunity to valorize organic waste materials and contribute to a circular economy. By converting organic waste streams into bioenergy, biofuels, or biochemicals, biomass-based processes can simultaneously address waste management challenges and provide renewable energy solutions. These global trends in biomass as renewable energy underscore the increasing recognition of biomass as a viable and sustainable energy source. By leveraging biomass's versatility and potential, researchers, policymakers, and industry stakeholders are actively working toward a cleaner and more sustainable energy future. Continued research and innovation in biomass utilization will be crucial in unlocking its full potential and accelerating the transition to a low-carbon economy.

This chapter provides an overview of the global research trends in biomass as renewable energy, highlighting key areas of focus and advancements in the field. It aims to shed light on the current state of biomass research, outline emerging technologies, and discuss the challenges and opportunities associated with biomass utilization. By exploring these key research trends, this paper aims to contribute to the understanding of the current advancements, challenges, and opportunities in the field of biomass as a renewable energy source. It underscores the importance of continued research and innovation in biomass utilization to achieve a sustainable and low-carbon energy future.

18.2   AVAILABILITY AND BIOMASS UTILIZATION

The availability of biomass as a renewable resource is significant and diverse. Biomass can be found in various forms and sources, including agricultural residues, forest residues, specific energy crops, algae, and organic waste materials. Its availability depends on factors such as geographical location, climate, land

TABLE 18.1
Major Biomass Waste Production Globally

S. No.	Biomass Waste	Annual Production (Million Tons)
1	Wheat straw, corn straw	529
2	Sugarcane bagasse	513
3	Rice straw	520 (Gummert et al., 2019)

availability, and agricultural practices. In agricultural regions, crop residues such as straw, husks, and stalks are abundant, while forestry areas have access to forest residues such as branches and wood chips. Urban areas can tap into biomass resources through municipal solid waste and industrial waste. Furthermore, the cultivation of dedicated energy crops, such as switchgrass or *Miscanthus*, can be pursued specifically for biomass production. The wide range of biomass sources and their availability throughout different regions make biomass a versatile and plentiful resource for various applications, including energy production, bioproducts, waste management, and carbon sequestration. Biomass was utilized as an energy source till 1950. In the 1800s, the use of fossil fuel was negligible which is evident from the literature. Saleem (2022) produced biofuel from a plant named *Euphorbia abyssinica* in the 1830s.

The commercial global energy supply was increased from 14 to 60 GJ with an increase in population from 1.6 to 6.1 billion from 1900 to 2000 (Burke and Innes, 2000). Among the total global energy supply, fossil fuel contributes about 81% and renewable energy sources of about 14% of which biomass contributes about 70%. Biomass waste is generated in huge quantities among which corn straw, wheat straw, sugarcane bagasse, rice straw, soybean, rapeseed, barley, and so on is generated in large amount globally. These are the major global crops produced in different countries such as the United States, China, Argentina, and India (Tripathi et al., 2019). The annual production of wheat straw and corn straw is about 529 million tons (Kapoor et al., 2016) and of sugarcane bagasse is 513 million tons produced per year (Allende et al., 2023). The highest biomass production is contributed by wheat straw and corn straw listed in Table 18.1.

18.3 BIOMASS AS AN ENERGY SOURCE

Biomass has been utilized as a fuel resource for centuries, and its use as a sustainable energy source is still significant today. When compared to fossil fuels, biomass has various advantages as a fuel, including its renewable nature, carbon neutrality, and ability to reduce greenhouse gas emissions. Biomass fuels can be obtained from a variety of sources, each with its own set of properties and applications. Some common forms of biomass used as fuel materials include wood, agricultural residue, and organic waste.

Agriculture biomass is a potential source for producing sustainable energy. Biomass from various agricultural sources has different energy content. The energy content of different biomass depends on the sources of biomass. That is for rice straw the energy content is about 18 MJ/Kg dry matter (Mahesh and Mohini, 2013), and for sugarcane bagasse, the energy content is about 18.25 MJ/kg (León-Niño et al., 2013). Approximately the amount of crop residue produced in the United States and the world is listed in Table 18.2.

The fuel value of total annual residue produced in the world is 1.5×10^{15} kcal which is equivalent to 1 billion barrels (bbl) of diesel (Lal, 2005). Agricultural biomass is popular in the majority of developing countries due to its favorable economic characteristics and easy accessibility. Figure 18.1 depicts a significant discrepancy in biomass consumption across regions, ranging from 80% in Cambodia to 5% in Indonesia in Asian countries (Tun, 2015). Statistics show a significant difference in biomass energy utilization between developed and developing countries, estimated at 4% and 22%, respectively (Ramanathan et al., 2001). This disparity is mostly due to environmental rules enacted by wealthy countries.

Despite the abundance of biomass and its utilization as a fuel source in numerous countries, wood remains the preferred choice for cooking and heating in many regions. Consequently, there is a need to further examine the economic and environmental factors associated with biomass to compare its viability with other available sources of biomass.

TABLE 18.2
Different Crop Residues and Their Annual Production

Crop Residue Type	United States	World
Cereals	367×10^6 Mg/year	$2,802 \times 10^6$ Mg/year
Cereals and legumes	450×10^6 Mg/yea	$3,107 \times 10^6$ Mg/year
Food crops	488×10^6 Mg/year	$3,758 \times 10^6$ Mg/year

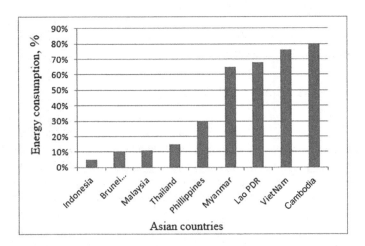

FIGURE 18.1   Biomass conversion technologies.

18.4 APPLICATION OF BIOMASS AS A RENEWABLE ENERGY SOURCE

Agricultural biomass refers to organic materials derived from plants and crops that can be used as a source of fuel. It is a renewable and sustainable energy resource that can help reduce greenhouse gas emissions and dependence on fossil fuels. The various form of agricultural biomass used as fuel includes crop residues from rice, wheat, corn, and sugarcane. These residues are collected, processed, or directly burned in biomass boilers to produce fuel. Among the array of energy resources available worldwide, biomass stands tall as a remarkable contender, securing the prestigious fourth position. Its significant role in meeting the world's energy demands cannot be undermined, as it diligently fulfills approximately 14% of the global energy requirements (León-Niño et al., 2013). Undoubtedly, biomass emerges as the paramount source of renewable energy, surpassing all others in importance.

18.4.1 BIOMASS AS SUSTAINABLE COMBUSTION FUEL

Biomass has gained recognition as a sustainable combustion fuel due to its renewable nature and potential to mitigate greenhouse gas emissions. Numerous studies and reports support the use of biomass as a viable and environmentally friendly energy source. There are many biomass sustainability associations and agencies to contribute biomass as sustainable combustion fuel. Renewable Energy and Sustainability Center (RESC), University of Texas at Austin conducted a study on the environmental benefits of biomass combustion. The research found that biomass combustion can reduce net carbon dioxide emissions compared to fossil fuel combustion. The utilization of agricultural residues and energy crops for biomass combustion helps to close the carbon cycle and reduce overall carbon emissions. International Energy Agency (IEA) Bioenergy: IEA Bioenergy is a research program that assesses sustainable biomass utilization for energy purposes. Their reports highlight the potential of biomass combustion to reduce greenhouse gas emissions and promote sustainable development. Biomass combustion can contribute to climate change mitigation by providing carbon-neutral or low-carbon energy options. National Renewable Energy Laboratory (NREL): NREL, a research institute funded by the U.S. Department of Energy, has extensively studied the environmental impacts of biomass combustion. Their research indicates that sustainable biomass combustion can significantly reduce greenhouse gas emissions compared to fossil fuel combustion. By replacing fossil fuels with biomass, emissions of carbon dioxide, sulfur dioxide, and nitrogen oxides can be reduced, thus improving air quality and mitigating climate change. European Biomass Industry Association (EUBIA) promotes the use of biomass as a sustainable energy source in Europe. According to their reports, biomass combustion offers several advantages, including reduced greenhouse gas emissions, improved waste management, and increased energy security. Biomass combustion technologies, such as fluidized bed boilers and advanced combustion systems, can achieve high energy efficiency and minimize environmental impacts. Sustainable Development Goals (SDGs): The United Nations' Sustainable Development Goals recognize the importance of sustainable energy production. Biomass combustion aligns with several SDGs, including affordable and clean energy (Goal 7), climate action (Goal 13), and responsible consumption and production (Goal 12). The use of biomass as a renewable energy source contributes to the transition toward a sustainable and low-carbon society (Mantlana and Maoela, 2020).

18.4.2   Biomass Utilization as Liquid Fuel

An important biofuel derived from biomass is biodiesel. Biodiesel is produced from vegetable oils or animal fats through a process called transesterification. It can be used in diesel engines with little or no modification, making it a viable alternative to fossil diesel. Biodiesel has lower emissions of sulfur, particulate matter, and carbon monoxide compared to conventional diesel, contributing to improved air quality. In addition, biodiesel production supports the agricultural sector by providing an additional market for oilseed crops, creating a diversified income source for farmers. Biofuels have been used since from 1970s due to the crude oil crisis. Biofuels are found to be alternate transportation fuels for petroleum-based fuels. The first biofuel from peanut oil was tested by Rudolph and the author found it was unsuitable. After the 1990s, with the increase in crude oil crisis, the United States and many European countries showed interest in biofuel production from biomass. Many countries have launched biofuel programmers to reduce greenhouse gas emission. According to the above discussion, biofuel production increased to the extent of 110 billion liters from 2008 to 2018 (Richards et al., 2017). Morales et al. studied biofuel (ethanol) production from lignocellulosic biomass and its environmental impact and energy balance (Morales et al., 2015). Sieverding et al. carried out the meta-analysis of nine peer-reviewed soybean life cycle analysis (LCA) biodiesel studies (Sieverding et al., 2015). Globally, 94 GL of ethanol was produced in 2015 from different biomass (Lehtikangas, 2001), and it was also noted that the United States and Brazil contribute 85% of the entire ethanol production.

18.4.3   Biomass as Gaseous Fuel

Biomass utilization extends beyond liquid biofuels to include solid biomass and biogas. Solid biomass, such as wood pellets and agricultural residues, can be burned directly or converted into heat and electricity through combustion or gasification processes. Biogas, on the other hand, is produced from the anaerobic digestion of organic waste materials, such as food scraps, animal manure, and sewage sludge. Biogas can be used as a renewable energy source for heating, electricity generation, and even as a replacement for natural gas in some applications. The conversion of biomass into biogas results in high carbon conversion and high calorific value than other biomass conversion processes. In addition, gasification technology and anaerobic digestion are more suited and extensively used conversion processes for single-source and hybrid energy generation. Rodriguez et al. experimented with mechanical pretreatment of waste paper to produce biogas and found that 21% methane gas was produced within 60 min (Rodriguez et al., 2017). Countries such as Europe have set

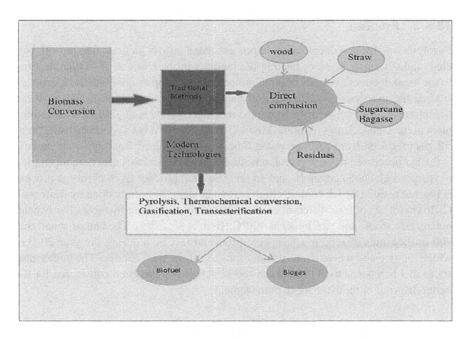

FIGURE 18.2 Global biomass energy consumption.

a target of 20% renewable energy and 10% renewable fuels in total EU transport. Furthermore, Cozier et al. underline the vast scope of biofuels and bioenergy, as well as their potential to replace fuels from petro products and that biofuel production can lead the "resource-productivity revolution," in which the agriculture sector will play a key part (Cozier, 2014). Figure 18.2 shows the conversion technologies for biomass into biofuels and biogas.

18.5 EMERGING TECHNOLOGIES FOR BIOFUEL PRODUCTION

The global trend in technology for the conversion of biomass into biofuels is focused on improving efficiency, reducing costs, and increasing the utilization of diverse feedstocks. There are several chemical and biological technologies used for the conversion of various feedstocks into biofuels and the most commonly used thermochemical conversion of biomass is gasification, pyrolysis, and combustion. The biological conversion technologies are fermentation, enzymatic hydrolysis, and microbial conversion.

18.5.1 TRANSESTERIFICATION

This process is used to produce biodiesel from vegetable oils or animal fats. Transesterification involves reacting the oil or fat with an alcohol, usually methanol, in the presence of a catalyst to produce fatty acid methyl esters, which is biodiesel. Sidohounde et al. (2018) studied the production of biodiesel from vegetable oil.

18.5.2 Pyrolysis

Pyrolysis is a thermal decomposition process that converts biomass into bio-oil, bio-char, and syngas. It involves heating the biomass in the absence of oxygen, leading to the breakdown of complex organic molecules into simpler compounds. Torri et al. studied the pyrolysis of softwood and hardwood to produce bio-oil and analyzed the intermediates using chromatographic technique (Torri et al., 2016). A bioturbine plant in the United States (Ankara-Mamak) produces 10 MW of power by using land-fill gas produced from garbage waste. The bio-oil produced from pyrolysis process is further processed to yield petrol, gasoline, and so on, and this also contains highly oxygenated alcohols, ethanols, and so on. The final product from pyrolysis process is biochar residue which contains high carbon content. In a study done by Bian et al. (2016) pyrolysis of four different crop residue were carried out by the author (wheat, maize, rice straw, and rice husk) at 400°C (Bian et al., 2016). The author stated that 700 million tons of crop residues were produced in China during the year 2010 in which the rice and wheat straw account for 75% of the total residue. The author also states that biochar is used to improve soil structure and has been considered for the better disposal of biomass waste management.

18.5.3 Gasification

Gasification is a process where biomass is converted into synthesis gas (syngas), which is a mixture of carbon monoxide, hydrogen, and other gases. The syngas can be used as a fuel or further processed to produce liquid biofuels such as ethanol or synthetic diesel. Gasification is considered an important method for the utilization of biomass. Advanced, cost-effective, and highly efficient gasification methods and systems are necessary to promote this technology in the future. A wide range of gasification processes, such as plasma gasification and wet biomass gasification in supercritical water, have been developed (Chaimaa et al. 2021). Van Oost et al. (2008) have made a detailed study on thermal plasma gasification. The author also showed that thermal plasma gasification is an environmental friendly treatment of waste bio-mass for sustainable energy and also studied the several types of plasma arches (Van Oost et al., 2008).

18.5.4 Fermentation

Fermentation is widely used for the production of biofuels such as ethanol. Biofuels derived from biomass through the fermentation process offer a promising solution for reducing greenhouse gas emissions and achieving sustainable energy production. This renewable energy source is produced by converting organic materials, such as crops, agricultural residues, and algae, into usable fuels. The fermentation process plays a crucial role in transforming biomass into valuable biofuels, such as ethanol and biogas. Fermentation is a metabolic process that occurs in the absence of oxygen and involves the conversion of complex organic compounds into simpler molecules. In the context of biofuel production, fermentation primarily refers to the conversion of sugars or carbohydrates present in biomass into ethanol or other biofuels through the action of microorganisms. One of the most widely used biofuels produced through fermentation is ethanol. Ethanol is a type of alcohol that can be blended with gasoline or used as a standalone fuel in vehicles. The process of ethanol production begins with the breakdown of complex carbohydrates into simple sugars, such as glucose and fructose. This can be achieved through various methods, including enzymatic hydrolysis or acid treatment. The resulting sugar solution is then fermented by yeast or bacteria, which convert the sugars into ethanol and carbon dioxide through a series of enzymatic reactions. The ethanol is separated and purified, making it suitable for use as a fuel.

The fermentation process can also be employed to produce biogas, which consists primarily of methane (CH_4) and carbon dioxide (CO_2). Biogas is generated through the anaerobic digestion of biomass, such as animal manure, food waste, or dedicated energy crops. During anaerobic digestion, microorganisms break down the organic matter, releasing methane-rich gas as a byproduct. Biogas can be utilized directly for heating or electricity generation, or it can be further processed to remove impurities and increase the methane content, resulting in biomethane. Biomethane has similar properties to natural gas and can be injected into the existing natural gas grid or used as a transportation fuel.

The use of biomass-based biofuels produced via fermentation offers several advantages. First, these biofuels are considered carbon-neutral because the carbon dioxide released during combustion is approximately equivalent to the carbon dioxide absorbed by the biomass during growth. This makes them an environmentally friendly alternative to fossil fuels, as they contribute to a net reduction in greenhouse gas emissions. Second, the use of biomass as a feedstock provides an opportunity to utilize agricultural and forestry residues, reducing waste and potentially offering additional income sources for farmers. Furthermore, biofuels derived from fermentation are compatible with existing fuel infrastructure and can be easily integrated into conventional engines and distribution systems (Liao et al., 2016).

18.5.5 HYDROTHERMAL LIQUEFACTION

This technology converts wet biomass, such as algae or sewage sludge, into a liquid bio-oil through a high-pressure, high-temperature process. The bio-oil can be further upgraded and used as a fuel or feedstock for other processes. The process takes place under high temperature and pressure conditions in the presence of water, typically between 250°C and 400°C and pressures of 10–25 MPa. The combination of high temperature and pressure, along with the presence of water, promotes the breakdown of complex organic compounds in the biomass into simpler, more energy-rich molecules. The solid residue left after Hydrothermal liquefaction (HTL), known as biochar, is a carbon-rich material that can be utilized as a soil amendment or as a precursor for the production of activated carbon. The gases produced during the process, mainly carbon dioxide and methane, can be used for energy generation or captured for other applications.

Hydrothermal liquefaction offers several advantages as a biomass conversion technology. It can utilize a wide range of feedstocks, including wet and high-moisture content biomass, which reduces the need for extensive biomass drying. Furthermore,

HTL can process biomass feedstocks with high water content, which makes it suitable for certain types of biomass, such as algae, sewage sludge, and food waste. The process also has the potential to convert non-food feedstocks into valuable energy products, minimizing competition with food crops (Elliott et al., 2015).

18.5.6 ALGAL BIOFUEL PRODUCTION

Algae can be cultivated to produce biofuels. Various methods are used, such as open ponds, closed photobioreactors, or raceway ponds, to grow and harvest algae. The harvested algae can be processed through techniques such as transesterification or hydrothermal liquefaction to obtain biofuels. Hossain et al. studied the microalgae biofuel production with nanoadditive since microalgae are rich in energy content, inflated growth rate, better CO_2 fixation, and O_2 addition (Hossain et al., 2019).

18.5.7 ENZYMATIC HYDROLYSIS

Enzymatic hydrolysis is employed in the production of cellulosic ethanol. It involves breaking down the complex carbohydrates present in cellulosic biomass, such as agricultural residues or dedicated energy crops, into simple sugars using enzymes. The resulting sugars can then be fermented to produce ethanol. Brummer et al. studied the enzymatic hydrolysis of waste paper using novoenzymes (Brummer et al., 2014). The author used cellulose paper, filter paper, and cardboard for carrying out the enzymatic hydrolysis and found that the highest yield was achieved for cardboard.

These are just a few examples of the technologies used for the conversion of biomass into biofuels. The choice of technology depends on the feedstock, scale of production, economic viability, and other factors. Ongoing research and development efforts continue to improve existing technologies and explore new avenues for efficient and sustainable biofuel production.

There are advance technologies for the conversion of biomass into biofuels that focus on improving efficiency, reducing costs, and increasing the utilization of diverse feedstocks. Here are some key trends in this field.

18.5.8 ADVANCED BIOCHEMICAL PROCESSES

Advanced biochemical processes, such as consolidated bioprocessing (CBP) and synthetic biology, are gaining attention. CBP aims to combine multiple steps of biomass deconstruction, enzyme production, and biofuel production into a single microorganism or enzymatic system, thereby reducing costs and increasing efficiency. Synthetic biology involves engineering microorganisms to enhance their ability to convert biomass into biofuels more effectively.

18.5.9 CELLULOSIC ETHANOL

Cellulosic ethanol production is gaining momentum globally. This technology focuses on converting non-food-based feedstocks, such as agricultural residues, forestry residues, and dedicated energy crops, into ethanol. Advances in enzymatic hydrolysis and fermentation techniques are improving the efficiency and commercial viability of cellulosic ethanol production. The biomass feedstock is collected and processed to remove any impurities and to break down the complex structure of cellulose. This can be achieved through various pretreatment methods such as mechanical grinding, steam explosion, or acid hydrolysis.

Once the biomass has been pretreated, enzymes are added to the mixture to break down the cellulose into its constituent sugars, primarily glucose. These enzymes, known as cellulases, are naturally occurring biological catalysts that can break the glycosidic bonds in cellulose. The resulting sugar-rich solution is known as cellulose hydrolysate.

The cellulose hydrolysate is then fermented by adding specific strains of microorganisms, typically yeast or bacteria, which convert the sugars into ethanol through a process called fermentation. The microorganisms consume the sugars and produce ethanol as a metabolic byproduct. This step is similar to the fermentation process used in the production of traditional ethanol from corn or sugarcane. The resulting ethanol solution undergoes further purification steps, such as distillation and dehydration, to remove impurities and increase the ethanol concentration. The final product is high-quality cellulosic ethanol that can be used as a fuel additive or blended with gasoline for use in transportation.

Cellulosic ethanol offers several advantages over traditional ethanol produced from food crops such as corn or sugarcane. It utilizes non-food biomass sources, reducing competition for resources between fuel and food production. In addition, cellulosic ethanol has a lower carbon footprint and can contribute to reducing greenhouse gas emissions compared to fossil fuels (Ronces, 2021).

Several countries, including the United States, have implemented policies and incentives to promote the production and use of cellulosic ethanol as part of their renewable energy strategies. Ongoing research and development efforts aim to improve the efficiency and cost-effectiveness of the biomass-to-ethanol conversion process, making cellulosic ethanol a viable and sustainable alternative to fossil fuels.

18.5.10 WASTE-TO-BIOFUELS

There is a growing emphasis on converting waste materials into biofuels. Technologies such as anaerobic digestion and gasification are being used to convert organic waste, such as food waste, agricultural waste, and municipal solid waste, into biogas or syngas, which can be further processed into biofuels.

18.5.11 ALGAE-BASED BIOFUELS

Algae-based biofuels have the potential to offer high yields and can be cultivated on non-arable land. Research and development efforts are focused on improving algae cultivation techniques, optimizing lipid content in algae strains, and developing efficient extraction methods to obtain biofuels from algae. These systems provide optimal conditions for algal growth, including sufficient sunlight, temperature, and nutrient availability. Algae can be cultivated using freshwater, seawater, or wastewater, depending on the species and the intended application (Chye et al., 2019).

Once the algae have reached their desired biomass concentration, they are harvested. Various methods can be employed for algae harvesting, including sedimentation, flocculation, centrifugation, or filtration. Harvested algae are then processed to separate the desired components, such as lipids, for biodiesel production or carbohydrates for bioethanol production.

For biodiesel production, the extracted algal lipids are converted into biodiesel through a process called transesterification. Transesterification involves reacting the lipids with an alcohol, typically methanol or ethanol, in the presence of a catalyst to produce fatty acid methyl or ethyl esters, which are the main components of biodiesel (Chisti, 2007).

In the case of bioethanol production, the extracted algal carbohydrates are subjected to enzymatic hydrolysis, which breaks down the complex carbohydrates into simple sugars. These sugars are then fermented by adding specific strains of microorganisms, such as yeast, to convert them into ethanol. The ethanol is further purified through distillation and dehydration processes.

Algae-based biofuels offer several advantages over traditional biofuels derived from food crops. Algae can be grown on non-arable land, including wastewater and saline environments, minimizing the competition for resources between fuel and food production. Algae also have a higher lipid content compared to oilseed crops, which makes them a more efficient source for biodiesel production. In addition, algae cultivation can help capture and utilize carbon dioxide, reducing greenhouse gas emissions.

Although algae-based biofuels show great potential, there are still challenges to overcome. These include improving the efficiency of algal cultivation, developing cost-effective harvesting and extraction methods, and optimizing the conversion processes. Ongoing research and development efforts are focused on addressing these challenges and scaling up algae-based biofuel production for commercial viability.

18.5.12 CATALYTIC CONVERSION

Catalytic conversion processes are being explored to convert biomass feedstocks into biofuels. These processes involve the use of catalysts to chemically convert biomass components into biofuels or intermediate chemicals. Catalytic pyrolysis, hydrothermal liquefaction, and Fischer–Tropsch synthesis are some examples of catalytic conversion technologies. This process utilizes various catalysts and reaction conditions to facilitate the conversion of complex biomass components into simpler molecules through chemical reactions such as pyrolysis, gasification, and hydrothermal processing.

In pyrolysis, biomass is heated in the absence of oxygen to break it down into pyrolysis oil, gas, and char. Catalysts can be employed to enhance the efficiency and selectivity of this process, promoting the desired reactions and minimizing unwanted side reactions. These catalysts can influence product distribution, leading to a higher yield of desired biofuels or chemicals.

Gasification involves the partial oxidation of biomass at high temperatures to produce syngas, a mixture of carbon monoxide and hydrogen. Catalysts play a crucial role in controlling the gasification reactions, improving the syngas composition, and reducing tar and methane formation. They can also enable the direct conversion of syngas into specific chemicals through Fischer–Tropsch synthesis or other catalytic processes.

Hydrothermal processing uses high-pressure and high-temperature water to break down biomass into various products. Catalysts can be employed to enhance the reaction rates and selectivity in hydrothermal liquefaction, resulting in the production of bio-oils, gases, and solid residues. These catalysts can also promote the deoxygenation and stabilization of bio-oils, making them suitable for further processing or utilization.

The choice of catalysts depends on the specific biomass feedstock and desired products. Various catalysts, including metals, metal oxides, zeolites, and supported catalysts, have been investigated for biomass conversion. Catalyst development focuses on improving their activity, selectivity, stability, and resistance to deactivation (Dowson et al., 2013).

The catalytic conversion of biomass is an active area of research and development, with the goal of achieving more efficient and sustainable processes for biofuel and chemical production. Advances in catalyst design and reaction engineering are crucial for optimizing biomass conversion and enabling the transition to a bio-based economy.

18.5.13 Electrocatalysis

Electrocatalysis, which utilizes electricity to drive chemical reactions, is an emerging trend for biomass conversion. This technology has the potential to convert biomass-derived molecules into biofuels or other value-added chemicals efficiently. Electrocatalysis plays a crucial role in the conversion of biomass into biofuels through electrochemical processes. Biomass electrocatalysis involves the use of catalysts to facilitate the electrochemical reactions involved in the transformation of biomass-derived compounds into biofuels. These reactions include electrochemical oxidation, hydrogenation, and electroreduction.

Electrocatalysis offers several advantages for biomass conversion compared to traditional thermochemical or biochemical methods. It enables selective and controllable transformations, operates under milder conditions, and can be integrated into renewable energy systems. In addition, electrocatalysis can improve the efficiency and selectivity of biomass conversion, leading to higher yields of biofuels.

One area of electrocatalytic biomass conversion is the electrochemical oxidation of biomass-derived molecules. Catalysts are employed to facilitate the oxidation reactions of carbohydrates, lignin, and other biomass components, leading to the production of valuable chemicals and intermediates. These catalysts can enhance the kinetics and selectivity of the oxidation process, allowing for targeted production of specific biofuel precursors (Page et al., 2023).

Another important aspect is the electrocatalytic reduction of biomass-derived molecules, such as carbon dioxide and biomass-derived platform chemicals, into biofuels. Catalysts are utilized to promote the electroreduction reactions, which convert these molecules into fuels such as ethanol, formic acid, or methane. These catalysts can enhance the efficiency and selectivity of the electrochemical reduction process, contributing to the development of sustainable biofuel production (Thangaraj and Solomon, 2021).

The design and development of efficient and cost-effective electrocatalysts for biomass conversion are active areas of research. Researchers are exploring various catalyst materials, including transition metals, metal complexes, metal nanoparticles, and carbon-based materials, to improve the performance and stability of electrocatalytic systems.

18.5.14 INTEGRATED BIOREFINERIES

The concept of integrated biorefineries is gaining traction. These biorefineries aim to optimize the utilization of different biomass feedstocks by integrating multiple conversion processes. They produce a range of biofuels, chemicals, and other bio-based products in a sustainable and efficient manner. Integrated biorefineries are facilities that utilize a variety of biomass feedstocks to produce multiple value-added products, such as biofuels, biochemicals, and biomaterials. They aim to maximize the efficiency and sustainability of biomass utilization by integrating multiple conversion processes within a single facility. Integrated biorefineries play a crucial role in the transition from a fossil fuel-based economy to a bio-based economy.

The concept of integrated biorefineries is inspired by traditional petroleum refineries, which extract multiple products from crude oil. However, instead of relying on fossil resources, integrated biorefineries focus on the utilization of renewable biomass resources. These facilities employ various conversion technologies, including biochemical, thermochemical, and physicochemical processes, to transform biomass into a wide range of valuable products.

In an integrated biorefinery, different feedstocks, such as agricultural residues, dedicated energy crops, and forest residues, can be processed simultaneously or sequentially. The feedstocks undergo various conversion pathways, including fermentation, enzymatic hydrolysis, pyrolysis, gasification, and extraction, to produce biofuels, chemicals, and other value-added products. The integration of these processes allows for the efficient utilization of the entire biomass feedstock and the generation of multiple revenue streams.

The integration of biorefinery processes offers several advantages. It enhances resource efficiency by utilizing all components of biomass, minimizing waste, and maximizing product yields. It also enables the production of a diversified portfolio of products, reducing reliance on a single market and improving economic viability. In addition, integrated biorefineries promote sustainability by reducing greenhouse gas emissions, lowering dependence on fossil fuels, and supporting rural development through the utilization of locally available biomass resources (Takkellapati et al., 2018).

The design and operation of integrated biorefineries require careful consideration of process integration, logistics, and optimization. The selection of appropriate conversion technologies, feedstock choices, and product portfolios is critical for the success of integrated biorefinery operations.

Process Optimization and Digitalization: There is a growing focus on process optimization and digitalization to improve the efficiency, monitoring, and control of biofuel production processes. Advanced analytics, machine learning, and automation technologies are being employed to optimize process parameters, reduce energy consumption, and enhance overall process performance. These trends reflect the ongoing efforts to develop and deploy technologies that enable the sustainable production of biofuels from diverse biomass sources while addressing economic, environmental, and social considerations. As the world aims to transition to a low-carbon economy, the global trends in technologies for biomass conversion to biofuels align with the goals of sustainability, energy security, and reduced greenhouse gas emissions. These trends reflect a growing emphasis on efficient feedstock utilization, process optimization, and the development of innovative and integrated approaches to biofuel production. By leveraging these technologies and embracing ongoing research and development efforts, the biofuel industry continues to evolve, offering promising solutions to address the energy and environmental challenges of the future.

18.6 ENVIRONMENTAL AND SOCIOECONOMIC IMPACTS OF BIOMASS UTILIZATION

Biomass-based energy projects offer a range of socioeconomic benefits while also presenting certain challenges that need to be addressed for their successful implementation. These projects have the potential to contribute to local economies, create employment opportunities, promote rural development, and enhance energy security. However, careful consideration of social and economic factors is essential to ensure biomass-based energy projects deliver maximum benefits to communities and minimize any adverse impacts.

One of the key socioeconomic benefits of biomass-based energy projects is job creation. These projects require a significant workforce for various activities, including biomass collection, transportation, processing, and plant operation. This can generate employment opportunities, particularly in rural areas where biomass resources are often abundant. In addition, biomass projects can stimulate local industries such as agriculture, forestry, and waste management, leading to additional job creation and economic grow

Biomass-based energy projects can also enhance energy security by diversifying energy sources and reducing dependence on imported fossil fuels. This can strengthen the resilience of communities and reduce vulnerability to energy price fluctuations. Moreover, biomass resources are often locally available and renewable, providing a sustainable energy solution that can contribute to a more reliable and decentralized energy system.

However, biomass-based energy projects also face certain challenges that need to be addressed. One challenge is ensuring the sustainable sourcing of biomass feedstocks. It is crucial to establish robust supply chains that prioritize responsible biomass production, considering factors such as land use, biodiversity conservation, and competition with food production. Proper governance and regulations are needed to prevent negative environmental and social impacts, such as deforestation or displacement of local communities.

Another challenge is the economic viability of biomass projects. Biomass supply, conversion technologies, and infrastructure development require significant investments. The availability of affordable financing and supportive policies are essential to attract investments and create a favorable business environment. In addition, biomass projects must compete with other renewable energy sources and fossil fuels in terms of cost competitiveness and market dynamics.

18.7 FUTURE RESEARCH DIRECTIONS AND EMERGING TECHNOLOGIES

Future research directions and emerging technologies in the field of biomass energy hold great promise for advancing sustainable and efficient bioenergy systems. As the world strives to transition toward renewable energy sources and reduce greenhouse gas emissions, ongoing research efforts focus on enhancing biomass conversion processes, improving feedstock availability, and integrating biomass with other renewable energy technologies. The following are key areas of interest for future research and development in biomass energy.

Advanced Conversion Technologies: Continued research is being conducted to enhance the efficiency and sustainability of biomass conversion technologies. This includes developing advanced pyrolysis and gasification techniques, improving enzymatic hydrolysis for bioethanol production, and exploring new catalysts for biofuel synthesis. Integration of emerging technologies such as plasma gasification, supercritical water gasification, and microbial fuel cells also hold potential for efficient biomass conversion.

Bioenergy from Lignocellulosic Biomass: Lignocellulosic biomass, such as agricultural residues, forest residues, and energy crops, offers abundant and sustainable feedstock for bioenergy production. Future research aims to optimize pretreatment methods, develop novel enzymes for efficient biomass degradation, and overcome challenges related to recalcitrance and inhibitor formation. Advances in lignocellulosic bioenergy will help unlock the full potential of this resource and enable the production of advanced biofuels and biochemicals.

Waste-to-Energy and Circular Economy: Research in biomass energy is increasingly focusing on waste-to-energy conversion technologies. This includes anaerobic digestion of organic waste for biogas production, thermochemical conversion of municipal solid waste, and utilization of agricultural and forestry residues for energy generation. Future research aims to improve waste management practices, develop efficient waste-to-energy conversion systems, and explore synergies between biomass energy and other waste valorization processes.

Integration with Renewable Energy Systems: Biomass energy can be integrated with other renewable energy systems to create hybrid energy solutions. Research is exploring the integration of biomass with solar and wind energy, as well as energy storage technologies, to address the intermittent nature of renewable energy sources. Combined heat and power (CHP) systems that utilize biomass for both electricity and heat generation are also gaining attention. Future research will focus on optimizing system integration, maximizing energy efficiency, and developing smart grid solutions for effective renewable energy integration.

Sustainability and Environmental Impact: As biomass energy continues to expand, it is crucial to address sustainability and environmental concerns. Future research aims to improve life cycle assessment (LCA) methodologies for better evaluation of the environmental impact of biomass energy systems. This includes assessing indirect land-use change, biodiversity impacts, and water footprint. In addition, the research will explore sustainable biomass sourcing, land management practices, and socioeconomic implications of biomass energy projects.

Digitalization and Data Analytics: The advent of digitalization and data analytics has the potential to revolutionize biomass energy systems. Future research will explore the application of artificial intelligence, machine learning, and big data analytics to optimize biomass feedstock selection, process control, and system performance. Digital tools can assist in real-time monitoring, predictive maintenance, and decision-making for biomass energy operations, leading to improved efficiency and cost-effectiveness.

In conclusion, future research in biomass energy will focus on advancing conversion technologies, utilizing lignocellulosic biomass, promoting waste-to-energy solutions, integrating with other renewable energy systems, addressing sustainability challenges, and harnessing the power of digitalization. These research directions and emerging technologies will contribute to the development of more sustainable, efficient, and economically viable biomass energy systems, helping to achieve global energy transition goals and reduce carbon emissions.

In conclusion, biomass-based energy projects offer socioeconomic benefits such as job creation, rural development, and enhanced energy security. However, challenges related to sustainable biomass sourcing and economic viability need to be addressed. Through appropriate governance, supportive policies, and stakeholder engagement, biomass projects can maximize their positive socioeconomic impacts while minimizing any potential drawbacks. It is crucial to consider the specific local context and conduct comprehensive socioeconomic assessments to ensure the successful implementation of biomass-based energy projects that benefit both the environment and communities.

18.8 SUMMARY AND CONCLUSION

The use of agricultural biomass as an alternative energy supply appears to be a feasible choice, as evidenced by several nations such as the United States, Europe, and so on. Agriculture biomass finds its primary application in the production of biofuels, which can undergo multiple stages of processing using different thermochemical, biochemical, and other chemical conversion methods. Furthermore, it offers a sustainable fuel source that has the potential to replace diminishing fossil fuel reserves and reduce greenhouse gas emissions and solid waste.

Moreover, there is diversity among countries in biofuel production from biomass since biomass waste from different sources especially agriculture depends on climatic conditions and economical facilities. To maximize the efficient and cost-effective utilization of agricultural biomass for energy production, the establishment of clear directives is crucial. The utilization of agricultural biomass for energy holds the potential to create new avenues and possibilities for farmers, enabling them to adapt their agricultural practices by cultivating energy-efficient crops. This presents an opportunity for farmers to diversify their activities and tap into the growing market for renewable energy sources derived from agriculture. By aligning agricultural practices with energy production, farmers can contribute to sustainable development while harnessing the economic benefits associated with the utilization of agricultural biomass for energy purposes.

In conclusion, global research trends in biomass as renewable energy highlight the significant potential of biomass utilization in addressing the world's energy challenges and advancing sustainable development. The research efforts focus on various aspects, including biomass feedstock availability and assessment, advanced conversion technologies, and integration with other renewable energy systems. Biomass-derived biofuels, such as bioethanol and biodiesel, offer lower carbon footprints and promote rural development through job creation and economic opportunities. Solid biomass utilization and biogas production from organic waste contribute to heat and electricity generation, waste management, and reduced emissions. However, future research should prioritize sustainability considerations, such as responsible biomass sourcing and environmental impact assessments. By embracing these research trends and emerging technologies, biomass-based energy systems can play a vital role in the global transition toward a cleaner and more sustainable energy future.

REFERENCES

Allende, S., Brodie, G., Jacob, M.V., 2023. Breakdown of biomass for energy applications using microwave pyrolysis: A technological review. *Environmental Research* 226, 115619. doi:10.1016/j.envres.2023.115619

Bian, R., Ma, B., Zhu, X., Wang, W., Li, L., Joseph, S., Liu, X., Pan, G., 2016. Pyrolysis of crop residues in a mobile bench-scale pyrolyser: Product characterization and environmental performance. *Journal of Analytical and Applied Pyrolysis* 119, 52–59. doi:10.1016/j.jaap.2016.03.018

Brummer, V., Jurena, T., Hlavacek, V., Omelkova, J., Bebar, L., Gabriel, P., Stehlik, P., 2014. Enzymatic hydrolysis of pretreated waste paper - Source of raw material for production of liquid biofuels. *Bioresource Technology* 152, 543–547. doi:10.1016/j.biortech.2013.11.030

Burke, A., Innes, P., 2000. Entomology in the twentieth century. *Advances* 70, 261–285.

Chaimaa, A., Malika, A., Mohammed, A., Loulidi Ilyasse, B.F., 2021. Analysis of the biomass behavior in the combustion process. *American Journal of Engineering Research (AJER)* 10, 26–35.

Chisti, Y., 2007. Biodiesel from microalgae. *Biotechnology Advances* 25, 294–306. doi:10.1016/j.biotechadv.2007.02.001

Chye, J.T.T., Jun, L.Y., Yon, L.S., Pan, S., Danquah, M.K., 2019. Biofuel production from algal biomass. *Bioenergy and Biofuels*. doi:10.1201/9781351228138-3

Cozier, M., 2014. Business highlights: Collaboration: Bigger and beta. *Biofuels, Bioproducts and Biorefining* 8, 743. doi:10.1002/BBB

Dowson, G.R.M., Haddow, M.F., Lee, J., Wingad, R.L., Wass, D.F., 2013. Catalytic conversion of ethanol into an advanced biofuel: Unprecedented selectivity for n-butanol. *Angewandte Chemie-International Edition* 52, 9005–9008. doi:10.1002/anie.201303723

Elliott, D.C., Biller, P., Ross, A.B., Schmidt, A.J., Jones, S.B., 2015. Hydrothermal liquefaction of biomass: Developments from batch to continuous process. *Bioresource Technology* 178, 147–156. doi:10.1016/j.biortech.2014.09.132

Gummert, M., Van Hung, N., Chivenge, P., Douthwaite, B., 2019. *Sustainable Rice Straw Management*. Springer, New York. doi:10.1007/978-3-030-32373-8

Hossain, N., Mahlia, T.M.I., Saidur, R., 2019. Latest development in microalgae-biofuel production with nano-additives. *Biotechnology for Biofuels* 12, 1–16. doi:10.1186/s13068-019-1465-0

Kapoor, M., Panwar, D., Kaira, G.S., 2016. *Bioprocesses for Enzyme Production Using Agro-Industrial Wastes: Technical Challenges and Commercialization Potential, Agro-Industrial Wastes as Feedstock for Enzyme Production: Apply and Exploit the Emerging and Valuable Use Options of Waste Biomass.* Elsevier, Amsterdam. doi:10.1016/B978-0-12-802392-1.00003-4

Lal, R., 2005. World crop residues production and implications of its use as a biofuel. *Environment International* 31, 575–584. doi:10.1016/j.envint.2004.09.005

Lehtikangas, P., 2001. Quality properties of pelletised sawdust, logging residues and bark. *Biomass and Bioenergy* 20, 351–360. doi:10.1016/S0961-9534(00)00092-1

León-Niño, A.D., Camargo, J.M.O., Neta, A.M.P., Toneli, J.T.C.L., Nebra, S.A., 2013. Sugarcane residual biomass briquetting aiming energetic use. In: *2nd International Conference on Wastes: Solution, Treatments and Opportunity.*

Liao, J.C., Mi, L., Pontrelli, S., Luo, S., 2016. Fuelling the future: Microbial engineering for the production of sustainable biofuels. *Nature Reviews Microbiology* 14, 288–304. doi:10.1038/nrmicro.2016.32

Mahesh, M.S., Mohini M., 2013. Biological treatment of crop residues for ruminant feeding: A review. *African Journal of Biotechnology* 12, 4221–4231. doi:10.5897/ajb2012.2940

Mantlana, K.B., Maoela, M.A., 2020. Mapping the interlinkages between sustainable development goal 9 and other sustainable development goals: A preliminary exploration. *Business Strategy and Development* 3, 344–355. doi:10.1002/bsd2.100

Morales, M., Quintero, J., Conejeros, R., Aroca, G., 2015. Life cycle assessment of ligno-cellulosic bioethanol: Environmental impacts and energy balance. *Renewable and Sustainable Energy Reviews* 42, 1349–1361. doi:10.1016/j.rser.2014.10.097

Page, J.R., Manfredi, Z., Bliznakov, S., Valla, J.A., 2023. Recent progress in electrochemical upgrading of bio-oil model compounds and bio-oils to renewable fuels and platform chemicals. *Materials* 16. doi:10.3390/ma16010394

Ramanathan, V., Crutzen, P.J., Kiehl, J.T., Rosenfeld, D., 2001. Atmosphere: Aerosols, climate, and the hydrological cycle. *Science* 294, 2119–2124. doi:10.1126/science.1064034

Richards, M., Pogson, M., Dondini, M., Jones, E.O., Hastings, A., Henner, D.N., Tallis, M.J., Casella, E., Matthews, R.W., Henshall, P.A., Milner, S., Taylor, G., McNamara, N.P., Smith, 2017. High-resolution spatial modelling of greenhouse gas emissions from land-use change to energy crops in the United Kingdom. *GCB Bioenergy* 9, 627–644. doi:10.1111/gcbb.12360

Rodriguez, C., Alaswad, A., El-Hassan, Z., Olabi, A.G., 2017. Mechanical pretreatment of waste paper for biogas production. *Waste Management* 68, 157–164. doi:10.1016/j.wasman.2017.06.040

Ronces, E.A., 2021. Process design. Life cycle of a process plant 59-74. doi:10.1016/B978-0-12-813598-3.00005-7

Saleem, M., 2022. Possibility of utilizing agriculture biomass as a renewable and sustainable future energy source. *Heliyon* 8, e08905. doi:10.1016/j.heliyon.2022.e08905

Sidohounde, A., Pascal Agbangnan Dossa, C., Nonviho, G., Papin Montcho, S., Codjo Koko Sohounhloue, D., 2018. Transesterification reaction and comparative study of the fuel properties of biodiesels produced from vegetable oils: A review. *Chemistry Journal* 4, 79–90.

Sieverding, H.L., Bailey, L.M., Hengen, T.J., Clay, D.E., Stone, J.J., 2015. Meta-analysis of soybean-based biodiesel. *Journal of Environmental Quality* 44, 1038–1048. doi:10.2134/jeq2014.07.0320

Takkellapati, S., Li, T., Gonzalez, M.A., 2018. An overview of biorefinery-derived platform chemicals from a cellulose and hemicellulose biorefinery. *Clean Technologies and Environmental Policy* 20, 1615–1630. doi:10.1007/s10098-018-1568-5

Thangaraj, B., Solomon, P.R., 2021. Biodiesel production by the electrocatalytic process: A review. *Clean Energy* 5, 19–31. doi:10.1093/ce/zkaa026

Torri, I.D.V., Paasikallio, V., Faccini, C.S., Huff, R., Caramão, E.B., Sacon, V., Oasmaa, A., Zini, C.A., 2016. Bio-oil production of softwood and hardwood forest industry residues through fast and intermediate pyrolysis and its chromatographic characterization. *Bioresource Technology* 200, 680–690. doi:10.1016/j.biortech.2015.10.086

Tripathi, N., Hills, C.D., Singh, R.S., Atkinson, C.J., 2019. Biomass waste utilisation in low-carbon products: Harnessing a major potential resource. *NPJ Climate and Atmospheric Science* 2. doi:10.1038/s41612-019-0093-5

Tun, M.M., 2015. *Asean Strategy on Sustainable Biomass Energy for Agriculture Communities and Rural Development in 2020-2030*. ASEAN.

Van Oost, G., Hrabovsky, M., Kopecky, V., Konrad, M., Hlina, M., Kavka, T., 2008. Pyrolysis/gasification of biomass for synthetic fuel production using a hybrid gas-water stabilized plasma torch. *Vacuum* 83, 209–212. doi:10.1016/j.vacuum.2008.03.084

19 Biomass Energy for Sustainable Development
Opportunities and Challenges

Kamla Malik, Monika Kayasth, Sujeeta Yadav,
Nisha Arya, Shweta Malik, Rekha Chahar,
Kashish Sharma, Dandu Harikarthik,
Shikha Mehta, and Meena Sindhu
CCS Haryana Agricultural University

19.1 INTRODUCTION

Due to the rapid depletion of energy resources, rising prices and environmental pollution have been major problems that the world is currently facing, and it is important to move towards sustainable sources of alternative energy, or in other words, renewable energy sources. Currently, about 90% of the world's energy demands are met by fossil fuels, which are expected to reduce to around 50% by 2040 with the advent of the above-mentioned renewable energy sources (Hussein, 2015; Palaniappan, 2017). One of the great technological challenges of today is the development of renewable energy technologies due to serious problems related to the production and use of energy. Renewable energy, defined as energy that comes from resources that are naturally replenished on a human timescale, is vital to the much-needed socio-economic development of developing countries. Biomass is the term for energy from plant matter grown to generate electricity or produce, for example, trash such as dead trees and branches, yard clippings, and wood chips for biofuel; it also includes plant or animal matter used for the production of fibres, chemicals, or heat (Bandyopadhyay, 2015). It may also include biodegradable waste that can be burned as fuel. This process releases large amounts of carbon dioxide gas into the atmosphere and is a major contributor to unhealthy air in many areas. Some of the more modern forms of biomass energy are methane generation, the production of alcohol for automobile fuel, and fuelling electric power plants. There is a dire need to develop eco-friendly, cost-effective technologies for the utilisation of biomass in value-added products.

Depending on where it comes from and how it is processed, biomass is characterised as traditional or modern. Traditional or conventional biomass is connected to the creation of energy-utilising materials from improper management and methods that are distinguished by poor efficiency and excessive pollution emissions. Modern biomass is produced through proper management, the application of technologies that ensure high productivity and conversion rates, and the production of high-quality

DOI: 10.1201/9781003406501-19

biofuels such as ethanol, biogas, and bio-oil from vegetable oils, reforested wood, municipal and industrial waste, etc. (Goldemberg, 2009).

Presently, there are no exact data available on the utilisation of biomass as energy source, but it is estimated that one-third of the world's population uses traditional biomass (which includes wood, agricultural, livestock, and forestry residues, among other sources) as their primary source of energy, making up about 90% of the world's total consumption of biomass. Families use traditional biomass to meet their energy needs in several parts of Africa, Asia, and Latin America, primarily for cooking. In these situations, the use of biomass is routinely inefficient, depleting natural resources and harming the operators of the cooking systems.

In addition, the production of fuels from conventional biomass sources has the potential to exacerbate the issue of deforestation, putting extra strain on the surrounding ecology and increasing greenhouse gas emissions. Despite these drawbacks, billions of people still use conventional sources of biomass to supply their energy needs because they are easily available and cost-effective. In addition, many nations would have to raise their energy imports without this provision, and many low-income families would have to pay more to buy other kinds of energy (Goldemberg, 2009).

19.1.1 CURRENT SCENARIO AND AVAILABILITY OF BIOMASS

Biomass has always been an important energy source for the country considering the benefits it offers. It is renewable, widely available, and carbon-neutral and has the potential to provide significant employment in rural areas. Biomass is also capable of providing firm energy. About 32% of the total primary energy used in the country is still derived from biomass and more than 70% of the country's population depends upon biomass for its energy needs.

The current availability of biomass in India is estimated to be 500 million metric tons per annum. Studies sponsored by the ministry have estimated surplus biomass availability at about 120–150 million metric tons per annum covering agricultural and forestry residues. This corresponds to a potential power generation of about 18,000 MW. Table 19.1 depicts the different biomass with potential of India.

Globally, because of the serious concern on the use of fossil fuels, it is important to start using renewable energy sources. Among the renewable energy sources, biomass plays a vital role especially in rural areas, as it constitutes the major energy

TABLE 19.1
Biomass Energy Potential of India

Biomass Type	Exploitable Potential (Mt)	Currently Exploited (Mt)
Woody biomass	422	75
Crop residue	180	90
Cattle dung	248	80
Urban waste	62	5
Total	912	250

source for the majority of households in India. Agricultural biomass is a relatively broad category of biomass that includes the food-based portion of crops (such as rice, corn, sugarcane, and beets), the non-food-based portion of crops (such as corn stover, orchard trimmings, and rice husks), etc. Ministry of New and Renewable Energy has realised the potential and role of biomass energy in the Indian context. Hence, the ministry has initiated a number of programmes for the promotion of efficient technologies to use in various sectors. For efficient utilisation of biomass, bagasse-based cogeneration in sugar mills and biomass power generation have been taken up under biomass power and cogeneration programme.

One of the great challenges in the upcoming decade is the constant thirst for energy sources and that too renewable since they are not only eco-friendly/cost-effective but human-friendly. There is an instant need to dwindle our high dependency on fossil fuels which till date are used for power generation, transportation, and industrial applications which are destroying our mother earth at an alarming rate. This is not the fault of fossil fuels but mankind who are overexploiting them for their development without giving a serious thought to their negative effects. India is an agrarian society, and there is a vast generation of biomass in every state of the country. A lot of biomass is generated annually which is either being improperly used or underutilised or else is being simply burnt creating further environmental problems such as emission of greenhouse gases, huge loss of plant nutrients, organic matter, and degradation of soil properties due to wastage of residue. The management of such a vast quantity of biomass/waste is the need of the hour.

In the world, energy consumption falls due to the economically restored and imbalanced supply of energy, which can be contented by accepting carbon-neutral technologies and mitigating the adverse effects of greenhouse gases (Liu et al., 2021). Carbon neutrality is one of the processes in which the balance between CO_2 emissions and atmospheric carbon counterbalances through the favourable technologies that aid in carbon sequestration and utilisation, in addition to the value-added products generation (Kazemifar, 2022). For example, microorganisms (macro and microalgae) could use atmospheric carbon dioxide for their growth and metabolism and further conversion by using technologies into biodiesel (biofuels), thereby reducing their carbon dioxide footprint at atmospheric levels as well (Kumar et al., 2022). Recently, research is more emphasised on carbon dioxide capture, biochar and activated carbon that have been produced from agricultural residues, thus exploiting the underutilised, easily available biomaterials (biomass) to achieve global sustainable development goals (SDGs) (Yaashikaa et al., 2020). Currently, nanoparticles such as metallic oxides and CNTs have been in the attention due to their properties and exceptional adsorption capabilities for carbon sequestration (Bathla et al., 2022). However, nanomaterials and nanoparticles could catalyse carbon dioxide conversion into carbon-neutral fuels (Dongare et al., 2021). This would help in attaining a balance between carbon dioxide generation and abatement/reduction that results in a zero carbon footprint. Several literatures reported in the discussion of various aspects of carbon neutrality, namely policy and carbon peaking (Wei et al., 2021), advanced engineering solutions for CO_2 capture, proper utilisation and storage (Kazemifar, 2022), and country-based specific challenges and opportunities for carbon neutrality (Liu et al., 2021). To achieve the world's SDGs, it is essential to review the need for a prototype move away from fossil or conventional fuels and acquire knowledge on the latest, advanced, and environmental friendly carbon sequestration techniques to eliminate emitted carbon dioxide emissions for a better future for next generation.

Likewise, biomass has gained various benefits – first, it cannot be depleted like fossil fuels because it is renewable, abundant, cost-effective, and eco-friendly that's used as a sustainable alternative source to fossil fuels. Second, as per sustainability management, it is considered carbon-neutral, whereas the burning of fossil fuels releases harmful gases such as CO_2 and other greenhouse gases and creates heat in the atmosphere. These fossil fuel sources have been responsible for hazardous impacts on the environment, human, and animals.

19.2   ENERGY ISSUES

The majority of human activities depend on energy, which is also crucial for the advancement of the economy and of society. Since the formation of fossil fuels (coal, gas, and oil) required millions of years of natural processes, they are regarded as non-renewable energy sources. Fossil fuels are very simple and accessible to employ for the production of heat and electricity when compared to other types of energy carriers. Since the Industrial Revolution, fossil fuels have gained popularity as a source of energy. As a result, seemingly unaffected by this, the world continues to advance with a growing reliance on fossil fuels, rapid growth in energy-intensive businesses and automobiles, population growth, and urbanisation. Therefore, there is a need of hour, how we can sustain the supply of reliable and affordable, eco-friendly energy sources.

Biomass – the fourth largest energy source after coal, oil, and natural gas – is the largest and most important renewable energy option at present and can be used to produce different forms of energy. As a result, it is, together with other renewable energy options, capable of providing all the energy services required in a modern society, both locally and in most parts of the world (Long and Wang, 2013). Renewability and versatility are, among many other aspects, important advantages of biomass as an energy source. Moreover, compared to other renewables, biomass resources are common and widespread across the globe.

In the recent era, human beings face three challenges:

- the greenhouse gas (GHG) emissions and global warming
- renewable or sustainable forms of energy
- energy security.

This led to an EU target of at least 40% domestic reduction in GHG emissions by 2030 compared to 1990, which the European Council agreed to in October 2014. The majority of non-renewable resources, such as coal, natural gas, oil, and nuclear energy, are over-exploited as a result of the need to meet constantly rising energy demands brought on by global urbanisation and industrialisation, which has a significant negative impact on climate change (Williams et al., 2019). Due to their propensity to absorb sunlight, GHGs such as carbon dioxide (CO_2), methane (CH_4), and nitrous oxide (N_2O) are recognised to have a significant role in global warming (Feng et al., 2022). There is an urgent need for mitigation, as evidenced by the average worldwide CO_2 emissions from 2000 to 2016 of 29364.4 Mt CO_2 (Londono-Pulgarin et al., 2021).

19.2.1 ENERGY FROM BIOMASS

In contrast to other renewable energy sources, biomass (the biodegradable portion of biogenic goods, such as plants and their waste) is readily accessible locally and technically adaptable in the energy conversion process. Biomass energy has the potential to be modernised worldwide, that is, produced and converted efficiently and cost-competitively into more convenient forms such as gases, liquids, or electricity (IBEP, 2006). The fact that biomass is the only renewable energy source that can be used to create liquid gasoline makes it even more alluring. These factors have led to an increase in interest in biomass energy worldwide, particularly in the form of oil. Biomass is naturally composed of the elements carbon, hydrogen, and oxygen. It first forms through a biological process using energy from the sun and carbon dioxide (CO_2) from the atmosphere (Dongare et. al., 2021). One of the simplest ways to use biomass energy is by combustion, which converts the energy in biomass into heat.

In addition, biomass power is mentioned as a significant alternative for supplying energy to the rural sector, and a few methods for biomass conversion are briefly reviewed. Biomass and renewable energy derived from biomass are regarded as end products, involving cutting-edge technologies to increase the effectiveness of power generation. There is evidence that biomass offers a clean, renewable energy source that could benefit the environment, economy, and energy.

Biomass power is an important alternative for providing energy in the rural sector. The inherent advantages of utilisation of biomass are that employment opportunities are created even for the cultivation, collection, transportation, and storage of biomass. In evaluating biomass energy chains, it is clear that a simple cost-benefit analysis does not capture a range of 'external' costs and benefits that arise from the supply of energy services. Figure 19.1 provides a schematic representation of biomass fuel chains (Bauen et al., 2004; Singhal et al., 2021). The technologies for biomass conversion mainly consist of direct combustion processes, thermochemical processes, biochemical processes, and agrochemical processes (Demirbas, 2001; IBEP, 2006).

19.2.2 BIOMASS WASTES AS ENERGY SOURCES

- Wastewaters and industrial wastes.
- Food industry wastes.
- Solid wastes are obtained from different industries such as fruits and vegetable scrap, food, pulping, and fiber from the extraction of sugar and starchy materials, and so on. All these solid wastes make a potential feedstock for energy generation.
- Liquid wastes are generated by fruit and vegetables, the meats washing process, cleaning of poultry and fish, and winemaking process. The wastewater contains dissolved organic matter (DOM) such as sugar, starch, etc. These

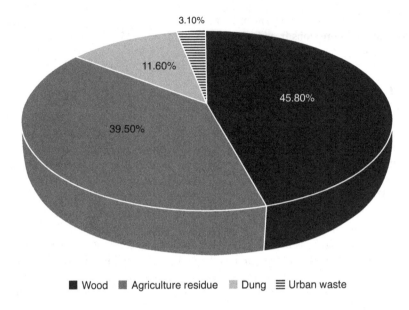

FIGURE 19.1   Biomass used as fuel in India.

industrial wastes are anaerobically digested to produce biogas and fermented to produce ethanol.

- Animal wastes: Mainly consist of organic material, moisture, and ash. Decomposition of these wastes may be in aerobic or anaerobic conditions. Under aerobic conditions, carbon dioxide and stabilised organic materials (SOM) are produced, whereas under anaerobic conditions, methane is produced.

19.3   BIOMASS WASTE ENERGY CONVERSION TECHNOLOGIES

Large quantities of various biomass wastes are available in India. These wastes can be converted to energy fuels called biofuels by biochemical as well as thermochemical conversion processes. Non-edible lignocellulosic biomass materials are increasing attention as renewable, economical, and abundant resources to reduce dependency on petroleum resources and minimise energy and material feedstock costs (Liu et al., 2022). These resources do not cause an additional increase in the carbon dioxide level in the earth's atmosphere compared to fossil-based energy fuels such as coal, gasoline, or natural gas. The carbon dioxide captured in biomass growth mostly balances with the release of carbon dioxide from bioenergy/biofuel. Therefore, the use of biomass energy has the potential to reduce greenhouse gas emissions. These biomass materials are the largest carbon sources for the production of various fuels, chemicals, and platform compounds and bioproducts. However, due to their heterogeneous, complex, and rigid structures, it is hard to break down these materials into smaller components and/or convert to a wide range of value-added products (Sikarwal et al., 2017). Biomass has a relatively low energy density; therefore, it requires more biomass feedstocks to supply the same amount of energy as a traditional hydrocarbon fuel. High oxygen contents of biomass materials can also negatively affect their conversion to various products such as fuels. For instance, to produce hydrocarbons fuels that can be comparable with petroleum-based one's oxygen should be removed from biomass structure. The efficiency of conversion processes can also vary depending on the biomass types (hardwood, softwood, grass, etc.).

In the current scenario, fossil fuels (petrol, diesel, coal, natural gas) are broadly used for power generation or mechanical energy in different sectors such as transport, commercial, agriculture, domestic, and industrial (Demirbas, 2009). In the transport sector, about 98% of the total energy is consumed which is derived from fossil fuel. Nowadays, there are several technological challenges towards the development of renewable energy production in a sustainable manner. There is an urgent need to search for an alternative fuel that is renewable, cost-effective, eco-friendly, and has fewer greenhouse gas emissions. The potential alternative for fossil fuels is biofuel production. In this direction, developing countries have taken initiatives to replace conventional fuels with biofuels as a sustainable substitute. Biofuels are renewable solid/liquid fuels that are produced from biomass used as biological raw materials (Bandyopadhyay, 2015).

There are several existing and developing routes to convert biomass and other feedstocks/raw materials into different value-added products. Basically, there are four routes which could be further sub-classified into different categories:

1. Physical route/processes
    a. Briquetting
    b. Palletisation
2. Agrochemical route/processes
3. Thermochemical route/processes
    a. Combustion in excess air
    b. Carbonisation
    c. Pyrolysis
    d. Gasification
4. Biochemical route/processes
    a. Fermentation
    b. Anaerobic digestion

The conversion of biomass or feedstock into liquid biofuels is based on two main processes that is, biochemical and thermochemical conversion and products are bioethanol, biogas, biodiesel, bio-oil, bio-syngas, biohydrogen, transport fuels, and heat generation through fermentation, pyrolysis, liquefication, transesterification, chemical synthesis, and combustion process (Figure 19.2).

Thermochemical conversion process is characterised by higher temperature and high conversion rates which is suitable for dry biomass/feedstock with lower moisture and is generally less selective for limited products. On the other hand, biochemical technologies are more suitable for wet wastes which are rich in organic matter (Sohel and Jack, 2011). Biofuels such as bioethanol, biodiesel, and biogas are important for future energy sources because they replace fossil fuels. It is derived from renewable

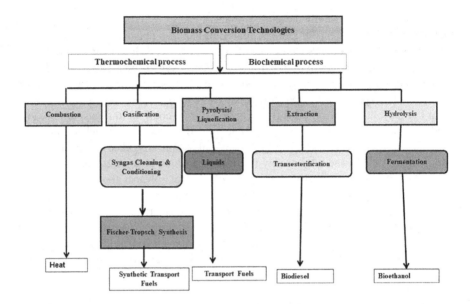

FIGURE 19.2 Biomass conversion technologies for biofuels production (Sikarwal et al., 2017).

sources of biomass such as plants, wheat, sugar beet, corn, straw, and wood used as a feedstock for bioethanol production. Similarly, biodiesel is derived from vegetable oils and animal fats through a process called transesterification. Renewable biofuels can be produced from different feedstock via conversion technologies such as liquefaction, enzymatic hydrolysis, saccharification, fermentation, transesterification, anaerobic digestion, and dark fermentation (Kamla et al., 2019). Table 19.2 shows the overview of conversion technologies for biofuel production from various feedstock/substrates/biomass (Mussatto, 2010).

19.3.1 THERMOCHEMICAL CONVERSION

19.3.1.1 Combustion

Burning of biomass in the air and used to transform the chemical energy held in the fuel into thermal energy, mechanical power, and electricity through a variety of processes and devices, such as furnaces, stoves, steam turbines, boilers, etc. Any kind of biomass can be burned, however in reality combustion is only practical for biomass with moisture content around 50%, unless the biomass is pre-dried. Biochemical conversion techniques are better adapted to biomass with high moisture content. Net bioenergy conversion efficiencies for biomass combustion power plants range from 20% to 40%. The higher efficiencies are obtained with systems over 100 MWe or when the biomass is co-combusted in coal-fired power plants. One heat engine cycle, the Stirling cycle, uses combustion to provide shaft power directly but the development of the cycle is presently limited to small power outputs.

TABLE 19.2

Feedstock Conversion Technologies for Biofuel Production

Type of Biofuels	Feedstock/Substrate	Conversion Technologies
Bioethanol	Cellulosic biomass: starch and sugars crops (grain, corn, potatoes, sugarcane, sugar beet, etc.)	Liquefaction, hydrolysis, saccharification, and fermentation
	Lignocellulosic biomass: Forest and wood waste resources (such as poplar); agricultural residues (such as corn, sorghum, oat, barley, wheat, soybean, cotton, bagasse and rice straws); and energy crops (hybrid sorghum, energy cane, miscanthus, switchgrass, eucalyptus and pine)	Pre-treatment, hydrolysis, saccharification, and fermentation
Biodiesel	Oil crops: rapeseed, oil palm, soybean, canola, jatropha, castor, etc.	Cold and warm pressing extraction, purification, and transesterification
	Waste cooking/frying oil	
	Macro and microalgae	Cultivation, harvesting, extraction, and transesterification
Biogas	Cattle dung, agricultural residue, sewage sludge, municipal waste, green waste and crops	Anaerobic digestion
Biohydrogen	Lignocellulosic biomass, biowaste, algae, and food waste	Dark fermentation, photo fermentation, bio photolysis

19.3.1.2 Biomass Gasification

Gasification is the conversion of biomass into a combustible gas mixture by the partial oxidation of biomass at high temperatures, typically in the range of 800–900 LC. The low calorific value (CV) gas produced can be burnt directly or used as a fuel for gas engines and gas turbines. The application of this produced gas can be used as a feedstock (syngas) for the production of chemicals such as methanol and hydrogen, which can be potential fuels for transportation and others.

19.3.1.3 Pyrolysis

It is a thermal chemical conversion process of biomass into biofuels that occurs in the absence of oxygen and converted to liquid, that is, bio-oil or bio-crude, solid (biochar), and gaseous fractions, by heating at high temperatures (500–1,000°C).

19.3.2 Biological Conversion

19.3.2.1 Fermentation

Fermentation is used commercially on a large scale in various countries to produce ethanol from sugar crops (e.g., sugar cane, sugar beet) and starch crops (e.g., maize, wheat). The biomass is ground down and the starch is converted by enzymes to sugars, with yeast then converting the sugars to ethanol. Purification of ethanol by distillation is an energy-concentrated step, with about 450 L of ethanol being produced by 1,000 kg of dry corn. Solid residue obtained from this process can be given to cattle to feed and bagasse which is obtained from sugar cane can be used for next gasification or as a fuel for boilers. The conversion of lignocellulosic biomass (such as wood and grasses) is more complex, due to the presence of longer-chain polysaccharide molecules and requires acid or enzymatic hydrolysis before the resulting sugars can be fermented to ethanol (Nizami and Rehan, 2018; Singh et al., 2021). Such hydrolysis techniques are currently at the pre-pilot stage.

19.3.2.2 Anaerobic Digestion

In anaerobic digestion (AD), organic material is directly converted to a gas which is termed as biogas. It is a mixture of mainly methane and carbon dioxide with small quantities of other gases such as hydrogen sulphide. The biomass is converted in anaerobic environment by bacteria, which produces a gas with an energy of about 20%–40% of the lower heating value of the feedstock. AD is a commercially proven technology and is widely used for treating high moisture content organic wastes, that is, 80%–90% moisture. Biogas can be used directly in spark ignition gas engines and gas turbines and can be upgraded to higher quality, that is, natural gas quality, by the removal of CO_2. The overall conversion efficiency can be 21% (Sohel and Jack, 2011; Noor, 2011). As with any power generation system using an internal combustion engine as the prime mover, waste heat from the engine oil and water-cooling systems and the exhaust could be recovered using a combined heat and power system.

19.3.3 Mechanical Extraction

Extraction is a mechanical conversion process in which oil is produced from the seeds of various biomass crops such as groundnuts, cotton, etc. The process produces not only oil but also a residual solid or 'cake', which is suitable for animal fodder. Three tons of rapeseed is required per ton of rapeseed oil produced. Rapeseed oil can be processed further by reacting it with alcohol using a process termed esterification.

19.4 BIOENERGY'S CONTRIBUTION TO SUSTAINABLE DEVELOPMENT SUSTAINABILITY

Bioenergy resources take many forms, which can be broadly classified into three categories:

a. Residues and wastes
b. Purpose-grown energy crops
c. Natural vegetation

In addition, a number of variables combine to make bioenergy an important issue for achieving the Millennium Development Goals (MDGs). On the basis of a few sustainability principles and metrics, the contributions of bioenergy to the sustainable development of humanity are also examined.

The progress of any nation today is measured in terms of its efforts towards the achievement of the MDGs. Many factors converge in making bioenergy a key component and a viable opportunity in the great effort towards the achievement of the MDGs (Table 19.3).

TABLE 19.3
Bioenergy's Contribution to Sustainability

Component of Sustainability	How Does Bioenergy Help Accomplish the MDGs?
a. Social development	• Bioelectricity production has the highest employment-creation potential among renewable energy options • Participation of representatives of socio-environmental organisations, participation of communities as decision-makers rather than just being consulted, the degree of inclusion of the local population in project design, and knowledge of the proposal and alternatives are all indicators of social accountability • Improve the quality of life by increasing access to and reliability of energy services for rural households • Encourage governance options, equity, and gender equality, particularly given women's important role in household energy management • Among renewable energy sources, bioelectricity production has the greatest potential for job creation • Attracts investment to rural areas, creating new business opportunities for small and medium-sized enterprises in biofuel production, preparation, transportation, trade, and use, as well as providing income to the people who live in and around these areas • Bioenergy has the potential to be a driver of rural development and regeneration in areas where investment is most needed and job creation is most difficult.
b. Environment	• CO_2 emissions have been reduced by using cleaner fuels such as ethanol and biodiesel • Reduced indoor air pollution from wood energy combustion in low-income households linked to favourable features of cooking equipment • Sustainable biomass production in marginal lands conserves resources and restores ecosystems.
c. Economics	• Improve access to energy services for small rural industries. • Increase national energy security while lowering the cost of oil imports • Save foreign currency by substituting imported fossil fuels • Agriculture, agro-industries, and forestry will benefit from increased diversification and income opportunities • Increased use of biomass for energy leads to improved economic development and poverty alleviation, particularly in rural areas (from sustainable resource management) • Increase the value of rural resources by encouraging participation and investment from the private and public sectors.

19.5   BIOMASS ENERGY SYSTEMS AND LINKAGES TO SUSTAINABLE HUMAN DEVELOPMENT

The uncontrolled and reckless human activities since the Industrial Revolution have caused a number of environmental problems such as fossil fuel depletion, climate change, and environmental pollution. Gradually, the importance of moving towards "sustainable development" was realised throughout the world. Sustainable development is the development that meets the needs of the present without compromising the ability of future generations to meet their own needs.

It is evidenced that biomass provides a clean, renewable energy source, which could improve economic, energetic, and environmental sectors. Also, many factors converge in making bioenergy a key issue toward the achievement of the MDGs. Contributions of bioenergy to the sustainable development of humanity are also discussed, based on some sustainability considerations and indicators. The future projection on the use of renewable energy resources points toward actions for economic development where renewable energy from biomass will play more and more a growing role, without affecting the community's food security.

19.5.1   Socio-Economic Implications of Biomass Usage

The industrial usage of biomass as a raw material for new products can potentially have multiple socio-economic benefits for farmers and their communities, especially if the supply chain is built with aspects of fairness and inclusion in mind.

Indian agriculture is characterised by several environmental and social sustainability challenges, which are linked, for example, to water use, soil quality, access to food, and farmers' income. At the same time, there are several opportunities to improve farmers' socio-economic status, but the challenging socio-economic context can also create and maintain risks that ought to be tackled once they set up new supply chains around the industrial usage of biomass. In this section, we introduce some of these opportunities and risks and point out some ways to cope with the difficulties and sustainability criteria and indicators (Table 19.4).

19.5.1.1   Economic Benefits

It is very important that the well-being of farmers is considered when developing new value chains for biomass usage. Poverty alleviation and creation of decent working conditions are highlighted in the UN's SDGs and this aspect should be emphasised in this context as well. Small-scale marginal farmers often face economic hardship and struggle to survive with their low income level. Any improvement in the profitability of their work is a possibility to enhance their socio-economic situation. Therefore, it is critically important that prices paid for the residue are fair and in line with the possible changes in their farming and household operations and workload caused by the collection of biomasses.

There is also a possibility for increased economic activity and income for communities around the areas of residue collection. Much of the residue is often collected by the local workforce, although this is not always the case. Whenever setting up new operations and supply chains it is important to look at who is doing what now, and do

TABLE 19.4
Sustainability Criteria and Indicators

Criteria	Desirable and Pre-Requisites	Indicators
a. Social		
• Social responsibilities • Gender equality • Active-Participation in decision-making • Social inclusion • Type of management • Job creation and management	Building capacity and information Recognition of women as key decision-makers at all stages of the process Education Political forums for participation with real influence over decisions, information, and training Benefits of the project are shared with the local community. Training for management Training for the creation of cooperatives, awareness and training of families with technical and political information	Participation of local residents and national socio-environmental organisations in project planning Existence of programmes and policies for women and youth Numerous sites, the nature and types of consultations, the publicity methods, information access, language, and material's accessibility Measures of quality and compliance with accepted standards of the involuntary resettlements, when necessary and accepted Contribution to access to services and infrastructure on the part of local populations to education, energy, waste and sewage services Number of families previously without access to energy who benefit from the project Epidemiological assessment and monitoring Impact on quality of life of the communities, Social programmes, especially for health and education Organisation structures and forms of decision-making, number of participants /decision-makers, involvement of organisation representing local workers, participation of women Number of jobs per unit of energy (production chain, implementation, and operation), profit sharing, generation of new local opportunities and sources of income, relation between local jobs before and after the project, indexes of increase in acquisitive power of the local population

(Continued)

TABLE 19.4 (*Continued*)
Sustainability Criteria and Indicators

Criteria	Desirable and Pre-Requisites	Indicators
b. Environment • Environmental management • Land use	Use of best practices available Crop diversification Agroforestry strategies Agroecology Pesticide use should be reduced or eliminated. Reducing soil loss Producer development Protection of natural areas Diversification and decentralisation of economic activities Regions classified as suitable by strategic environmental assessment Comply with economic/ ecological zoning Defined limits for occupation of biomass	Monoculture area Soil depletion Emissions from the atmosphere and effluents into bodies of water Sizes of continuous areas of monocultures Time necessary to manage crops Distance from energy source to consumer Decentralisation and diversification of production systems in an area/ region Distance travelled and time spent by workers in project area
c. Economic • Production organisation/labour relations • Technology • Use of bioenergy • Financing	Cooperatives Family agriculture New technology that can reduce the impact of energy production on ecosystems Contribution to energy matrix diversification Horizontal transfer of technologies and knowledge Decentralised generation and production Adoption of technology by the local population Promotion of energy efficiency Developing more efficient transportation systems Credits, access to land	Profit sharing among family farmers in the biofuels production chain Level of satisfaction with existing contracts Relationship between local workers and project maintenance outsiders Clean technology application Technological advancement Capacity of technology reproduction Origin of equipment Existence of technology licences International technical assistance is required. Changes in the use of renewable energy, cogeneration Consumption reduction rates Enhanced end-of-life conservation Capability to reduce, reuse, and recycle inputs in the final activities for which the energy is intended Demand management should be included in the project planning horizon. Conditions for government financing Programmes and lines to credit

they stand to lose out in novel supply chains. The inclusion of locals and those now employed by residue chains is thus encouraged. To gain socio-economic benefits for whole communities, it should be made sure that the local workforce is used to the extent possible, with formal, fair contracts that do not leave the workers in precarious positions. Otherwise, there is a risk of losing the economic benefits from biomass collection that has brought income to village communities besides farmers.

19.5.1.2 Health Improvements

The burning of straw has many negative side effects that can affect the health and well-being of large amounts of people in the countryside. If this residue can be used in industrial applications instead of burning, then multiple social benefits can occur.

19.5.1.3 Other Externalities

There are multiple other uses for the biomass (straw). For the best outcome for farmer and the environment alike, most likely only a portion of the residue should be bought per farmer/field. This way some of the straw can be left as mulch and used for other purposes as necessary. Some farmers use the residue as fodder, which at least in the case of rice, is not high-quality fodder. If farmers can sell their rice straw and enhance their possibility to buy good quality fodder, then this can in turn improve their livestock productivity, which then again can improve farmers' livelihoods. For this to happen, the price paid for the residue ought to be decent, to cover for the previous uses of the residue.

We believe taking the above considerations into notion, it is possible to create a fair and socio-economically beneficial supply chain for straw usage, in which farmers and communities alike can enhance their well-being and livelihoods (Figure 19.3).

19.6 OPPORTUNITIES, CHALLENGES, AND PROSPECTS OF BIOMASS ENERGY

Biomass is a natural resource used for various purposes, including energy, all around the globe (Perea-Moreno, 2019). In 2010, bioenergy accounted for 12% of the world's total final energy consumption, with 9% coming from traditional sources and 3% from modern bioenergy (IRENA, 2016). Therefore, to meet international goals to double the global share of renewables by 2030, a rapid increase in the use of modern biomass is necessary.

Some challenges need to be considered in the effort to use biomass energy in Ethiopia, which includes the lack of comprehensive national biomass policy and regulation: there is a lack of well-thought and comprehensive policies that direct activities in the biomass energy sector. When there is a requirement to promote the growth of particular renewable energy technologies, policies might be declared that do not adhere to the plans for the development of renewable energy. There is no defined framework for the biomass sector (Kaygusuz, 2012).

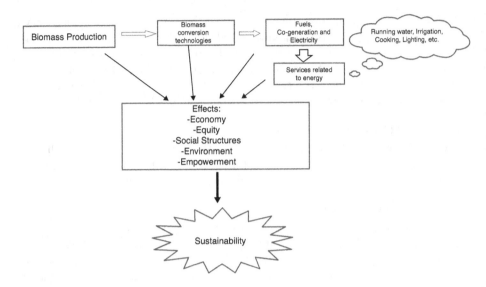

FIGURE 19.3 A conceptual picture of biomass energy systems and their relationships to human development.

Opportunities and challenges related to biomass have to do with greenhouse gas emissions (biomass can contribute to reducing carbon emissions, but emissions may not be fully accounted for); resource availability (biomass can contribute to energy security, but its sources are finite); environment and human health (increased use of biomass for energy can have adverse effects on air quality, soil properties, and biodiversity). To address sustainability concerns, different responses have been put forward, including the principle of the cascading use of biomass, whereby it is used more than once, with energy conversion typically as the last step. Biomass – the fourth largest energy source after coal, oil, and natural gas – is the largest and most important renewable energy option at present and can be used to produce different forms of energy. As a result, it is, together with the other renewable energy options, capable of providing all the energy services required in a modern society, both locally and in most parts of the world. Biomass has proven to be an effective energy carrier capable of fulfilling the growing demand for clean and everlasting energy sources for the sustainable development of society.

19.6.1 CHALLENGES RELATED TO BIOMASS

The existing challenges of the biomass supply chain related to different feedstock can be broadly classified into operational, economic, social, policy, and regulatory challenges (Figure 19.4).

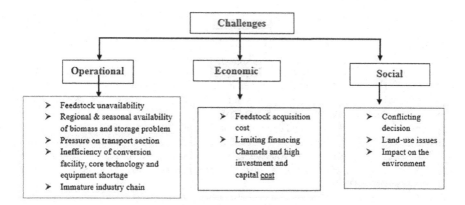

FIGURE 19.4 Classification of challenges of biomass. (*Source*: Raychaudhuri and Ghosh, 2016.)

19.7 CONCLUSION

Due to the depletion of fossil fuel, high oil prices, huge population, and limited energy resources in mind, the nation is searching for alternative renewable sources of fuels that are clean, non-toxic, and eco-friendly. India is one of the fast-growing economies with an increase in demand for fuel and energy. In this context, biomass industries are rapidly growing with a promising role in producing renewable energy and tackling climate change. In the present scenario, biomass plays a vital role as a source of energy with potentially address to several aspects such as environmental degradation, global warming, and climate change. The use of biomass, along with other renewable energy sources, can help to meet the world's growing energy demand. Biomass-based energy technologies are the largest source of renewable energy as well as a vital part of waste management and energy as well as fuel requirement. A greater emphasis is being laid on the promotion of bioenergy in the industrialised as well as developing world to counter environmental issues. These biomass conversion technologies could play an important role in making more sustainable and environmentally friendly fuels.

REFERENCES

Bandyopadhyay, K. R. 2015. *Biofuel Promotion in India for Transport: Exploring the Grey Areas*. The Energy and Resources Institute, TERI, India. https://petroleum.nic.in

Bathla, A., Lee, J., Younis, S. A. and Kim, K. H., 2022. Recent advances in photocatalytic reduction of CO2 by TiO2-and MOF-based nanocomposites impregnated with metal nanoparticles. *Materials Today Chemistry*, 24, 100870.

Cao, Y. and Pawłowski, L., 2013. Effect of biofuels on environment and sustainable development. *Ecological Chemistry and Engineering*, 20(4), 799–804.

Demirbas, A. 2010. Social, economic, environmental and policy aspects of biofuels. *Energy Education Science and Technology Part B: Social and Educational Studies*, 2, 75–109

Demirbas, A., 2008. Biofuels sources, biofuel policy, biofuel economy and global biofuel projections. *Energy Conversion and Management*, 49(8), 2106–2116.

Dongare, S., Singh, N. and Bhunia, H., 2021. Electrocatalytic reduction of CO2 to useful chemicals on copper nanoparticles. *Applied Surface Science*, 537, 148020.

Feng, H., Sun, C., Zhang, C., Chang, H., Zhong, N., Wu, W., Wu, H., Tan, X., Zhang, M. and Ho, S. H., 2022. Bioconversion of mature landfill leachate into biohydrogen and volatile fatty acids via microalgal photosynthesis together with dark fermentation. *Energy Conversion and Management*, 252, 115035.

Goldemberg, J. 2009. Biomass and energy. *Química Nova*, 32(3), 582–587.

Hussein, A. K. 2015. Applications of nanotechnologyin renewable energies-A comprehensive overviewand understanding. *Renewable Sustainable Energy Reviews*, 42, 460–476.

IRENA, 2016. *Renewable Energy Statistics 2016*. The International Renewable Energy Agency, Abu Dhabi, UAE.

Joshi, P. and Visvanathan, C., 2019. Sustainable management practices of food waste in Asia: Technological and policy drivers. *Journal of Environmental Management*, 247, 538–550.

Kaygusuz, K., 2012. Energy for sustainable development: A case of developing countries. *Renewable and Sustainable Energy Reviews*, 16(2), 1116–1126.

Kazemifar, F., 2022. A review of technologies for carbon capture, sequestration, and utilization: Cost, capacity, and technology readiness. *Greenhouse Gases: Science and Technology*, 12(1), 200–230.

Kumar, A., Venkatkarthick, R., Jayashree, S., Chuetor, S., Dharmaraj, S., Kumar, G., Chen, W. H. and Ngamcharussrivichai, C., 2022. Recent advances in lignocellulosic biomass for biofuels and value-added bioproducts: A critical review. *Bioresource Technology*, 344, 126195.

Kurchania, A. K., 2012. Biomass energy. In: Baskar, C., Baskar, S., and Dhillon, R. S. (Ed.), *Biomass Conversion: The Interface of Biotechnology, Chemistry and Materials Science*, pp. 91–122. Berlin Heidelberg: Springer-Verlag.

Liu, F., Dong, X., Zhao, X. and Wang, L., 2021. Life cycle assessment of organosolv biorefinery designs with the complete use of biomass. *Energy Conversion and Management*, 246, 114653.

Liu, Z., Deng, Z., He, G., Wang, H., Zhang, X., Lin, J., Qi, Y. and Liang, X., 2022. Challenges and opportunities for carbon neutrality in China. *Nature Reviews Earth & Environment*, 3(2), 141–155.

Londono-Pulgarin, D., Cardona-Montoya, G., Restrepo, J. C. and Munoz-Leiva, F., 2021. Fossil or bioenergy? Global fuel market trends. *Renewable and Sustainable Energy Reviews*, 143, 110905.

Long, H., Li, X., Wang, H., Jia, J., 2013 Biomass resources and their bioenergy potential estimation: A review. *Renewable and Sustainable Energy Reviews*, 26, 344–352.

Majid, M. A., 2020. Renewable energy for sustainable development in India: Current status, future prospects, challenges, employment, and investment opportunities. *Energy, Sustainability and Society*, 10(1), 1–36.

Malik, K., Ahlawat, S. and Malik, K. 2021. Nanotechnology: A sustainable solution for bioenergy and biofuel production. *Journal of Nanoscience and Nanotechnology*, 21(6), 3481–3494.

Ministry of New and Renewable Energy. 2009. National Policy on Biofuels. https://mnre.gov.in/file-manager/UserFiles/biofuel_policy.pdf

Mussatto, S. I., Dragonea, G., Guimarãesa, P. M. R., Silva, J. P. A., Carneiro, L. M., Roberto, I. C., Vicentea, A., Domingues, L. and Teixeira, J. A. 2010. Biotechnol. *Advances*, 28(6), 817–830.

Nizami, A. S. and Rehan, M., 2018. Towards nanotechnology-based biofuel industry. *Biofuel Research Journal*, 5(2), pp.798–799.

Noor, S., Latif, A. and Jan, M., 2011. Overview of biomass conversion technologies. *Science Vision*, 16, 17.

NREL, 2009. Biomass Research. www.nrel.gov/biomass/biorefinery.html

Pachauri, S. and Jiang, L. 2008. The household energy transition in India and China. *Energy Policy*, 36, 4022–4035.

Palaniappan, K., 2017. An overview of applications of nanotechnology in biofuel production. *World Applied Sciences Journal*, 35(8), 1305–1311.

Perea-Moreno, M.-A., Esther Samerón-Manzano, A. J. P. M. 2019. Biomass as renewable energy: Worldwide research trends. *Sustain Artic*, 11, 863.

Raychaudhuri, A. and Ghosh, S. K., 2016. Biomass supply chain in Asian and European countries. *Procedia Environmental Sciences*, 35, 914–924.

Saleem, M., 2022. Possibility of utilizing agriculture biomass as a renewable and sustainable future energy source. *Heliyon*, e08905.

Saleh, N. M., Saleh, A. M. and Mahdi, H. H., 2022. Production of biofuels from biomass as an approach towards sustainable development: A short review. *NTU Journal of Renewable Energy*, 3(1), 9–21.

Sikarwal, V. S., M. Zhao, P. Fennell, N. Shah. and E. J. Anthony. 2017. Progress in biofuel production from gasification. *Progress Energy Combustion* 61, 189–248.

Singh, J. and Gu, S., 2010. Biomass conversion to energy in India: A critique. *Renewable and Sustainable Energy Reviews*, 14(5), 1367–1378.

Singhal, B. S., Jain, A. K. and Jain, S. 2021. Biomass energy potential in India: A review. *Journal of Renewable and Sustainable Energy Reviews*, 215, 125–138.

Sohel, M. I. and M. W. Jack. 2011. Thermodynamic analysis of lignocellulosic biofuel production via a biochemical process: Guiding technology selection and research focus. *Bioresource Technology*, 102, 2617–2622.

Stout, B. A., 2012. *Handbook of Energy for World Agriculture*. Elsevier, Amsterdam. https://books.google.com/books

Sudheer, P. Suresh Kumar, and P. K. 2020. Viswanathan agricultural residues for bioenergy in India: Status, challenges, and policy implications. *Journal of Renewable and Sustainable Energy Reviews*, 347, 625–675.

Wei, S., Heng, Q., Wu, Y., Chen, W., Li, X. and Shangguan, W., 2021. Improved photocatalytic CO_2 conversion efficiency on Ag loaded porous Ta_2O_5. *Applied Surface Science*, 563, 150273.

Williams, E. A., Raimi, M. O., Yarwamara, E. I. and Modupe, O., 2019. Renewable energy sources for the present and future: An alternative power supply for Nigeria. *Energy and Earth Science*, 2, 2.

Yaashikaa, P. R., Kumar, P. S., Varjani, S. and Saravanan, A. 2020. A critical review on the biochar production techniques, characterization, stability and applications for circular bioeconomy. *Biotechnology Reports*, 28, e00570.

Zhang, L., Ling, J. and Lin, M., 2023. Carbon neutrality: A comprehensive bibliometric analysis. In: *Environmental Science and Pollution Research*, 30(16), 1–17.

20 Global Status of Biomass Energy Programmes– Challenges and Roadmap

Priyanka Devi, Palvi Dogra,
Prasann Kumar, and Joginder Singh
Lovely Professional University

20.1 INTRODUCTION

Without a doubt, the greatest challenge facing humanity now is climate change, which poses the greatest danger to international security. Since the Industrial Revolution, unchecked and irresponsible human behaviour has contributed to several environmental issues, including the depletion of fossil fuels, climate change, and environmental degradation. Throughout the world, the significance of pursuing "sustainable development" has gradually become apparent. There must be a significant reduction in greenhouse gas emissions to build a sustainable future and stimulate active economic development. One of the primary challenges to achieving carbon neutrality is the efficient use of natural resources for energy applications. This waste material is collectively referred to as biomass, which is the term used to refer to the whole category. Any commercial or industrial product (derived from food or feed) that utilizes biological products or domestic, renewable agricultural (plant, animal, or marine). Forestry materials are referred to as a "bio-based product" throughout this context (ABB, 2003; Industry Report, 2008; OCAPP, 2007). Different sectors must undergo long-term reform to thrive, resulting in more sustainable and balanced patterns of energy supply and demand. Modern bioenergy offers advantageous chances to create energy and valuable products from renewable sources for effective negotiation of conventional reactionary energies. Sustainable development satisfies the needs of the present without jeopardizing the ability of future generations to satisfy their own needs. It includes two essential generalizations the idea of restrictions imposed by the level of technology and social association on the ability of the landscape to meet present and future needs, and in especially the belief that the basic needs of the world's impoverished should be given priority. It is projected that biomass will continue to provide the majority of human energy demands. The adoption of renewable energy is essential, especially to the ever-increasing energy demand as well as the moment's increasingly pressing need for resources. In addition, the question of biomass application and eventuality is extremely important for the production of food and feed. To preserve the Earth's natural resources and physical infrastructure, one of the difficulties associated with the Sustainable Development Goals (SDGs) of the

DOI: 10.1201/9781003406501-20

United Nations (UN) needs to address feeding a growing population. Although biomass for energy operations is widely acknowledged and carefully researched, producing energy from biomass is still quite expensive due to both technological and logistical limitations. The expenses associated with scaling up and expanding biomass applications be reduced, which calls for stronger public cooperation and innovation. This special issue aims to compile research on the opportunities, difficulties, and unintended consequences of using biomass fuels as sustainable energy sources for energy production. This special issue will provide a forum for in-depth discussions about how to overcome the current obstacles of using biomass for energy, such as functional, financially beneficial, social and policy issues, and nonsupervisory concerns. Applicable pretreatment techniques are required to promote biodegradation and boost the original energy content to break through the toughest technical barriers to efficient biomass application. Some researchers suggest that we are in the middle of a broad, sustainable development revolution that is influencing every area of our lives. Due to its multidimensional nature, the world must be seen as a system that links space and time. We need to change our existing unsustainable production along with consumption patterns and embrace a better approach that takes into account political, social, specialized, and environmental concerns if we are to achieve sustainable development. In light of Pawowski's reorientation, a comparable change is especially crucial and critical for the achievement of sustainable energy, which is necessary for making the idea of intergenerational justice a reality. Numerous publications have examined various aspects of the biomass energy situation and the technologies that can be used to improve it. Further research must be done to determine the total effects of biomass force chain operations. In addition, it is necessary to handle updated technology outcomes suitable for the efficient and alluring utilization of biomass funds. It is important to present and value international case studies of the effective integration of biomass-based energy outcomes as priceless reference illustrations for other nations and diligent industries. To hasten technological advancement toward a sustainable future, knowledge of these motifs should be shared widely. The prospects of biomass-derived energy to comprehensively fulfil rising energy demand and effectively contribute to the energy transition will be the focus of this special issue. We especially encourage research papers that address implicit biomass energy upgrades and biomass-powered combined heat and power (CHP), co-firing, and combustion facilities. These factors could be essential in determining the future of biomass energy usage. There will be presentations on social, nonsupervisory, and policy-related concerns.

20.2   ENERGY ISSUES

Energy is necessary for almost all aspects of human life and is an essential factor in the development of both the economy and society. Since the formation of fossil fuels (oil, gas, and coal) is the result of natural processes that take millions of years, these types of energy sources are not considered to be renewable. For the production of both heat and electricity, the use of fossil fuels is comparatively straightforward and convenient in comparison to the use of other kinds of energy carriers. Since the beginning of the Industrial Revolution, people have favoured the utilisation of fossil fuels as a primary energy source is a widely employed practise in contemporary society an increasing amount. Therefore, it would seem as if nothing has changed, yet the whole globe is moving ahead with rising dependence. The increased utilization of fossil fuels, coupled with the rapid expansion of energy-intensive industries and automobiles, as well as the proliferation of urbanization and population growth, have contributed to this phenomenon. Unfortunately, our planet is unable to provide us with the quantity of fossil fuels that we need regularly. According to Pawłowski's (2010) estimates, it is projected that the global reserves of coal, natural gas, along with oil will be depleted within the next 150–200 years. The global demand for fossil fuels is increasing concurrently. The International Energy Agency's World Energy Outlook report of 2009, predicts that if governments continue with their current policies, the world's primary energy demand will increase by 30% between 2010 and 2035, with fossil fuels continuing to dominate the world's energy mix by that point, accounting for 75% of it. Particularly for China, the greatest developing nation and home to approximately a quarter of the world's population, the energy crisis is evident. China has sustained high economic growth over the past four decades, with an average annual gross domestic product (GDP) growth of roughly 9.8% from 1980 to 2009 (Ma, 2010). China, now the world's biggest energy user and the second-largest oil consumer after the United States, is sadly largely reliant on the expansion of energy-intensive manufacturing and heavy industries, exports, and fixed asset investment. With an average annual growth rate of 5.8% between 1980 and 2009, China's total energy consumption in 2010 reached 3.25 billion tons of coal equivalent (Ma, 2010). More than 90% of China's overall energy demands are fulfilled by fossil fuels, with coal accounting for 71% of that, oil for 19%, and natural gas for 3% (as of 2008) (Odgaard & Delman, 2014). Coal reserves of 176.8 billion tonnes, crude oil reserves of 21.2 billion tonnes, and natural gas reserves of 22.03 trillion cubic metres are all recoverable the nation is not particularly resource-rich, though (Zhang, 2009). The majority of developed nations have comparable energy issues. Fossil fuel consumption currently accounts for over 80% of the entire energy mix in the European Union (EU), with over 50% of that coming from renewable sources. The percentage of imports needed for fossil fuels increased significantly from 51% in 2000 to 54% in 2005 (EEA, 2008) and was projected to reach 70% by 2030 (Romero, 2012). Fossil energy is growing more expensive on a worldwide scale, which is causing a narrowing bottleneck to economic growth and wealth creation. We are also warned that the world is getting warmer in addition to the security of our energy supply. Unfortunately, there is a strong correlation between the two problems, and both are due to our constant use of fossil fuels. The next two points served as the foundation for our argument. First, there is a growing consensus that global warming is occurring and those greenhouse gas emissions, such as those from carbon dioxide and methane, are the primary culprits. Second, the analysis shows that fossil fuels, which presently account for the bulk of the world's energy supply, are the largest generators of greenhouse gases, creating 74% of all carbon dioxide emissions worldwide (Moriarty & Honnery, 2011). Combining the two, we may infer. It is widely acknowledged that an increase in energy consumption is directly proportional to the emission of greenhouse gases into the atmosphere, thereby increasing the likelihood of global warming. By taking into account the circumstances in China, the following is an illustration of how the first two components correlate: As we noted earlier, China's rapid economic growth, which is fuelled by fossil fuels, makes it the world's largest energy consumer while also making it the largest producer of greenhouse gases. Even while some scientists have published findings casting doubt on the origin and severity of climate changes (Pawłowski, 2006; Lindzen, 2010; Sánchez, 2008), the precautionary principle requires us to take some action concerning the association between the last two components. Perhaps even unfavourably, while reducing greenhouse gas emissions can lessen global warming, the current control mechanisms frequently impede or even halt economic growth. It is remarkable that the biggest source of air pollutants, such as particles in the air, tropospheric ozone components, acidic gases such as SO_2 and NOx, CO, CH_4, and volatile organic molecules other than methane are produced by burning fossil fuels. For instance, the European Environment Agency reports that in 2005, energy production and consumption accounted for roughly 55% of acidification-causing substances emitted in the EU, 76% of precursors for tropospheric ozone emitted in the EU, and 67% of particles emitted in the EU (EEA, 2008). In conclusion, finding a solution to keep the lights on while also protecting the environment is a pressing problem that has to be addressed as soon as possible.

20.3 BIOMASS: AS A SOURCE OF ENERGY

At the start of the twenty-first century, regional, national, and local governments all over the world adopted several laws to encourage the use of renewable energy. For instance, by 2010, the EU aims to have renewable energy account for 12% of gross final energy consumption (Directive, 2001). Following this, Directive 2009/28/EC was adopted by the EU, with the aim of achieving a 20% increase in the proportion of renewable energy in gross final energy consumption by 2020 as the overall target of the Community (Union (2009) relative to the level in 1990); Poland's target being a decrease to 15%. Ethical considerations dictate that only non-food biomass should be used to produce things such as fuels, chemicals, electricity, and heat. Industrialised countries have almost 1,500,000,000 acres of crop, forest, and woodland, with 460,000,000 acres devoted to agriculture. To achieve the 15% reduction, it may be required to convert 1.25 million hectares of farmland into energy plantations per year. This accounts for little over 2% of all land in industrialised countries (Bauen et al., 2004). During photosynthesis, plant tissues absorb sunlight and store it as biomass energy. Bioenergy resources come in a wide variety, but they may be broken down into different three categories (Rosillo-Calle et al., 2007):

1. Waste,
2. Energy crops are grown specifically,
3. Organic foliage.

Traditionally, wood, garbage, and alcohol fuels have identified as the three main sources of traditional biomass (Figure 20.1).

Some examples of renewable non-fossil fuel sources that can power facilities such as landfill gas and wastewater treatment plants are biomass. Biomass (the

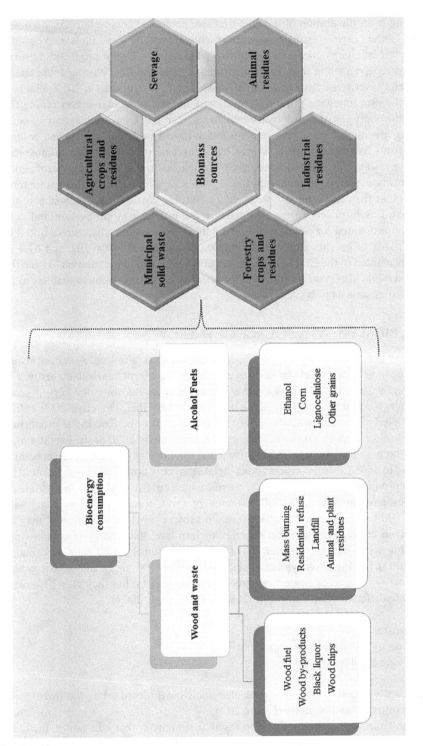

FIGURE 20.1   Resources for producing bioenergy from biomass.

biodegradable element of biogenic commodities, such as plants and their waste) is distinct from other renewable energy sources in that it is easily the energy conversion process can benefit from the utilisation of locally available and technically adaptable resources. Biomass is an attractive option due to its unique characteristic of being the sole renewable energy source that is capable of being utilised. As a result, there has been a global uptick in enthusiasm for biomass energy, especially in the form of oil. Carbon, hydrogen, and oxygen are the three main components of biomass. At first, it is formed biologically, with the help of the sun's energy and atmospheric carbon dioxide (CO_2). Combustion is one of the most basic forms of biomass energy utilization since it directly transforms the biomass's energy content into heat. Using solar energy in the form of biomass is a way to do so indirectly and under control. The generation process is optimized. The utilization of biomass as a source of energy has been found to have positive ecological effects and holds significant promise for further advancement in the field of sustainable energy production, provided that the energy generation process is suitably refined is transferred into more contemporary energy carriers such as electricity and liquid and gaseous fuels (IBEP, 2006). It is important to keep in mind that there are many different types of biomass, and they are classified based on their origin (plant or animal) or their phase (solid, liquid, or gaseous) (Bauen et al., 2004; IBEP, 2006). It is possible to burn biomass directly, as well as to convert it into solid, liquid, or gaseous fuel. This can be used for generating heat and power. Although all organic materials have the potential to be transformed into usable energy sources, the advantage of contemporary biomass-to-electricity systems is that they are capable of handling a variety of lignocellulose-rich materials. In contrast, production chains for liquid biofuels are generally reliant on annual, oil-, sugar-, or starch-rich crops. Below is a description of the options for converting biomass to energy. The bioenergy sector's biomass resources can be used in one of three ways (EREC, 2007; IBEP, 2006): – Biomass for transportation fuels, – Electricity generation using biomass, heat generation using biomass (transport biofuels), and so on. The potential for the expansion of China's biomass energy sector is considerable, given that it is the world's biggest agricultural country. Tables 20.1 and 20.2 provide a historical look at China's biomass energy production. Burning biomass wastes for heat and power generation and biofuel production (including bioethanol and biodiesel) make up the bulk of China's biomass energy consumption. Sorgo, which is adapted to colder northern temperatures and can survive periods of drought,

TABLE 20.1
Goals for Biomass Energy Production in China

Energy from Biomass	Unit	Years		
		2005	2010	2020
Bioelectricity	MV	2,000	5,500	3,0000
Bioethanol	1,000 tons	1,020	3,000	1,0000
Biodiesel	1,000 tons	50	200	2,000
Solid biomass fuel	1,000 tons		1,000	5,0000

shows promise as a biofuel crop in China's northern areas. Several different types of bio-oils are being studied for potential production in the arid southern areas, including colza oil, cottonseed oil, wood oil, and even Chinese tallow oil (Zhang, 2010).

There's little doubt that switching to biomass energy instead of fossil fuels would improve both the reliability and variety of the global power grid. It might also decrease. One possible approach to mitigate greenhouse gas emissions is to sequester carbon dioxide from the atmosphere and store it in crop biomass along with soil. The format of our debate is as follows: First, we categorize the biomass used for energy production into two groups: biomass residues/wastes and commercial energy crops. Then, we go over each of their drawbacks and difficulties separately. The rivalry between energy crops and food crops is a serious issue. Food crops (such as maize, sugarcane, soybeans, and palms) account for the vast majority of biofuels today, likely because of their convenience, high energy content, and amenability to widespread application. This raises the age-old biofuels vs. food argument over the best use of these crops. Since arable land is limited and the world's population is increasing, a quick transfer from edible oil to energy supplied by food crops might worsen food shortages. Furthermore, there is a chance that dense plantations of energy crops will harm the ecology. Large-scale cropping in places where crops are grown for energy purposes may cause significant biodiversity loss, soil erosion, and nutrient leakage, according to a warning from the UN. Even a wide diversity of crops could be harmful if they replace natural grasslands or woods (UN, 2007). However, our rationale does imply that the current energy crop-based biofuels are problematic, even though we do not oppose research for alternative fuels such as energy crops. Looking at the issues as a system and managing the system sustainably while considering both positive and negative impacts are necessary for finding solutions. Bio wastes, which are described as leftovers or remainders of the agricultural and forestry sectors, consist of items such as tree branches, grain husks, and straw, while biomass residues and wastes include things such as organic matter in municipal solid wastes including sewage sludge, as well as animal wastes. It acknowledged that using energy from biomass waste or residues can be beneficial in the fight against energy and climate change challenges. In addition, it gives a chance for waste management, especially for MST and sewage sludge, which account for the bulk of the ever-increasing garbage we generate – to be done sustainably. The development or use of dependable, economical, and energy-efficient conversion technology is the fundamental problem facing energy from biomass residues or trash. A barrier to the widespread use of energy generation is created by the dispersed and challenging-to-collect nature of this category of biomass. In terms of technology, the environment, society, and the economy, this calls on us to be more efficient with our resources while also potentially requiring the integration and diversity concerning the gathering and processing of feedstock, the techniques of energy conversion, and the applications of the energy that is created.

20.4 CONVERSION OF BIOMASS INTO ENERGY

The possibility exists for biomass energy to be modernized globally, that is, produced and transformed into more opportune forms such as gases, liquids, or electricity, cost-effectively and efficiently (IBEP, 2006; Larson and Katha, 2000). Today, industrialised countries get just 3% of their primary energy from modern biomass, and this fraction has been the same over the last several years. However, a sizeable section (approximately half of the global population residing in developing nations, primarily in rural areas) continue to depend on conventional biomass sources, particularly wood, as their primary energy source. In developing countries, biomass is conventionally employed as a primary energy source, contributing to 35% of the total energy supply, bringing the total worldwide contribution to 14%. The use of biomass power plants is becoming an increasingly viable option for the provision of electricity in rural regions. The intrinsic benefits of using biomass include the creation of job possibilities for biomass collection, storage, transportation, and horticulture. According to Demirbas (2001) and IBEP (2006), the most common kinds of technologies used for converting biomass include techniques of the utilisation of direct combustion, thermochemical processes, and biochemical processes, as well as agrochemical processes, which are commonly employed in various applications. As per the Environmental Protection Agency's report in 2007, various technologies have the potential to convert biomass into inert gases, organic oils, gases, and fuels. These converted products can be further utilized to produce desired energy products.

20.5   THERMOCHEMICAL TECHNOLOGIES

Biomass feedstocks are typically subjected to elevated temperatures to facilitate their conversion into energy, which is commonly manifested in the form of thermal and electrical energy. The utilisation of various technologies has the potential to produce thermal energy, electrical power, fuel sources, and bioproducts – biochemical technologies: biological agents transform substances' energy from biomass (liquid and gaseous fuels) and chemical technologies: liquid fuels typically produced from biomass feedstock using chemical agents. These methods of converting biomass might also yield by-products that can lead to useful bio-based goods (EPA, 2007). Several technologies are included in Table 20.2 that can transform biomass in its solid form

TABLE 20.2
Different Technology for Biomass Production

Technology	Scale	Energy Services Provided
Biogas	Small	Cooking, heating, and local electricity (perhaps distributed through the utility grid for local pumping, mining, lighting, communications, and refrigeration)
Producer gas	Small to medium	Cooking, heating, and local electricity (perhaps distributed through the utility grid for local pumping, mining, lighting, communications, and refrigeration)
Ethanol	Medium to large	Transportation by vehicle Cooking
Steam turbine	Medium to large	Heating process heat Electricity (used in industrial processing and grid distribution)
Gas turbine	Medium to Large	Heating process heat Electricity (used in industrial processing and grid distribution)

into energy sources that are both clean and usable (Demirbas, 2001; Kaltschmitt & Weber, 2006).

Although some are used more frequently than others most of these technologies are already in use commercially. The descriptions of the various technologies include a general overview of the implementation of gasification anaerobic digestion and ethanol production, steam turbine, and gas turbine technologies in a project that requires the resolution of fundamental technical challenges. These are the technologies being discussed. In addition to this, it provides a comprehensive analysis of the topics related to the operation and maintenance of a system, including its basic operating principles, input requirements, maintenance challenges, and cost considerations. In addition, environmental factors are also considered.

20.6   CHALLENGES RELATED TO BIOMASS

Operational, economic, social, legislative, and regulatory issues can be used to broadly categorize the current challenges facing the biomass supply chain associated with various feedstocks.

20.7   OPERATIONAL CHALLENGES

Feedstock scarcity: The inefficient management of resources and a hands-off approach to regulation taken by the government are the two primary factors holding back the expansion of the biomass industry.

The price of fuel is affected by seasonal shifts due to factors such as the availability of biomass according to area and season, as well as a problem with storage. Because of its low energy density, biomass is difficult to collect and store, making it difficult to purchase land for these purposes.

Transport section under pressure: As distance increases, it becomes more expensive and energy-unfavourable to transport wet biomass from the plantation to the manufacturing site.

The absence of standards for bioenergy systems and equipment: In particular when considered in light of the vast number of potential energy sources, resulted in the creation of technical obstacles in the conversion facility, core technology, and a scarcity of equipment. The appropriate pretreatment, which is required to put a halt to biodegradation and the reduction in heating value, results in an increase in production costs as well as investment expenses for new equipment.

Immature industrial chain: Obtaining long-term contracts for dependable feedstock supply at reasonable costs is practically difficult due to the immaturity of the sector. One further reason why many upstream businesses do not possess the driving reasons that are behind the technological revolution is because of their limited potential for profit.

20.8   ECONOMIC DIFFICULTIES

Feedstock acquisition costs: Due to the dispersion of biomass resources, many projects exhibit a preference for proximity to the source as a cost-saving measure for transportation expenditures. The aforementioned phenomenon has resulted in the process of centralization of activities.

Because of scattered challenges faced by the industry including capital constraints, reduced profitability, volatility in the global crude oil prices, and substantial market risk, investors have seldom taken the initiative to join in the biomass power generating business. This is mostly due to the fact that the industry carries a high level of market risk. The generation of energy from biomass is hampered by high operational costs and exorbitant investment prerequisites. It's possible that scattered farmers and small fuel firms won't be able to afford the extra costs that come with technologies involved with biomass pretreatment.

20.9   SOCIAL PROBLEMS

When making choices that are in conflict with one another, proper communication is necessary, regardless of whether one is selecting a supplier, location, route, or technology. By strengthening leadership and carrying out commitments, the many stakeholders should be made fully aware of the positive effects on society, the environment, and the economy that may be derived from the utilisation of resources.

Problems with land use: Problems with land use lead to the destruction of the houses of native people and the ecosystems on which they depend. The biomass plantation has a negative impact on the environment because it removes nutrients from the soil, encourages the deterioration of the aesthetic quality of the area, and quickens the rate at which biodiversity is being lost. The building of energy farms in rural areas will have additional consequences on society, such as increased demand for goods and services and an increase in the volume of transportation. The possibility of unfavourable consequences on society is sufficiently high to outweigh the benefits of generating new, long-term employment opportunities.

20.10   CHALLENGES IN REGULATION AND POLICY

Policies: Now, the government is subsidizing the price of domestic fuel, which lowers the cost of producing electricity from traditional sources than from renewable fuels.

System: There are no particular regulations governing the use of biomass resources. It is not clear whether a sanction can imposed if someone behaves in a manner that should be fully utilized.

Regulation: There is no specific mechanism or department to oversee the execution of pertinent national standards and policies or the growth of the biomass resources business.

20.11   THE ISSUES WITH RENEWABLE ENERGY

The field of renewable energy is now beset by a great number of challenges. Political pressures, governmental rules, corporate influence, antiquated infrastructure, a lack of sufficient battery storage, and the present status of the business are some of the factors that prevent greater worldwide adoption. In spite of these challenges, the use of renewable energy as a weapon in the fight against climate change has become widespread around the globe. Direct pollution as well as carbon emissions are absent. It is the most feasible alternative to using fossil fuels in the long run. It is steadily gaining importance as a component in the overall mix that is used to generate electricity.

The installation's huge initial cost: Carbon emissions are the primary contributor to the phenomenon of global warming. Numerous attempts have been conducted, with the goal of increasing its use. However, one factor that plays a crucial role in determining its level of adoption is the price. One of the challenges that stand in the way of its further growth is the very high initial cost of installation and effective sources of renewable energy, surpassing traditional fossil fuels in terms of affordability and sustainable forms of energy that are accessible. However, the initial installation costs of a gas-fired power plant and a solar power system are very different from one another. The installation of a large-scale solar power system has an average cost of 2,000 dollars per kilowatt. In a manner comparable to how the price per kilowatt-hour for a new gas-fired plant is just $1,000, yet the cost of a small-scale residential system is around $3,700. There is little question that the substantial range in pricing for installations is significant. Investors and lenders consider renewable energy sources to be high risk owing to the high initial installation costs associated with these sources. This is in contrast to the cheaper construction costs associated with fossil fuel plants.

Inadequate infrastructure: Taking into consideration the total cost of ownership throughout their lifetime investments. As the world seeks to reduce its dependence on non-renewable sources of energy hampered by a lack of adequate infrastructure, the utilisation of solar and wind power is a crucial area of focus in contemporary times, because of its potential to yield greater profitability. The infrastructure that is now in place was primarily developed for establishments dealing with nuclear power and fossil fuels. Since the present energy system is incapable of managing considerable quantities of renewable energy, it needs to be updated as soon as possible. It is crucial that some of its finest sources have no infrastructure at all, since this fact describes them. An important concern is the increasing age of the nation's electrical system. The bulk of electric transmission and distribution lines were constructed in the 1950s and 1960s. These decades saw this construction. It has already reached the end of its expected lifetime of fifty years. As a consequence of this, they are unable to respond adequately to the critical demands and significant changes in the weather. It is an extremely challenging task to determine the size of the solar system. It will be difficult to develop an energy-producing system if the system in question is too small. If the solar power generation system is too huge, then a significant energy storage system is necessary. If there is not a storage system with a large capacity, the energy that is created will be wasted.

Power reserves: The absence of power storage at a cost that is affordable is the most major and significant drawback. At some times of the day, renewable energy sources supply the bulk of the energy that is needed. Its electricity production cannot keep up with demand during the busiest hours of the day. Since sunlight and wind are both erratic phenomena, they are not suitable candidates for a reliable supply of on-demand electricity seven days a week. Both wind power and solar electricity are subject to variation. There is potential for instability in both the generation and the loads. The generation of energy by the combustion of fossil fuels is considered to be a more dependable method. However, owing to the intermittent nature of renewable energy sources, it is necessary to have a battery storage system that is both efficient and effective. Through the use of a battery storage device, the surplus of energy may be saved for later consumption. It is possible to prevent blackouts by lowering the grid's instability. Technology has enabled improvements in both the durability of the storage system and the battery capacity it has. Its very expensive nature makes broad implementation of it unlikely. Battery costs need to come down to make the storage of solar energy more financially viable.

Monopoly on non-renewable energy: The bulk of the world's energy demand is now satisfied by sources of non-renewable energy, such as fossil fuels, which gives these types of fuels a monopoly on the market. This presents an additional challenge that must surely be surmounted. It is necessary to develop several types of renewable energy to compete with the well-established business of fossil fuels. Some examples of these forms of renewable energy are solar and wind power. The fossil fuel sector continues to get major support from the government, despite the fact that the government supports solar energy via the provision of subsidies and other sorts of aid. The business that relies on fossil fuels is making concerted efforts to undermine the scientific consensus on climate change to slow down the shift toward renewable energy sources. Using energy that comes from sources that can be renewed is the most effective tactic that can be used to fight climate change. People have been making use of fossil fuels ever since the beginning of recorded human history. Because of this, it has a significant bearing on the economy of the nation. This is a direct result of the situation.

Lack of knowledge and unawareness: People are reluctant to adopt renewable energy technology because they lack the knowledge and understanding necessary to do so. Plants that use fossil fuels are almost always situated in densely populated areas. Because it needs a huge staff to function properly, it hires people from the surrounding community. The large property tax generated by fossil fuel facilities is going to be to the neighbourhood's advantage. The fact that Australia is the world's leading exporter of coal is hampering attempts to reduce the effects of climate change. Since the signing of the Paris Agreement in 2015, a number of countries have pledged support for investments in renewable energy. As a direct consequence of this, investments in renewable energy are currently ahead of those countries' investments in fossil fuels in terms of legislation, subsidies, and other such things. The broad adoption of renewable energy technology is hampered by the absence of supportive regulations, subsidies, incentives, and limits. To attract investors, the market for renewable energy has to have rules and regulations that are clear. To put it another way, for the government to achieve its goal of expanding the market for renewable energy sources, it must initiate and carry out assistance programmes. There is still a significant obstacle in the way of a smooth transition from the time-honoured fossil fuel industry to the renewable energy sector. This obstacle is caused by lobbying by corporations, political pressure, and a basic dependency on fossil fuels. The threats that are presented by climate change have started having a substantial influence on the lives of humans. Utilising renewable sources of energy is the only and most important answer. As a result of the multiple incentives and subsidies offered by the government, the cost of renewable energy sources has drastically fallen. The difference in cost per kWh between them and fossil fuels has shrunk significantly in recent years. This unquestionably aided in the company's establishment as a leader in the energy industry. Numerous Australians have switched to solar energy because of rising power costs, falling solar energy costs, and rising feed-in tariffs. Over 2.3 million solar rooftop installations may be found in Australia. As a result, solar energy is advantageous to the majority of Australians.

Large-scale biomass supply difficulties: Energy density is one of the primary concerns about the provision of biomass on a big scale. To put it simply, if the biomass moisture content of average wood is 30%, then it will be necessary to convey 300 kilograms of water for every ton of wood. In addition, the form of the biomass feedstock – whether it is chipped, pelletized, rounded, or bale – has a considerable influence on the bulk density as well as the costs associated with transportation. Because of this, compaction as well as densification are considered to be crucial components of a dependable biomass supply. The supply of biomass on a large scale is hampered by a number of obstacles, some of which include the initial raw material prices, the engagement of biomass producers, environmental laws, and the need to maintain sustainability, in addition to bulk and energy density. Finding answers to all of these issues involves figuring out how to produce biomass that will be a valuable commodity in Europe well beyond.

Biomass in the future: today, tomorrow, and beyond: The only sustainable, reliable, and dispatchable option for CHP at this time is woody biomass. The climate crisis calls for widespread participation. To be clear, the purpose of the world's industries was never to produce less carbon; hence, more effort will be required to achieve net zero and decarburization than simply banning plastics and promoting recycling. The global community must work together to solve the climate catastrophe, and governments, NGOs, society, and enterprises must all think differently about how they conduct business. Why now? The current trajectory of the world economy does not support the Paris Agreement's goal of gradually phasing out fossil fuels. The global carbon budget will be used up by 2040, Even if every country in the world implemented the most stringent emission reduction policies, the Intergovernmental Panel on Climate Change (IPCC) reports that global warming would continue. As the recent rise in peak power prices throughout Europe demonstrates, the majority of the globe is still dependent on coal to meet the need for energy. This is shown by the fact that the majority of the world still relies on coal. The demand for dispatchable power keeps fossil fuels, especially coal, competitive rather than pricing them out of the market. So why do we use such a large amount of coal and fossil fuels? It's because it's difficult to sustain baseload, dispatchable electricity. Managing the erratic nature of solar and wind energy is just as challenging as developing batteries and other electricity storage technologies. In the medium and long terms, batteries and hydrogen provide some hope, but the deadline for reducing EU emissions by 55% is only nine years away. We do not have enough time to wait for such solutions to become cost-competitive or to rise up to the challenge. If we want to live in a world that is carbon neutral or carbon negative tomorrow, all governments and businesses must take aggressive steps to cut their carbon emissions today. We must take immediate action to identify options based on renewable energy sources that are capable of addressing the climatic crisis of this century.

The sustainability of biomass: The entire forest products industry is highly regulated, but the biomass industry is the most so. Biomass must be developed and used in accordance with a network of regulations that intertwine at the international, federal, and state levels to be considered "sustainable" and qualify for subsidies. According to our standards, biomass must be obtained from sources where forest carbon stores are stable or growing, be replanted after harvest, not be made from wood that could be used for another function with high value, and harvesting should take into account the importance of biodiversity and conservation, respectively. We think that these standards should be followed in all wood fibre purchases and that compliance should be independently confirmed. The IPCC of the UN, the International Energy Agency, the United Kingdom Committee on Climate Change, the United States Department of Agriculture, the United States Department of Energy, the EU, and a great number of other leading academic and governmental bodies from around the world have acknowledged the use of biomass as a climate change mitigation measure and have continued to support its utilisation. When generated sustainably, the underlying science of wood bioenergy's benefits for forestry and carbon storage is well-established and unchanging. It is bolstered by an ever-expanding collection of scholarly writing, with substantial new studies due out in the years 2019 and 2020 and has been attested to by the top academics and experts in this subject. Today, biomass can deliver stable, dispatchable baseload power that is low in carbon. Therefore, biomass provides a quick low-carbon answer, and bioenergy needs to be included when we think about the global energy transition. We need to use alternative energy sources right away. When wind and solar energy are in limited supply, replacing coal with sustainably produced biomass can keep the lights on while reducing carbon emissions by 85% over the course of a year of lifecycle (Reid et al., 2020).

20.12   PROSPECTS FOR LONG-TERM LAND-INTENSIVE BIOENERGY

With 9.5% to the International Energy Agency (2017b, 2019), bioenergy accounts for a significant portion not just of the world's primary energy supply but also of the approximately 70% of all renewable energy that is now being used. Over 50% of the bioenergy utilised is derived from the traditional application of biomass. This practise is predominantly observed in domestic environments, especially in kitchens and heating systems, and also in commercial settings, including brick kilns and charcoal kilns. While acknowledging the pressing need for enhancing the sustainability, effectiveness, and medical safety of traditional biomass utilisation (Creutzig et al., 2015), this study confines its investigation to contemporary bioenergy due to the latter's potential for the rapid increase in the decades to come. Modern bioenergy also referred to as "bioenergy," contributed four times as much to the total amount of renewable energy used in 2017 than a combination of wind energy and solar photovoltaics (PV), according to the International Energy Agency (2018). According to the International Energy Agency (2018), the bulk of bioenergy is used toward the heating of homes and industry. However, it is expected that by the year 2023, bioenergy will also create 3% of electricity and account for around 4% of the need for energy to power transportation. Before 2010, the growth rate of liquid biofuel production for transportation exceeded 10% per year; however, from 2010 to 2016, that growth slowed to 4% yearly. From 2010 to 2016, the capacity of bioenergy electricity increased by 6.5% year on average (International Energy Agency, 2017b). Most current scenarios for addressing climate change include bioenergy significantly. According to the IPCC's most recent Special Report on Global Warming of 1.5°C, which analysed 85 1.5°C scenarios, the median share of primary energy (154 EJ/year) provided by biomass increased from 10% in 2020 to 26% in 2050 (range 10%–54%) (Rogelj et al., 2018). Conversely, it is projected that the combination of solar and wind energy sources will yield a median contribution of 22% toward the overall primary energy output by the year 2050. Rogelj et al. (2018) reported that a significant proportion of the simulated pathways analysed in their study indicate a persistent reliance on high levels of bioenergy without CCS (carbon capture and storage) until the conclusion of the current century, while there will be a significant rise in the use of bioenergy with CCS. To attain a future with a temperature increase of 1.5°C without the implementation of Bioenergy with Carbon Capture and Storage (BECCS) and with reduced bioenergy, certain modelled pathways necessitate significant reductions in overall energy demand, amounting to a 32% decrease by 2050 in comparison to 2010 levels. These pathways also require substantial modifications in human behaviour, such as dietary changes, along with rapid technological advancements and a relatively small global population along with a constructive outlook that considers utilizing enormous areas of land for the production of bioenergy. What makes bioenergy and BECCS so important in these scenarios is that they are both low-cost, low-emission energy sources that don't have to deal with the intermittent nature of other renewables such as solar and wind power. Furthermore, even if all forms of bioenergy were completely free of carbon emissions, it is very doubtful that land-intensive forms of bioenergy would continue to be an economically desirable source of energy beyond the year 2050 due to increased market rivalry and technical improvement. Humanity's dependence on ecosystem services has already been affected. According to recent studies (Daz et al., 2019; Reid et al., 2005), the conversion of natural habitats and ecosystems into controlled scenes, agricultural land, as well as urban areas has resulted in the endangerment of approximately one million species. As a result, it is essential to protect the natural ecosystems that are still standing and, if it is feasible to do so, to revive ecosystem services on lands that have been abandoned for the production of food or fibre or that have been degraded. The spread of land-intensive bioenergy on a wide scale is essentially incompatible with the need for conservation and restoration. According to Arneth et al. (2019), the IPCC found an increase in land demand. The implementation of bioenergy, reforestation, along with afforestation initiatives at a significant scale may lead to adverse impacts on desertification, land degradation, and food security. These factors must be considered in the pursuit of reducing greenhouse gas emissions and removing carbon dioxide. The land is limited, so if we must utilize it to produce energy, we should do so as effectively as we can. For instance, PVs can significantly increase the efficiency of how land is used to produce energy. According to the European Academies Science Advisory Council (2019), PV technology has the potential to produce somewhere between fifty and one hundred times more power from a single hectare of land than biomass does. Despite its somewhat inefficient use of land and likely rivalry with other land uses, bioenergy is a significant component of the majority of energy models that project the world's energy landscape beyond the middle of this century. This is due to the combination of three different causes. First, unlike intermittent energy sources, bioenergy can offer baseload electrical power. This ability is one that is expected to become increasingly important as the current thermal capacity that is dependent on fossil fuels is decommissioned. Second, the production of biofuels can be done at a cheap cost while satisfying the high energy density standards for use in airplanes and ships. This is an advantage for both of these industries. Third, BECCS has the potential to provide a source of energy that really reduces carbon emissions. Because they are able to maintain a steady stream of emissions in the short term while simultaneously removing greenhouse gases in the long term, negative emission technologies (NETs) are viewed favourably by integrated assessment models. NETs play a large role in the second half of the century in nearly all of the climate mitigation scenarios that fulfil the 2-degree aim and almost all of the scenarios that meet the Paris goal of significantly below 2°C.

The current state of bioenergy: Although bioenergy may only contribute a small portion, according to the International Energy Agency (2019), traditional and contemporary forms of bioenergy now provide 9.5% of the world's primary energy supply. In 2100, this percentage is expected to increase. The proportion of bioenergy, as well as its volume, is expected to rise dramatically. Demand for forest biomass is expected to rise over the next few decades as governments and asset owners in the power sector try to retain coal-powered infrastructure while moving away from it. According to Tran et al. (2017), the production of wood pellets intended for utilisation in biomass energy has increased by over four times between 2006 and 2015, reaching a total of 26 million tons (MT). In the EU, over half of all renewable energy comes from solid biomass, which accounts for 44.7% of the total (40% of this biomass is utilized for domestic heating). East Asian markets for new biomass are also growing quickly and may soon catch up to European demand. For example, the Japanese government has approved 11.5 GW of biomass electricity projects, 40% of which Obayashi (2017) and Watanabe (2017) suggest might be powered by palm oil. Many countries' carbon pricing regimes recognise biomass as a zero-carbon fuel, and its used to meet national (and corporate) climate targets is a driving factor in this growth. This assumption leads to excessive usage of bioenergy (European Academies Science Advisory Council, 2019) as only a fraction of the available biomass may provide a climate benefit over a 10-year period. We consider a period of 10 years to be the most pertinent to actual climate consequences; if the use of bioenergy causes a rise in CO_2 over that time. The carbon absorption by the fuel source notwithstanding, the exacerbation of climate impacts is an inevitable consequence of the aforementioned action. The utilisation of forest biomass has the potential to pose a unique "double climate problem" as it may lead to a substantial rise in near-term emissions that surpasses those of the majority of fossil fuels. In addition, the carbon payback periods associated with forest biomass utilisation may extend from decades to over a century. These assertions are supported by various sources (Brack, 2017; Buchholz et al., 2016; Cornwall, 2017; Sterman et al., 2018). Managing this difficult contrast between expected longer-term decline and projected near-term expansion in the usage of bioenergy poses several special issues. To commence the analysis, it is advantageous to consider three discrete categories of biomass provision, each of which can be sourced from ecosystems with diverse capacities and durations for carbon sequestration both on and off the site. Biomass is a potential feedstock that can be derived as a byproduct or residual from various processes, including crop and wood harvesting, as well as cooking oil utilisation. Ecosystems may experience a reduction in biomass as a means of increasing carbon storage or improving the quality of the habitat in various other manners. Strategies, such as biomass removal, aimed at mitigating the risk of wildfires, promoting tree growth, and incentivizing the utilisation of wood fibre in durable goods, have the potential to enhance carbon sequestration while simultaneously functioning as a bioenergy source. Ecosystems that are intentionally managed for energy generation have the potential to offer biomass (what we refer to as land-intensive bioenergy). The desirability, sustainability, and prospects vary greatly for each of these categories.

20.13 CONCLUSIONS

It is most expected to rise during the next 10 years or more, based on existing patterns and policy. However, there are fewer bioenergy resources available than most models and scenarios suggest that can be produced sustainably while still providing net climate benefits. By the end of the century, the incorporation of land-intensive bioenergy into the energy mix is improbable to constitute a substantial proportion. In the forthcoming decades, policymakers ought to bear in mind the ensuing objectives when contemplating the utilization of bioenergy. Initially, the utilization of bioenergy as a substitute for non-renewable energy sources ought to result in a significant reduction in emissions within a brief timeframe, which is pivotal for climate-related ramifications, spanning years instead of decades. Biomass finds its optimal utilization in extended periods of preservation, such as in the erection of edifices, provided it is obtainable as a byproduct or due to suitable management techniques. The utilization of energy is considered to be the second most favourable application, provided that it does not result in any adverse effects on the quality of air or water, or exacerbate water scarcity concerns. It is highly desirable for energy generation to be equipped with CCS technology. It is likely that by mid-century, bioenergy that requires significant land use will be regarded as a fuel of the past. In light of this, it is recommended that governing bodies impose limitations on the provision of short-term benefits for land-intensive bioenergy. Rather, they should provide incentives regarding the upcoming range of technologies that have the potential to facilitate a carbon emission-free world.

ACKNOWLEDGEMENT

We would like to acknowledge the Department of Agronomy at the Lovely Professional University, Phagwara, Punjab, India, for their consistent moral support and encouragement throughout the writing process.

REFERENCES

Arneth, A., Barbosa, H., Benton, T., Calvin, K., Calvo, E., & Conners, S. (2019). *Summary for Policymakers: Climate Change and Land.* Geneva: Intergovernmental Panel on Climate Change.

Bauen, A., Woods, J., & Hailes, R. (2004). *Biopowerswitch: A Biomass Blueprint to Meet 15% of OECD Electricity Demand by 2020.* London: Imperial College London, Center for Energy Policy and Technology, E4tech Ltd..

Brack, D. (2017). *The Impacts of the Demand for Woody Biomass for Power and Heat on Climate and Forests.* https://www.chathamhou se.org/publi cation/impacts-demand-woody-biomass-power and heat-climate-and-forests

Buchholz, T., Hurteau, M. D., Gunn, J., & Saah, D. (2016). A global meta-analysis of forest bioenergy greenhouse gas emission accounting studies. *GCB Bioenergy, 8*(2), 281–289. doi:10.1111/gcbb.12245

Cornwall, W. (2017). The burning question. *Science, 355*(6320), 18–21, doi:10.1126/science.355.6320.18

Creutzig, F., Ravindranath, N. H., Berndes, G., Bolwig, S., Bright, R., Cherubini, F., & Masera, O. (2015). Bioenergy and climate change mitigation: An assessment. *GCB Bioenergy, 7*(5), 916–944. doi:10.1111/gcbb.12205

Demirbas, A. (2001), Biomass resource facilities and biomass conversion processing for fuels and chemicals, *Energy Conversion and Management, 42*, 1357–1378.

Díaz, S., Settele, J., Brondízio, E., Ngo, H. T., Guèze, M., Agard, J., & Vilá, B. (2019). *Summary for Policymakers of the Global Sssessment Report on Biodiversity and Ecosystem Services of the Intergovernmental Science-Policy Platform on Biodiversity and Ecosystem Services.* Bonn: Intergovernmental Panel on Biodiversity and Ecosystem Services.

EPA. (2007). *Biomass Conversion:Emerging Technologies, Feedstocks, and Products.* Washington, D.C: Sustainability Program, Office of Research and Development, EPA/600/R-07/144, U.S. Environmental Protection Agency.

EREC. (2007). *Bioenergy.* Brussels, Belgium: European Renewable Energy Council.

EU Directive. (2001). 77/EC on the promotion of the electricity produced from renewable energy source in the internal electricity market. *Official Journal of European Community, 283*, 33.

EU. (2009). Directive 2009/28/EC of the European Parliament and of the Council of 23 April 2009 on the promotion of the use of energy from renewable sources and amending and subsequently repealing Directives 2001/77/EC and 2003/30/EC. *Official Journal of the European Union, 5*, 2009.

European Academies Science Advisory Council. (2019). *Forest Bioenergy, Carbon Capture and Storage, and Carbon Dioxide Removal: An Update.* https://easac.eu/fileadmin/PDF_s/reports_statements/Negative_Carbon/EASAC_Commentary_Forest_Bioenergy_Feb_2019_FINAL.pdf

European Environment Agency. (EEA). (2008). *Energy and Environment Report 2008.* Official Publications of the European Communities European Environment Agency.

Field, C. B., & Mach, K. J. (2017). Rightsizing carbon dioxide removal. *Science, 356*(6339), 706–707. doi:10.1126/science.aam9726

IBEP. (2006). *Introducing the International Bioenergy Platform.* Rome: Food and Agriculture Organization of The United Nations.

International Energy Agency (IEA). (2009). *The World Energy Outlook 2009.* https://www.iea.org.

International Energy Agency. (2017). *Technology Roadmap: Delivering Sustainable Bioenergy.* https://webstore.iea.org/technology-roadmap-delivering-sustainable-bioenergy

International Energy Agency. (2018). *Market Report Series: Renewables 2018.* https://web-store.iea.org/market-report-series-renewables-2018

International Energy Agency. (2019). *Key World Energy Statistics 2019.*

IPCC. (2014). Climate change 2014: Mitigation of climate change. In: O. Edenhofer, R. Pichs-Madruga, Y. Sokona, E. Farahani, S. Kadner, K. Seyboth, J. C. Minx (Eds.), *Contribution of Working Group III to the Fifth Assessment Report of the Intergovernmental Panel on Climate Change.* Cambridge: Cambridge University Press.

Kaltschmitt, M., & Weber, M. (2006). Markets for solid biofuels within the EU-15. *Biomass and Bioenergy, 30,* 897–907.

Larson, E. D., & Katha, S. (2000). Expanding roles for modernized biomass energy. *Energy for Sustainable Development, 4,* 15–25.

Lindzen, R. S. (2010). Global warming: The origin and nature of the alleged scientific consensus. *Problems of Sustainable Development, 5*(2), 13–28.

Ma, J. (2010). *China Statistical Yearbook.* China: National Bureau of Statistics of China.

Moriarty, P., & Honnery, D. (2011). The transition to renewable energy: Make haste slowly. *Environmental Science and Technology, 45*(7), 2527–2528.

Obayashi, Y. (2017). Japan fires up biomass energy, but fuel shortage looms, *Reuters,* September 22.

Odgaard, O., & Delman, J. (2014). China's energy security and its challenges towards 2035. *Energy Policy, 71,* 107–117.

Pawłowski, A. (2006). The multidimensional nature of sustainable development. *Problemy Ekorozwoju,* 1(1), 23.

Pawłowski, A. (2007). Barriers in introducing sustainable development. Ecophilosophical point of view. *Problemy Ekorozwoju,* 2(1), 59.

Pawłowski, A. (2010). The role of environmental engineering in introducing sustainable development. *Ecological Chemistry and Engineering,* 17(3), 263–278.

Reid, W. V., Ali, M. K., & Field, C. B. (2020). The future of bioenergy. *Global Change Biology, 26,* 274–286.

Reid, W. V., Mooney, H. A., Cropper, A., Capistrano, D., Carpenter, S. R., Chopra, K., & Zurek, M. (2005). *Ecosystems and Human Well-Being: Synthesis.* Washington, DC: Island Press.

Rogelj, J., Shindell, D., Jiang, K., Fifita, S., Forster, P., Ginzburg, & Vilariño, M. V. (2018). Mitigation pathways compatible with 1.5°C in the context of sustainable development. In: V. Masson-Delmotte, P.Zhai, H.- O. Pörtner, D. Roberts, J. Skea, P. R. Shukla, T. Waterfield (Eds.), *Global Warming of 1.5°C. An IPCC Special Report on the Impacts of Global Warming of 1.5°C above Pre-Industrial Levels and Related Global Greenhouse Gas Emission Pathways, in the Context of Strengthening the Global Response to the Threat of climate change, sustainable development, and efforts to eradicate poverty* (pp. 313–443). Geneva, Switzerland: Intergovernmental Panel on Climate Change.

Rosillo-Calle, F., de Groot, P., & Hemstock, S. L. (2007), *General Introduction to the Basis of Biomass Assessment Methodology.* In: F. Rosillo-Calle, P. de Groot, S. L. Hemstock, J. Woods (Eds.), *The Biomass Assessment Handbook. Bioenergy for a Sustainable Environment* (pp. 27–68). London, Sterling: Earthscan.

Ruiz Romero, S., Colmenar Santos, A., & Castro Gil, M. A. (2012). EU plans for renewable energy: An application to the Spanish case. *Renewable Energy, 43,* 322.

Sánchez, A. (2008). Perspectives and problems in sustainable development. *Problemy Ekorozwoju,* 3(2), 21.

Sterman, J. D., Siegel, L., & Rooney-Varga, J. N. (2018). Does replacing coal with wood lower CO_2 emissions? Dynamic lifecycle analysis of wood bioenergy. *Environmental Research Letters,* 13(1), 015007.

Thrän, D., Peetz, D., Schaubach, K., Mai-Moulin, T., Junginger, H. M., Lamers, P., & Visser, L. (2017). *Global Wood Pellet Industry and Trade Study 2017: IEA Bioenergy Task 40*. IEA Bioenergy.

UN. (2007). *Sustainable Bioenergy: A Framework for Decision Makers*. Santiagon de Chile: ECLAC.

Watanabe, C. (2017). Japan's green energy incentives cast spotlight on controversial use of palm oil. *The Japan Times*, November 9.

Zhang, G. (2009). *Report on China Energy Development for 2009*. Beijing: Economic Science Press.

Zhang, X., Ruoshui, W., Molin, H., & Martinot, E. (2010). A study of the role played by renewable energies in China's sustainable energy supply. *Energy*, *35*(11), 4392.

Index

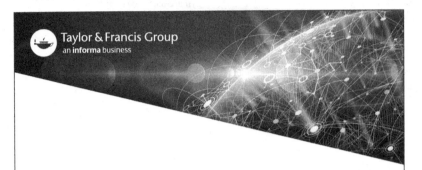

Printed in the United States
by Baker & Taylor Publisher Services